NATIONAL DEFENSE RESEARCH INSTITUTE

ENDING

THE FINAL TRANSITION,

THE U.S.

OPERATIONAL MANEUVER,

WAR IN

AND DISESTABLISHMENT OF UNITED STATES FORCES–IRAQ

IRAQ

Richard R. Brennan, Jr.,
Charles P. Ries, Larry Hanauer,
Ben Connable, Terrence K. Kelly,
Michael J. McNerney,
Stephanie Young, Jason Campbell,
K. Scott McMahon

Prepared for the United States Forces–Iraq
Approved for public release; distribution unlimited

The research described in this report was prepared for United States Forces–Iraq. The research was conducted within the RAND National Defense Research Institute, a federally funded research and development center sponsored by the Office of the Secretary of Defense, the Joint Staff, the Unified Combatant Commands, the Navy, the Marine Corps, the defense agencies, and the defense Intelligence Community under Contract W74V8H-06-C-0002.

Library of Congress Cataloging-in-Publication Data

Brennan, Richard, 1954-
  Ending the U.S. War in Iraq : the final transition, operational maneuver, and disestablishment of United States Forces-Iraq / Richard R. Brennan, Jr., Charles P. Ries, Larry Hanauer, Ben Connable, Terrence K. Kelly, Michael J. McNerney, Stephanie Young, Jason Campbell, K. Scott McMahon.
    pages cm
  Includes bibliographical references.
  ISBN 978-0-8330-8047-9 (pbk.)—ISBN 978-0-8330-8245-9 (hc)
  1. Iraq War, 2003-2011. 2. Disengagement (Military science) 3. Postwar reconstruction—Iraq. 4. United States—Armed Forces—Iraq. I. Title.
  DS79.76.B738 2013
  956.7044'310973—dc23
                                                          2013032286

The RAND Corporation is a nonprofit institution that helps improve policy and decisionmaking through research and analysis. RAND's publications do not necessarily reflect the opinions of its research clients and sponsors.

Support RAND—make a tax-deductible charitable contribution at www.rand.org/giving/contribute.html

RAND® is a registered trademark.

Cover: image by Reuters; design by Dori Gordon Walker

© Copyright 2013 RAND Corporation

This document and trademark(s) contained herein are protected by law. This representation of RAND intellectual property is provided for noncommercial use only. Unauthorized posting of RAND documents to a non-RAND website is prohibited. RAND documents are protected under copyright law. Permission is given to duplicate this document for personal use only, as long as it is unaltered and complete. Permission is required from RAND to reproduce, or reuse in another form, any of our research documents for commercial use. For information on reprint and linking permissions, please see the RAND permissions page (www.rand.org/pubs/permissions.html).

RAND OFFICES
SANTA MONICA, CA • WASHINGTON, DC
PITTSBURGH, PA • NEW ORLEANS, LA • JACKSON, MS • BOSTON, MA
DOHA, QA • CAMBRIDGE, UK • BRUSSELS, BE
www.rand.org

Foreword

In March 2003, the United States and its coalition partners began Operation Iraqi Freedom. The military campaign leading to the destruction of Saddam Hussein's military, the capture of his seat of power in Baghdad, and many other tasks associated with the invasion phase of the operation were complete by April 30; a mere six weeks after the start of the war.

As military history demonstrates, wars rarely end as first planned, for reasons that may not have been considered when crafting war plans. For example, war changes a country's internal political and social dynamics, affecting its internal security, economic development, and governance. In addition, countries and nonstate actors that were not part of the initial conflict may pursue their own interests in ways that bring new challenges to ending a war. Further, the initial political goals may expand in light of changing situations on the ground, causing major shifts in the military campaign.

From the beginning to the end of a war, all participants operate under a cloud of uncertainty. Military leaders and national security experts use history as the foundation of their professional knowledge. Military history is primarily concerned with facts, figures, and lessons learned about how to fight and win battles and campaigns. However, the history of war suggests that how a war ends is at least as important as how it is waged in establishing a given postwar environment. Despite this importance, military practitioners, strategists, and historians sometimes pay less attention to understanding how wars end than how they are fought.

This book is the product of a two-year RAND Corporation effort not only to create a historical record of the retrograde of military forces and the transitions that occurred during Operation New Dawn (OND) but also to provide an independent and objective analysis including key insights and recommendations on how to end large-scale military operations. In collaboration with the U.S. Embassy in Iraq, the United States Forces–Iraq (USF-I) provided RAND access to plans, operations orders, internal staff deliberations, strategic and operational assessments, and a host of other contemporaneous information on how U.S. forces completed, transferred, transformed, or terminated all activities being conducted in Iraq. In addition, a RAND research team spent two weeks in Iraq in 2011, interviewing the leaders and staffs of both Embassy Baghdad and USF-I.

OND was the last campaign of the U.S. military's eight-year effort in Iraq, as the U.S. forces completed their withdrawal and the United States transitioned from a military-led relationship to a comprehensive partnership that not only encompasses security but also commerce, culture, education, and economics. The overall political mission in Iraq during OND was to produce a long-term and enduring strategic partnership between the United States and a sovereign, stable, and self-reliant Iraq that contributes to the peace and security in the region. With U.S. assistance, Iraq has been given an opportunity for a sovereign and stable future, possessing the tools necessary to maintain internal security and the foundation necessary for external defense. The United States and Iraq should continue to work together to develop a government that is answerable to its people and their elected representatives, with a growing economy that is capable of continued growth and development.

This partnership is the same the United States seeks to share with all nations governed by principles of freedom, that respect the rights of their citizens, and that ensure the benefits of this freedom for all. This is the future the United States desires with Iraq. It is a future of mutual respect and mutual benefit. This opportunity has come at great cost and sacrifice, both by the people of Iraq and all who have served there. It should not be squandered.

The transition process, which is a focus of this book, included efforts to increase the capacity of the Iraqi security forces, strengthen the embassy and its newly formed Office of Security Cooperation to operate independently of U.S. military support, hand over responsibility for the accomplishment of activities to other DoD entities operating outside of Iraq, and responsibly withdraw forces and equipment in a manner that allowed U.S. forces to depart Iraq "with honor and success."

However, just as waging war is not the sole purview of deployed military forces, neither is ending a war. Both the Executive Branch and Congress play important roles, not only in determining how to end the U.S. military's effort but also in preparing for the transformational relationship with the host nation once military forces have departed. Perhaps more importantly, the host nation—in this case, the government of Iraq—plays an important role in the transition from a military-led effort to new, enduring partnership. Consequently, any study of how a war ends must examine the interaction of these key players throughout the transition, a period that will ultimately transform the relationship between two countries.

While every conflict has its own set of dynamics that are unlikely to be replicated elsewhere, the historical lessons learned from Iraq and elsewhere are valuable to inform policymakers and military planners as they examine options for future transitions that plan the end of future conflicts.

—James F. Jeffrey
U.S. Ambassador to Iraq, 2010–2012

Preface

This book is being published ten years after the U.S. invasion of Iraq in March 2003. During the intervening years, many studies have been conducted about the U.S. involvement in Iraq, covering the intelligence failures that led to the initial invasion, the rapid success of the invasion, the failure to adequately plan and resource operations necessary to maintain security during the immediate aftermath of the invasion, the conduct of the Coalition Provisional Authority, the establishment of an interim government, the Iraqi civil war and long counterinsurgency, and the efforts undertaken in what must be viewed as the largest nation-building effort since World War II. It is a history of U.S. military transitions in Iraq that culminated in the final transition as it divested responsibility for over 30,000 tasks, activities, and relationships through a systematic effort made in conjunction with the Embassy of the United States in Baghdad, Iraq, to terminate or transfer responsibility to the embassy, the newly formed Office of Security Cooperation–Iraq, the Iraqi Security Forces, U.S. Central Command, or some other government department or agency. This complex transition was conducted as part of a larger operation that involved the operational maneuver of forces and equipment out of Iraq. While the bulk of the book details the final 15 months of Operation New Dawn, culminating with the disestablishment of U.S. Forces–Iraq (USF-I), it also describes the events that occurred in the 12 months after the departure of U.S. forces in an effort to develop observations, insights, and recommendations that policymakers and military planners should find useful when considering how to end a war involving a large-scale U.S. military involvement.

During the conduct of the Iraq war, the RAND Corporation deployed many analysts to support the Coalition Provisional Authority; Embassy Baghdad; Multi-National Force–Iraq; and its successor organization, USF-I. One of the principal authors of this study, Richard R. "Rick" Brennan, served as a member of the Multi-National Force–Iraq Joint Interagency Task Force–Iraq from September 2008 to October 2009, where he was a primary planner for Balancing Iranian Influence and Countering Violent Extremist Organizations. Subsequent to that, he served as the Director of the Checkmate Team and Senior Advisor to the USF-I Director of Operations (J3) from November 2009 through December 16, 2011. As such, he participated in much of the planning efforts related to the transition and was an eyewitness to the final operational maneuver of forces. As a participant in the final stages of the operation, Dr. Brennan took contemporaneous notes during the last three years of the mission, which this book references to fill gaps in the historical record.

In March 2011, then-MG William B. "Burke" Garrett, the USF-I Chief of Staff, asked Dr. Brennan to develop a proposal for RAND to conduct an independent and objective analysis of the transition process and the final operational maneuver of U.S. forces out of Iraq, both to record the history of Operation New Dawn and to draw strategic- and policy-level lessons that would be useful for future transitions. On June 1, 2011, GEN Lloyd J. Austin III, Commander of USF-I, was briefed on the project and agreed to provide RAND unfettered access to planning and operational details necessary to conduct an unclassified study whose results could be widely distributed. Two weeks later, Ambassador James Jeffrey was briefed on the study and also made a commitment to provide necessary support for the research team to collect information necessary to publish this book. This research could not have been done without the full support of these two leaders and their organizations.

The research in this book is based largely on primary sources, including unclassified portions of plans, reports, assessments, and briefings developed by USF-I and Embassy Baghdad. While some of the information relating to the transition and operational maneuver of forces out of Iraq was made available by USF-I for this report, a fuller history will have to await the future release of classified USF-I and Embassy Baghdad documents now in the archives of the U.S. military historian in Washington, D.C.; the U.S. Central Command historian in Tampa, Florida; and at the National Archives. A substantial amount of material was also available to the RAND study team from official reports developed by government agencies, such as the Special Inspector General for Iraq Reconstruction and the Army Audit Agency, and through congressional testimony and reporting. For discussions of enduring threats emanating from Sunni and Shi'a extremist groups, the authors chose to rely largely on unclassified research conducted by noted subject-mater experts and such organizations as the Counter Terrorism Center at West Point to preclude the possibility of inadvertent release of information developed by the U.S. Intelligence Community. In some cases, however, internal assessments developed by USF-I J3 and J5 (Plans) have been used with the approval of USF-I as the originating authority. Finally, a substantial amount of information was gained through interviews with military officers, Foreign Service officers, Department of Defense and Department of State officials, and government contractors serving in Iraq. Where possible, the accuracy of these interviews was verified either through multiple interviews or, where possible, through documentary evidence.

In addition to the primary source documents from USF-I, a RAND research team consisting of the principal authors of this book visited Iraq from June 24 to July 1, 2011. During this visit, senior leaders and staff from both Embassy Baghdad and USF-I provided briefings and interviews covering all aspects of the transition and the planned operational maneuver of forces out of Iraq. The information provided and the candid discussions were crucial in developing the richness of detail contained in this book.

On December 13, 2011, less than a week before the departure of U.S. forces from Iraq, General Garrett, exercising his authority as the Chief of Staff of USF-I and the originating authority for classified and "unclassified for official use only" material originating within that command, approved the release of the details relating to the Joint Campaign Plan, operations plans, operation orders, internal staff processes and procedures, internal USF-I assessments regarding the Iraq operational environment, and a host of other details necessary to accurately document the history of Operation New Dawn in an unclassified report. Great care has been used to ensure that sensitive or classified information derived from other sources, data related to current plans and operations, and information related to sensitive programs and activities were excluded from the report. Consequently, some transitions are not addressed in this report. During this meeting, then-BG Jeff Snow, the J5 of USF-I, was designated to have oversight of the study to ensure compliance with the agreed-on protocols. In February 2013, General Snow personally reviewed the document to ensure the report did not divulge sensitive internal deliberations or classified material. He also provided valuable comments to ensure the nuances of the transition and operational maneuver were accurately reported. His continuous support and assistance throughout the research and writing process was crucial to the final product.

This research was sponsored by USF-I and conducted within the International Security and Defense Policy Center of the RAND National Defense Research Institute, a federally funded research and development center sponsored by the Office of the Secretary of Defense, the Joint Staff, the Unified Combatant Commands, the Navy, the Marine Corps, the defense agencies, and the defense Intelligence Community.

For more information on the International Security and Defense Policy Center, see http://www.rand.org/nsrd/ndri/centers/isdp.html or contact the director (contact information is provided on the web page).

Contents

PART IV
The Aftermath, Conclusions, and Recommendations

CHAPTER TEN
After the Transition

CHAPTER ELEVEN
Conclusions and Recommendations

Figures

Tables

Executive Summary

On December 18, 2011, the final U.S. forces stationed in Iraq under Operation New Dawn (OND) crossed the Iraq-Kuwait border, bringing to a close the transition and withdrawal of U.S. Forces–Iraq (USF-I) that is the focus of this study. The U.S. war in Iraq is over. As characterized by USF-I Commanding General Lloyd J. Austin III, U.S. forces completed one of history's most complex handovers of authority and retrograde of U.S. personnel and equipment with "honor and success." The extraordinary effort, which began with detailed military and civilian planning in early 2009, was over. For Iraqis, of course, elevated levels of violence and political instability unfortunately continue.

Among Iraq's new challenges, the most immediate is the civil war in Syria, which has the potential to engulf Iraq in a broader regional sectarian conflict fuelled both by Shi'a and Sunni hatred and mistrust and by activities of such regional and global powers as Turkey; Saudi Arabia; Russia; Iran; and the Iranian proxies, Lebanese Hezbollah and Iraqi Kata'ib Hezbollah and Asa'ib al-Haq, which have recently entered the war to support the Asad regime. Experienced fighters from al Qaeda in Iraq have joined the Syrian rebels, aligning themselves with the Islamist "Al Nusra Front." The escalation of violence in Syria poses numerous threats to Iraq and has the potential to place Baghdad once more at the epicenter of a regional Sunni-Shi'a conflict.

Moreover, while the war involving the United States and Iraq has ended, conflict in Iraq is not over. Tensions over power-sharing, the remnants of the Sunni insurgency, other unresolved issues that aggravated the Iraqi civil war, and external support for violent actors remain challenges the Iraqis will have to solve themselves. Thus, the Iraqi transition from wartime activities to peace and stability will take much longer to realize, and could yet spiral downward into a new civil war.

At its peak in 2011, the transition involved virtually every military and civilian American stationed in Iraq, as well as hundreds in Washington, in Kuwait, and at U.S. Central Command (USCENTCOM) headquarters. The result—the retrograde of U.S. forces after an eight-year presence during which both Americans and Iraqis worked diligently to place the country on a path toward stability—marked a pivotal point in U.S.-Iraqi relations. While the future of the relationship is uncertain, all U.S. officials who contributed to the successful implementation of the transition and the safe departure of U.S. forces can take satisfaction in their contributions to U.S. national interests.

A History of Transitions

The study summarized here focuses on the story of how U.S. forces stepped back from broad engagement in Iraq, drew down force levels, and handed off the myriad activities for which they had become responsible. Even though the Iraq conflict was among the longest the United States has engaged in, the legal basis for the presence of forces and nature of our interactions with Iraqi authorities was in constant flux throughout.

In this overall context, *transition* has meant many different things, to different stakeholders, at different times. The term has referred to institutional transitions, such as those from the Office of Reconstruction and Humanitarian Assistance to the Coalition Provisional Authority (CPA), from CPA Combined Joint Task Force 7 to Multi-National Force–Iraq (MNF-I), from MNF-I to USF-I, or from USF-I to the U.S. embassy. For the U.S. military, *transition* has also referred to functional transitions—for example, from combat to counterinsurgency, to training and advising, and to broad-based reconstruction and economic development.

Over more than eight years, the functions and activities the military undertook in support of Iraqi government and U.S. civilian authorities variously expanded, transformed, changed hands, and were reclaimed by the Iraqis and other U.S. agencies. The success of the final transition was a tribute to extraordinarily productive working relationships between the U.S. military and the U.S. embassy. However, a unified interagency approach was not the norm for earlier transitions. The first truly joint campaign plan (JCP) in Iraq was not developed until 2006. Joint civilian-military planning improved substantially after the 2008 security agreement (SA) and its companion Strategic Framework Agreement (SFA) specified an end to the U.S. military presence,[1] at which point both organizations became increasingly focused on the shared goal of preparing the U.S. embassy to operate successfully after the departure of U.S. troops.

In almost no case, however, did Iraqi transitions proceed as planned. In part, this was because much U.S. transition planning relied on assumptions and performance measures that ultimately proved overly optimistic. Transitions were also affected by the highly dynamic internal situation in Iraq; as security deteriorated, the nature of key transitions evolved, and the time lines for performance goals slipped.

---

[1]   United States of America and the Republic of Iraq, "Agreement Between the United States of America and the Republic of Iraq on the Withdrawal of United States Forces from Iraq and the Organization of Their Activities During Their Temporary Presence in Iraq," November 17, 2008a, and United States of America and the Republic of Iraq, "Strategic Framework Agreement for a Relationship of Friendship and Cooperation Between the United States of America and the Republic of Iraq," November 17, 2008b.

The war began with U.S. forces conducting traditional combat operations against Saddam's army but with relatively limited plans for postconflict activities. Over time, however, U.S. forces would come to perform a remarkably broad array of functions, shouldering substantial responsibility for the development of Iraqi security, political, and economic institutions and establishing the foundation for a long-term strategic partnership with the United States.

This volume focuses on the transition as U.S. forces drew down force levels, stepping back from these responsibilities and handing them off to Iraqi and embassy civilian authorities.

USF-I planners established several novel internal organizations and processes to develop and adjudicate the command's positions and policies related to the transition of military activities to other U.S. government agencies and to the government of Iraq. For civil-military planning and implementation of OND transitions, the command utilized existing multilevel interagency structures. Interagency "rehearsal of concept" drills were crucial to the command's ability to test the efficacy of its major operations order and to synchronize it with other U.S. military and civil agency plans. The uncertain political environment in Iraq and assumptions about a residual U.S. force in Iraq posed serious challenges to USF-I planners.

Security Challenges

Transition planners expected that Iraq would face a number of enduring security threats that would test its ability to stand on its own without U.S. military assistance and the ability of the U.S. embassy to operate effectively in what would remain a hostile environment. These challenges were well known, although the specifics were tough to address definitively prior to the departure of U.S. forces. Planners assessed that there were four interrelated drivers of instability that would affect Iraq after the transition: communal and factional struggles for power and resources within Iraq, insufficient capacity of the government of Iraq, the activities of violent extremist groups, and external interference from such countries as Iran and Syria. In particular, USF-I considered that violent extremist groups, if left unchecked, posed an existential threat to emerging democracy in Iraq. Almost to its final days in country, USF-I continued to assist Iraqi counterparts to address this threat through partnered operations and the provision of key enablers. Contingency plans envisioned the U.S. military continuing to provide the Iraqi Security Forces (ISF) with some level of technical support to fill gaps beyond 2011.

In addition, the transition planners appreciated that one of the most important lingering challenges would be managing Arab-Kurdish tensions. The concern was that, although some of the structures USF-I and the embassy had put in place to manage Arab-Kurd frictions might endure, their long-term viability would be in question. In particular, planners asked, would the Iraqis or the U.S. diplomatic contingent be able to manage force confrontations with no one filling the "honest broker" role U.S. forces had played?

By the time U.S. troops withdrew, ISF had sufficient capabilities to handle most of the internal security threats violent extremist organizations posed. This is especially true for Sunni violent extremist groups, which the government of Iraq has shown a willingness and ability to target. However, Sunni nationalist extremist groups, as well as al-Qaeda in Iraq, retain the capacity to strike at government and Shi'a civilian targets, posing a long-term risk to Iraqi stability. The government of Iraq must address these groups in a manner that does not increase the level of popular support they have with the Sunni population. Indeed, one of the successes the U.S. military began to achieve in 2007 was to drive a wedge between Sunni extremists and the Sunni population, largely eliminating the base of support necessary for a successful insurgency. Unfortunately, actions Prime Minister Maliki has taken since the departure of U.S. forces have weakened this wedge and provided more fertile ground for an insurgency to regenerate.

An equal security challenge for the Iraqi government is to garner the political will to take on Iraqi Shi'a extremist groups, such as Kata'ib Hezbollah and Asa'ib Ahl al-Haq, both of which continue to receive weapons, equipment, training, and funding from Iran. Doing so, however, would both threaten the political coalition that keeps the Maliki government in power and antagonize Tehran, which uses the groups to alter Iraqi political dynamics by dialing the level of violence up or down. Although the groups appear to have been less active inside Iraq since U.S. troops—their principal target—withdrew, they will no doubt remain capable to engage in sectarian attacks as long as they continue to receive support from Iran. The recent involvement of both Kata'ib Hezbollah and Asa'ib al-Haq in Syria working at the behest of the Islamic Revolutionary Guards Corps–Quds Force suggests that these organizations will remain violent Iranian proxies for the foreseeable future.

A Diplomatic Outpost Like No Other

The Department of State (DoS) broke new ground by undertaking the transition to a diplomatic outpost that would be self-sufficient and that would assume some of the functions similar to those a U.S. military force ten times its size had undertaken. U.S. transition plans envisaged an "expeditionary embassy" of unprecedented scope and scale to maintain U.S. influence and help the Iraqis maintain their security, political, and economic gains. Many embassies have operated in difficult security environments, depending on armored vehicles and security guards. But never before had an embassy managed support functions of this size and scale: field hospitals, a small airline, military-style life-support mechanisms, and a small army of security guards to protect facilities and ensure secure movements of personnel.

At a tactical and operational level, many aspects of the transition to a stand-alone embassy went well. Despite earlier challenges attracting qualified personnel to Iraq, DoS had relatively little difficulty finding Foreign Service staff willing to take positions in 2011 and 2012. The department established comprehensive medical and air transportation capabilities on schedule, working through obstacles created by the need to let large new contracts, unclear legal authorities, and uncertain funding. It secured aviation and support agreements with multiple countries in the region. Despite doubts about whether DoS would be able to manage large-scale contracts for security, life support, medical operations, and other necessary functions, the department improved its contract oversight capabilities, trained all deployed Diplomatic Security officers in contract oversight, and made short-term use of Department of Defense (DoD) contract managers to fill gaps. Transition planners—particularly those on the ground in Iraq—demonstrated considerable flexibility by recognizing midstream that the embassy would benefit from a robust knowledge management initiative that would ensure it could benefit from DoD's considerable collections of information.

Transition planners encountered tactical challenges as well. Although DoS and USF-I had agreed at senior levels to transfer excess DoD equipment to the embassy (such as housing units and generators), the uneven implementation of these agreements created problems for DoS that led to construction delays and short-term operational and security shortcomings. Compounding these challenges was DoS's contracting process, which is far less flexible than DoD's, making it difficult to adapt rapidly to changes. Congress's unwillingness to fully fund the Police Development Program required DoS's Bureau of International Narcotics and Law Enforcement Affairs to scale the program back dramatically. After the departure of USF-I, increased security threats limited the ability of even the reduced numbers of trainers to get to training sites or demonstrate their effectiveness to U.S. or Iraqi stakeholders, and the program was even further reduced. It is important to note that the Coalition Police Assistance Training Team that the CPA started in 2003 was also unable to effectively train the numbers of police required. Taken together, the failure of both the Police Development Program and the Coalition Police Assistance Training Team calls into question whether civilian organizations have the capacity to manage a training program as large as what was envisioned in Iraq. In retrospect, the skepticism of appropriators about plans for a large civilian footprint seems prescient. More than officials directly involved in the military-to-civilian planning, appropriators were in tune with the concerns of influential Iraqis and doubted the feasibility of an extraordinarily robust U.S. civilian presence in Iraq from 2012 forward.

However, the single biggest obstacle to standing up an effective diplomatic mission in Baghdad may have been the lack of political support on the banks of both the Potomac and the Tigris. In Baghdad, the Iraqi government quickly made clear that it did not want a large-scale, highly visible official American presence in the country. Senior Iraqi officials objected to the embassy's proactive security initiatives and pressured the United States to reduce the embassy's 17,000-person footprint. As a result, mere weeks after the military's departure, DoS was forced to plan drastic cuts to embassy staffing and consider whether to close some embassy facilities. This raised concerns in Washington about whether the transition process needlessly spent billions of dollars on construction, security measures, and outside contracting for a presence that was not sustainable.

Moreover, after eight years of outsized U.S. military influence, Iraqi officials in 2012 eagerly asserted Iraq's sovereignty in ways that complicated U.S. goals. For example, the shift of the embassy personnel and logistic operations from being under the umbrella of a large-scale military presence operating under very liberal "wholesale" status-of-forces-agreement procedures to being under a traditional Vienna Convention legal regime did not go smoothly. At the time U.S. forces left Iraq, Embassy Baghdad was ten times larger than even a "normally" huge American embassy, conducting extraordinary operations, such as air movement, security, and convoy logistics. The Iraqis were not prepared for instituting new procedures for requesting visas, obtaining contractor work permits, clearing imports, and dealing with other routine matters from which the embassy was previously exempt. A more-concerted effort to engage Iraqi officials in transition planning might have generated a greater and quicker measure of host-nation support (or at least alerted the United States to the potential for future hurdles), although the contemporaneous political gridlock at senior levels of the Iraqi government meant there were no clear, empowered interlocutors with whom the embassy could collaborate. The focus of both sides on the political debate over a possible residual force made Iraqi planning all but impossible. Eventually, literally facing a possible administrative meltdown of the U.S. mission in late December 2011 through February 2012, the Iraqis (with much embassy support and assistance) managed to put workable, if cumbersome, procedures in place.

Notwithstanding these challenges, the embassy was generally prepared to assume the lead U.S. role in Iraq in December 2011. However, the long-term success of the "expeditionary embassy" created by this transition process is not guaranteed as long as Iraq remains dangerous and politically unstable. Unless the embassy's operating environment improves, DoS may have to revisit a number of central tenets of the transition plan.

The Final Retrograde of U.S. Forces

Starting in July 2011, USF-I began the operational maneuver of forces out of Iraq. This final operation entailed closing over 50 bases, including the seven large bases and logistical hubs at Taji, Victory Base, Al Asad, Kalsu, Echo, Basra, and Adder. To complete the transition, USF-I had to move 1.8 million pieces of equipment out of Iraq, requiring the movement of over 20,000 truckloads of equipment. Simultaneously, approximately 50,000 military and more than 56,000 contractors had to safely and responsibly exit Iraq by December 31, 2011. To accomplish the extremely complex mission, USF-I developed an operational maneuver plan whereby the force would slowly collapse on itself as forces departed, much like a rearward passage of lines. Each unit conducting the tactical movement was proceeded by engineers providing route clearance. Units then received ground escort from either U.S. or Iraqi forces that controlled the geographic area; priority for intelligence, surveillance, and reconnaissance support within Iraq; and aerial escort from attack helicopters. Fixed-wing air cover was available within minutes of units' locations, should they be attacked.

The 1-8 CAV Squadron was the first of 24 battalions to make this tactical movement out of Iraq on October 6, 2011. The three-day movement began in Northern Iraq in western Diyala province and continued south through Baghdad and Nasiriya through Khabari border point with Kuwait (K-Crossing), after which it transitioned into an administrative convoy movement. During the three-day movement toward Camp Buehring, Kuwait, all U.S. and Iraqi units along the route were given the mission to protect the force moving through their respective sectors. After 1-8 CAV crossed into Kuwait on October 9, it assumed the USF-I strategic reserve mission, which it and its parent organization 2/1 Cavalry Division maintained until the USF-I operational maneuver was completed on December 18, 2011. The tactical movement would be repeated another 23 times. By the end of the operational maneuver, all forces successfully completed the retrograde, suffering only one fatality, the result of an improvised explosive device on November 14, 2011, just one month before the end of the operation.

Departing Iraq with honor and success was defined to mean that military bases and facilities were turned over to the Iraqis in better condition than when they were initially occupied. It also meant transfer and transit of remaining military equipment would be completed in a manner consistent with U.S. laws and best practices of property accountability. The orderly movement of forces was completed on a time line the United States determined unilaterally. From General Austin's perspective, it meant the last soldier had to leave Iraq safely, with his or her head held high, knowing that the United States did all it could to help Iraq secure a better future for its people.[2]

The Transition's Aftermath

Iraq began its post-transition future with a raucous political system, an uneven security force structure, and (fortunately) a growing economy. Iraq's situation reflects the myriad contributions Americans and other coalition allies have made since 2003.

---

[2] GEN Lloyd J. Austin III, oral guidance given to staff, al Asad Air Base, October 15, 2011.

Nevertheless, the Iraqi people and their elected representatives must address a wide range of challenges in the years ahead, a task made more difficult by the internal struggle for power among competing groups and factions; high levels of corruption; acquiescence to Tehran on some issues affecting regional security; and an unwillingness on the part of the Maliki government and the Iraqi political system to address critical issues associated with reconciliation, reintegration, and repatriation of the Sunni minority. The growing conflict in Syria is presenting the government of Iraq with a new set of challenges as Iraqi Sunni refugees who fled to Syria to escape the violence of the Iraqi civil war are now returning to their homeland to escape the violence in Syria. These returning refugees have returned as internally displaced Sunnis, a community that could serve as a breeding ground for a future insurgency should the Maliki government not take proactive steps to reintegrate them into Iraqi society. This problem has recently been compounded by Syrian Sunni refugees, who are fleeing to Iraq to escape the civil war in their country.

The Uncertainty of the End State

The final transition for U.S. forces in Iraq was hindered by uncertainties, which made it difficult for agencies to ensure DoS would be ready to assume sole leadership of the mission on January 1, 2012. Key decisions were frequently revisited or left open-ended; necessary Iraqi approvals could not be secured because empowered Iraqi interlocutors did not exist or were not disposed to make decisions in politically uncertain times; and funds were appropriated months after programs were supposed to have started. None of these factors, by themselves, posed insurmountable obstacles, but the Iraq transition was so complex and so hemmed in by deadlines that it was difficult for agencies to adapt.

On one level, the desired end state of the transition was clear from the outset: The SA, signed by the United States and Iraq in 2008 and ratified by the Iraqi Council of Representatives shortly thereafter, provided that "all United States Forces shall withdraw from all Iraqi territory no later than December 31, 2011."[3] The companion SFA anticipated a long-term strategic relationship between the two nations, with economic and political components, scientific research, cooperation between cultural and educational institutions, and the many other aspects of a normal bilateral relationship between two friendly nation-states.[4] Yet the two agreements were ambiguous about whether there would be an enduring security relationship beyond 2011 and what form it would take after January 1, 2012.

---

[3]   United States of America and the Republic of Iraq, 2008a.

[4]   United States of America and the Republic of Iraq, 2008b.

The transition was thus dogged by uncertainty of Iraqi and U.S. interest in a follow-on U.S. troop presence. Neither the U.S. administration nor the Iraqi government had a clear position on the desirability of a follow-on presence when transition planning began in earnest in 2009 and 2010. Moreover, neither side could agree on the missions any enduring force might undertake. Most important, it was unclear whether Iraq would be prepared to enter into a second security agreement to provide any remaining U.S. troops with legal protections, a U.S. precondition for any enduring military presence.

The necessity for a two-track effort to plan for both possible outcomes complicated U.S. transition planning. Given the President's clear statement at Camp Lejeune that "I intend to remove all U.S. troops from Iraq by the end of 2011,"[5] USF-I and embassy planners prepared for a complete withdrawal of U.S. forces. At the same time, however, compartmented contingency planning efforts considered a range of possible follow-on options to provide the President with flexibility should the United States and Iraq agree on a post-2011 mission.

Iraqi officials similarly tried to straddle the fence. Driven by overwhelming popular opposition to an enduring U.S. military presence, Iraqi leaders from all parts of the political spectrum issued public statements opposing the continued deployment of U.S. forces—even though many of these same leaders privately confided to U.S. officials that they believed an enduring presence would contribute positively to internal security and hoped the two governments would find a way to extend the mission.[6] Ultimately, in August 2011, all major political factions (except the Sadr Trend) agreed in principle to support an appropriately scaled "training" mission but refused to grant immunities and protections, on which U.S. officials had insisted.[7] Unable to reach an agreement on immunities that both sides could accept, President Obama announced definitively on October 21, 2011, that "the rest of the troops would come home by the end of the year."[8]

The uncertainty about whether there would or would not be a follow-on presence affected transition planning in important ways. For example, planning guidance initially established June as the decision point where U.S. forces would have been directed to initiate the withdrawal of U.S. forces beginning in September and the transition of responsibilities to Embassy Baghdad and the Office of Security Cooperation

---

[5] Barack Obama, "Responsibly Ending the War in Iraq," remarks, Camp Lejeune, N.C., February 27, 2009.

[6] Marisa Cochrane Sullivan, "Obama's Iraq Abdication," *Wall Street Journal*, July 28, 2011, and Qassim Abdul-Zahra and Rebecca Santana, "Iraqis Want U.S. Trainers, Without Immunity," Associated Press, October 4, 2011.

[7] Tim Arango and Michael S. Schmidt, "Despite Difficult Talks, U.S. and Iraq Had Expected Some American Troops to Stay," *New York Times*, October 21, 2011.

[8] Barack Obama, "Remarks by the President on Ending the War in Iraq," Washington, D.C.: The White House, Office of the Press Secretary, October 21, 2011.

in Iraq (OSC-I). This timing would have enabled U.S. military and civilian officials to work through any unanticipated challenges while USF-I remained in Iraq. This earlier timing would also have focused the entire USF-I and DoS effort toward the transition instead of continuing to balance competing efforts with limited staffs and resources. But to allow time for the U.S. and Iraqi governments to negotiate a follow-on agreement, the decision point to execute the final phase of the overarching Operations Order (OPORD) 11-01,[9] continued to shift "to the right," with October 15—the date by which "the laws of physics" required the redeployment to begin for it to be completed by December 31—becoming the final deadline for a decision. When time for negotiations ran out and the President decided to proceed with the redeployment, what had initially been envisioned as a gradual withdrawal of forces became a steep "waterfall." When USF-I, OSC-I, and the embassy identified unanticipated transition-related challenges in November and December, there was little USF-I could do to assist in resolving these issues because of the immense requirement to reposture the force and exit Iraq in a responsible manner. Perhaps more important, given the requirements for USF-I to conduct the operational maneuver out of Iraq, USCENTCOM should have assumed responsibility for providing assistance and support to OSC-I during the last months of the operation much earlier in 2011 rather than allowing USF-I to do so until its departure.

The waterfall of U.S. force departures was designed to keep as many forces as possible in country as long as possible, both to continue the advise and assist mission and to preserve options for the President.[10] USF-I identified forces, locations, and equipment that might be involved in a follow-on mission and released them for final disposition only in the last stages of the transition. As General Austin put it: "Quite frankly, we're not pushing the Iraqis to ask us for help. All we're saying is if they are going to ask us for help, [they should know] that sooner is better for us because it will not cause us to disassemble things that we might have to spend money to reassemble at a later date."[11]

The uncertainty about an enduring presence meant that planning for Embassy Baghdad OSC-I was somewhat delinked from other transition planning processes and did not receive much senior-level attention until mid-year 2011. A security cooperation plan—crucial for guiding normal defense relationships with partners around

---

[9]   The first two numbers of an operations order indicate the year it was published, and the second set of numbers refer to the sequence. Thus, OPORD 11-01 is the first operations order published in 2011. An operations order can be modified and published as "Change 1 to OPORD 11-01." Alternatively, a new operations order can be published, such as OPORD 11-02, indicating the second OPORD published in 2011.

[10]   Ambassador Jeffrey explained that USF-I also wanted to keep significant forces in the north committed to the Combined Security Mechanism along the line between Arabs and Kurds as long as possible to mediate any possible disputes. RAND team interview with Ambassador James Jeffrey, Arlington, Va., March 12, 2012.

[11]   Jim Garramone, "Austin Gives Insight into Drawdown, Possible Aid to Iraq," American Forces Press Service, July 11, 2011.

the world—was not finalized until late in 2011.[12] Moreover, a USCENTCOM exercise program that included Iraq was not developed until late 2011—well outside the normal window for developing, scheduling, and funding this portion of a security and cooperation program.[13]

The Political Transition

There have been few, if any, comparable military-to-civilian transitions on such a scale as OND, and none was completed under such less-than-ideal security conditions. The careful process of "binning" activities to be transitioned, the 2010 JCP, and the execution of activities under OPORDs 10-01, 10-01.4, 11-01, and 11-01 Change 1 ensured comprehensive oversight of the activities being transitioned, adjustment to changing circumstances and opportunities, and fostered close civil-military cooperation throughout.

It is perhaps ironic, therefore, that a consequential factor related to the success of the "transition" was an erstwhile "line of operation" that was not transferred at all: the political aspect of the U.S.-Iraq relationship. The transition plan did not include elements of the political line of operation because the 2010 JCP made clear that they were already the responsibility of the embassy, leaving no political tasks to transition.[14] The embassy had the lead for political engagement with the government of Iraq. However, the decision to delink these activities from the transition process had unintended consequences: Transition plans failed to identify and assign measures to support this critical component of the overall mission, which likely contributed to the U.S. failure to anticipate the full effects of the rapid withdrawal of U.S. forces on the already fragile Iraqi political arena.

That is not to say that the embassy was unaware that the withdrawal of forces would likely have unpredictable political consequences within Iraq. For example, in January 2011 Ambassador James Jeffrey delivered a presentation at USCENTCOM's Washington Transition Conference in which he discussed what he considered the "five Ms of transition": money (budget and authorities), missions (what the embassy would have to do with a focus on USF-I's binning process), months (time available before December 2011), management (the tools available to do this along with the overall magnitude of the operation), and Maliki (shorthand for the actions the government

---

[12] Interview with former DoD official, January 17, 2012.

[13] Interview with USF-I J5 staff officer, Baghdad, July 1, 2011.

[14] As we describe in the body of this book, the 2010 JCP and OPORD 10-01 actually dropped references to the political relationship as a line of operation, on the basis that USF-I had little direct effect on it. Instead, the 2010 OPORD approved three lines of operations: strategic partnership, operations, and civil support and theater sustainment.

of Iraq would have to take to transition from an embassy operation under what was essentially a status-of-forces agreement to an embassy operating under the Vienna Convention).[15] However, planning for this political transition was not a focus either in Baghdad or in Washington.

The primary U.S. political objectives for Iraq in 2010 and 2011 were interrelated: to help ensure the success of government formation following the 2010 national elections with a broad-based, stable government; to ensure Iraq's security and territorial integrity; and to preserve and enhance a strategic U.S. relationship with Iraq.

It had been evident to planners that the transition, especially the planned complete departure of U.S. military forces, would influence the Iraqi political process, affecting various groups in divergent ways. For the United States, the challenge was, as Ambassador Jeffrey stated at his confirmation hearing, to "reinforce in words and deeds that the withdrawal of U.S. combat forces in no way signals a lessening of our commitment to Iraq."[16] Vice President Joe Biden's November trip to Iraq for a meeting of the U.S.-Iraq Higher Coordinating Committee (as provided for under the SFA) and meetings with Iraqi political leaders were part of this reassurance effort,[17] a process that culminated with President Obama's invitation to Prime Minister Nouri al-Maliki to visit Washington December 12–13, 2011.[18]

Yet the day after the last U.S. soldier departed, a prosecuting judge and Interior Ministry personnel precipitated one of Iraq's most significant political crises of the last few years by issuing an arrest warrant for Vice President Tariq al-Hashimi and other Sunni leaders, which in turn led Sunni leaders to boycott Iraqi political institutions. While Prime Minister Maliki disclaimed responsibility for the arrest warrant, he was most likely aware of its significance. An important question about this crisis is whether it was precipitated or aggravated by the final withdrawal of U.S. forces and, if so, whether the transition could have been managed in such a way to attenuate such a political effect. As is typical in such political analysis, it is not possible to know for sure what the counterfactual would have been. And in Iraq, even when U.S. forces were at peak levels, crippling political crises had emerged. Nevertheless, the dramatic Iraqi political events that followed the transition were sobering reminders of the limitations of the transition.

---

[15] Email correspondence between Ambassador James Jeffrey and Charles Ries, January 11, 2013.

[16] James F. Jeffrey, "Statement by Ambassador James F. Jeffrey: Senate Foreign Relations Committee," July 20, 2010.

[17] Amy Dudley, "Vice President Joe Biden: 'In America, and in Iraq, the Tide of War Is Receding,'" blog, Washington, D.C.: The White House, December 2, 2011. See also United States of America and the Republic of Iraq Higher Coordinating Committee, "Joint Statement by the United States of America and the Republic of Iraq Higher Coordinating Committee," Washington, D.C.: The White House, Office of the Press Secretary, November 30, 2011.

[18] "Obama and Maliki Back Iraq Post-War Future," *BBC News*, December 12, 2011.

Iraqis and the Scale of the U.S. Civilian-Led Presence

Complicating the transition in Iraq was the evolution in Iraqi attitudes toward the U.S. presence. The U.S. "occupation" of Iraq and its symbols, such as large military convoys, aerostats around U.S. facilities, and the ever-present armored High Mobility Multi-purpose Wheeled Vehicles, had always been distressing to the Iraqis, who are strong nationalists across the political spectrum.[19] Memories of the 2007 Nisour Square incident, in which a U.S. contractor protective detail killed Iraqi bystanders, hardened Iraqi opposition to the aggressive and highly visible U.S. security posture.

While the formal occupation of Iraq legally ended in 2004 with the establishment of the interim Iraqi government, U.S. and British forces continued to operate in Iraq with the legal authorities of occupation forces in accordance with United Nations Security Council Resolutions through the end of 2008. For Americans, the sense of the U.S. military as an occupying force ended *de jure* on January 1, 2009, with the implementation of the SA, and to a more tangible extent on July 1, 2009, when U.S. forces moved out of Iraqi cities in accordance with the SA's provisions. Even so, U.S. officials and many classes of contractors, as well as equipment and supplies, routinely entered and exited the country without inspection by Iraqi authorities. Helicopters and aerostats remained highly visible. To most Iraqis, therefore, Iraq did not fully regain its sovereignty until the last U.S. forces left the country in December 18, 2011.

As December 2011 approached, senior American and Iraqi policymakers were focused on discussions about the scope, privileges, and immunities for a possible follow-on U.S. military training mission. But also on the to-do list was the need to secure land-use agreements for the U.S. government to be able to construct facilities at a few locations chosen to support a civilian-led presence. Yet Iraqi policymakers—distracted by internal political crises, without permanent ministers of Defense and Interior, and under no deadline pressure themselves—were unwilling to authorize the U.S. diplomatic mission the land use it needed. Such a bifurcation of incentives could also apply in future stabilization mission transitions.

The lack of land-use agreements, however, was but a symptom of a broader political problem affecting the follow-on U.S. civilian presence: a widespread Iraqi allergy to the scale of the envisaged U.S. civilian footprint. Ambassador Jeffrey bluntly described this public sentiment in March 2012 by stating, "Iraqis hate us for having occupied the country for eight years, and they don't want to see us around anymore."[20]

---

[19] RAND interview with Ambassador James Jeffrey, Arlington, Va., March 12, 2012.

[20] RAND interview with Ambassador James Jeffrey, Arlington, Va., March 12, 2012.

Posttransition Goals Were Overly Optimistic

The signing of the SFA and the associated SA can be viewed as the final terms of settlement of the U.S. war in Iraq. The SA established the terms of reference guiding the conduct of the U.S. military in Iraq, including ending the combat mission and moving out of the cities in 2010 and the retrograde of US forces in 2011. However, what both policymakers in Washington and planners in Iraq largely missed was that the years following the signing of the SFA and the associated SA should have been viewed as a period of political and diplomatic change that would result in a new U.S.-Iraq relationship, requiring a fundamental reassessment of U.S. policy and strategic goals in Iraq.

At the macro level, the transition was a carefully planned deliberate handover of responsibilities for activities that the U.S. military had previously conducted. The overarching strategic and policy goals that the 2006 to 2010 JCPs established remained constant. It was not until October 2011, during a war termination assessment, that military planners fully recognized that the goals and objectives contained in the JCPs would not be viable once USF-I was no longer on the scene.

Just as the redeployment of USF-I had unforeseen political ramifications within Iraq, the end of U.S. military presence in Iraq would also have a critical impact on what the United States would be able to achieve in Iraq, especially in the short term. What gradually occurred between 2008 and 2011 was a widening gap between established strategic goals and the means and resources available to achieve them. A fundamental reassessment of the U.S. strategy for Iraq given the overall situation in Iraq and the region may have identified the resource/objectives mismatch. Such a strategic review was not undertaken, leaving Embassy Baghdad and USF-I the challenge of attempting to accomplish overly optimistic strategic and policy goals with insufficient resources. The new OSC-I was tasked with a mission set for which it was woefully under resourced. Moreover, OSC-I lacked the necessary authorities to accomplish its assigned tasks because it had to operate under the Vienna Convention framework. While an October 2011 USF-I J5 assessment identified a large number of authorities that would end with the departure of U.S. forces, there was insufficient time left to develop a transition plan for authorities. If viewed as a process, the transitions that accompanied the end of OND were part of a phase change that should have required a new strategic vision with achievable goals developed by the interagency and USCENTCOM. That type of fundamental strategic reassessment did not take place until well after the departure of USF-I.

Implications for Future Military to Civilian Transitions: Key Insights and Recommendations

The following insights and recommendations are presented as strategic- and policy-level lessons learned that military planners and policymakers should consider when crafting strategies for transitions and posttransition relationships. These insights and recommendations relate to relations with host countries and priorities for security assistance and include more-technical recommendations on civil-military coordination and cooperation, planning horizons, contracting, and knowledge management. To be successful, all these elements of successful transitional planning require long lead times and high-level commitment.

Recommendation 1
Policymakers should initiate a multiagency planning process under the direction of the White House national security staff well in advance of the anticipated transition to (1) define enduring U.S. interests in the country, (2) establish realistic goals and objectives that an embassy operating under the requirements and limitations of the Vienna Convention can achieve, (3) assess follow-on military presence and resources required to achieve desired objectives, and (4) identify authorities that the embassy and its Office of Security Cooperation will require to operate within the country.[21]

An embassy-led presence is fundamentally different from a military-led mission and must be designed to be consistent with global U.S. foreign and security policy interests and with the requirements and limitations of the Vienna Convention operating framework. The Iraq experience illustrated that a transition from a U.S. presence dominated by a major military command to one managed by a U.S. embassy is not just a matter of scale but also of kind. A fundamental transformation of the mission took place in Iraq. While a programmatic approach to what can and should be transitioned from military to civilian organizations (including an Office of Security Cooperation, within the embassy) is necessary, planning should start by identifying U.S. strategic goals for the era after the transition and only then considering how a civilian-led embassy can be set up to accomplish these goals. An approach that transfers functions "as is" from the military to the embassy may not be as effective as an approach that plans a fundamentally new mission from scratch and only looks at functional transfers once this new planning foundation is established.

---

[21] After reviewing the draft manuscript, Ambassador Jeffrey prepared a personal assessment for us, which we have enclosed as Appendix A at his request. Email correspondence between Ambassador James Jeffrey and Charles Ries, January 11, 2013.

Recommendation 2

Policymakers should secure support from relevant congressional committees on the nature and likely cost of an enduring civilian-led mission well in advance of the departure of military forces, in the context of U.S. foreign and security policy goals and in conjunction with normal budget planning cycles.

Transitions in force posture do not always imply transitions from war to peace. After the U.S. military departure, Embassy Baghdad was expected to begin performing functions that no other U.S. diplomatic post in the world must undertake. The embassy and OSC-I sites needed to operate in an insecure environment with limited force-protection capabilities and restricted movement options. Of more than 180 bilateral U.S. embassies in the world, Embassy Baghdad is the only one to have a "sense and warn" radar system, aerial surveillance drones, or a fleet of mine-resistant, ambush-protected vehicles for quick-reaction rescue of personnel *in extremis*.

In July 2010, the independent, congressionally chartered Commission on Wartime Contracting identified 14 "lost functionalities" to be expected with the departure of U.S. forces. USF-I and Embassy Baghdad identified an additional seven critical functions that the military performed that the U.S. embassy would need to assume. While such requirements and related funding authorizations were small from a DoD perspective, they were enormous from a DoS perspective because of the department's more modest resources.

Normally, U.S. executive branch agencies begin developing their budgets two years ahead of time. Furthermore, DoS and the U.S. Agency for International Development face far greater political obstacles than DoD does in getting large appropriations or supplemental appropriations to cover contingencies. Moreover, if planning efforts do not anticipate all costs, DoS and the U.S. Agency for International Development do not have anywhere near the flexibility that DoD has to reprogram funds within existing budgets to meet needs. Planning should, therefore, include options driven by different potential funding levels, and budget proposals should incorporate foreseeable requirements.

As responsibility shifts from DoD to DoS, it is also important to make sure the U.S. embassy has all the legal authorities it needs to operate after the drawdown. This did not occur during the Iraq transition and caused a number of problems that came to the surface immediately after the departure of USF-I. These included the initial inability of contractors to enter Iraq to support embassy operations, the requirement to gain Iraqi government approval for the movement of food and other goods into Iraq, the inability of the chief of OSC-I to obligate funds, and the requirement to license embassy vehicles to operate in Iraq—to name just a few. These challenges regarding authorities highlight the transformational nature of the transition from a DoD to a DoS mission.

In the Iraq transition, congressional committees consistently rejected the Obama administration's requests for increased funding to support expanded embassy operations in Iraq. In retrospect, congressional committees appear to have correctly antici-pated U.S. domestic—and Iraqi public opinion—constraints on U.S. posttransition presence better than the administration (and Baghdad-based) planners did.

Recommendation 3

Policymakers and military transition planners should initiate work early with the host nation to identify posttransition requirements and to reach firm agreements with the host nation to ensure the smooth transition and success of posttransition U.S. presence. The parameters of the scope and functions of the U.S. presence should be identified early, and, when possible, agreements should be crafted to support U.S. and host-nation needs, possibly even accommodating future varia-tions in the footprint to build flexibility into plans and programs. Such dialogues should be buttressed by outreach to other political interest groups and should be integrated with public diplomacy efforts.

In the future, posttransition circumstances and programs will depend on the security situation, U.S. objectives, and agreements with the host nation regarding the residual U.S. footprint after the military force draws down. This will require extensive engagement with senior host-nation officials and other political interests and extensive public-diplomacy efforts to ensure wider understanding of U.S. goals and objectives. For a variety of political and practical reasons, such consultations do not appear to have been systematically undertaken in the case of Iraq.

Recommendation 4

Military and civilian planners both in theater and in Washington should make a fundamental reassessment of campaign goals and objectives well before the departure of forces, recognizing that previously established campaign goals likely will not have been achieved by the end point of the transition process. Therefore, planning should rigorously prioritize efforts in advance to set the critical con-ditions for the success of the organizations that will assume some of the mili-tary force's responsibilities rather than aim to achieve all the goals and objectives established during the campaign planning process. In particular, with respect to the crucial task of training security forces, *minimum essential capability* for host country forces is the "good enough" functionality required to fulfill basic responsibilities, not equivalence to U.S. forces capabilities.

In Iraq, the mantra was that OND's JCP was conditions based and time con-strained. In reality, the JCP was conditions based and resource constrained, with time being the limiting factor. In building the ISF, the successive U.S. military transition commands operated on the basis of defining, then seeking to help the Iraqis achieve, a "minimum essential capability" for each discrete function (air, naval, special forces, combined arms, etc.) it was assumed that Iraq would need to maintain its security and sovereignty. While USF-I stopped using the term *minimum essential capability* in late

2010, when it assessed that the goals were unachievable in the time remaining, the focus remained on providing a minimal capability deemed as a necessary foundation for future development of an external defense capability. Once the security agreement made it clear that U.S. training would come to a close (or at least change significantly) at the end of 2011, the U.S. changed its aspirations for the ISF. Instead of striving to develop requirements-based competencies, U.S. officials worked to develop the minimum capabilities that would permit U.S. forces to depart, defined in practice as whatever was possible by the end of the time-constrained U.S. military presence. In the end, the capabilities identified were driven more by the reality of time available rather than the achievement of the goals established in the JCP (and approved by both DoS and DoD). However, at no time were the JCP goals and objectives modified to meet the time and resource constrains. Moreover, neither policymakers nor planners conducted the type of campaign plan reassessment that was necessary to establish achievable goals and objectives during and after the period of transition.

Recommendation 5

Military planners should make institution-building a priority effort to ensure that the progress made through training, advising, and assisting will be sustained after the transition. In planning for sustainable host-country posttransition security, the human resource functions of recruitment, training, and professionalization are more important than providing equipment and modernization. Institutional capacity must ensure that the equipment provided can be successfully used and maintained after the departure of U.S. forces.

Of greater concern than achieving tactical and operational skills competencies, however, was whether the ISF would continue on the path to professionalization. The U.S. military training and advisory mission focused significant effort on individual and small unit skills necessary to conduct tactical operations. However, much less effort was placed on creating the type of institutional capacity that would ensure the continuation of this training by the ISF after the departure of U.S. forces. While there were some success stories, at the time of the transition all ISF elements had serious institutional deficiencies in their training capabilities and thus in their abilities to sustain the process of recruiting, training, and fielding professional military and police forces.

Recommendation 6

Prior to fielding equipment packages for a host nation, military planners should critically assess the long-term capacity of the partner nation to independently sustain the equipment and systems after the departure of U.S. military, contractors, and funding. Planning for sustainable host-country posttransition security, the life-cycle management of the equipment, and the capacity and capabilities of the host country are just as important as the intended purpose of that equipment.

The gap in the ISF components' ability to sustain their equipment and systems with a mature logistics system was well known. While the USF-I Deputy Commanding General for Advising and Training, ISF, and other agencies worked to close this gap prior to USF-I's transition, the complexity of the equipment, numerous variants, and logistics management programs and processes made the task that much greater. The departure of U.S. military advisors, contractors, and funding exposed a lack of Iraqi capacity to independently sustain much of the modern equipment and systems the United States had provided. Nowhere is this more evident than in logistics and maintenance, where both institutional failings and cultural norms have worked together to impede progress.

Recommendation 7

Pretransition planning should be launched several years ahead of the transition deadline, led jointly by a general officer and a senior civilian, staffed with capable planners who are not involved in current operations, and granted all necessary authorities. Moreover, effective transition planning must proceed on the basis of seamless top-level collaboration between the senior military and senior civilian in country working together in partnership.

Civil-military cooperation in Iraq throughout the last transition was exceptionally good, and this is clearly one of the major reasons it went as smoothly as it did. The cooperation was due in large part to the commitment of the ambassador and the USF-I commanding general. They made it clear that they would take all key decisions together and demanded comparable cooperation from their subordinates. The cooperation was also a result of the increasing convergence of the core missions of USF-I and Embassy Baghdad. In particular, after U.S. forces left the cities in summer 2009, USF-I's combat mission was largely limited to counterterrorism and force protection. As a result, the primary mission of USF-I from that point forward was to set the conditions for the Iraqi government and U.S. embassy to succeed after USF-I departed. Such unity of effort is critical to a successful transition.

The Iraq security assistance transition effort began as a small cell in Multi-National Security Transition Command–Iraq well in advance of the actual transition, with a planning team that was not involved in current operations and could therefore focus on the long-range planning. However, the planning team did not have access to senior-level officials, the authority to task MNF-I (later USF-I) or USCENTCOM staffs for support and information, or a direct civilian counterpart at the embassy. These shortcomings caused challenges and delays in the cell's ability to plan for post-transition security assistance and security cooperation at the level of detail required.

Recommendation 8

A single office to manage all contracts and contractors should be established in theater early in the operation. The USCENTCOM Contracts Fusion Cell estab-

lished for Iraq is a model that could usefully inform other U.S. efforts to develop and maintain a common operating picture for the state of contracts, as well as to coordinate with and among contract owners. In addition, a mechanism must be developed to ensure individual accountability of all contractors in country to help facilitate their departure along with the military forces they support.

Since the U.S. military relied heavily on contractors, it spent a great deal of time planning for the demobilization and redeployment of tens of thousands of civilians, as well as uniformed military. Planners need to account for contract provisions for ending services and getting contractors and their equipment out of the host nation in ways that support and do not hinder the military drawdown. However, at no point in the eight-year operation did the military ever have an accurate accounting of the individual contractors who were in country. This was largely due to the fact that contractors are paid for services provided, not on the basis of the number of individuals it takes to provide the services.

Thus, the military headquarters from the initial outset of the contingency operation must oversee, manage, and prepare to terminate or hand over contracts managed by several organizations (e.g., Corps of Engineers, Logistics Civil Augmentation Program, USCENTCOM Contracting Command) both during and after the transition. Doing this well requires in-depth knowledge of major U.S. government contracts in the host nation, a designated staff lead, and a commitment to transfer that knowledge to the succeeding embassy-led team.

Recommendation 9

Transition planners should engage host-nation officials in planning for use of third-country contractors following departure of U.S. forces because immigration restrictions and political constraints may limit an embassy's ability to use contractors for specific support functions.

The remaining civilian presence in Iraq is also dependent on contractors for security and base support and, if agreed, to provide training and other services to the Iraqis. New contracts had to be in place to support the embassy well before the transition. Embassy Baghdad's reliance on private-sector support also required awareness of host-nation political sensitivities to large numbers of foreign contractors. Contractors cannot be a staffing solution unless the host nation agrees.

Recommendation 10

Future transition efforts should undertake a systematic knowledge management survey and ensure that all databases (military and contracted civilian), key leader engagement logs, assistance project files, and other vital information remain accessible to the follow-on civilian mission.

As responsibility for many functions is handed over from U.S. forces to civilian officials, there is a danger that critical information could be lost. Furthermore, it is important for staff planners and operators to get the right information when short-falls cannot be made up with manpower and money. As a result, a robust knowl-edge management effort is very important during and after the drawdown period. The knowledge management effort that Embassy Baghdad and USF-I developed provided a means of identifying and ensuring access to a wide variety of data without collecting it all in one location.

Recommendation 11

Policymakers and commanders in future transitions should resist the temptation to delay final decisions on ending operations to such an extent that rapidly ret-rograding forces create a power vacuum like the one that may have occurred in Iraq. A more-gradual "waterfall" of troops, contractors, and equipment not only would have been more logistically manageable but might also have contributed to greater political stability in Iraq.

The delay in making the final decision regarding residual forces had multiple causes, the most important being the inability to reach a U.S.-Iraq consensus on the mission of, size of, and protections afforded to any U.S. Title 10 military personnel who might remain beyond 2011. Moving the decision point forward from June to October 2011 not only created a monumental logistical challenge associated with the sharp retrograde of military personnel, contractors, and equipment and the accelerated handover of military bases to the Iraqis but also likely exacerbated a power vacuum in Iraq that Prime Minister Maliki and others immediately exploited to gain power over political adversaries. Although the exact motivations are not known, the government of Iraq initiated preemptive measures that had not been attempted previously (e.g., arrest warrants for Vice President Tariq al-Hashimi, Deputy Prime Minister Saleh al-Mut-laq, and others) in the immediate aftermath of the retrograde of U.S. forces. A more-gradual departure of U.S. forces might have reduced both the opportunity and incen-tives to make such sudden and destabilizing moves.

Recommendation 12

Policymakers, commanders, and planners should use the lessons derived from the final two years of USF-I and its transition efforts to inform critical decisions and time lines required to end large-scale military operations successfully in the future.

Making a decision to go to war is profound. Wars often change combatant coun-tries' internal political and social dynamics and affect regional and international secu-rity. How a war is fought will contribute to the postwar security environment. Finally, history shows us that the most important part of a war is how it ends, for that will set the stage for what is to follow. Despite the importance of understanding how wars end, this topic has received far less attention from historians, social scientists, and military strategists than other phases of war. The preponderance of literature about war focuses on how and why wars begin and, once initiated, how battles and campaigns are fought. This study on how the United States ended the war in Iraq is a first attempt to bridge the gap in strategic and policy thinking regarding how wars end. DoD, the broader national security community, and academia should use these lessons learned to conduct policy relevant research and analysis, including the development of joint doctrine that focuses on the strategic and operational aspects of how wars end.

Summary

It took roughly two years to wrap up a long-term, countrywide military presence in Iraq that, at its peak, involved more than 170,000 American troops; a comparably sized army of supporting contractors; and 505 bases and outposts. Political, operational, bureaucratic, and fiscal challenges arose from both the U.S. and Iraqi sides, but Embassy Baghdad was prepared to undertake its primary diplomatic missions when U.S. forces departed. More than a year later, and despite hardships and political upheaval, the embassy continues to manage a multifaceted bilateral relationship that advances the shared political, economic, and strategic interests of both nations—something that was not possible a mere ten years earlier.

It has often been said that all conflicts are *sui generis*. Each conflict has its own set of dynamics that are unlikely to be replicated elsewhere. Each transition therefore must be planned for given the unique opportunities and constraints associated with the particular conflict at hand. However, while the transition process will vary, the key lessons learned from Iraq should be used to inform policymakers and military planners as they devise future transition plans for operations given the particulars of the specific conflicts; U.S. interests; and a broader assessment of the ends, ways, and means necessary and/or available to advance the U.S. interests. The USF-I transition process was uniquely developed for Iraq, but the policy and strategic lessons learned provide important data points that can inform how to end future conflicts.

Acknowledgments

Collecting the necessary information for a study of this depth would not have been possible without the strong support and assistance RAND received from both Embassy Baghdad and United States Forces–Iraq. The authors would like to thank both Ambassador James Jeffrey and GEN Lloyd J. Austin III (USA) for their leadership and proactive support for this study. Much assistance was also received from the USF-I Chief of Staff, then-MG William B. "Burke" Garrett (USA), who was actively involved in all aspects of the development of the research plan and facilitated access to information not readily available to the general public. The authors would also like to thank then-BG Jeff Snow (USA) for making USF-I J5 staff available to support the collection of usable information from the various staff elements of USF-I and subordinate units. Much of the planning related to the RAND research team's visit to Iraq was conducted by Maj Craig Tayman (USAF) of USF-I J5 and by Kevin Lyon at Embassy Baghdad. Their assistance was crucial to the successful collection of information, coordination of staff presentations, scheduling of interviews, ground and aerial movements to various locations in Baghdad, and the provision of accommodations necessary to support the research team visit.

The authors would like to thank all those who participated in interviews or reviewed early drafts of this work—and, in many cases, did both. These include Ambassador James Jeffrey, Ambassador Larry Butler, Ambassador Peter Bodde, Ambassador Patricia Haslach, Kathleen Austin-Ferguson, LTG William B. "Burke" Garrett (USA), LTG Michael Ferriter (USA), LTG Robert "Bob" Caslen (USA), MG Edward Cardon (USA), MG Arthur Bartell (USA), MG Jeff Snow (USA), MG Jeff Buchanan (USA), MG Thomas Richardson (USA), BG Rock Donahue (USA), BG Robin Mealer (USA), Brig Gen Russell Handy (USAF), RDML Hank Bond (USN), RDML Kevin Kovacich (USN), Col Frank Rossi (USAF), Col Darryl Stankevitz (USAF), COL Steve Smith (USA), COL Dave Boslego (USA), Col Jeff Hupy (USAF), COL Mike Culpepper (USA), COL Doug Harding (USA), COL Chris Bado (USA), COL Matt Dawson (USA), COL Brian Cook (USA), COL Rich Kramer (USA), COL Jorge Rangel (ARNG), COL Tommy Steele (USA), Lt Col Randy Oakland (USAF), LTC Jerry Brooks (USA), LTC Peyton Potts (USA), CPT Richard R. Brennan III (USA), Paul Carter, James Cooke, Tim Hoffman, Anthony Rosello, and Agnes Schaffer. The authors would also like to thank the former USF-I J4 staff members who prepared an unclassified record of the how logistics were integrated to support all aspects of the operational plan, including the transition and the final operational maneuver out of Iraq. While this information was crucial to the study, only a small portion could be included in the text. The three unclassified reports they provided to RAND in December 2011 are included in this book as Appendixes D, E, and F to provide a historical record for future research. The authors would also like to the hundreds of military and civilian personnel who were interviewed for this project, both within USF-I and Embassy Baghdad. The insights they provided on the various aspects of the transition enabled the authors to capture details that otherwise would have been lost. Finally, the authors would like to thank Nora Bensahel, Douglas Ollivant, and Olga Oliker for their careful and thoughtful reviews.

Abbreviations

AAB	advise-and-assist brigade
AAFES	Army and Air Force Exchange Services
ACCE	Air Component Coordination Element
AFCAP	Air Force Contract Augmentation Program
AFSB	Army field support brigade
AMB	ambassador
ARCENT	U.S. Army Central
ARNG	Army National Guard
BCT	brigade combat team
BCTC	Besmaya Combat Training Center
BII	Balancing Iranian Influence
BOS-I	Base Operations Support–Iraq
bpd	barrels per day
BPAX	Baghdad Police College Annex
C3	U.S. Central Command Contracting Command
CAV	cavalry
CCC	combined coordination center
CCP	combined checkpoint
CD	cavalry division
CERP	Commander's Emergency Response Program

CFLCC	Coalition Forces Land Component Command
CG	commanding general
CJTF-7	Combined Joint Task Force 7
CMATT	Coalition Military Assistance Training Team
COB	contingency operating base
COIN	counterinsurgency
COS	contingency operating station
CPA	Coalition Provisional Authority
CR	continuing resolution
CSA	combined security area
CSM	Combined Security Mechanism
CT	counterterrorism
CTS	Counterterrorism Service
CWC	Commission on Wartime Contracting
D2D	door-to-door
DCG	deputy commanding general
DCG (A&T)	Deputy Commanding General for Advising and Training
DCMA	Defense Contract Management Agency
DCOS	Deputy Chief of Staff
DJ3	Deputy Director, Joint Operations
DJ5	Deputy Director, Joint Strategy and Plans
DLA	Defense Logistics Agency
DoD	U.S. Department of Defense
DoS	U.S. Department of State
DoJ	U.S. Department of Justice
DS	Diplomatic Security
EBO	embassy branch office

ECG	Executive Core Group
ePRTs	embedded PRTs
ESC	Expeditionary Sustainment Command
FAM	Foreign Affairs Manual
FEPP	Foreign Excess Personal Property
FM	field manual
FMF	Foreign Military Financing
FMS	Foreign Military Sales
FOB	forward operating base
FRAGO	fragmentary order
FY	fiscal year
GAO	Government Accountability Office
HMMWV	High Mobility Multipurpose Wheeled Vehicle
IAW	in accordance with
IEA	International Energy Agency
IED	improvised explosive device
IG	Inspector General
IHEC	Independent High Electoral Commission
IMET	International Military Education and Training
INCLE	International Narcotics, Counterterrorism, and Law Enforcement
INL	Bureau of International Narcotics and Law Enforcement Affairs (Department of State)
INSCOM	U.S. Army Intelligence and Security Command
IOT	in order to
IqAAC	Iraqi Army Air Corps
IRAM	improvised rocket-assisted mortars
IRGC-QF	Islamic Revolutionary Guards Corps–Quds Force

ISAM	Iraq Security Assistance Mission
ISCI	Islamic Supreme Council of Iraq
ISF	Iraqi Security Forces
ISFF	Iraqi Security Forces Fund
ISOF	Iraq's special operations forces
ISR	intelligence, surveillance, and reconnaissance
ITAM	Iraq Training and Advisory Mission
J1	Directorate of Joint Personnel
J2	Directorate of Joint Intelligence
J3	Directorate of Joint Operations
J35	Joint Future Operations
J4	Directorate of Joint Logistics
J5	Directorate of Joint Strategy and Plans
J6	Directorate of Joint Communications and Information
J7	Directorate of Joint Engineering
J8	Directorate of Resource Management/Comptroller
J9	Directorate of Strategic Effects
JAM	Jaysh al-Mahdi
JCC-I	Joint Contracting Command–Iraq
JCP	joint campaign plan
JCWG	Joint Campaign Working Group
JFSOCC-I	Joint Forces Special Operations Component Command–Iraq
JIATF-I	Joint Interagency Task Force–Iraq
JLOC	Joint Logistics Operations Center
JOC	joint operations command
JP	joint publication
JPT	joint planning team

JRTN	Jaysh Rijal al-Tari al Naqshabandi
JSS	joint security station
KBR	Kellogg Brown &Root
KM	knowledge management
KRG	Kurdistan Regional Government
LN	local national
LNO	liaison officer
LOE	line of effort
LOGCAP	Logistics Civil Augmentation Program
LOO	lines of operation
M-RPAT	Mobile Redistribution Property Assistance Team
MCWP	Marine Corps warfighting publication
MEC	minimum essential capability
MND-B	Multinational Division–Baghdad
MNF-I	Multi-National Force–Iraq
MNSTC-I	Multi-National Security Transition Command–Iraq
MOD	Ministry of Defense
MOI	Ministry of the Interior
MRAP	mine-resistant ambush protected vehicle
MRT	Mobile Redistribution Team
NATO	North Atlantic Treaty Organization
NDAA	National Defense Authorization Act
NSC	National Security Council
NSDD	national security decision directive
NSPD	national security Presidential directive
NTM-I	NATO Training Mission–Iraq
OBO	Bureau of Overseas Building Operations (Department of State)

OCO	Overseas Contingency Operations
OIG	Office of the Inspector General (Department of State)
OND	Operation New Dawn
OPORD	operations order
ORHA	Office of Reconstruction and Humanitarian Assistance
OSB	Operations Synchronization Board
OSC-I	Office of Security Cooperation in Iraq
OSD	Office of the Secretary of Defense
PDP	Police Development Program
PIC	Provincial Iraqi Control
PJCC	Provincial Joint Coordination Center
POL-MIL	political-military
PRT	Provincial Reconstruction Team
PSD	personal security detail
PSG-I	Partnership Strategy Group–Iraq
R&R	rest and relaxation
RGB	Regional Guard Brigade
ROC	rehearsal of concept
RPAT	Redistribution Property Assistance Team
RSO	regional security officer
SA	security agreement
SASO	stability and support operations
SAT	security assistance team
SDDC	Surface Deployment and Distribution Command
SCO-I	Senior Contracting Officer–Iraq
SDDC	Surface Deployment and Distribution Command
SFA	Strategic Framework Agreement

SIGIR	Special Inspector General for Iraq Reconstruction
SJA	Staff Judge Advocate
SLF	Senior Leader Forum
SOF	U.S. Special Operations Forces
SOFA	status-of-forces agreement
SOI	Sons of Iraq
SPOT	Synchronized Predeployment and Operational Tracker
SRD	self-redeployment
SWG	Senior Work Group
TAC	tactical command post
TAL	Transitional Administrative Law
TCN	third-country national
TRM	tactical road march
TSC	theater support command
UAV	unmanned aerial vehicle
UN	United Nations
UNSCR	United Nations Security Council Resolution
USA	U.S. Army
USAAA	U.S. Army Audit Agency
USACE	U.S. Army Corps of Engineers
USAF	U.S. Air Force
USAID	U.S. Agency for International Development
USEMB-B	U.S. Embassy Baghdad
USC	U.S. Code
USCENTCOM	U.S. Central Command
USD-N	U.S. Division–North
USETTI	U.S. Equipment Transfer to Iraq

USF-I	U.S. Forces–Iraq
USG	U.S. government
USN	U.S. Navy
VBC	Victory Base Complex
WG	working group
WMD	weapons of mass destruction
WPS	Worldwide Protective Services

PART I

Setting the Stage

CHAPTER ONE

Introduction: How Wars End

Let us learn our lessons. Never, never, never believe any war will be smooth and easy, or that anyone who embarks on that strange voyage can measure the tides and hurricanes he will encounter. The Statesman who yields to war fever must realize that once the signal is given, he is no longer the master of policy but the slave of unforeseeable and uncontrollable events. Antiquated War Offices, weak, incompetent or arrogant Commanders, untrustworthy allies, hostile neutrals, malignant Fortune, ugly surprises, awful miscalculations—all take their seats at the Council Board on the morrow of a declaration of war. Always remember, however sure you are that you can easily win, that there would not be a war if the other man did not think he also had a chance.

—Winston Churchill[1]

Overview

In March 2003, the United States and a number of important allies invaded Iraq in what they envisioned as a war with limited objectives. On April 4, National Security Advisor Condoleezza Rice stated that the United States sought to

> help Iraqis build an Iraq that is whole, free and at peace with itself and with its neighbors; an Iraq that is disarmed of all WMD [weapons of mass destruction]; that no longer supports or harbors terror; that respects the rights of Iraqi people and the rule of law; and that is on the path to democracy.

She continued to pledge that the United States would work with coalition partners and international organizations to rebuild Iraq and then leave Iraq "completely in the hands of the Iraqis as quickly as possible."[2]

---

[1] Winston Churchill, *My Early Life: A Roving Commission*, London: Thornton Butterworth, Ltd., 1930, p. 246.

[2] Condoleezza Rice, "Dr. Condoleezza Rice Discusses Iraq Reconstruction," press briefing, April 4, 2003.

By the end of April 2003, the Iraqi Army had been defeated, and coalition forces exercised control of Iraq. With military objectives met, it was time to shift emphasis from combat operations to civilian-led reconstruction activities with the hopes of establishing the foundation for a democratic and prosperous Iraq that would be free from the brutal totalitarian rule of Saddam Hussein.

Unfortunately, the accomplishment of the military objectives did not rapidly translate to the achievement of the broader political goals President George W. Bush had established. Not all involved in the operation shared the expectation of a quick transition to Iraqi sovereignty that Rice had articulated. Rice and other senior members of the Bush administration initially believed that the Iraqi people would welcome "an opportunity to build a better future" and would not do anything to "blow it up" after suffering years of brutality at the hands of Saddam Hussein.[3] Such a belief was an integral assumption of U.S. Central Command's (USCENTCOM's) invasion plan, but planners in the Coalition Forces Land Component Command (CFLCC)—who feared the possibility of a potential insurgency in the aftermath of destroying the Iraqi armed forces—were far less optimistic. In fact, a month prior to the initiation of hostilities, CFLCC planners concluded that "the joint campaign was specifically designed to break control mechanisms of the regime and that there would be a period following regime collapse in which we would face the greatest danger to our strategic objectives," likely to be caused by "an influx of terrorists to Iraq, the rise of criminal activity, the probable action of former regime members, and the loss of control of WMD that was believed to exist."[4]

When these concerns were brought to the attention of the CFLCC commander, LTG David R. McKiernan, the planners "failed to persuade the Commanding General [CG] and dropped these issues with little resistance."[5] Indeed, as a U.S. military officer wrote, "both the [CFLCC] planners and commander had been schooled to see fighting as the realm of war and thus attached lesser importance to postwar issues."[6] The focus of the U.S. military education system had been to fight military campaigns, not to worry about the aftermath. As GEN Tommy Franks stated to Undersecretary of Defense Douglas J. Feith in the lead-up to the invasion, "You pay attention to the day after the war, and I'll pay attention to the day of."[7] Secretary of Defense Donald

---

[3] Rice, 2003.

[4] Stephen W. Peterson, "Central but Inadequate: The Application of Theory in Operation Iraqi Freedom," research paper, Washington, D.C.: National War College, 2004, p. 10. For a more detailed assessment of military planning for postwar military operations in Iraq, see Nora Bensahel, Olga Oliker, Keith Crane, Richard R. Brennan, Jr., Heather S. Gregg, Thomas Sullivan, and Andrew Rathmell, *After Saddam: Prewar Planning and the Occupation of Iraq*, Santa Monica, Calif.: RAND Corporation, MG-642-A, 2008, pp. 5–14.

[5] Peterson, 2004, p. 11.

[6] Peterson, 2004, p. 10.

[7] General Tommy Franks and Matthew McConnell, *American Soldier*, New York: HarperCollins, 2004, p. 441.

Rumsfeld also shared this focus on the military campaign to defeat the Iraqi Army and remove Saddam Hussein and specifically provided guidance that the military plan would not address issues associated with reconstruction, nation-building, or other aspects of what are commonly called stability and support operations (SASO).[8]

Unfortunately, the end game in Iraq did not go as planned. After toppling the Saddam Hussein regime, the United States found itself embroiled in what would become an eight-year war. Iraqis' initial euphoria regarding their country's liberation soon turned into frustration over an extended foreign military presence. As time and opportunities passed, the United States faced a situation for which it had deliberately not planned—criminal acts of retribution against former regime officials, al-Qaeda sponsored terrorism, an insurgency, and finally civil war.

This book is primarily an analysis of how the United States ended its large-scale military involvement in Iraq. Operation New Dawn (OND), the final chapter of the military operation in Iraq, lasted from January 1, 2010, through December 18, 2011. During this time, the U.S. military shifted many of its missions to the Iraqi government (particularly to the Iraqi Security Forces [ISF]), USCENTCOM, the Intelligence Community, and the interagency team at the U.S. Embassy in Iraq,[9] which would come to include a new entity, the Office of Security Cooperation–Iraq (OSC-I), to oversee continued military training and assistance. To place the complexity associated with this phase of the operation into context, however, we will first review early efforts to transition responsibility for security and governance to nascent Iraqi institutions, the subsequent surge and the negotiation of the two key U.S.-Iraq agreements providing for the end of the U.S. occupation, and the longer-term framework for U.S.-Iraqi relations.

The book will then assess the three years of developments that led up to the reposturing of U.S. forces in Iraq; the formulation and implementation of policy decisions that defined the nature of the post-2011 U.S. presence; the planning for security challenges that would endure after U.S. forces departed; U.S. military training efforts, intensified by a looming withdrawal deadline, to ensure that Iraqi forces could maintain internal security on their own; and preparations to ensure that Embassy Baghdad could operate successfully and safely without the logistical support and safety net U.S. forces provided. The book will detail military and civilian planning processes—most of which were characterized by extensive civil-military collaboration—for the transition and its aftermath.

---

[8] For a more complete discussion on the failure of the Bush administration to plan for the aftermath of the invasion, see Bensahel et al., 2008.

[9] The term *U.S. Mission in Iraq* refers to the interagency team that works within the Embassy of the United States in Baghdad, under the authority of the chief of mission, the U.S. Ambassador to Iraq. The term *Embassy Baghdad* is often used as shorthand when referring to the U.S. Mission in Iraq.

The challenges associated with ending the war were, in many respects, more daunting than those involved with the initial invasion. Not since Vietnam had the U.S. military withdrawn its forces from a zone of conflict and left civilian diplomats to lead the U.S. presence in such a hostile security environment. There were no textbooks to guide planners through the complexities of this transition. Policy guidance coming from Washington would prove too broad, sometimes contradictory, and often ambiguous. While continuing to operate in a combat setting, officials within Embassy Baghdad and U.S. Forces–Iraq (USF-I) nonetheless sought to develop, monitor, and execute the transition of military activities in a manner designed to ensure honorable, successful transitions and set the stage for the success of the organizations that would exercise responsibility for these activities once USF-I departed.

As Chapter Four will discussed in more detail, the working relationships among the leadership and staffs of Embassy Baghdad and USF-I were critical to the transition. The transition was both planned and conducted by USF-I and embassy personnel working as partners with a common purpose that would result in a more-capable ISF and an embassy that would be able to conduct its diplomatic mission in a zone of conflict without the direct support of the U.S. military.

How Wars End

During the movement toward Baghdad in March 2003, then-MG David Petraeus raised the question, "Tell me, how does this [war] end?"[10] The question he asked was the question that all statesmen and policymakers should ask *before* initiating hostilities. Making a decision to go to war is profound. Wars often change combatant countries' internal political and social dynamics and affect regional and international security. How a war is fought will contribute to the postwar security environment. Finally, history shows us that the most important part of a war is how it ends, for that will set the stage for what is to follow.

Despite the importance of understanding how wars end, this topic has received far less attention from historians, social scientists, and military strategists than other phases of war. The preponderance of literature about war focuses on how and why wars begin and, once initiated, how battles and campaigns are fought. A very small body of literature exists that specifically examines what political scientists and military strategists call *war termination*.[11] The dearth of research relating to war termination has a

---

[10] This quote was the inspiration for the title of the book detailing the history of the Iraq war through the surge: Linda Robinson, *Tell Me How This Ends: General David Petraeus and the Search for a Way Out of Iraq*, New York: PublicAffairs, 2008.

[11] For more information on war termination, see Stewart Albert and Edward C. Luck, *On the Endings of Wars*, Port Washington, N.Y.: Kennikat Press, 1989; Stephen J. Cimbala, ed., *Strategic War Termination*, New York: Praeger, 1986; H. E. Goemans, *War & Punishment: The Causes of War Termination & The First World War*,

profound effect on the way both policymakers and military decisionmakers approach the decision to go to war. Policymakers can rely on a vast amount of historical information about how others have come to the decision to make war, drawing lessons learned from those who have gone before them. Likewise, military leaders have been schooled in the art of war, and they, too, draw lessons learned from military strategists and historians who have fought or wrote about historic campaigns. Unfortunately, there is little written to inform either about how to draw a war to successful conclusion.

What is it about war termination that makes it such a challenging topic for both scholars and practitioners? First of all, the bulk of prewar military planning is dedicated to perfecting the campaign plan, leaving far too little time to contemplate how even the achievement of the military objectives will advance political goals and even less on how to turn unanticipated setbacks into acceptable outcomes. Part of the reason for this is that military plans are amenable to evaluation using well-established processes and procedures, including modeling, war games, rehearsal-of-concept (ROC) drills,[12] and other forms of assessment. In contrast, the achievement of a political goal, such as establishing a democratic government that abides by the rule of law, may require incremental steps that could take a generation to achieve. Moreover, assessing how military objectives might lead to the accomplishment of overarching political goals is difficult at best. Further, while there is a certain calculus associated with the art of conducting military operations, the nature of the steps necessary to attain broad political objectives is far more ambiguous.

In the case of Iraq, the military plan assumed that coalition forces would be able to conduct operations without interference once major combat operations were over.[13] Over time, many of the initial planning assumptions proved to be false, but no plans were on the table for responding quickly to such situations.[14] For example, when

---

Princeton, N.J.: Princeton University Press, 2000; Michael I. Handel, *War Termination—A Critical Survey*, Jerusalem: Hebrew University, 1978; Jane Holl Lute, *From the Streets of Washington to the Roofs of Saigon: Domestic Politics and the Termination of the Vietnam War*, dissertation, Stanford, Calif.: Stanford University, 1989; Fred Iklé, *Every War Must End*, New York: Columbia University Press, 1971; Paul Pillar, *Negotiating Peace: War Termination as a Bargaining Process*, Princeton, N.J.: Princeton University Press, 1983; Gideon Rose, *How Wars End: Why We Always Fight the Last Battle*, New York: Simon & Schuster, 2012; and Elizabeth Stanley, *Paths to Peace: Domestic Coalition Shifts, War Termination and the Korean War*, Stanford, Calif.: Stanford University Press, 2009.

[12] Military commanders use ROC drills to ensure that all aspects of a plan are integrated with the others, appropriately resourced, and synchronized to achieve the mission and commander's intent. For a large organization, a ROC drill can take weeks to plan and a full day (or more) to execute.

[13] All military plans include assumptions around which the plan is based. When time is available, contingency plans are typically developed to address the situation if one or more of the assumptions prove to be wrong.

[14] The fact that the United States did not develop contingency plans to draw on if key assumptions embedded within the war plan failed to hold was a result of both poor policy guidance coming from the Secretary of Defense and a failure on the part of the USCENTCOM commander, GEN Tommy Franks, to even consider the possibility that Iraqi civil authorities would not continue to run essential services and that police would not continue to maintain law and order. While there was concern about the potential for lawlessness, commanders did not develop plans to address the possibility of a strong insurgency. See Bensahel et al., 2008, pp. 5–14. In fact, even as

criminal activity occurred in the wake of the invasion, the military was not prepared to respond. Likewise, when the insurgency started, U.S. forces had no plan on the shelf to counter it. And when Iran became engaged in training, equipping, and funding extremist groups to counter the U.S. occupation, the United States had no plan for how to respond.

Planning for ending a war and for transitions is also difficult because policymakers may alter the overall purpose of the war as it takes unanticipated twists and turns. One reason for this is that policymakers have great difficulty exiting a war even after it becomes apparent that the costs of the war will be greater than anticipated and are likely to adjust the ends to justify the actual costs of the war.[15] In Iraq, as the insurgency grew in late 2003 and 2004, the Bush administration chose to embark on a large-scale reconstruction program that included rebuilding host-nation security forces; developing the capacity of the government of Iraq; and instituting a Western-style democracy in Iraq that would be, it was thought, inherently aligned with the United States and other democratic countries.

However, history has clearly shown that a population increasingly sours on wars as costs increase and as fighting drags on. In Iraq, the realization that the cost of the war would be much higher than anticipated affected the American public after the first battle of Sadr City in April 2004. As fighting intensified throughout Iraq in 2004 and 2005, the strength of the domestic opposition to the war intensified.

Once it becomes clear that military forces will remain involved in a conflict for an extended period, political leaders often find themselves confronting a dilemma: Do they acquiesce to domestic political demands or intensify the effort to justify the increasing costs? Policymakers become heavily invested in the decision to go to war; once they have made a decision to start a war, it is very difficult to reverse that decision before being able to proclaim success. In such a situation, policymakers are likely to commit additional resources to implement their earlier decision rather than reverse course and acknowledge either that the war was a mistake or that the achievement of partial success is sufficient to end the conflict. As Fred Iklé notes in his book *Every War Must End*, for this reason, "fighting often continues long past the point where a 'rational' calculation would indicate that the war should be ended—ended, perhaps even at the price of major concessions."[16] A purely rational calculus would assess the current

---

the insurgency was growing, Secretary of Defense Donald Rumsfeld continued to publicly refute the existence of any insurgents, calling them "dead-enders," and banned the use of the word *insurgent* from use by DoD personnel. See James Joyner, "Rumsfeld Bans Word 'Insurgents,'" Outside the Beltway blog, November 30, 2005.

[15] Examples of the difficulty of exiting wars include the Japanese War in the Pacific with the United States, the Hungarian Revolution in 1956, and the Tibetan independence movement in 1959. See Iklé, 1971, pp. 8–15. After serving as the Under Secretary of Defense in the Reagan administration, Fred Iklé served as a distinguished scholar at the Center for Strategic and International Studies. His 1971 book was one of the first analytical efforts to explore the challenges of bringing a war to an end.

[16] Iklé, 1971, p. 16.

situation, future costs, potential risks, and expected outcomes without considering past expenditures. However, no politician wants "to ask a man to be the last man to die for a mistake," as a young John Kerry famously testified to a congressional committee during the Vietnam War.[17] Having directed the expenditure of national treasure—both human and financial—on a war, political leaders are resistant to end a conflict unless they can claim that the results justified the costs.

As a war continues to drag on with increasing costs and risks, new plans must be developed to guide military operations. Should national policymakers conclude that a war is spiraling out of control, the increasing fear associated with the costs and risks of the war is most likely to cause them to demand a new approach. While policymakers always have the option of engaging an adversary to develop terms of settlement, uncertainty about the enemy's intentions and the likely costs of potential terms creates a perverse incentive to continue the war in the hopes of gaining a better position for negotiations. Consequently, to ensure that the war does not end on unfavorable terms, national policymakers will often approve new military plans that have the effect of expanding or prolonging the conflict.[18]

This is what occurred in Iraq. The early decision the Bush administration made to keep a small postcombat footprint in Iraq resulted in the deployment of far fewer troops than were necessary to constrain criminal activity and, later, to counter a growing insurgency.[19] As noted in previous RAND Corporation reports, "[t]he United States entered Iraq with a maximalist agenda . . . and a minimalist application of money and manpower."[20] But in 2006, as Iraq slipped deeper and deeper into a civil war and domestic opposition grew, President Bush directed that new options be presented to him to address the increasingly perilous situation. While the majority of senior officials within the Department of Defense (DoD) supported the continuation of the existing plan to transition responsibility to the ISF as the middle ground between escalation and withdrawal,[21] others in the administration concluded that this was not a feasible option.[22] During a December 2006 NSC meeting, President Bush stated:

---

[17] John Kerry, *Testimony of John Kerry, Legislative Proposals Relating to the War in Southeast Asia, Hearings Before the Committee on Foreign Relations*, U.S. Senate, 92 Cong., 1st Sess., Washington, D.C.: Government Printing Office, April 22, 1971.

[18] For a discussion of historical examples where this form of "peace by escalation" occurred, see Iklé, 1971, pp. 38–58.

[19] Michael R. Gordon, "Troop 'Surge' in Iraq Took Place Amid Doubt and Public Debate," *New York Times*, August 31, 2008, p. 41.

[20] James Dobbins, Seth G. Jones, Benjamin Runkle, and Siddharth Mohandas, *Occupying Iraq: A History of the Coalition Provisional Authority*, Santa Monica, Calif.: RAND Corporation, MG-847-CC, 2009, pp. xxix.

[21] Donald Rumsfeld, "DoD News Briefing with Secretary Rumsfeld and North Korean Minister of National Defense Yoon Kwang-Ung at the Pentagon," October 20, 2006.

[22] Interview with a senior official within the Bush administration National Security Council (NSC), January 4, 2007.

I want to make clear what I see as the options here. We can hold steady. None of you say it is working. We can redeploy for failure. . . . Or, we can surge for success.[23]

In the end, President Bush's decision to pursue the surge expanded the scope, scale, and commitment of the United States to the rebuilding of Iraq. In essence, President Bush abandoned the short war he had approved in 2002 for a "long-war" plan that had been largely rejected during prewar deliberations. The new counterinsurgency (COIN) strategy required a temporary increase in the number of troops to establish and maintain security as a prelude to the most expansive nation-building effort the United States has undertaken since the reconstruction of Germany and Japan after World War II.[24]

As domestic debate grows more acrimonious, national leaders struggle to find an acceptable way to stop the fighting and end the war. As fighting continues, government leaders engage in negotiations to end the conflict, each nation seeking to solidify gains that will assure them of a better future. In Iraq, the bulk of these negotiations took place in 2007 and culminated with the signing of the Strategic Framework Agreement (SFA) and the associated security agreement (SA).[25] The latter agreement established conditions governing the future conduct of U.S. forces and established a time line for the gradual withdrawal of U.S. forces from Iraq. Pending any subsequent agreements between Iraq and the United States, the SFA and SA would serve as the final negotiated settlement ending the U.S. military's involvement in Iraq.

The Final Transition in Iraq

The existing literature on war termination highlights the fact that how a war ends is at least as important as how it started because the end of the war establishes conditions governing the new peace.[26] This lesson of history was not lost on the U.S. military and civilian leaders in Baghdad. The signing of the SA codified the broad parameters of

---

[23] Bob Woodward, *The War Within: A Secret White House History 2006–2008*, New York: Simon & Schuster, 2008, p. 292.

[24] For a detailed discussion of U.S. involvement in nation-building, see James Dobbins, John G. McGinn, Keith Crane, Seth G. Jones, Rollie Lal, Andrew Rathmell, Rachel Swanger, and Anga Timilsina, *America's Role in Nation Building: From German to Iraq*, Santa Monica, Calif.: RAND Corporation, MR-1753-RC, 2003.

[25] United States of America and the Republic of Iraq, "Strategic Framework Agreement for a Relationship of Friendship and Cooperation Between the United States of America and the Republic of Iraq," November 17, 2008b (the SFA), and United States of America and the Republic of Iraq, "Agreement Between the United States of America and the Republic of Iraq on the Withdrawal of United States Forces from Iraq and the Organization of Their Activities During Their Temporary Presence in Iraq," November 17, 2008a (the SA). For the sake of brevity, the remainder of this report uses these abbreviations to refer to and to cite these documents.

[26] For a discussion on this issue, see Iklé, 1971; Rose, 2012; and Goemans, 2000.

how the U.S. military would exit Iraq, unless modified by a subsequent agreement.[27] While history records this as the last agreement between Iraq and the United States regarding the end of the war, the reality is that the SFA and SA were just the beginning of the bargaining and negotiations that would govern the final years of the U.S. military in Iraq. Indeed, while the SFA and SA established the framework for a future U.S.-Iraqi relationship, the details of this new relationship would be worked out by the ambassador and embassy staff, along with the commander of the U.S. forces and his staff, starting in January 2009.

The lesson from history of the criticality of how a war ends was always a factor governing the conduct of USF-I. GEN Raymond Odierno, commander of Multi-National Force–Iraq (MNF-I) from October 2010 until the end of August 2011,[28] would frequently tell his subordinate commanders and staff to ensure that the conduct and actions of all units and soldiers comported with both the letter and the spirit of the SFA and SA.[29] Starting in December 2008, MNF-I and Embassy Baghdad personnel began negotiations with their Iraqi counterparts to address the conduct of U.S. military operations with a sovereign Iraq governed by Iraqi law. By March 2009, processes and procedures in place dictated that the U.S. military could only conduct targeted operations pursuant to a warrant issued by an Iraqi court, using Iraqi rules of evidence.[30] The result was that, even though the American military might know through intelligence that a particular person was involved in conducting acts of violence, U.S. forces were unable to do anything toward that person except for immediate self-defense.[31] While this was frustrating for military personnel who were being targeted by extremists using Iraqi law to shield themselves from U.S. and Iraqi military actions, General Odierno would constantly remind his subordinates that the United

---

[27] The possibility of entering into a status-of-forces agreement (SOFA) or, more likely, some other type of formal bilateral agreement approved by the Iraqi Council of Representatives that would govern a follow-on military force in Iraq beyond 2011 remained until October 2011, when it became clear that the United States and Iraq could not find mutually acceptable terms regarding either the size of the force package or the legal protections afforded to military and DoD civilian personnel serving in Iraq.

[28] While MNF-I was disestablished on January 1, 2010, General Odierno continued to command its successor organization, USF-I, through the end of August 2010.

[29] Notes from a deployed RAND analyst serving on the MNF-I/USF-I staff between September 30, 2008, through December 16, 2011.

[30] The rules of evidence in Iraq differ from those in the United States in many ways. Of critical importance in Iraq is eyewitness testimony, without which it was extremely difficult to obtain a warrant, regardless of the other forms of direct evidence. Between 2009 and 2011, Iraqi judges became more open to issuing warrants based on direct evidence other than eyewitness accounts. However, unlike in the United States, circumstantial evidence is not acceptable in an Iraqi court. Interview with USF-I Staff Judge Advocate (SJA) staff officer, Baghdad, January 15, 2011.

[31] U.S. forces always retained the right to maintain force protection and take actions necessary to defend U.S. personnel, bases, and facilities. However, such action could only be taken to eliminate an imminent threat, not a potential threat sometime in the future. Interview with USF-I SJA staff officer, Baghdad, January 15, 2011.

States was a "guest in Iraq" and had to act in a manner that was conducive to building a "long-term strategic partnership" with Iraq.[32] This point was repeatedly made as U.S. forces moved out of Iraqi cities in June 2009 and during the subsequent disestablishment of MNF-I on January 1, 2010.

On September 1, 2010, GEN Lloyd J. Austin III assumed command of USF-I. Like his predecessor, General Austin was aware that the actions of the military conducted during the last months of the war were critical to the long-term success of the eight-year war in Iraq. While General Austin did not select the USF-I motto, "Honor and Success," it was consistent with his intent that U.S. forces would always interact with their Iraqi counterparts and the Iraqi people in a manner that was respectful of Iraqi culture, rule of law, and the spirit of the SFA and SA, carefully balanced with the requirements for force protection and the accomplishment of mission requirements. Negotiations and bargaining with the government of Iraq continued throughout the tenure of USF-I to both address immediate challenges and establish the conditions that would govern post-2011 U.S. military involvement with Iraq.[33] Unfortunately, because of the inability of Prime Minister Nouri al-Maliki to form a government in the aftermath of the 2010 national elections, the Obama administration delayed initiation of the necessary negotiations and bargaining involving the possibility of a follow-on military force that would continue to conduct training and advisory missions in Iraq beyond 2011.[34] Regardless, leaders in USF-I concluded that, should a residual force be agreed on, it would have the following three missions: (1) train and assist Iraq's special operations forces (ISOF) so they could continue to target al-Qaeda in Iraq; (2) train and assist the Iraqi Army, Air Force, and Navy so they could develop the capabilities necessary to maintain internal security and defend themselves against an external threat; and (3) assist in preventing the Kurdish *peshmerga* and the Iraqi military from engaging each other in combat along the contentious disputed boundaries between the Kurdish region and the remainder of Arab Iraq. Of course, sufficient U.S. forces would need to be available to provide self-protection and support to Embassy Baghdad in the event of an emergency.

Negotiations between the United States and Iraq to replace the SA with a new bilateral agreement authorizing a residual force took on new emphasis during summer 2011. In mid-July, a U.S. delegation consisting of Deputy Secretary of State Thomas

---

[32] Notes from a deployed RAND analyst serving on the MNF-I/USF-I staff from September 30, 2008 through December 16, 2011.

[33] A particularly daunting challenge was how to address threats to U.S. personnel, bases, and facilities emanating from Shi'a extremist groups supported by Iran. The concern over force protection was an issue of repeated, and sometimes intense, discussion with Prime Minister Maliki and other senior Iraqi government and military officials.

[34] It should not be presumed that, given more time, the final outcome would have been any different. However, the delay in starting the negotiation process most likely delayed the final decision to end the mission, and this delay clearly affected the conduct of the transition and redeployment of military forces.

Nides, Deputy National Security Advisor Denis McDonough, and National Security Advisor to Vice President Biden Antony Blinken met with Prime Minister Maliki and informed him that time was running out to request a residual U.S. force, setting out the four conditions President Barack Obama had established:

1. Iraq had to make a formal request for such assistance.
2. The prime minister would have to get parliamentary approval for a SOFA governing the residual force.
3. Prime Minister Maliki would need to appoint a Minister of Defense.
4. Prime Minister Maliki had to make a commitment that he would protect this force against Iranian-backed extremist groups.[35]

As the departure of U.S. forces became imminent, Prime Minister Maliki took the matter of a residual military presence to the Iraqi Parliament in early October. Although the parliament approved the presence of residual military trainers, it would not approve legal immunities for military personnel in Iraq.[36] On October 21, President Obama held a video teleconference with Prime Minister Maliki and informed him of his decision that negotiations were over and that all U.S. forces would depart Iraq by the end of 2011.[37] The role of the United States military in Iraq would revert to the type maintained with other countries in the region, and the U.S. military's role in the Iraq war would end no later than December 31, 2011.

Starting with the U.S. withdrawal from Iraqi cities in June 2009 and continuing until the departure of the last U.S. soldier in December 2011, USF-I was not only advising and assisting Iraqi troops in a combat environment; it was also conducting a deliberate transition of responsibilities. This so-called transition may therefore also be viewed as a military operation conducted pursuant to a national decision to end a war. The transition of USF-I consisted of the

> transfer, transformation, completion or termination of tasks, programs, projects or relationships that are owned, performed or managed by a military organization engaged in combat, COIN, SASO, or any other military operation to the host nation, U.S. embassy and OSC, follow-on DoD organizations operating in country, or any other [U.S. government] or international entity.[38]

This book details this transition process in an effort to develop strategy and policy lessons that may facilitate any future transition of U.S. military missions to civilian entities—including such efforts in Afghanistan.

---

[35] Michael R. Gordon and Bernard E. Trainor, *The Endgame: The Inside Story of the Struggle for Iraq, from George W. Bush to Barack Obama,* New York: Pantheon, 2012, p. 667.

[36] Gordon and Trainor, 2012, p. 670.

[37] Gordon and Trainor, 2012, p. 670.

[38] Interview with USF-I Chief of Staff MG William B. "Burke" Garrett, USA, Baghdad, September 15, 2011.

It is important to note that, while the U.S. military's transition and operational maneuver from Iraq was an important part of ending the war, so too was the management of the political transition that occur when the forces leave a country. Using the word *transition* to address the departure of forces is instructive because it does not suggest an end point. Rather, the withdrawal of forces from a zone of combat establishes the baseline for the next transition. For example, while USF-I ended its advise-and-assist mission and departed Iraq, the United States remains committed to continuing the development of ISF through the efforts of the embassy-based OSC-I and routine interactions between the Department of State (DoS) and DoD and their counterparts at the Iraqi ministries of Foreign Affairs and Defense. The partnership the SFA established serves as the conduit from one transition to the next. While it is still not possible to know "how this ends," it is clear that the ongoing transition will continue for years to come.

The Three Faces of the War in Iraq

The conflict in Iraq started as an interstate war between the United States and Iraq. The scholarly literature on war termination discussed earlier directly relates to how nations end wars with one another. It is important to highlight the fact that the transition discussed here ends only the interstate portion of the war in Iraq. Two other types of conflict emerged after the 2003 invasion: an insurgency in 2004 followed by a bloody civil war beginning in 2005. While the interstate war between the United States and Iraq has ended and while the U.S. military has departed following its mission to advise and assist the ISF, what remains are the remnants of an insurgency and the unresolved issues that caused the start of the Iraqi civil war.

Prior to the 1980s, very little scholarly work had been done that related to how to end civil wars. Since that time, however, a robust body of literature has been written as a result of the internal conflicts that characterized countries around the world during the late 1980s and the 1990s.[39] A special subset of that field of study looked at the role of outside mediators and international organizations, such as the United Nations (UN), to help warring factions draw a civil war to an end.[40] Combined, these bodies of literature examined the conditions that lead to a political settlement and conflict reso-

---

[39] See, for example, Roy Licklider, ed., *Stopping the Killing: How Civil Wars End*, New York: New York University Press, 1993; I. William Zartman, *Ripe for Resolution: Conflict and Intervention in Africa*, New York: Oxford University Press, 1989; and Carnegie Commission, *Preventing Deadly Conflict: Final Report*, New York: Carnegie Commission, 1997.

[40] In 1992, the UN Secretary-General published a report entitled, "An Agenda for Peace: Preventative Diplomacy, Peacemaking, and Peacekeeping." That report spawned substantial academic and policy analysis that examined how the capabilities of the UN might be increased to enable it to conduct these types of operations more effectively. See, for example, Carnegie Commission, 1997.

lution. For example, one side to the conflict could realize that it will lose the war and will seek to end it while still having leverage in negotiations. Or there could emerge a "mutually hurting stalemate," in which the costs of war have become so great that a general war weariness causes both sides of the conflict to seek acceptable accommodations to end the conflict.[41] In any event, civil wars tend to end in negotiated settlements between belligerents, often accompanied by terms of settlement that include disarmament, demobilization, and resettlement. The key to ending a civil war is the institution of formal processes and procedures that ensure reconciliation and reintegration. It is often important to have a neutral third party facilitate and later monitor compliance with the terms of settlement. This never took place in Iraq. Moreover, unless there is an outright victor, the terms of settlement in a civil war often require rewriting the country's constitution, bringing together the power brokers representing the various sides involved in the fighting. This also did not happen in Iraq. As will be discussed in Chapter Two, while the Coalition Provisional Authority (CPA) did press the Iraqis to quickly write a constitution and hold national elections, the "de-Ba'athification" process ensured that the group of people who formerly ruled Iraq would have little voice in the future, setting the stage for the civil war that erupted in 2005.[42]

The third type of war that existed in Iraq was a widespread Sunni insurgency fueled by multiple external actors, including al-Qaeda. While the United States had largely ignored COIN since the end of Vietnam, the Sunni insurgency that began to emerge in late 2003 resulted in a fundamental reassessment of this type of warfare. Embedded within the COIN doctrine developed in 2006 was the assessment that, to defeat an insurgency, the host nation needed to gain and maintain legitimacy with the population.[43] Consequently, building the political, economic, and security capacity of the host nation was viewed as central to conducting a successful COIN operation. While building the capacity and legitimacy of the government, a COIN strategy must also seek either to delegitimize the insurgents and separate them from the population or to find a mechanism to being the insurgents back into the political process. However, as the host government increases its capacity, especially in the security sector, the less likely it is that the government will seek a negotiated settlement with insurgents. Thus, a successful COIN campaign that strengthens the capacity of the government to conduct COIN operations may actually undermine the long-term resolution of a civil war because those who hold governmental power have little incentive to establish and enforce reconciliation and reintegration programs. And once the third-party military

---

[41] For a more complete discussion of this concept, see Licklider, 1993, p. 72.

[42] See Toby Dodge, *Iraq: From War to a New Authoritarianism*, London: International Institute for Strategic Studies, December 2012.

[43] See for example, U.S. Army and U.S. Marine Corps, *Counterinsurgency Field Manual*, FM 3-24/MCWP 3-33.5, Chicago: University of Chicago Press, 2007; Joint Publication 3-24, *Counterinsurgency Operations*, Washington, D.C.: Joint Staff, October 5, 2009; and Ben Connable and Martin Libicki, *How Insurgencies End*, Santa Monica, Calif.: RAND Corporation, MG-965-MCIA, 2010.

departs, the "winners" are free to seek retribution and/or consolidate power in structural ways to ensure their long-term dominance.

Thus, what is known from these three bodies of literature is that actions and activities that are necessary to bring an end to one type of conflict may actually create new challenges that can cause another type of war to emerge.[44] So, for example, by strengthening the Shi'a-led government in Iraq and helping it fight an insurgency, the United States was doing exactly the opposite of what the literature on peace operations and civil war resolution requires, which would have been using neutral third-party mediators to facilitate reconciliation and reintegration of the portion of society opposed to the government.[45]

What is clear is that the conflict in Iraq is not over. What is over is the interstate war between the United States and Iraq. The bargaining and negotiation regarding a potential residual military presence, as well as the discussions regarding enduring bases and facilities for the Embassy Baghdad, mark the last chapter of the interstate war that began in 2003. The remaining remnants of the Sunni insurgency and the unresolved issues that caused the Iraqi civil war remain elements of the conflict that the Iraqis will have to solve themselves.

Overarching Themes

Several overarching themes recur throughout the book and will be elaborated on in the conclusion. The first is that military was in constant state of transition starting in 2003 with the invasion and continuing until USF-I departed Iraq in December 2011. While the word *transition* has meant many things over time, it has always reflected changes in roles, functions, and responsibilities in response to changes in the security situation and mission. The second major theme is that, from the very beginning of the Iraq war, planners and policy developed overoptimistic goals and objectives based on planning assumptions that would later prove inaccurate. A third theme derives from the fact that U.S. plans and objectives for Iraq were often made without buy-in from the Iraqi government. That is not to say that the leadership of Iraq did not "agree"

---

[44] This observations was also made in Heather S. Gregg, Hy S. Rothstein, and John Arquilla, eds., *The Three Circles of War: Understanding the Dynamics of Conflict in Iraq*, Herndon, Va.: Potomac Books, 2010, pp. 1–10.

[45] A key concept of peacekeeping is that a neutral third party should practice impartiality when dealing with either side of the conflict. As the civil war in Iraq grew in 2006 and 2007, the U.S. military began to employ what was called *balanced targeting*. At the tactical level, this was consistent with the concept of impartiality. Unfortunately, the United States was also a major party to the conflict, and although it later attempted to assume the role of a neutral third party to facilitate reconciliation and reintegration, it was simultaneously building Iraqi military capacity, causing both the Sunni and Kurdish minorities to fear what would occur after the U.S. military departed. These observations and a number of others throughout this book are from contemporaneous notes author Richard R. ("Rick") Brennan, Jr., took while serving as senior advisor to the MNF-I/USF-I Directorate of Joint Operations (J3), Baghdad, 2009–2011.

in face-to-face discussions. However, these personal discussions actually reflected not buy-in but temporary acquiescence. Finally, as can be seen throughout this book, all U.S. participants, both in Iraq and in Washington, were focused on the operational aspects of the transition. Unfortunately, U.S. policymakers and planners did not proactively consider the transformative nature of the withdrawal of U.S. military forces and the effects that transformation would have on strategic- and policy-level issues for both Iraq and the region.

Organization of the Report

This book is organized in four parts. Part I establishes the framework and background for the final transition and redeployment of U.S. forces. Chapter One has provided an overview of war termination to provide a context for how the study of the USF-I transition fits into what we know about the complexities associated with ending wars. The chapter also highlighted the fact that, once a decision is made to start a war, the conflict is likely to take many unexpected twists and turns, often ending in a way unimagined at its inception. The remainder of the book will tell this story. Chapter Two provides a historical overview of the first six years of the war. It focuses on the various attempts to transition responsibility for security from the U.S. military to the Iraqis and the eventual decision to surge U.S. forces and expand the mission to include far-reaching nation-building goals. A central thread that will be introduced in this chapter is that, from 2003 onward, the U.S. military repeatedly sought to transition responsibility to the newly established government of Iraq but was continually thwarted by a combination of increased security threats and a lack of Iraqi capacity to maintain their own security. As a consequence, U.S. forces were in a constant state of transition. Thus, while the original plan developed in 2002 envisioned a quick end to the war and the redeployment of all U.S. forces by December 2003, the war took a turn that resulted in a significantly longer, costlier, and more deadly conflict than planners and policymakers had anticipated. Chapter Three highlights the planning and initial transitions that took place during first eight months after the establishment of USF-I in January 2010. Such preliminary transition planning served as the foundation that would later be built on when General Austin assumed command in September 2010. For those familiar with the details of the Iraq War, much of what is contained in Chapters Two and Three will serve as a review. However, these two chapters provide important details for those who are not familiar with the Iraq War. In either case, understanding the conditions that led up to the beginning of OND in January 2010 is crucial to understanding the transitions that occurred during the following two years.

Part II details the operational planning and management of the final transition, including impediments to the transition. The largest and most complex portion of the transition in Iraq took place during OND, during the tenure of USF-I. To manage the complexities of the transition process and the retrograde of forces, USF-I and Embassy Baghdad established a transition management process that will be described in Chapter Four. As will be discussed, this process was independent of, but in parallel with, internal USF-I planning processes. Chapter Five provides a detailed assessment of policy decisions in both Washington and Baghdad that influenced transition planning and execution.

Part III provides the details associated with how USF-I and Embassy Baghdad executed the transition and retrograde of U.S. forces. As introduced in Chapters Two and Three, the inability of the ISF to address new and emerging threats forced the United States initially to provide security for the Iraqi populace. U.S. forces would later partner with and then advise and assist the Iraqi forces as the Iraqis increasingly took the lead in providing security. Chapter Six will elaborate on the enduring security threats Iraq faces, providing some historical background on the source and strength of the threats that the Iraqis will now have to confront on their own. This chapter will also describe how USF-I transferred its responsibilities for security-related activities during the final years of the operation. Given these enduring threats, Chapter Seven details the extensive efforts made to equip and train ISF so they could effectively maintain internal security. This chapter will also highlight the efforts made to provide the Iraqi military with a foundation so that it will one day also have the capabilities necessary for external defense. Chapter Eight details the myriad challenges associated with the establishment of an "expeditionary" embassy that could conduct normal diplomatic activities in an extremely hostile security environment. This chapter also examines the fiscal and political challenges associated with the unprecedented security and life-support infrastructure such a diplomatic mission required. Chapter Nine establishes the historical record of how USF-I conducted the operational maneuver and redeployment of forces out of Iraq, focusing on the period between September and December 2011. This chapter not only addresses the movement of forces and contractors out of Iraq but also explains how base closures and the final disposition and accountability of equipment was integrated into the operational reposturing of the force.

Part IV seeks to capture the strategic and policy lessons learned from U.S. efforts to end the war in Iraq. Many events occurred in 2012, after the departure of U.S. forces, that can be used to illuminate some of the lessons learned. Therefore, Chapter Ten examines the aftermath of the transition, depicting the subsequent political crises and continued security challenges that took place in 2012, as well as the ways in which the U.S. embassy adjusted its presence and operations to accommodate unanticipated security challenges and political pressures from governments in both Washington and Baghdad. Finally, Chapter Eleven offers our overall assessment of the entire transition continuum and identifies lessons that could be applied to future similar transitions. It is often said that conflicts are *sui generis*. Similarly, the lessons learned and recommendations developed in this book are derived from the unique situation in Iraq. Consequently, the transition process in Iraq cannot be simply replicated and improved on. Rather, the lessons learned from Iraq should be considered as important factors that policymakers should carefully consider when confronting the next challenge of ending a war and establishing normal diplomatic relations with a host country.

CHAPTER TWO

The First Six Years

Overview

Since 2003, the term *transition* has meant different things to different stakeholders at various times. Indeed, words used publicly to describe U.S. efforts in Iraq and the nomenclature of U.S. military organizations suggest a perpetual state of transition throughout the eight years of war. For example, prewar planning anticipated a "transition to democracy" in Iraq; assessments of Iraqi capacity were done by means of "transitional readiness assessments"; military "transition teams" were embedded with Iraqi units and ministries; and the major U.S. command with responsibility for training ISF for much of the war was the "Multi-National Security Transition Command." This chapter will examine the military transitions conducted from prewar planning through the end of 2008.

Prewar Planning

The early history of transitions in Iraq is a story about how the military mission was originally envisioned, how it expanded and evolved over time, its repeated attempts to transfer military activities to the Iraqis, and the changing relationships between military and civilian authorities. This dynamic began even before the U.S. invasion of Iraq as the prewar planning for combat and postcombat operations in Iraq set the stage for the military transitions to follow. As we will discuss here, prewar assumptions about military roles for securing the postcombat environment proved to be wildly optimistic.

As with most wars, the war in Iraq rarely proceeded as planned. This did not reflect a lack of forethought on the part of the U.S. government so much as planning for a postconflict environment that never emerged. Previous studies have documented that this failure was due largely to the adoption of a set of planning assumptions that went largely unchallenged and the failure to develop contingency plans in the event the assumptions failed to hold up.[1] For the purposes of this study, we will focus on two key elements: interactions between DoS and DoD regarding responsibilities for postwar activities and the assumed postcombat environment, which largely failed to materialize. These two critical aspects governed the conduct of the final transition in 2011.

The march to war began in fall 2001, when Secretary of Defense Donald Rumsfeld tasked USCENTCOM with developing a plan to remove Saddam Hussein from power.[2] From the outset, DoD planners assumed that U.S. forces would not face steep opposition after combat operations ceased and would quickly be able to transfer governing responsibility to an interim Iraqi authority.

DoS officials, however, developed a more cautious view, suggesting instead that regime change in Iraq would be significantly more complex and challenging.[3] In October 2001, DoS's Bureau of Near Eastern Affairs launched the "Future of Iraq Project," a broad effort to anticipate and plan for postwar challenges. The initiative recruited professional (e.g., nonpolitical) Iraqi exiles and organized them into 17 working groups charged with developing ideas for managing critical sectors of government and society—such as defense, public health, energy, education, and justice—during a transition to a post-Saddam Iraqi government.[4] Its final product—a 1,200-page, 13-volume report—was less a detailed transition plan than a series of concept papers and recommendations that might have helped inform a national planning process about the transition from war to peace.[5]

While DoS continued these planning efforts through 2002 and into 2003, its work did not shape or even influence the postwar planning effort then under way at DoD. According to the SIGIR, DoS's "insights and suggestions were not used as a basis for postwar planning efforts within the interagency process."[6]

The question of which of the conflicting visions for postwar reconstruction that DoD and DoS were planning simultaneously was going to guide the war planning effort was firmly settled in January 2003, 14 months after planning for combat operations started in earnest and only two months before the start of the war. National Security Presidential Directive (NSPD) 24 consolidated responsibility for postwar Iraq in a DoD institution called the Office of Reconstruction and Humanitarian Assis-

---

[1] Bensahel et al., 2008, p. 1. To the extent challenges were raised, they were largely unsuccessful. See, for example, Special Inspector General for Iraq Reconstruction (SIGIR), *Hard Lessons: The Iraq Reconstruction Experience,* Washington, D.C.: U.S. Government Printing Office, 2009a.

[2] Bensahel et al., 2008, pp. 6–7.

[3] SIGIR, 2009a, p. 3.

[4] DoS, "The Future of Iraq Project: Overview," Washington, D.C., May 12, 2003.

[5] Bensahel et al., 2008, pp. 31–33. Most of the papers were declassified in September 2006 and are available via National Security Archive, "New State Department Releases on the 'Future of Iraq' Project," Electronic Briefing Book No. 198, September 1, 2006.

[6] SIGIR, 2009a, p. 30.

tance (ORHA), which commenced staffing and planning in January 2003.[7] Secretary of Defense Rumsfeld selected retired Army LTG Jay Garner to lead ORHA. General Garner had been the Deputy Commander of Operation Provide Comfort, which led humanitarian and reconstruction efforts in Northern Iraq following the first Gulf War.[8] While Garner compiled an interagency team—including representatives from the departments of Agriculture, Commerce, State, and Treasury; the U.S. Agency for International Development (USAID); and other agencies that would undertake postwar assistance missions—he reported directly to the Secretary of Defense.

Unlike the broad postwar planning vision then under way in DoS, ORHA was tasked with the relatively narrow task of preparing to address anticipated humanitarian crises in a postconflict Iraq, particularly by providing emergency relief aid and overseeing repairs to war-damaged Iraqi infrastructure, such as oil fields, roads, and communications networks. The primary planning assumption was that the U.S. military would be able to decapitate the regime by removing Saddam Hussein and his loyalists but that the government of Iraq would continue to function and provide many necessary services to the population with only limited disruption from combat damage.

In the end, prewar interagency planning and collaboration for the postcombat environment fell far short of what was necessary. The failures of the interagency process, however, do not explain why the military was ill prepared to respond to security concerns that arose in the immediate aftermath of major combat operations. The success of the ground campaign in Operation Iraqi Freedom demonstrates the importance of military planning as an effective guide to how battles should be prosecuted and war fought. However, a preference for planning the major combat operations first and foremost, leaving stabilization efforts (some of which have to be undertaken during the war) to be planned afterward, left the military unprepared for the postwar task of winning the peace—the ultimate object of war.

Combat Operations (March 20 to June 23, 2003)

On March 20, 2003, coalition ground forces entered Iraq. Prewar military planning focused on combat phases of war against Iraq.[9] Indeed, the execution of the combat phase of the war went remarkably well. However, unanticipated security challenges arose that quickly pushed U.S. forces into nontraditional military roles for which they had not planned. Forces anticipating a relatively permissive security environment instead confronted widespread looting and lawlessness. The Iraqi police units that U.S. planners assessed would remain capable of maintaining law and order lacked

---

[7] NSPD 24, "Post-War Iraq Reconstruction," January 20, 2003; SIGIR, 2009a, p. 33.

[8] Dobbins et al., 2009, p. 3.

[9] Bensahel et al., 2008, pp. 5–10.

the capacity (and perhaps the will) to carry out such functions. Moreover, whatever political and societal cohesion existed prior to the invasion evaporated once the fear of Saddam Hussein and the Ba'athist regime was removed.[10]

As the war quickly proceeded, ORHA scrambled to staff its efforts on an aggressive time line and began deploying to Kuwait in April 2003.[11] As an early priority, ORHA sought to ensure that Iraqi ministries continued to perform basic governance functions.[12] To provide some continuity of leadership in the ministries in advance of (what was assumed to be a relatively near-term) transition to Iraqi sovereignty, ORHA moved to establish an advisory team for each ministry. These teams would consist of senior advisors from the U.S. or coalition partner to guide ministry decisions, expatriate Iraqi technocrats, and the most senior non-Ba'athist member of the ministry staff.[13] While the intent was to provide continuity, ORHA staffing problems and its short lifespan meant that few of these senior advisors joined ORHA in March 2003, before it deployed to theater, and some had still not arrived once ORHA roles and missions transitioned to the CPA in June. Yet the concept of the senior advisor for each ministry, and their fundamental roles, continued under the CPA.[14] Broader reconstruction efforts were to be led by USAID and facilitated by ORHA, yet USAID's plans anticipated that Iraqi ministries would again be in the lead within a relatively short time.[15] This assumption also proved incorrect.

Once ORHA was in theater, little effort went into coordinating the organization's activities with military operations. An early sign that the military and civilian efforts were not integrated came as soon as ORHA landed in the theater. General Garner and his staff discovered that they would not be collocated with CFLCC. The CFLCC Commander, LTG David McKiernan, did not want a mostly civilian organization housed within the perimeter of his military headquarters at Camp Doha. As a result, for the first weeks of the war, ORHA was based in a Hilton Hotel 45 minutes away from CFLCC—making effective communication and collaboration almost impossible.[16] More broadly, the two organizations lacked clear command relationships; they also had different priorities, with CFLCC naturally placing the greatest emphasis on

---

[10] Bensahel et al., 2008, p. 81. See also USCENTCOM, planning slides, August 2002; Andrew J. Enterline, J. Michael Greig, and Yoav Gortzak, "Testing Shinseki: Speed, Mass, and Insurgency in Postwar Iraq," *Defense and Security Analysis*, Vol. 25, No. 3, September 2009.

[11] Bensahel et al., 2008, p. 66.

[12] Bensahel et al., 2008, p. 58. While it was assumed that the bureaucratic machinations of the ministries would largely survive an overthrow of the Ba'athist regime, this assumption proved incorrect.

[13] Bensahel et al., 2008, p. 59.

[14] Bensahel et al., 2008, p. 107.

[15] Bensahel et al., 2008, p. 65.

[16] Bensahel et al., 2008, p. 66.

combat operations. When ORHA requested security, logistics, or communication support from CFLCC, these capabilities were often unavailable.

On May 1, 2003, President Bush stood on the deck of the USS *Abraham Lincoln* before a banner reading "Mission Accomplished" and announced, "Major combat operations in Iraq have ended. In the battle of Iraq, the United States and our allies have prevailed. And now our coalition is engaged in securing and reconstructing that country."[17] However, the transition between war and peace the President anticipated proved elusive. As the scope of the reconstruction project became clearer, the administration replaced ORHA with the CPA, an organization with a much broader and more ambitious mandate.

Combined Joint Task Force 7 in Support of the Coalition Provisional Authority (April 23, 2003, to June 28, 2004)

On April 11, 2003, Garner arrived in Baghdad, just two days after the statue of Saddam Hussein had been toppled. Less than two weeks later, on April 24, Secretary Rumsfeld notified him that President Bush planned to appoint former Ambassador L. Paul (Jerry) Bremer as the head of the new CPA.[18] Bremer arrived in Baghdad less than two weeks after the "Mission Accomplished" speech. The establishment of the CPA as the occupation authority for postwar Iraq marked a transformation in the scope and scale of postwar U.S. military involvement, going from what most expected to be a four-month limited transition to an explicit occupation with an indeterminate time frame and no prior planning. No longer could U.S. forces expect a relatively limited mission involving combat phases of the war and subsequent humanitarian assistance; instead, the CPA and the military authority, Combined Joint Task Force 7 (CJTF-7), became involved in a remarkable range of security, economic development, and governance functions. As the SIGIR noted:

> ORHA was designed as a short-term holding mechanism, to be followed by a rapid shift to an interim Iraqi authority. . . . But the deteriorating situation in Iraq apparently had caused the White House to change plans . . . [the CPA] rapidly and massively expanded, far eclipsing the minimalist liberation vision that had set planning for war in motion a year and a half earlier.[19]

---

[17] George W. Bush, "Address to the Nation on Iraq from the U.S.S. Abraham Lincoln," The American Presidency Project website, May 1, 2003.

[18] Dobbins et al., 2009, p. 8.

[19] SIGIR, 2009a, p. 64.

Legal authority for the occupation relied on UN Security Council Resolution (UNSCR) 1483, dated May 22, 2003, which named the United States and coalition partners as "occupying powers" in Iraq, in accordance with international law.[20]

Transition from the Office of Reconstruction and Humanitarian Assistance to the Coalition Provisional Authority

While ORHA had existed for only five months before Bremer's arrival in Baghdad, some ORHA staff resented that the CPA leadership did not seek to utilize the expertise ORHA had generated since January, concluding that many of their initiatives would soon be forgotten.[21] While General Garner remained in country for three weeks after Ambassador Bremer's arrival, the transition from ORHA to the CPA was not smooth.[22] By some assessments, it also proceeded without adequate guidance from Washington.[23] In one of the earliest (and subsequently most controversial) CPA actions, Ambassador Bremer signed CPA Order 1 on May 16, 2003, removing Ba'athists from senior government positions. Undersecretary of Defense for Policy Douglas Feith had drafted the de-Ba'athification policy earlier that month. General Garner was shown the draft order shortly after Ambassador Bremer's arrival in Baghdad. Apprised of the plan, General Garner appealed to Ambassador Bremer to instead adopt a more limited de-Ba'athification order instead, but he was rebuffed.[24]

The leaders of the two organizations also expressed a disjointed vision regarding the pace and process of transition. In engagements with Iraqi leaders, General Garner had articulated a swift path to transition to Iraqi control; Ambassador Bremer, however, articulated a more qualified vision in his first meeting with Iraqi leaders. "We expect the progress along this path to be incremental," Ambassador Bremer told the Iraqi Interim Authority, "but we are prepared to cede increasing responsibility to responsible Iraqi leaders."[25] Operating with the conviction that he had President Bush's unqualified support, Ambassador Bremer developed an 18-month plan to rebuild Iraq by having the Iraqis draft a constitution, conduct a referendum on the final text, establish election law, and then hold national, regional, and local elections. Rather than a quick handoff to the Iraqis, Ambassador Bremer established a plan that started a long

---

[20] Dobbins et al., 2009, p. 13. The UN Security Council reaffirmed the authority, responsibilities, protections, and designation of the United States as the occupation power each year through the end of 2008, when U.S. and Iraq signed the SFA and Iraq regained sovereignty. From this point until the final departure of forces in December 2011, the actions and activities of U.S. military forces were constrained by Iraqi law, the SFA, and the related SA.

[21] Bensahel et al., 2008, p. 72.

[22] Bensahel et al., 2008, p. 71.

[23] Gordon W. Rudd, *Reconstructing Iraq: Regime Change, Jay Garner, and the ORHA Story*, Lawrence, Kan.: University of Kansas Press, 2011, p. 304.

[24] Dobbins et al., 2009, p. 114.

[25] Rudd, 2011, pp. 322–323.

transition to Iraqi sovereignty by establishing himself as the head of an occupation authority to facilitate the emergence of a Western-style democracy in much the same way General Douglas MacArthur had in Japan after World War II.

Military Activities in Support of the Coalition Provisional Authority

CPA Order 1 specified that the military organization, CJTF-7, would support CPA activities. Yet the chain of command remained ambiguous in practice. Although both Ambassador Bremer and CJTF-7 commander LTG Ricardo Sanchez reported to Secretary of Defense Rumsfeld, they did so through different chains of command, which constrained institutional interaction. On a personal level, working relationships between Ambassador Bremer and General Sanchez were reportedly tense.[26]

As the legally recognized governing authority in Iraq, the CPA had a broad mandate and large aspirations. In its year of existence, the CPA took on a remarkable range of functions for governing Iraq, far beyond ORHA's narrow charter.[27] The CPA "Vision Statement" of July 2003 established four principal objectives for itself: security (establishing a secure and safe environment), governance (enabling the transition to transparent and inclusive democratic governance), economy (creating the conditions for economic growth), and essential services (restoring basic services to an acceptable standard).[28] CJTF-7 supported these aims, although it was increasingly occupied with battling criminality, violence, and a growing insurgency. As was perpetually the case in Iraq, the deteriorating security situation in the CPA years shaped military activities and the nature and timing of transitions.

Iraqi Army

Training and equipping the Iraqi military to provide for the nation's security was not a function the U.S. military sought. In fact, General Sanchez identified his mission as defeating "internal armed threats," disarming the militias, and locating the caches of weapons of mass destruction that intelligence analysts thought might still remain.[29] On May 23, 2003, Ambassador Bremer signed CPA Order 2, which dissolved the Ministry of Defense (MOD) and announced that the coalition intended to develop a

---

[26] Rudd, 2011, p. 379.

[27] While Ambassador Bremer technically reported to the Secretary of Defense, the CPA was not a military organization. In fact, its exact administrative status remained fairly unclear; as the Congressional Research Service noted: "No explicit, unambiguous and authoritative statement has been provided that declares how CPA was established, under what authority, and by whom, and that clarifies the seeming inconsistencies among alternative explanations for how CPA was created" (L. Elaine Halchin, "The Coalition Provisional Authority [CPA]: Origin, Characteristics, and Institutional Authorities," Washington, D.C.: Congressional Research Service, RL32370, June 6, 2005, p. 39). In any case, this section will focus on CJTF-7 activities in support of the CPA, not the history of CPA activities in themselves.

[28] Bensahel et al., 2008, p. 104.

[29] Gordon and Trainor, 2012, pp. 13–15, 18.

New Iraqi Army instead of attempting to recall members of the Iraqi Army that had collapsed when the United States invaded.[30] While CPA Order 2 dissolved the MOD, it did not do the same thing to the Ministry of Interior (MOI) or the Iraqi police. As will be discussed, the decision to essentially disband the Army had unanticipated consequences that would not only fuel the emergence of a Sunni insurgency but also make the establishment of a future Iraqi military force much more difficult. Moreover, the decision to rebuild the Iraqi Army from the ground up marked a transition of great consequence for the evolving role of the U.S. military in Iraq. From this moment forward, the United States would not only have the responsibility to train and equip the New Iraqi Army but would also assume responsibility for Iraq security against both internal and external threats until that new military was sufficiently trained, organized, and equipped.

Responsibility for training the Iraqi Army fell to MG Paul Eaton, who arrived in June 2003 to take command of the CPA's Coalition Military Assistance Training Team (CMATT).[31] Departing from the Saddam-era precedents, the United States decreed that the New Iraqi Army, as it was then called, was to be an all-volunteer force and, as a truly national force, should reflect the ethnic, regional, and religious diversity of the nation.[32] The first battalion of the New Iraqi Army recruits began training on August 2, 2003, and graduated on October 4, 2003, after successfully completing a nine-week course of instruction at a training base located in the Kurdish region of northern Iraq.[33]

Police

Unlike the Iraqi military forces and intelligence services, which CPA Order 2 disbanded, police forces and the MOI remained intact. A prewar DoD planning assumption held that the security environment following major combat operations would be relatively permissive and that existing Iraqi police would be able to maintain law and order.[34] The existing MOI was expected to take responsibility for the civil security mission. The extent to which this assumption proved invalid became clear as criminality and violence continually increased through 2003 and 2004. Frank Miller, a senior member of the NSC planning staff explained: "We believed that the Iraqi police were a corrupt, but generally efficient police force. It turns out they were both corrupt and not a particularly efficient police force."[35] CPA police trainers were caught off guard by the

---

[30] Dobbins et al., 2009, pp. 15, 57. CPA "orders" were directives to the Iraqi people, which altered Iraqi law.

[31] SIGIR, 2009a, p. 128.

[32] Dobbins et al., 2009, p. 63.

[33] U.S. Department of the Army, *The Iraq War: 2003–2011*, U.S. Government Printing Office, May 10, 2012.

[34] SIGIR, 2009a, pp. 124–125.

[35] Quoted in SIGIR, 2009a, p. 125.

Iraqi police's poor skills, equipment, and infrastructure. As civil society began to fracture, the inability of police forces to maintain order became increasingly problematic.

Police training was initially considered part of the civilian mission. Before the war, DoS proposed that 5,000 armed civilian police might be necessary to maintain order in the postcombat environment.[36] In August 2003, Ambassador Bremer signed an agreement with King Abdullah of Jordan to build a police academy for Iraqi forces in Jordan.[37] The CPA needed a facility with requisite space and security to support training of large numbers of Iraqi recruits. Large-scale police training at the Jordanian Police Academy did not begin until late November 2003.[38]

Assessing the Effectiveness of Training

After a January 2004 assessment found the CPA's resourcing for security force training efforts to be lacking, Secretary Rumsfeld sent MG Karl Eikenberry to Iraq that month to assess the situation. In the first of what was to become a pattern for civilian-military transitions during the early years, General Eikenberry found CPA efforts underresourced and disorganized and recommended giving CJTF-7 the responsibility for training both the New Iraqi Army and Iraqi police forces.[39]

In March 2004, Secretary Rumsfeld transferred CMATT from the CPA to CJTF-7 as Eikenberry had recommended. The move reflected concern that the CPA lacked the capacity to manage the effort and a sense that the scope and scale of the required training was appropriately the purview of the military, a sense Ambassador Bremer shared. In the face of rising violence in Iraq, policymakers in Washington wanted to generate large numbers of trained Iraqi forces and wanted them trained quickly. CMATT did succeed in ramping up the training pipeline, yet operational demonstrations of the capabilities of trained recruits left many observers concerned that quality was being sacrificed in the name of quantity and expediency.[40] Desertion was a problem both because salaries were relatively low and because security in the country was increasingly perilous. The New Iraqi Army also lacked a clear mission. The question of whether Iraqi military power should focus primarily on maintaining internal security amidst a growing insurgency or on protecting the country from external threats remained unresolved.[41] This question would persist through subsequent military transitions.

---

[36] Dobbins et al., 2009, p. 73.

[37] Dobbins et al., 2009, p. 74.

[38] SIGIR, 2009a, pp. 127.

[39] Dobbins et al., 2009, pp. 62–69, 71. In 2009, MG Paul Eaton, the head of CMATT, stated, "In the beginning, there was no, zero, urgency on the part of the Secretary of Defense to provide the requisite resources to truly develop the Iraqi security force."

[40] Dobbins et al., 2009, pp. 70–71.

[41] Dobbins et al., 2009, p. 71.

The security assessment Eikenberry conducted also found significant shortfalls in the quality of police training. The training program was reportedly plagued by poor coordination among U.S. agencies involved.[42] The challenge of training the Iraqi police was compounded by inadequate vetting of cadets and an overemphasis on quantities of trainees rather than quality of training.[43] The Eikenberry assessment concluded that the scope and scale of police training required in Iraq exceeded the capability of the CPA and, consequently, recommended giving the military responsibility for police training and the supervision of the MOI. Ambassador Bremer opposed this recommendation, however, and argued that the "Chief of Mission . . . must have policy supervision of the police program."[44]

Secretary Rumsfeld ultimately accepted the recommendation of the Eikenberry assessment and directed that the military immediately assume responsibility for the MOI. In March 2004, CJTF-7 took on responsibility for training police from the CPA's MOI team, as it had for the military training mission. The military's accretion of police training activities, placed under the purview of the new Coalition Police Assistance Training Team, reflected a judgment that civilian agencies lacked capacity for building requisite police numbers as quickly as they were needed to quell rising violence.

Governance

The CPA supported the development of nascent democratic Iraqi political institutions, including appointment of the Iraqi Governing Council, which, while not elected or much empowered, reflected an effort to put an Iraqi face on the occupation.[45] This CPA-appointed council consisted of former exiles and individuals who could be proven to have no connection to the Ba'ath party. Consequently, most Sunnis viewed the council as illegitimate. One of the council's key tasks was development of the interim constitution, known as the Transitional Administrative Law (TAL),[46] which established a process and a time line for establishing a permanent democratic government in Iraq. Shoring up the *political process* became both a means to the end of building Iraqi government legitimacy and an end itself as military forces worked to ensure that elections proceeded with minimal violence and intimidation.

---

[42] The DoS Bureau of International Narcotics and Law Enforcement Affairs (INL), DoD, the Federal Bureau of Investigation, and others all shared responsibility for the police training program.

[43] Anthony H. Cordesman, *Iraqi Security Forces: A Strategy for Success*, Washington, D.C.: Center for Strategic and International Studies, 2006, pp. 109–116.

[44] L. Paul Bremer, "Security Assessment," memorandum for Secretary Rumsfeld, Office of the Secretary of Defense document number 01562-04, Rumsfeld Library, February 3, 2004.

[45] Dobbins et al., 2009, pp. 31–49. The Governing Council, which formed July 13, 2003, was a largely advisory body of émigré and internal Iraqis whom Ambassador Bremer had chosen with the help of the UN and U.S. and British regional experts.

[46] CPA, *Law of Administration for the State of Iraq for the Transitional Period*, March 8, 2004.

On November 15, 2003, the CPA and the Governing Council agreed that sovereignty would be transferred to the interim Iraqi government by July 1, 2004—a shorter time frame than any CPA officials initially thought they had or needed. The November agreement gave the CPA and the interim Iraqi government a clear time line for the transfer of sovereignty. In so doing, the November agreement launched the CPA into a period of heightened emphasis on transitioning tasks as quickly as possible, both to Iraqis and to a new U.S. embassy, to be opened in June 2004.

In anticipation of that imminent change, the November 2003 agreement called for drafting of the TAL, consisting of basic laws in advance of adoption of a permanent constitution. On March 8, 2004, the CPA and the Governing Council approved the TAL, which outlined an 18-month period of transition. The CPA transfer of sovereignty would hand power to the Iraqi interim government, which would hold elections for provincial councils and the National Assembly by January 31, 2005. The National Assembly would be responsible for drafting a permanent constitution by August 15, 2005, followed by a referendum by October 15, 2005. By December 15, 2005, Iraqis would hold national elections to choose a Council of Representatives, which would be seated by the end of the year.[47]

Economy and Essential Services

CPA work on economic development included efforts to reform the banking sector, issue a new currency, secure debt relief, promote foreign investment, reduce subsidies, and privatize inefficient state-owned enterprises. CJTF-7 supported this work by providing security for major infrastructure sites, such as oil fields, and at banks from the day the currency exchange began.[48]

The CPA also took on hefty responsibilities for developing the Iraqi economy and for reconstructing infrastructure necessary for the provision of basic services. The poor state of Iraqi infrastructure for oil production, energy generation, health care, and education shocked many CPA officials, and these became major high priority areas for CPA investment. CJTF-7 supported this work in part by manning distribution centers and guarding payrolls and benefit deliveries to civil servants, pensioners, and teachers.[49]

In May 2003, the CPA authorized U.S. field commanders to use seized Iraqi assets to fund urgent humanitarian, relief, and reconstruction requirements within a commander's area of operations through a program called the Commander's Emergency Response Program (CERP).[50] By the end of 2003, Congress began appropriating

---

[47] SIGIR, 2009a, p. 155.

[48] Bensahel et al., 2008, p. 208.

[49] Bensahel et al., 2008, p. 199.

[50] Funds were drawn from the Development Fund for Iraq, which was established in May 2003 through UNSCR 1483, as a means to direct seized Iraqi funds toward relief and reconstruction efforts in Iraq. See SIGIR, "Com-

U.S. funds to CERP, instead of relying solely on seized assets.[51] Its first appropriation called for CERP funds to be spent on the humanitarian, relief, and reconstruction projects that will "immediately assist the Iraqi people."[52] Over the course of military operations in Iraq, CERP became a major source of military support for civil capacity building, eventually funding a remarkably wide array of projects and activities. As the Government Accountability Office (GAO) noted, CERP became an exceptionally flexible tool commanders could use to affect projects in their areas of operations.[53] Its inception during the CPA reflected the broadening scope of military activities in the reconstruction of Iraq.

Transfer of Sovereignty, Transition to the U.S. Embassy

The end of the CPA on June 28, 2004, and the transfer of sovereignty to the Iraqis marked the end of the formal occupation of Iraq.[54] Marking the transfer ceremony, President Bush stated:

> Earlier today, 15 months after the liberation of Iraq, and 2 days ahead of schedule, the world witnessed the arrival of a free and sovereign Iraqi government. Iraqi officials informed us that they are ready to assume power, and Prime Minister Allawi believes that making this transition now is best for his country.[55]

This political transition had important consequences for the U.S. military mission in Iraq. The transfer of sovereignty ended the official U.S. occupation of Iraq, although continuing military and civilian activities in the country did not cease. In fact, they would grow and change through additional transitions in the following seven

---

mander's Emergency Response Program in Iraq Funds Response Program in Iraq Funds Many Large-Scale Projects," Washington, D.C., SIGIR-08-006, January 25, 2008, p. i.

[51] SIGIR, 2008, p. i.

[52] Public Law 108-106, Emergency Supplemental Appropriations Act for Defense and for the Reconstruction of Iraq and Afghanistan, 2004, November 6, 2003.

[53] GAO, *Military Operations: Actions Needed to Better Guide Project Selection for Commander's Emergency Response Program and Improve Oversight in Iraq*, Washington, D.C., GAO-08-736R, June 23, 2008b. The flexibility and fungibility of CERP has made it the subject of numerous government audits by, for example, SIGIR and GAO.

[54] While the formal occupation ended with the establishment of the Iraqi interim government, UNSCR 1483 officially recognized the United States and Great Britain as occupying powers. UNSCR 1546 afforded MNF-I the authorities and responsibilities of an occupation military until December 31, 2008. Although the U.S. military remained in Iraq for an additional three years as a guest of the government of Iraq, consistent with a bilateral security agreement, polling data through the end of 2011 consistently showed that upwards of 80 percent of Iraqis viewed U.S. military forces as an occupying army. Brennan notes, 2009–2011.

[55] George W. Bush, "The President's News Conference with Prime Minister Tony Blair of the United Kingdom in Istanbul," The American Presidency Project website, June 28, 2004. Notably, the transition two days earlier than anticipated created some turbulence as the announcement, which surprised even CPA staff, undercut CPA officials who had been planning to use the full period to let contracts. Interview with Ronald Neumann, September 2011.

years of war. The 2004 transition, however, was not simply a transfer from occupation to a newly sovereign Iraq; in addition to being a transition to the Iraqis, it was a transition from CJTF-7 to its successor, MNF-I, and from the civilian CPA to a new U.S. embassy. This marked an important transition to a newly crowded landscape of institutions with responsibilities for carrying out governance and security functions in Iraq. Nevertheless, despite this "transfer of authority," U.S. military forces retained immunities and authorities of an occupation force under the UNSCR.[56]

On May 11, 2004, President Bush issued NSPD 36, which formalized the institutions charged with continuing the U.S. mission in Iraq following the transfer of sovereignty.[57] DoS, through the Chief of Mission, would lead all U.S. activities except for military operations and the development of the ISF, which DoD would conduct through MNF-I.[58] The directive also established a new DoS-led organization, the Iraqi Reconstruction Management Office, to help support the reconstruction program. The directive put the new office, led by Ambassador William B. Taylor, in charge of developing reconstruction policy.[59] The office also absorbed responsibility for managing the CPA's former senior advisors.[60] Within the office, the advisors were to continue to provide technical assistance to the Iraqi ministries and to help them coordinate with reconstruction projects.

Multi-National Force–Iraq in Support of the Iraqi Interim Government and the U.S. Embassy (June 28, 2004, to May 3, 2005)

With the transfer of sovereignty, the military mission transitioned to the newly established MNF-I.[61] This important political transition for Iraq was accompanied by a transition in U.S. military and civilian leadership. GEN George Casey assumed command of MNF-I from LTG Ricardo Sanchez, and Ambassador John Negroponte arrived in June to take control of the civilian mission.

The security situation MNF-I inherited was more challenging than the one CJTF-7 had faced. In July, two weeks after the transfer of sovereignty, a MNF-I assessment found that "the insurgency is stronger than it was nine months ago and could

---

[56] U.N. Security Council, "The Situation Between Iraq and Kuwait," Resolution 1546, 2004.

[57] NSPD 36, "United States Government Operations in Iraq," May 11, 2004; SIGIR, 2009a, p. 157.

[58] SIGIR, 2009a, p. 157.

[59] SIGIR, 2009a, p. 165.

[60] SIGIR, 2009a, p. 157.

[61] On May 15, 2004, CJTF-7 was redesignated MNF-I, the intent of which was to stand up a strategic-level headquarters to support the June 30 transfer of sovereignty. USF-I, *OIF/OND Chronology*, 2011a.

deny the IIG [Iraqi interim government] legitimacy over the next nine months."[62] The paper expressed concern about the ability of the fledgling Iraqi government to maintain control over a simmering insurgency, perform basic governance functions, and protect infrastructure. It also pointed to a transition in the operational campaign that was reverberating through the activities of the military authorities: "Now that the warfighting and pre-insurgency phases of the conflict are over, military power must be employed differently . . . . The coalition's military capability (MNF-I) has to assume a lower profile, push the ISF forward when trained and ready."[63] This elusive transition from warfighting to building capacity for Iraqis to fight for themselves would remain on the horizon for military forces for many years. As the insurgency grew, combat operations swelled to occupy an ever-greater portion of the U.S. effort, and the time frame for transition to Iraqi security self-reliance continued to slip.

Multi-National Force–Iraq Campaign Plan

In August 2004, a year and a half into the conflict, General George Casey released the first military campaign plan.[64] *Operation Iraqi Freedom: Partnership, From Occupation to Constitutional Elections*, emphasized that the U.S. military now operated in a fundamentally different context. Unlike CJTF-7, the plan noted, "MNF is no longer a force of occupation."[65] In part, this meant that MNF-I now operated in a landscape defined by other institutional stakeholders. According to the campaign plan,[66]

> As the campaign has progressed, the balance of authority has changed. In order for the triangle of interests [between the United States, coalition partners, and the Iraqi government] to remain stable it is important to keep a balance between the three players.

It called for coordination, consultation, and partnership between the U.S. mission, military forces, and the interim Iraqi government as the "glue" binding the working relationships together. A few weeks later, General Casey and Ambassador Negroponte signed a joint mission statement predicated on the assumption that "the Multinational Force's effort in Iraq is inextricably linked to that of the U.S. embassy."[67] The need to manage the evolving relationship between the embassy, military, and the

---

[62] MNF-I, *Building Legitimacy and Confronting Insurgency in Iraq: End States and Bottom Line*, July 15, 2004b, p. 1.

[63] MNF-I, 2004b, p. 3.

[64] Ricks, Tom, *Fiasco: The American Military Adventure in Iraq*, New York: Penguin Books, 2006, pp. 392–393.

[65] MNF-I, *Campaign Plan: Operation Iraqi Freedom; Partnership: From Occupation to Constitutional Elections*, Baghdad, 2004a, p. 17.

[66] MNF-I, 2004a, p. 3.

[67] George Casey and John Negroponte, "MNF-I-Embassy Joint Mission Statement," August 18, 2004, p. 1.

Iraqi government was an outcome of the transition of sovereignty, and it would continue to shape subsequent transitions. However, for the next few years, the nature of the competition and resource imbalances made MNF-I the dominant partner in the relationship. MNF-I narrowed its focus to its urgent kinetic mission rather than to leading a significant interagency effort. MNF-I and Embassy Baghdad would not issue the Joint Campaign Plan (JCP) until April 2006.

The campaign plan of 2004 articulated diverse objectives for the military mission in Iraq. It described four lines of operation (LOOs): security, governance, economic development, and strategic communication. The four LOOs, separate but deeply intertwined, reflected the large scale and scope of MNF-I's engagement in Iraq. While the security LOO continued to occupy the preponderance of MNF-I's efforts, it is notable that Operation Iraqi Freedom was not being conceptualized as narrowly focused on traditional military roles. This was neither the narrow humanitarian ORHA mission nor the CJTF-7 military occupation in advance of the transfer of sovereignty; this was a military plan with a broad vision of U.S. military responsibilities for building security institutions, rule of law, economic development, and even a communication strategy that would support a positive U.S. relationship with Iraq. Yet as concerns about a growing insurgency increasingly threatened the mission, the security LOO became the leading priority.

The security LOO called for continued development of ISF. Reflecting concern over training under the CPA/CJTF-7, the campaign plan called for an approach with a "[f]ocus on building quality, not quantity."[68] But it called for committing ISF to operations as soon as possible without mission failure. On the governance LOO, the campaign plan placed a significant emphasis on supporting the national elections in January 2005. Support to the political process would be a significant focus of subsequent campaign plans. The elections and other activities aimed at transitioning increased responsibility to Iraqis were aimed at bolstering the legitimacy of Iraqi political institutions. The campaign plan exhorted the forces to always remember that the "visible legitimacy and success of the Iraqi Government is the key to mission success."[69]

Five months later, MNF-I released a campaign plan progress update to assess progress along the four LOOs. The update made clear the extent to which the deteriorating security situation was disrupting the execution of the plan. It noted that, while progress was being made, the "insurgency has not been neutralized. Levels of violence and insecurity are higher than when the Campaign Plan was signed."[70] The growing intensity of the insurgency, it warned, threatened mission success; defeating the insurgency must be MNF-I's top priority. Along the security LOO, progress in developing

---

[68] MNF-I, 2004a, p. 6.

[69] MNF-I, 2004a, p. 3.

[70] MNF-I, *Five Month Campaign Progress Review*, Baghdad, December 12, 2004c, p. 5.

the ISF varied. "Numerically," ISF development was "broadly on track," the assessment found, but "their operational record has been uneven."[71]

Along the governance LOO, the campaign plan update continued to build anticipation for the elections. "The key Decisive Point in the near term," it stated, "and perhaps of the whole campaign, is the holding of inclusive, free and fair elections on 30 January 2005."[72] It explained that elections were essential to building the kind of "government legitimacy" the interim government could not achieve.

The economic development assessment found that,

> [i]n general, the basic needs of the Iraqi people are being met, although their expectations of the Coalition are high, and probably unrealistic. That said, the economic LOO is not yet achieving its full potential, and the apparent lack of progress on reconstruction, employment and economic recovery is reducing Iraqi tolerance of the Coalition presence.[73]

The review pointed specifically to challenges the Iraq Relief and Reconstruction Fund programs faced, many of which were planned under the assumption of a relatively permissive security environment. Nevertheless, the review asserted, the "situation has changed. Attacks on contractors are widespread, internal movement is subject to restrictions, and security costs consume an increasing proportion of project funds."[74]

The campaign plan had emphasized the importance of a whole-of-government approach to the U.S. mission in Iraq, but the update found that interagency coordination was not yet functioning as needed. It found

> Unity of Effort is a central pillar of counter-insurgency doctrine. Within MNF-I, top level liaison and understanding between Force and State appears excellent but is based on personalities rather than structures. At the working level there are few structures for joint co-ordination, and those that exist are largely *ad hoc*.[75]

Developing Iraqi Security Forces

On June 28 2004, MNF-I stood up the Multi-National Security Transition Command–Iraq (MNSTC-I), centralizing and adding emphasis to efforts to train and equip ISF.[76] MNSTC-I took command of two existing entities that had previously transitioned from the civilian CPA to CJTF-7: the Civilian Police Assistance Training

---

[71] MNF-I, 2004c, p. 8.

[72] MNF-I, 2004c, p. 10.

[73] MNF-I, 2004c, p. 10.

[74] MNF-I, 2004c, p. 11.

[75] MNF-I, 2004c, p. 14.

[76] USF-I, 2011a.

Team, responsible for training the police and border police, and CMATT, responsible for training members of the Iraqi military.[77]

Tapped to lead MNSTC-I was LTG David Petraeus, who already enjoyed an unusually high profile after commanding the 101st Airborne Division in Mosul during the first year of the war. In Mosul, General Petraeus had won accolades for an innovative approach to security and reconstruction that relied on engaging the Iraqi population to develop productive relationships.[78] A 2004 profile remarked,

> It's not hard to see why President Bush may have chosen LTG Petraeus as the commander to turn things around on the Iraq security front. The general had emerged from the tumultuous first year of occupation with a reputation as perhaps the finest American commander in Iraq, capable of making important Iraqi friendships in one of the country's most hostile regions, as well as maintaining fierce loyalty among his soldiers.[79]

He remained in the MNSTC-I job until September 2005, when he left Iraq to become commander of the Combined Arms Center at Fort Leavenworth. There, he would take the lead on drafting a new counterinsurgency manual, which would be credited by many with turning the war around in 2007.[80]

In fall 2004, efforts to enlist the North Atlantic Treaty Organization's (NATO's) support for training ISF resulted in the establishment of NATO Training Mission–Iraq (NTM-I); General Petraeus was named commander of this effort as well.[81] This program provided advisors and trainers at the MOD, three national command centers in Iraq, a joint headquarters, and a small team at the Iraqi Military Academy.[82] General Petraeus described the focus of NTM-I as being on institutional development, while MNSTC-I was "the train and equip piece."[83]

Soon after being established, MNSTC-I worked to improve on the ad hoc assessment mechanisms the CPA had used to track the progress of the development of the ISF. For example, the CPA counted the ISF "on duty," irrespective of whether or not these forces had received coalition training. Immediately after the transition, MNSTC-I

---

[77] Jeremy M. Sharp and Christopher M. Blanchard, *Post-War Iraq: Foreign Contributions to Training, Peacekeeping, and Reconstruction,* Washington, D.C.: Congressional Research Service, June 6, 2005, p. 2.

[78] Tom Ricks, *The Gamble: General Petraeus and the American Military Adventure in Iraq,* New York: Penguin Books, 2009a, pp. 20–22.

[79] Vivienne Walt, "Can Petraeus Salvage Iraq?" *Time,* June 19, 2004.

[80] Robinson, 2008, p. 75.

[81] Sharp and Blanchard, 2005, p. 3.

[82] David Petraeus, "DoD Briefing on NATO Training Mission in Iraq," February 10, 2005.

[83] Petraeus, 2005.

replaced this metric with the more stringent "trained and equipped" count.[84] Yet General Petraeus was careful to note that the count would not be simply about quantity; quality would be the ultimate metric. "This is not about just numbers of trained and equipped," General Petraeus emphasized in a 2005 press conference. "I want to be very clear that we have always specified that there's trained and equipped to a specific standard."[85] Yet questions soon arose regarding the meaningfulness of the "trained and equipped" metric. Anthony Cordesman noted in 2007 that measuring the number of Iraqis who had received basic training and equipment was no measure of their availability or operational effectiveness: "The US has reported Iraqi manning levels based on the number of men it has trained and equipped that bear no resemblance to the actual manning levels of men that are still in service."[86] The actual number of forces was lower because of desertions, authorized leave, and attrition.[87] In November 2006, DoD's quarterly *Measuring Stability and Security in Iraq* report similarly noted that "the trained-and-equipped number should not be confused with present-for-duty strength. The number of present-for-duty soldiers and police is much lower."[88] While the CPA had only counted the number of personnel that were present for duty regardless of whether they had been trained, MNSTC-I initially counted only the number of soldiers who had been trained, regardless of whether they were present for duty. Both measurements fell far short of accurately portraying the capacity of the Iraqi Army.

One means by which MNSTC-I sought to assess the quality of its trained and equipped troops was the Transition Readiness Assessment process, developed in the winter of 2004–2005. The Transition Readiness Assessment was designed to measure the training and equipping status, and the operational effectiveness, of Iraqi forces.[89] The process was based on the U.S. Army's readiness assessments and took into account such elements as personnel, command and control, training, sustainment and logistics, equipment, and leadership.[90] Military transition teams were responsible for conducting these assessments at the battalion, brigade, and division levels.

---

[84] SIGIR, 2009a, p. 198.

[85] Petraeus, 2005.

[86] Anthony H. Cordesman, *Iraqi Force Development and the Challenge of Civil War: The Critical Problems the US Must Address If Iraqi Forces Are to Do the Job*, Washington, D.C.: Center for Strategic and International Studies, April 26, 2007, p. ix.

[87] Anthony H. Cordesman and Emma R. Davies, *Iraq's Insurgency and the Road to Civil Conflict*, Washington, D.C.: Center for Strategic and International Studies, 2008, p. 447.

[88] DoD, *Measuring Stability and Security in Iraq*, November 2006, p. 3.

[89] SIGIR, 2009a, p. 199.

[90] DoD, *Measuring Stability and Security in Iraq*, July 2005, p. 13.

National Elections, January 2005

In January, Iraq held national elections for a 275-member National Assembly, provincial councils for the 18 provinces, and a Kurdistan regional assembly.[91] UNSCR 1511 had called for elections to be held no later than January 31, 2005. The lead-up to the elections was tumultuous, as it became clear that a Sunni boycott, other sectarian divisions, and growing violence threatened the legitimacy of electoral outcomes.[92] Despite violence, elections proceeded with relatively high voter turnout (60 percent of eligible voters) and few claims of vote rigging.[93] As would be the case after subsequent elections, it took months after the election to form a government. Months of wrangling led to the choice of Jalal Talabani as president and Ibrahim al-Jaafari as prime minister. On May 3, 2005, Prime Minister Jaafari and his cabinet were sworn in as the Iraqi transitional government.[94]

A week after the January elections, MNF-I and U.S. embassy released a plan for a two-pronged transition to self-reliance. In the year ahead, they would seek to "diminish the insurgency and prepare ISF and ITG [Iraqi Transitional Government] to begin to accept the counterinsurgency lead."[95] The COIN fight continued to dominate the U.S. military mission, but the joint mission statement asserted, U.S. forces also needed to focus on handing that fight over to capable Iraqi institutions. The joint mission statement also emphasized the central role of political process in successfully carrying out the transition: "The more political space the Transitional National Assembly and Iraqi Transitional Government occupy through their cooperative efforts, the less space will be available to insurgency and terrorism, and the sooner Iraq can forge a civic culture that will sustain the rule of law."[96]

By April 2005, as the National Assembly struggled to complete negotiations on forming a government, MNF-I released a new campaign action plan. Describing 2005 as "a period of key transitions," the plan noted the Multi-National Corps–Iraq

transfer of authority in February, the Iraqi Interim Government to the Iraqi Transitional Government transition running through April and potentially extending

---

[91] Kenneth Katzman, *Iraq: Reconciliation and Benchmarks*, Washington, D.C.: Congressional Research Service, 2005, p. 1.

[92] Ali A. Allawi, *The Occupation of Iraq: Winning the War, Losing the Peace*, New Haven, Conn.: Yale University Press, 2007, pp. 334–347.

[93] Allawi, 2007, p. 391.

[94] Allawi, 2007, pp. 391–397.

[95] George Casey and John Negroponte, "Joint Mission Statement, A Plan for the Year Ahead: Transition to Self-Reliance," memorandum, February 7, 2005.

[96] Casey and Negroponte, 2005, p. 3.

further, the transition of key embassy personnel, the shift in our main effort, and another Iraqi election period late in the year.[97]

These evolutions in Iraqi politics, American military and civilian leadership, and military activities suggest how interdependent different kinds of transitions in Iraq have been over time. Since elections had been held, the campaign action plan directed that

> MNF-I will progressively shift our main effort for fighting the counterinsurgency campaign to the Iraqi Transitional Government and the ISF. We will do this by building Iraqi capacity to conduct independent counterinsurgency operations at the local, provincial, and national levels.[98]

MNF-I's stated objectives for 2005 were continuing to support key Iraqi political transitions and transitioning the lead for the COIN fight to the Iraqis—as they would continue to be for years thereafter.

Multi-National Force–Iraq in Support of the Iraqi Transitional Government and the U.S. Embassy (May 3, 2005, to May 20, 2006)

The period of MNF-I support to the Iraqi transitional government was marked by steady Iraqi progress toward major political milestones—development of a permanent constitution and national elections—while, at the same time, the country spiraled into civil war.

In 2005, MNF-I was deeply involved in supporting the development of Iraqi political institutions and political processes. Building the capacity and legitimacy of Iraqi political leaders had been a key objective in previous MNF-I campaign plans. However, the focus on building numbers or checking off metrics, wrote Linda Robinson, "betrayed a mechanistic thinking about the problem."[99] Instead, in a country with deep sectarian, regional, and cultural divisions, the real challenge was the broader question of "how political and military power and economic resources were to be shared and used."[100] These fundamental divisions simmered beneath a percolating political process and boiled over into full-blown civil war in 2005 and 2006.

Despite progress in meeting targets for political milestones and elections, the security situation continued to deteriorate in summer and fall 2005. The campaign

---

[97] Commanding General, MNF-I, "MNF-I Campaign Action Plan for 2005—Transition to Self-Reliance," April 22, 2005, p. 2.

[98] Commanding General, MNF-I, 2005, p. 3.

[99] Robinson, 2008, p. 12.

[100]Robinson, 2008, p. 12.

plan update in December noted that, "[a]lthough the specific components of the Iraqi insurgency have changed little over the past six months, we have seen a shift in the nature of the insurgency. The level of insurgent attacks, and the terrorist capacity for spectacular acts of terrorism, remains stubbornly high." It also noted the troubling proliferation of more-potent improvised explosive device (IED) technology.[101]

While the security challenges built through the end of 2005, the situation imploded on February 22, 2006, when the golden dome of the revered Shi'a al-Askari Mosque in Samarra was destroyed in a bomb attack. MNF-I quickly moved to try to help Iraqi authorities project calm and confident leadership in the aftermath of the crisis.[102] General Casey's "Golden Mosque Directive" presciently recognized the potential danger the act held for inflaming sectarian tensions: "The completion of the political process and our recent operations have positioned us for decisive action in 2006," Casey wrote. However, "the fallout from this attack has the potential to disrupt our momentum and threaten the progress we can make."[103] He called for MNF-I engagement to prevent the development of sectarian divisions in the ISF or in communities. Despite such efforts, SIGIR found that, in the four days following the attack, at least 1,300 Iraqis, most of them Sunnis, were murdered.[104] The bombing of the al-Askari Mosque and the ensuing sectarian violence was a turning point in the Iraq war, causing the rapid spiral into a civil war that neither the United States nor MNF-I were prepared to counter.

Iraqi Security Force Development

An MNF-I assessment in September 2005 asserted that steady progress was being made in building the capabilities of ISF into self-reliance.[105] While the Iraqi Army was showing confidence in combined operations with coalition partners, there was less progress in developing ministerial capacity. "Iraqi Security Ministries [MOI and MOD] show limited progress toward self-reliance," it remarked, and "Iraq's other key ministries remain underdeveloped. All lack effective senior leadership, a professional civil service, and suffer from patronage."[106] MNSTC-I, led since September by LTG Martin Dempsey, was assigned new responsibilities for supporting security ministry institutional development. In October 2005, it would take over capacity building for

---

[101] MNF-I, *Campaign Progress Review, June 2005–December 2005*, Baghdad, December 20, 2005b, p. 9.

[102] George Casey, "Strategic Directive: Golden Mosque Bombing," memorandum to commanders and staff, February 24, 2006.

[103] Casey, 2006.

[104] SIGIR, 2009a, p. 274.

[105] MNF-I, *The September Assessment*, September 23, 2005a.

[106] MNF-I, 2005a, p. 5.

the development of the MOD and MOI from DoS's Iraq Reconstruction Management Office.[107]

In June 2005, MNSTC-I launched a concept for embedding military transition teams into Iraqi Army and police units. The idea was that embedded coalition personnel would enable Iraqi units by means of facilitating their access to coalition air support, logistical coordination, and intelligence. They would also, by their presence, deter detainee abuse or violence against civilians.[108] Citing positive Transitional Readiness Assessments, MNF-I announced a plan to embed transition teams into security ministries also to support institutional development and provide for continuity through the transition from the Iraqi transitional government to the government of Iraq following anticipated national elections in December 2005.[109]

MNF-I increased its focus on development of the Iraqi police, which were increasingly being associated with abusive practices. In May 2005, MNF-I established the Police Partnership Program to improve police capacity and ministerial development.[110] MNF-I declared 2006 the "Year of the Police," indicating the increasing focus on police forces. The campaign progress update noted, "[i]nfiltration of Iraqi Police by insurgents and militias remains a problem—Badr Corps have achieved influence over police in Basrah, and insurgents have infiltrated those in Kirkuk, Mosul, and Samarra."[111] An MNF-I and U.S. Embassy Baghdad joint mission statement explained,

> 2006 must be the year of the police and the rule of law. The Iraqi government's ability to protect its citizens and to evenhandedly enforce the laws will have a profound effect on developing the trust and support needed from its citizens to reject the insurgency and to stand up to coercion.[112]

As will be discussed in Chapter Seven, U.S. efforts to establish a nonsectarian Iraqi police force that operated in accordance with the Iraqi rule of law fell far short of achieving the desired goals.

Provincial Iraqi Control

In June 2005, Prime Minister Jaafari and MNF-I announced a new process for assessing the readiness of Iraqi provincial security institutions and for transitioning them

---

[107] MNF-I, 2005a, p. 9.

[108] MNF-I, 2005a, p. 12.

[109] MNF-I, 2005a, p. 9.

[110] MNF-I, 2005a, p. 9.

[111] MNF-I, 2005b, p. 9.

[112] George Casey and Zalmay Khalilzad, "MNF-I-Embassy Joint Mission Statement: Building Success: Completing the Transition," December 6, 2005, p. 5.

from coalition control to Iraqi control.[113] The Provincial Iraqi Control (PIC) process was intended to be a conditions-based means of assessing progress in security at the local and provincial levels.[114] However, the SIGIR noted ongoing challenges for PIC assessments. A year later it noted that

> not one province had attained what Iraqi and U.S. officials considered the requisite degree of stability for the transition to provincial control to proceed. Target dates for the PIC process shifted five more times over the next year and a half. By the end of 2007, handoff had occurred in only nine of eighteen provinces, and three of those handed off had long been under de facto Kurdish control.[115]

Anthony Cordesman of the Center for Strategic and International Studies noted that the PIC process had problems in addition to slippage in target transfer dates. In 2008, he reflected,

> [a] great deal of past US and MNF-I reporting has focused on the formal transfer of provinces from MNF-I to Iraqi control . . . it was never clear what this transfer really meant. In many cases, it was clearly more cosmetic than real.[116]

He also noted that, in certain cases, MNF-I transferred provinces to Iraqi control prematurely, which exacerbated problems and undermined the legitimacy of the PIC process. While the process was intended to be conditions based, MNF-I faced pressure to "PIC" provinces as quickly as possible, even if that assessment had little relationship to security on the ground, as justification for the redeployment of U.S. military personnel. This would not be the last time that time pressures, rather than the achievement of predetermined capabilities, would drive a transition in Iraq.

Developing Counterinsurgency Doctrine

As the security situation continued to deteriorate, thoughtful military experts—in many cases people with firsthand experience in Iraq—began to reconsider fundamental approaches to the war. With notable exceptions, the dominant strategy in Iraq through 2006 was to hunt down and killing insurgents—with traditional approaches to conducting patrols and raids.[117] As U.S. casualties continued to climb in 2005 and 2006, General Casey began moving forces out of population centers and began con-

---

[113] SIGIR, 2009a, p. 285.

[114] Anthony H. Cordesman, *Transferring Provinces to Iraqi Control: The Reality and the Risks*, Washington, D.C.: Center for Strategic and International Studies, September 2, 2008, p. 10.

[115] SIGIR, 2009a, p. 286.

[116] Cordesman, 2008, p. 10.

[117] Ricks, 2009a, p. 12.

solidating them onto big, isolated bases.[118] As the approach did not create the security gains MNF-I expected, reconsiderations about how to fight an insurgency proliferated. Reassessments led to a transition of sorts in military approaches to the war, which shaped fundamental conceptions about the effective use of military power. In doing, it gave the military new responsibilities, outside the limits of what many describe as "traditional military operations."[119]

In a widely read article in the July 2005 issue of *Military Review*, LTG Peter Chiarelli made a strong case that fighting an insurgency required military engagement in nontraditional roles.[120] Based on his experience leading the 1st Cavalry Division (CD) in Baghdad from early 2004 through 2005, Chiarelli argued that military objectives were attainable only if the requirements for security were understood more broadly than traditional military operations. He found that the campaign to "'win the peace' in Iraq has forced us, as an instrument of national power, to change the very nature of what it means to fight."[121] Activities in support of democratic institutions, employment, infrastructure, and economic development could not be an afterthought to traditional operations; such activities needed to be part of the full spectrum of military operations.

Symptomatic of new interest in different approaches to COIN operations, the use of CERP was also evolving. Initially, the program was to be linked to security efforts, but for two years, commanders had had significant ability to affect reconstruction and broadly defined civil affairs projects in their areas of responsibility through the allocation of CERP funds. In October 2005, MNF-I issued policy and procedural guidance for the use of CERP in a document called "Money as a Weapon System."[122] The new guidance reflected the increasingly broad range of uses for CERP funds. Indeed, the concept that money could be used as effectively as a weapon, and the very fact that

---

[118] Ricks, 2009a, pp. 12–13.

[119] The term *traditional military operations* is frequently used to address security-related functions that a military organization brings to a battlefield. These capabilities include such capabilities as intelligence, surveillance, logistics, and a host of lethal and less than lethal capabilities in all branches of the military. Conversely, nontraditional use of the military may include such activities as assisting in economic development, establishing the rule of law, assisting in government formation and effective governance—functions usually performed by other agencies of the U.S. government. However, the reality is that, throughout the history of the U.S. military, the Army, especially, has frequently been called on to perform such nontraditional activities in such places as Cuba, Haiti, the Dominican Republic, Germany during the Marshall Plan, and Japan during the aftermath of World War II—to name just a few. Whether or not the military should perform these missions remains a point of contention both within DoD and across government agencies. However, the fact remains that, during exigent situations, the nation often turns to the military to perform these missions because it is the only institution with capacity and organization that can rapidly be redirected to perform these functions.

[120] Peter W. Chiarelli and Patrick R. Michaelis, "The Requirements for Full-Spectrum Operations," *Military Review*, July–August 2005.

[121] Chiarelli and Michaelis, 2005, p. 4.

[122] SIGIR, 2008.

MNF-I needed to issue guidance, reflects the extent to which spending money for reconstruction and civil affairs had become a significant activity for the U.S. military. In January 2006, General Chiarelli became commander of MNSTC-I and deputy to MNF-I Commander General Casey. According to SIGIR, Chiarelli oversaw a "significant evolution" of CERP. What had originally been conceived of as "walking around money"—a flexible means of supporting relatively small and near-term projects that unit commanders could use—grew into a tool that would be used to help "win the hearts and minds" of the Iraqi populace by providing services that would help improve the quality of their lives. The use of CERP funds would hereafter be used as an integral element of the emerging COIN strategy and came to be viewed as an "invisible capacity" that could be leveraged to augment the mission.[123]

As noted earlier, after leaving his post at the helm of MNSTC-I in September 2005, General Petraeus returned to the United States to lead the Combined Arms Center, an organization focused on training and educating soldiers.[124] In February 2006, Petraeus began a nine-month effort to draft a new Army/Marine Corps COIN manual, widely known as FM 3-24, which was released at the end of the year.[125] The published manual presented both a strong critique of the current military strategy and a framework for a new approach. "The cornerstone of any COIN effort is establishing security for the civilian populace," it asserted.[126] This new doctrine required restraint on the part of U.S. troops on the ground to limit the growth of the insurgency. The new approach to war also required forces to come off secure bases to provide security for the Iraqi populace. In addition to facilitating population security, operating among the people allowed U.S. forces to maintain contact with the population rather than ceding initiative to the insurgents.[127] The new manual emerged from (and in turn formalized) contemporary conversations and practices about the way to fight an insurgency. After taking command of MNF-I in February 2007, however, General Petraeus would have an opportunity to implement—and develop—the new concept of COIN operations that he helped shape as doctrine.

---

[123]SIGIR, 2009a, pp. 280–281.

[124]Robinson, 2008, p. 76.

[125]Army Field Manual (FM) 3-24/Marine Corps Warfighting Publication (MCWP) 3-33.5, *Counterinsurgency*, December 15, 2006. This document was later published by the University of Chicago Press in 2007 to ensure the broadest public dissemination. The document was later revised and published as Joint Publication 3-24 (JP 3-24), *Counterinsurgency Operations*, October 5, 2009. JP 3-24 is not identical to FM 3-24. Consequently, unless JP 3-24 is used for a particular reason, all references to FM 3-24/USMC 3-33.5 will use the book published by the University of Chicago for consistency.

[126]FM 3-24/MCWP 3-33.5, 2007, p. 5-2.

[127]FM 3-24/MCWP 3-33.5, 2007, p. 5-2.

Civil-Military Relations
Efforts at Integrating the U.S. Mission

As the campaign progressed, efforts at forging more effective relationships between MNF-I and Embassy Baghdad continued, although not as robustly as might have been desired. General Casey and Ambassador Zalmay Khalilzad developed processes and institutions that, at least on paper, aimed to improve joint coordination. For example, in 2005, the Iraqi Reconstruction Management Office established coordinators tasked with bringing civilian reconstruction resources to an area once military operations had ceased. Civilian and military staff also held weekly discussions on reconstruction, economic policy, political issues, and public communications.[128] In November 2005, Ambassador Khalilzad approved the creation of the Joint Strategic Planning and Assessment office to conduct strategic planning, independent analyses, and act as a liaison to MNF-I's planning and assessment efforts.[129] The goal was to establish a joint civil-military strategic planning and assessment unit reporting directly to the Chief of Mission and the CG of MNF-I.[130] A key function of the new cell was to support the development of a joint civil-military strategic plan "that fully integrates *and* balances military, political, economic and communication plans that will be required to conduct a successful counterinsurgency in Iraq."[131] Although the Joint Strategic Planning and Assessment office was intended to be an interagency staff, in actuality it was an embassy office. MNF-I never integrated its efforts with the office, but the two worked together on joint efforts.

In April 2006, two months after the bombing of the Golden Mosque in Samarra, MNF-I and the U.S. embassy released their first JCP. In December 2005, shortly before the scheduled national elections, they had issued a joint mission statement on "Completing the Transition."[132] It noted, "At present, Iraq is going through a difficult transition," and called on all dimensions of coalition power—political, military, and economic—to help support the transition to Iraqi self-reliance.[133] The JCP embodied such a whole-of-government approach.[134] The plan emphasized the need to bolster the government of Iraq: "The strategic center of gravity is the legitimacy of the GoI [gov-

---

[128]SIGIR, 2009a, p. 237.

[129]SIGIR, 2009a, p. 237.

[130]Brad Higgins, "Joint Strategy, Planning & Assessments (JSPA)," memorandum to ambassador, Baghdad, November 18, 2005.

[131]Higgins, 2005.

[132]Casey and Khalilzad, 2005.

[133]Casey and Khalilzad, 2005.

[134]GAO, *Operation Iraqi Freedom: Actions Needed to Enhance DOD Planning for Reposturing of U.S. Forces from Iraq*, Washington, D.C., September, 2008c, p. 13.

ernment of Iraq]."[135] Yet it assessed that the explosive sectarian violence, militias, the presence of external fighters, and corruption threatened the capacity and legitimacy of the nascent government.[136] The plan stated that

> the fundamental conflict in Iraq is between and among its ethnic and sectarian groups over the distribution of political and economic power. The new government, which largely comprises sectarian parties pursuing parochial interests, will need to rise above their sectarian agendas.[137]

Provincial Reconstruction Teams

In fall 2005, Ambassador Khalilzad introduced Iraq Provincial Reconstruction Teams (PRTs), a concept first deployed in Afghanistan, where Khalilzad previously served as U.S. Ambassador. The U.S.-led civilian-military PRTs were announced in a joint message from MNF-I and the U.S. embassy.[138] The PRTs' mission was to

> assist Iraq's provincial governments with developing a transparent and sustained capability to govern, promote increased security and rule of law, promote political and economic development, and provide the provincial administration necessary to meet the basic needs of the population.[139]

Three proof-of-concept PRTs deployed to Mosul, Hillah, and Kirkuk in November 2005.[140] The interagency teams, led by senior Foreign Service officers, included personnel from the departments of State, Justice, and Agriculture; USAID; and MNF-I (among other partners). The military provided support capabilities, including logistical assets and liaison officers. The civilian-military teams, along with locally employed Iraqis, were to execute their mission through training, coaching, and mentoring of staff in the provincial governments. The effort was originally conceived as a two-phase program over four years at the three original sites, but it was soon accelerated and expanded. By early 2007, seven U.S.-led PRTs and three others that were led by coali-

---

[135] Zalmay Khalilzad and George W. Casey, "Joint Campaign Plan: Operation Iraqi Freedom, Transition to Iraqi Self-Reliance," April 28, 2006a, p. 14.

[136] Khalilzad and Casey, 2006a, p. 5.

[137] Khalilzad and Casey, 2006a, p. 11.

[138] SIGIR, "Status of the Provincial Reconstruction Team Program in Iraq," Washington, D.C., 06-034, October 29, 2006, p. i.

[139] SIGIR, 2006, p. 241.

[140] MNF-I, 2005b, p. 15.

tion partners were in operation.[141] At its height, there would be 32 PRT or PRT-like entities.

The PRTs reflected the first major operational collaboration in Iraq between MNF-I and the embassy, and the effort faced significant challenges. It took more than a year to resolve basic issues relating to resources, security, and command-and-control relationships. SIGIR reported that the delay limited the PRTs' early effectiveness in the field.[142] Once the teams were fielded, debate remained about how they could best be used. Certain MNF-I advisors, including LTG Raymond Odierno, then the assistant to the Chairman of the Joint Chiefs of Staff, and Dr. Philip Zelikow, DoS's Counselor, called for embedding small teams in every brigade, although the ambassador preferred fielding fewer large teams that remained under embassy control.[143] Notably, PRTs in Afghanistan had been military led. Ambassador Khalilzad's initiative in Iraq to have civilians lead them broke new bureaucratic ground but also created challenges because PRT leaders operating in an area of operations controlled by a brigade commander had to find ways to work together to achieve common U.S. goals. The results of this "partnership" were mixed and were highly dependent on the personalities of each PRT leader and his or her brigade commander counterpart.[144]

In addition to coordination challenges, the PRT program faced significant security risks. SIGIR noted in 2006 that

> [b]ecause of security concerns, face-to-face meetings between provincial government officials and PRT personnel are often limited and, in some cases, do not occur. PRT members are at particular risk when traveling to and from their engagements with their Iraqi counterparts, as are provincial government officials and local Iraqi staff working with the PRT. If identified as cooperating with the U.S. government, all are at risk of threats and attacks by anti-coalition elements.[145]

Also, the program faced resource turbulence, especially following a military SJA ruling in April 2006 limiting the availability of DoD funds for the PRT mission.[146]

Nevertheless, the SIGIR audit in 2006 saw significant reason for optimism about the program:

> The Provincial Reconstruction Team Program in Iraq provides the best opportunity for U.S. government experts to provide grassroots support in the development

---

[141] SIGIR, "Status of the Provincial Reconstruction Team Program Expansion in Iraq," Washington, D.C., 07-014, July 25, 2007a, p. 1.

[142] SIGIR, 2007a, p. 241.

[143] SIGIR, 2009a, p. 241.

[144] Interview with USF-I command historian, Baghdad, August 1, 2011.

[145] SIGIR, 2006.

[146] SIGIR, 2006.

of local governance capacity in Iraq . . . . Despite very difficult operating conditions, creating the PRTs in a short period of time is a noteworthy achievement.[147]

The SIGIR attributed the PRT success to effective senior leadership at the Iraqi Reconstruction Management Office and by the CG of Multi-National Corps–Iraq.

Strategic Direction

In November 2005, the NSC issued the *National Strategy for Victory in Iraq*, which defined success along political, economic, and security tracks.[148] It was the first published strategy in more than two-and-a-half years of war.[149] It described the strategy as "clear, hold, build": Clear territory of insurgents, hold the territory, and build rule-of-law institutions and economic development.[150] The strategy also emphasized an idea the President would give frequent voice to in coming months: "As the Iraqi security forces stand up, coalition forces can stand down."[151] The GAO found the document lacking in terms of specific plans or requisite resources.[152]

Transition from the Iraqi Transitional Government

Despite the continuing spiral into civil war and concern in the COIN community about strategic drift on the battlefield, the development of Iraqi political processes continued to hew to prescribed milestones. MNF-I had long seen adherence to agreed-on timetables as a way of shoring up the legitimacy of the Iraqi government, ultimately providing a way out for U.S. forces. In August 2005, the National Assembly drafted a permanent constitution; Iraqi voters ratified it in October, with 13 of the 18 provinces in approval.[153] On December 15, 2005, in accordance with the schedule articulated in the TAL, Iraq held elections for a full-term national government.[154]

Much of MNF-I's support for the political transition in Iraq was in the form of election security. MNF-I activities in support of the election focused on "creating the security conditions that made it safe for Iraqis to cast their ballots."[155] MNF-I prioritized election security because, it predicted in September, "[a] compelling electoral outcome will have observable follow-on political, security, and economic effects that MNF-I and the

---

[147] SIGIR, 2006.

[148] NSC, *National Strategy for Victory in Iraq*, November, 2005.

[149] Robinson, 2008, p. 19.

[150] Robinson, 2008, p. 19.

[151] Ricks, 2009a, p. 14.

[152] GAO, 2008c, p. 13.

[153] USF-I, 2011a.

[154] Kenneth Katzman, *Iraq: Politics, Elections, and Benchmarks*, Washington, D.C.: Congressional Research Service, March 1, 2011a, p. 3.

[155] MNF-I, 2005b, p. 20.

Iraqi Government must be prepared to capitalize on and to sustain the positive direction generated to date."[156]

The Bush administration heightened expectations for the outcome of the election. With the adoption of a permanent constitution and a national government, administration officials began to formulate plans for major drawdowns of U.S. forces in 2006.[157] After the December 2005 elections, Vice President Cheney stated that Iraq had "turned the corner" and predicted that "when we look back ten years hence, we'll see that the year 2005 was a watershed year here in Iraq."[158] Of course, 2005–2006 proved to be a watershed, but not of the sort the vice president had envisioned. Only two months later, the bombing of the Samarra Mosque would become both symptom and cause of explosive sectarian tensions. By the end of 2006, serious reevaluations of the failing strategy in Iraq would lead to a new approach.

Government formation following the December elections proved frustratingly slow. In March 2006, Iraq inaugurated its legislature, the Council of Representatives. In April, the council approved Talabani's continuation as president, and, after months of wrangling, approved Nouri Kamal al-Maliki as prime minister in May.[159] The new government was inaugurated on May 28, 2006. President Bush congratulated the prime minister on the event: "Iraqis now have a fully constitutional government, marking the end of a democratic transitional process in Iraq that has been both difficult and inspiring."[160] This transition would not be the end of the democratic transitional process, of course, but rather the beginning of a new phase, as the levels of violence pushed MNF-I in new directions.

Multi-National Force–Iraq in Support of the Government of Iraq and the U.S. Embassy (May 20, 2006, to December 31, 2008)

After the May 2006 seating of the Maliki government, MNF-I plans envisioned rapid progress toward transition to Iraqi self-reliance. "2006 will see the pace of transition accelerate in all areas of activity," the Joint Campaign Action Plan projected. "The ISF will continue to progressively take the lead, responsibility for security will begin

---

[156] MNF-I, 2005b, p. 20.

[157] Ricks, 2009a, p. 31.

[158] Ricks, 2009a, p. 31.

[159] Katzman, 2011a, p. 3.

[160] George W. Bush, "Statement on the Formation of Iraq's Government," The American Presidency Project website, May 20, 2006.

to transition to Iraqi civil authorities, and the new government of Iraq will exercise its sovereign powers as a democratically-elected govt."[161]

Unfortunately, such rapid progress toward this final transition continued to prove elusive, as Iraqi society began to unravel under the stress of the escalating civil war, facilitated in large part by an ISF that was too small, inadequately trained, and complicit in the sectarian violence in many cases. In a relatively somber campaign update in June, MNF-I reported on the risks and opportunities of that period of transition: "Levels of violence remain unacceptably high, sectarian violence is rife, economic development has been slow, basic needs are not meeting Iraqi aspirations, and the Iraqi population is beginning to polarize along ethnic and sectarian lines," the update admitted, and the long struggle to form a government of national unity had allowed space for the insurgency to strengthen.[162] The update continued to find that development of the ISF was proceeding apace, but that ministerial capacity was lagging at all levels of government.

By December 2006, the U.S. embassy and MNF-I's JCP update presented a truly sobering assessment of the current situation, and a dark vision of the way ahead:

> The situation in Iraq has changed considerably since the JCP was written. Many of the risks identified within the campaign plan have materialized. Many of the assumptions did not hold. We are failing to achieve objectives in the Economic Development, Governance, Communicating, and Security LOOs within the planned time frames. It is extremely unlikely that the End State will be achieved by 2009.[163]

Despite hopes that the national elections in 2005 would allow U.S. forces to lessen their commitments in Iraq, the campaign progress update found that

> [t]he 5 month delay in the formation of the new government created a political vacuum which meant the new administration struggled to gain traction from the start. Sectarian and partisan influences and corruption are adding friction and delay in creating capacity of governance which is progressing at Iraqi pace and will take longer than envisaged.[164]

Furthermore, trends going forward gave little reason for hope that the situation would change in the near term. Finding that "[t]he security situation has worsened

---

[161] Zalmay Khalilzad and George W. Casey, "2006 Joint Campaign Action Plan 'Unity, Security, Prosperity,'" 2006b, p. 1.

[162] U.S. Mission–Iraq and MNF-Iraq, *Campaign Progress Review, June 2004–June 2006*, June 14, 2006b, pp. 3, 9.

[163] U.S. Mission–Iraq and MNF-Iraq, *Campaign Progress Review, June 2006–December 2006*, Baghdad, 2006a, p. 2.

[164] U.S. Mission–Iraq and MNF-Iraq, 2006a, p. 3.

since the beginning of 2006 as sectarian violence has become self-sustaining," the review concluded, "[i]t may be that [coalition force] and ISF surge operations coupled with political and reconciliation progress will stabilize the situation in 2007 and prevent the level of violence increasing but confidence levels are low-medium to achieve stability in 2007."[165]

Given this reality, the JCP called for a new way forward:

> The unifying deduction from this analysis of the Campaign is that its ends, ways, and means are out of alignment . . . . If the risk is constant or even increasing, and the ends remain constant, then the means need to be increased, in addition to adjusting the ways. The single most important resource is time, but this is also fixed.[166]

By late fall 2006, high-level evaluations of progress in Iraq led the administration to an admission it had long avoided: The current approach was not working, and military forces needed to take a different path forward. The "surge," as the new approach has since become known, was unveiled in a White House address in January 2007. It consisted of varied and loosely associated military activities, some of which preceded the January speech; changes to leadership and leadership style; and the addition of tens of thousands of troops.[167] Above all, the surge reflected the administration's commitment to stay in Iraq in the face of vocal calls for an immediate withdrawal—a highly meaningful political signal.

The surge marked an important turning point for the concept of "transition" in Iraq. For years, MNF-I had made transition of security and governance functions to the Iraqis, as quickly as possible, the top priority. Perhaps second only to fighting the insurgency, building security forces was a central focus because Iraqi force development was considered the U.S. military's ticket out of the country. Yet the surge represented a move away from this kind of near-term transition. The new focus was less on near-term force development (although there was a doubling down by means of such approaches as partnered operations) and more on reducing the level of violence, then slowly and painstakingly building the conditions necessary for effective governance, economic development, political reconciliation, and sustainable security.

Announcing the Surge, Transition to a New Leadership Team

On January 10, 2007, President Bush stood in the White House library and addressed the nation on the status of the war. "It is clear," he said, "that we need to change our

---

[165] U.S. Mission–Iraq and MNF-Iraq, 2006a, p. 5.

[166] U.S. Mission–Iraq and MNF-Iraq, 2006a, p. 6.

[167] Kenneth Katzman, *Iraq: Politics, Governance, and Human Rights*, Washington, D.C.: Congressional Research Service, April 1, 2011b, p. 4.

strategy in Iraq."[168] He explained that the security gains anticipated to follow national elections had failed to materialize. "The violence in Iraq, particularly in Baghdad," he explained, "overwhelmed the political gains the Iraqis had made."[169] He called for an additional 20,000 troops (which ultimately became 30,000[170]) to support population security in the capital. The surge deployment would bring the total boots on the ground to a wartime peak of more than 170,000.[171]

The speech capped high-level discussions in the fall about the strategy in Iraq. In September 2006, a bipartisan commission of prominent policymakers, the Iraq Study Group, had sought to examine the unraveling war with a fresh set of eyes.[172] On December 6, the group released its report, which opened with a grim finding: "The situation in Iraq is grave and deteriorating."[173] Time was running out, and current U.S. policy was not working. Rather than continuing down the current path, the study group's report called on the President to increase pressure on the Iraqis to make significant progress and to make continued U.S. engagement contingent on such progress.[174]

A new leadership team for the United States in Iraq also transformed the command climate, especially with regard to civil-military coordination and integration. General Petraeus, one of the most thoughtful military officers on new approaches to

---

[168] George W. Bush, "Address to the Nation on the War on Terror in Iraq," The American Presidency Project website, January 10, 2007.

[169] Bush, 2007.

[170] See Tom Bowman, "As the Iraq War Ends, Reassessing the U.S. Surge," National Public Radio, December 16, 2011.

[171] Bush, 2007. See also Katzman, 2011b, p. 4. Determining the actual number of troops in Iraq at any given time was always difficult, in large part because of personnel accounting procedures. The number given for "boots on the ground" generally included only the service members who were deployed to Iraq and assigned to a unit or headquarters that was on deployment orders to Iraq. Consequently, the reported number often left out a substantial number of personnel. For example, the count did not include service members who were visiting Iraq on temporary duty for fewer than 90 days. Likewise, personnel who were stationed in Kuwait to support the war in Iraq were not counted, even if their duties required them to enter Iraq for operational purposes for short periods. Finally, because of the Office of the Secretary of Defense's "business rules," some organizations with missions that could extend outside the territorial boundaries of Iraq were also not counted. For example, at the peak of the surge in 2007, the count for MNF-I included none of the organizations that were either assigned or attached to Task Force 714 under the command of then-LTG Stanley McChrystal. Therefore, for the purposes of this book, we have not specified a particular number but simply state that, at the peak, more than 170,000 boots were on the ground (Brennan notes, 2006–2008). The largest DoD projection was 172,000. See Richard Sherlock, "DoD Operational Update Briefing with Maj. Gen. Sherlock from the Pentagon Briefing Room, Arlington Va.," September 6, 2007.

[172] Lyndsey Layton, "The Story Behind the Iraq Study Group," Washington Post, November 21, 2006.

[173] Iraq Study Group, The Iraq Study Group Report, December 2006, p. xiii.

[174] Bruce R. Pirnie and Edward O'Connell, Counterinsurgency in Iraq (2003–2006): RAND Counterinsurgency Study—Volume 2, Santa Monica, Calif.: RAND Corporation, MG-595/3-OSD, 2008, p. 18.

the war,[175] succeeded George Casey as CG of MNF-I. In March, Ryan Crocker arrived to succeed Zalmay Khalilzad as ambassador. Crocker was a DoS expert in the Arab world who had both worked on the Future of Iraq project and served briefly as Jerry Bremer's senior advisor on governance issues during the CPA period.[176]

By fall 2007, security trends in Iraq had improved for the first time in four years of war.[177] No single factor accounts for the reduction in violence in this period, although many developments redefined the scope and scale of military activities in Iraq. Robust military engagement in reconciliation efforts, new approaches to providing population security, a doubling down of current efforts to team with and train ISF and to refine counterterrorism efforts all contributed to improved security trends. These new strategies expanded and changed the nature of military activities in Iraq.

Population Security

A key element of the surge was providing security to the population, especially in Baghdad. Although the concept had been in development for months before Petraeus arrived, the Baghdad Security Plan, or "Fardh al-Qanoon," was an initiative aimed at making Baghdad secure enough to facilitate political reconciliation.[178] The fundamental idea embodied in this new plan has been traced back to LTC Doug Ollivant, the chief of plans (G-5) at Multinational Division–Baghdad (MND-B) during the surge, who had coauthored a widely read article in 2006 arguing for a new approach to combatting an insurgency.[179] Ollivant argued then that "the combined arms battalion," partnering with ISF, "living among the population it secures, should be the basic tactical unit of counterinsurgency warfare."[180] This was a rejection of the approach then in practice of consolidating U.S. forces on large isolated bases. MNF-I implemented the Baghdad Security Plan by dividing Baghdad into a grid of "security districts" and

---

[175] As noted earlier, while commanding the Army's Combined Arms Center, General Petraeus had sponsored a collaborative effort with the Marine Corps' Gen. James F. Amos that resulted in the publication of the Counterinsurgency Field Manual.

[176] Dobbins et al., 2009, p. 32. Crocker served with the CPA for only about three months.

[177] Robinson, 2008, p. 324.

[178] In 2006, General Chiarelli had tried a Baghdad security plan, which also aimed to reduce the violence in the capital by employing coalition and Iraqi forces working together to create the security conditions necessary for reconstruction (Michael R. Gordon, "To Stand or Fall in Baghdad: Capital Is Key to Mission," *New York Times*, October 22, 2006). In an October 2006 interview for the Council on Foreign Relations, an analyst from the Brookings Institution assessed that this security plan stalled because it "was not properly conducted or resourced the way that General Chiarelli wanted it to." Kenneth M. Pollack, "Pollack: Iraq May End Up Worse Off Than Under Saddam," interview with Bernard Gwerzman, Council on Foreign Relations, October 13, 2006.

[179] Douglas A. Ollivant, "Producing Victory: Rethinking Conventional Forces in COIN Operations," *Military Review*, October 2006. A significant amount of research about COIN was conducted between 2004 and 2006. What was unique about this situation was that an author of a professional journal article would be in a position to implement the ideas he had proffered months earlier.

[180] Robinson, 2008, p. 121.

embedding a U.S. battalion and an Iraqi brigade in each district. The embedded troops were first to clear areas of armed combatants, control the area, secure the population, and support governance and economic activities. Finally, only after Iraqis were capable of retaining security in the area would U.S. forces withdraw into advisory and over-watch roles.[181] As Linda Robinson explained,

> [b]y denying the enemy access to the population on a continuing basis, and provid-ing the population with security and basic services, the new plan [was] to shift the population's allegiance to the host-nation government and induce them to identify and expel or neutralize the armed insurgents.[182]

Unlike previous plans, it also sequenced the areas it would secure based on how Bagh-dad operated.

The surge in military forces that began in January 2007 was accompanied by a surge in civilian boots on the ground and joint civil-military teams focused on neigh-borhood reconstruction throughout Baghdad and the security belts located around the outskirts of Baghdad.[183] This civilian and military surge would double the number of reconstruction advisors serving outside the relatively safe area of the International Zone of Baghdad over the next nine months.[184] An element of this approach was the introduction of the concept of "embedded" PRTs (ePRTs), which would place civilian development experts directly inside brigade combat teams (BCTs). As with traditional PRTs, ePRTs were led primarily by DoS personnel and included staff from USAID and DoD (both military and civilian). The intent behind the ePRTs was to facilitate engage-ment between brigades and local government officials and to help integrate brigades' reconstruction efforts (largely through CERP funds) with embassy reconstruction plans and the actions of civil affairs teams.[185] Yet the ePRTs did not always share the military's priorities for the use of CERP funds. Reflecting a broader critique of CERP, some observers argued that the metrics the military used to assess CERP's effectiveness overemphasized getting money out the door at the expense of ensuring that the money was directed to the most effective projects. Quoting one ePRT leader, SIGIR noted, the military is "being graded on how many projects are being carried out, how much money is flowing to the districts." A better approach would have been to assess "how many projects are being turned over to the Iraqis and how much less money [military

---

[181] Robinson, 2008, p. 122.

[182] Robinson, 2008, p. 123.

[183] SIGIR, 2009a, p. 295.

[184] SIGIR, 2009a, p. 295. Before 2004, the area of Baghdad near the Republican Palace and the crook in the Tigris River now known as the International Zone was officially called the "Green Zone." The area was renamed after Iraqi independence in June 2004. This part of Baghdad remains heavily fortified and is the home for foreign embassies, the Iraqi Parliament, and ministries of the government of Iraq.

[185] SIGIR, 2009a, p. 300.

units] are spending."[186] While CERP remained an important tool, it was never integrated into the overall strategy but, rather, remained a capability employed at the discretion of local brigade (tactical) commanders.[187]

Sons of Iraq

One of the efforts most widely cited as responsible for the substantial security gains in 2007 was a program that brought Sunnis and other Iraqis, including former insurgents, onto U.S. payrolls to provide security functions. For months before the surge, commanders engaged influential local leaders, including tribal sheikhs and imams, to recruit Sunnis and put them to work in support of coalition objectives. This effort, which began in fall 2006 in al-Anbar province, came to be called the "Sunni Awakening" or the "Anbar Awakening."[188] In time, the Sunnis who participated in this program would be known as the "Sons of Iraq" (SOI). By June 2007, MNF-I was paying SOI members in Baghdad willing to pledge allegiance to the government of Iraq to provide security at checkpoints and key infrastructure $10 a day.[189] A few months later, thousands of mostly Sunni Iraqis had signed on, motivated by a range of incentives for turning to the Americans. Many members of the Awakening rejected the extreme tactics and violence of al-Qaeda in Iraq that not only targeted Sunnis who failed to follow their lead but also threatened the authority and power of the tribal structure in Iraq. An additional motivation was the opportunity for jobs funded through MNF-I CERP funds. While individual motivations varied, MNF-I viewed this as an important inflection point because of the potential for reconciliation and the potential end to large-scale sectarian violence.[190]

Critics charged that the United States was empowering dubious and little-understood individuals and might well be fanning an ongoing civil war. Supporting SOI might be a nearsighted and expedient approach to reducing violence, some argued, rather than a way to achieve sustainable stability.[191] Short-term solutions were particularly concerning to the Shi'a-led government of Iraq, which would be responsible for maintaining stability long after the Americans were gone. One Iraqi official likened SOI to raising a crocodile: "It is fine when it is a baby, but when it is big, you can't keep it in the house."[192] The program accelerated through spring 2007, eventually

---

[186] SIGIR, 2009a, p. 303.

[187] Ricks, 2009a, p. 203.

[188] Ricks, 2009a, p. 203.

[189] Robinson, 2008, p. 252.

[190] Interview with USF-I command historian, Baghdad, December 14, 2012.

[191] Robinson, 2008, p. 253.

[192] Saad Yousef Al-Muttalibi, quoted by David Enders, "Iraqi Tribes Reach Security Accords, *Washington Times*, July 23, 2007.

putting 103,000 Iraqis, largely Sunnis and many former insurgents, on the U.S. payroll to provide local security.

The SOI was widely credited with reducing violence in Sunni areas in 2007 and 2008, although other factors—such as the U.S. troop surge and a new U.S. approach to COIN—also helped bring unrest under control. Describing SOI's value, the SIGIR wrote:

> The SOI provided intelligence on the location of insurgent groups and weapons caches, acted as a force multiplier by freeing U.S. and Iraqi forces to perform other operations; denied insurgent groups a recruitment pool; and, in some cases, began to cooperate with the Iraqi Security Forces.[193]

Thus, SOI significantly augmented the coalition's ability to provide local security and access valuable intelligence that was used to systematically target the al-Qaeda in Iraq network and other "nonreconcilable" Sunnis.[194] General Petraeus touted the program's success in his September 2007 testimony on progress in Iraq. He described "the most significant development in the past six months" as "the increasing emergence of tribes and local citizens rejecting Al Qaeda and other extremists."[195] In the second half of 2007, the number of al-Qaeda in Iraq fighters reportedly fell by 70 percent, from 12,000 to 3,500.[196]

The integration of SOI into Iraqi institutions, such as the ISF or civil service, was a critical component of U.S. efforts to facilitate Sunni reconciliation. In fall 2007, MNF-I began transitioning responsibility for paying and directing SOI to the government of Iraq, a process that was not completed until April 2009. The government of Iraq committed to finding jobs for SOI members, including in the ISF and in government offices, although the integration process has been slow and uneven.[197] Unfortunately, the government of Iraq was continually late in paying the SOI, was slow to provided the promised employment and integration, and arrested SOI leaders and personnel far too often based on dubious arrest warrants generated by Shi'a political leaders who were seeking revenge and to consolidate power.[198] While the transition of the SOI program to the government of Iraq was completed in 2009, USF-I and Embassy

---

[193] SIGIR, "Sons of Iraq Program: Results Are Uncertain and Financial Controls Were Weak," SIGIR 11-010, January 28, 2011a, p 5.

[194] For a detailed discussion of how the U.S. military used intelligence to relentlessly target the al-Qaeda in Iraq network and its success in doing so, see Stanley McChrystal, *My Share of the Task: A Memoir*, New York: Penguin Group, 2013.

[195] David Petraeus, "Report to Congress on the Situation in Iraq," September 10–11, 2007, p. 5.

[196] Attorney General of Australia Robert McClelland, letter to Arch Bevis, Chair, Parliamentary Joint Committee on Intelligence and Security, document 08/11412, October 21, 2008.

[197] SIGIR, *Quarterly Report to the United States Congress,* April 30, 2009c, p. 8.

[198] Interview with USF-I command historian, Baghdad, August 1, 2011.

Baghdad remained engaged with the government of Iraq on this issue through the departure of U.S. forces in 2011.[199]

Civil Affairs

CERP spending peaked during the Petraeus-Crocker years. Some widely discussed projects aimed to support economic development in the capital. For example, CERP funds went to support the renovation of Mutanabi Street, a bookselling district in Baghdad that long served as the intellectual center of Iraq and, in earlier times, the wider Arab world.[200] In March 2007, a car bomb had killed 26 people on Mutanabi Street. As a result, T-walls were put up as barriers, and Awakening Councils manned checkpoints. A restoration project repaired storefronts and installed new water and sewage lines.[201] CERP funds also went to support the revitalization of Abu Nawas Park, a restaurant-lined stretch of park along the Tigris River. One journalist recalled the park in the August 2006 as "a grim, spooky, deserted place, a symbol for the dying city that Baghdad had become." Yet on a return in September 2008, he observed that Abu Nawas was filled with people: "It was an astonishing, beautiful scene— impossible, incomprehensible, only months ago."[202]

In April 2008, MNF-I and the government of Iraq signed a memorandum of understanding establishing Iraqi-CERP. Under this program, MNF-I would be responsible for executing reconstruction projects using remaining balances of blocked Iraqi funds but spending the funds on priorities the Iraqis established.[203] A SIGIR review of Iraqi-CERP in its first year cited a number of specific projects, including a warehouse refurbishment project in Anbar, a school refurbishment in Sadr City, and a water treatment plant in Salah al Din to provide local potable water and improve the water distribution infrastructure.[204] By September 2009, MNF-I had obligated about $229 million of the $270 million available for Iraqi-CERP.

Benchmarks

In May 2007, Congress passed a supplemental war-funding bill that required the administration to report on progress toward meeting certain political and security related benchmarks by September 2007.[205] The bill made "progress" along 18 bench-

---

[199] Interview with USF-I J3 staff officer, Baghdad, December 10, 2011.

[200] Damien Cave, "A Baghdad Book Mart Tries to Turn the Page," *New York Times*, September 15, 2007.

[201] Eric Owles, "Then and Now: A New Chapter for Baghdad Book Market," *New York Times*, December 18, 2008.

[202] Dexter Filkins, "Back in Iraq, Jarred by the Calm," *New York Times*, September 20, 2008.

[203] SIGIR, "Iraq Commander's Emergency Response Program Generally Managed Well, but Project Documentation and Oversight Can Be Improved," Washington, D.C., 10-003, October 27, 2009b, p. 1.

[204] SIGIR, 2009b, pp. 7–8.

[205] Robinson, 2008, p. 169.

marks a requirement for subsequent funding. Since August 2006, U.S. and Iraqi officials had been assessing progress in Iraq in relation to key benchmarks; the 2007 supplemental linked the release of funds to the administration's own indicators. The law required GAO to cross-check the administration's assessments. It also mandated an independent assessment of the capabilities of the ISF.[206]

The time line put pressure on the administration to demonstrate progress in concrete terms. In his high-profile September 2007 testimony, General Petraeus began, "the military objectives of the surge are, in large measure, being met."[207] He highlighted reduced levels of sectarian violence and of civilian casualties in general and the improved capacity of the ISF. He particularly praised the security gains the Sunni Awakening had wrought. In the same hearings, Ambassador Crocker complemented the security assessment with an update on the state of the domestic political situation and regional dynamics for Iraq. His measured findings concluded that, in his judgment, "the cumulative trajectory of political, economic, and diplomatic developments in Iraq is upwards, although the slope of that line is not steep."[208] He conceded that "2006 was a bad year in Iraq. The country came close to unraveling politically, economically, and in security terms," but asserted that "2007 has brought improvement."[209]

Petraeus concluded his testimony with a recommendation that U.S. troop levels begin to be drawn down. He called for the withdrawal of the first Army surge brigade in December 2007 and the subsequent withdrawal of four brigades and two Marine battalions by July 2008.[210] U.S. troop presence hit its peak of more than 170,000 in September 2007 and remained at this level until November 24, when 5,000 soldiers from the 3rd Brigade, 1st CD began their redeployment without replacement. Additional phased withdrawals would be considered after the troop presence returned to presurge levels. The logistical burden of safely reposturing thousands of troops and materiel required significant coordination for military forces, which would become a theme in the years ahead.[211]

Charge of the Knights

Some worried that premature drawdown of U.S. forces and transition to Iraqi forces not yet capable of maintaining security might jeopardize the substantial security gains. A key turning point in demonstrating Iraqi capacity and willingness to maintain secu-

---

[206]Katzman, 2011b, p. 3.

[207]Petraeus, 2007.

[208]Ryan C. Crocker, "Statement of Ambassador Ryan C. Crocker, United States Ambassador to the Republic of Iraq," before a Joint Hearing of the Committee on Foreign Affairs and the Committee on Armed Services, September 10, 2007.

[209]Crocker, 2007.

[210]Robinson, 2008, p. 299.

[211]GAO, 2008c.

rity came in a March 2008 operation known as the "Charge of the Knights." During the first months of 2008, intrasectarian violence increased between the followers of Muqtada al Sadr and other Shi'a political leaders in and around the city of Basra. This violence signaled Sadr's renewed commitment to using force to achieve political goals. While Sadr had previously encouraged violence against Sunnis, this time Sadr targeted other Shi'a leaders who were their political opponents. This direct challenge to the prime minister's authority not only posed a security challenge in southern Iraq but also created a political crisis that could have threatened the Maliki government. While MNF-I routinely worked with the Iraqi Army to provide assistance in such areas as intelligence, surveillance, movement, airpower, special operations, and planning, the prime minister decided to forgo discussions with the United States and launched the Iraqi Army south to Basra to take on major Shi'a militias—most notably Sadr's Jaysh al-Mahdi (JAM) militia. The operation was poorly planned, had insufficient resources, and had no backup or reserve positioned in the event the initial assault failed. Prime Minister Maliki personally led the forces into Basra and soon found himself and his forces surrounded by the elements of JAM. Fortunately for Maliki, MNF-I quickly dispatched surveillance and combat aircraft that were able to break the stranglehold and turn the tide in favor of the ISF.[212] One MNF-I official stated, "[t]hey went in with 70 percent of a plan. Sometimes that's enough. This time it wasn't."[213]

As a demonstration of capacity, the risky operation proved largely successful at pacifying militant factions in the city.[214] Moreover, because U.S. military support was not obvious to Iraqis, the Charge of the Knights operation bolstered the image of the prime minister at a time when his authority was being challenged. This assault served as a demonstration of Maliki's willingness to use Iraqi military power against Shi'a groups conducting violence against other Shi'a. As such, many lauded the prime minister's actions and raised Maliki's profile as a national, rather than sectarian or coalition-controlled, leader. Unfortunately, as will be discussed in subsequent chapters, the prime minister's actions could also be seen as a reflection of his willingness to use all means of national power to eliminate threats to his personal political survival.[215]

Security Agreement and the Strategic Framework

In November 2007, President Bush and Prime Minister Maliki agreed that, by July 2008, the two countries would complete a bilateral strategic framework agreement and

---

[212] James Glanz and Alissa J. Rubin, "Iraqi Army Takes Last Basra Areas from Sadr Force," *New York Times*, April 20, 2008; Interview with USF-I command historian, Baghdad, August 1, 2011.

[213] Michael R. Gordon, Eric Schmitt, and Stephen Farrell, "U.S. Cites Planning Gaps in Iraqi Assault on Basra," *New York Times*, April 3, 2008, p. 3.

[214] Katzman, 2011b, p. 4.

[215] Interview with USF-I J3 staff officer, Baghdad, December 10, 2011.

a related SOFA.[216] Significant pressure in both the United States and Iraq to define a timetable for U.S. withdrawal provided an impetus for developing such an agreement, as did the approaching expiration of the UNSCR that provided international authorities and protections for continued U.S. military presence in Iraq, which was due to expire on December 31, 2007.

Framing the Issues in the United States

Early in 2008, elements of the U.S. and Iraqi political leadership began defining the terms of the evolving bilateral agreement. In February 2008, Secretary of State Condoleezza Rice and Secretary of Defense Robert Gates penned an op-ed in the Washington Post outlining the U.S. objectives in negotiations.[217] They noted that, while the U.S. presence to date had been sanctioned by a series of UNSCRs, the Iraqis sought "an arrangement that is more in line with what typically governs the relationships between two sovereign nations." They identified a key objective for the agreement as the establishment of the jurisdiction and authorities for a continued U.S. presence to conduct combat operations, develop the ISF, and counter Iranian involvement in Iraq. "There is little doubt that 2008 will be a year of critical transition in Iraq as our force levels continue to come down, as our mission changes and as Iraqis continue to assert their sovereignty," Rice and Gates noted. The agreement they sought to forge with the Iraqis was to define the terms and the time line for that transition.

From the inception of negotiations in spring 2008 through the signing of a final deal that November, both U.S. and Iraqi negotiators worked to reconcile strategic objectives with the priorities of their divided domestic political constituencies.[218] Iraqis rejected the Bush administration's initial draft, which included provisions for unilateral U.S. control of continued military operations, as well as control of borders and airspace. Disagreement also arose over the time line for U.S. withdrawal. As the two

---

[216] Kenneth Katzman, *Iraq: Post-Saddam Governance and Security*, Washington, D.C.: Congressional Research Service, July 8, 2009, p. 40. Some observers note that the SA is more of a policy document than is standard for other SOFAs. A *SOFA* is defined as "an agreement that establishes the framework under which armed forces operate within a foreign country," and SOFAs tend to focus more on formal agreements regarding the legal treatment of military personnel than is evident in the wide-ranging policy agreements in the 2008 SA. The SA with the government of Iraq is in some ways unusual as a SOFA, such as in its provisions for the conduct of combat operations, but as a Congressional Research Service report notes, "[t]here are no formal requirements governing the content, detail, and length of a SOFA," so the issues different documents address varies significantly. Chuck R. Mason, *U.S.-Iraq Withdrawal/Status of Forces Agreement: Issues for Congressional Oversight*, Washington, D.C.: Congressional Research Service, July 13, 2009, pp. 5–7. See also Chuck R. Mason, *Status of Forces Agreement (SOFA): What Is It, and How Has It Been Utilized?* Washington, D.C.: Congressional Research Service, January 5, 2011.

[217] Condoleezza Rice and Robert Gates, "What We Need Next in Iraq," *Washington Post*, February 13, 2008.

[218] The two senior negotiators for the United States were David M. Satterfield of DoS and Brett McGurk of the NSC. Karen DeYoung, "Lacking an Accord on Troops, U.S. and Iraq Seek a Plan B," *Washington Post*, October 14, 2008.

sides worked to forge a compromise in subsequent months, they reached agreement for joint control of Iraqi airspace by May, and Secretary Rice reported in August that they had settled on the end of 2011 for the withdrawal of U.S. forces.[219]

Perhaps the most contentious issue arising during the security agreement negotiations was the framework for the legal treatment of U.S. personnel. The United States reluctantly agreed to lift immunity protections from Iraqi law for U.S. contractors but resisted calls from some Iraqi politicians to end immunity for U.S. military and civilian government personnel as well.[220] The immunities in place for contractors dated to a policy adopted during the days of the CPA.[221] In October 2008, Iraqi Foreign Minister Hoshyar Zebari stated that "[t]he hanging issue is the issue of immunity," with regard to the final brokering of a deal.[222] The Sadrist bloc had been among the most vocal opponents of the security agreement, objecting to any agreement that included a continued U.S. presence in Iraq.[223]

Notably, the key transition in Iraq that the development of the SA exemplified took place amid a major political transition in the United States: a presidential election. The U.S. President that had taken the country to war and spent most of his presidency overseeing its execution would not be on the ballot.

Elements of the Security Agreement and the Strategic Framework Agreement

The Strategic Framework Agreement (SFA) and the associated security agreement (SA) took about a year in the making.[224] On November 17, 2008, U.S. Ambassador Ryan Crocker and Iraqi Foreign Minister Hoshyar Zebari signed the documents that both (1) set the stage for the final transition of the U.S. military out of Iraq and established a framework for a long-term strategic relationship between the United States and Iraq and (2) created the legal framework governing U.S. military operations in Iraq through the end of 2011. The SA specified limits to the U.S. military operations in Iraq and provided limited Iraqi jurisdiction over U.S. troops who committed serious crimes

---

[219] DeYoung, 2008.

[220] Sabrina Tavernise, "U.S. Agrees to Lift Immunity for Contractors in Iraq," *New York Times*, July 2, 2008.

[221] James Risen, "End of Immunity Worries U.S. Contractors in Iraq," *New York Times*, November 30, 2008.

[222] Charles Levinson and Ali A. Nabhan, "U.S., Iraq Negotiate on Troop Immunity," *USA Today*, October 7, 2008.

[223] Stephen Farrell, "Protests in Baghdad on U.S. Pact," *New York Times*, November 21, 2008.

[224] Section III of the SFA specified creation of a companion document, stating that "security and defense cooperation shall be undertaken pursuant to the Agreement Between the United States of America and the Republic of Iraq on the Withdrawal of United States Forces from Iraq and the Organization of their Activities during their Temporary Presence in Iraq." This so-called security agreement provided U.S. military and civilian government personnel immunities and protections while serving in Iraq. However, from an Iraqi perspective, it also required the withdrawal of all U.S. military personnel by December 31, 2011. Any discussions regarding a long-term SOFA for a potential enduring U.S. military presence were postponed until after the next Iraqi government was formed in 2010.

while off duty and off a U.S. military installation.[225] It constrained the use of unilateral U.S. military power by stipulating that operations must be coordinated by means of a joint U.S.-Iraqi military committee. At the same time, it limited Iraqi leverage over U.S. government personnel by providing for substantial immunities for U.S. military and civilian personnel, but not for contractors. This had been a major sticking point during negotiations. The agreement held that U.S. military and civilian personnel who committed "grave premeditated felonies" (vaguely defined) while off base and off duty would be subject to Iraqi jurisdiction.[226] Yet determinations of the applicability of this standard would be subject to a joint U.S. and Iraqi committee. The agreement also outlined a phased process leading to the complete withdrawal of U.S. troops by December 31, 2011. Control of all 18 provinces would transition to Iraq by January 1, 2009. This superseded the PIC process, despite the fact that five provinces had failed to achieve the PIC thresholds.[227] Also on January 1, 2009, Baghdad's International Zone would return to Iraqi control. The SA provided that U.S. troops would cease patrolling Iraqi cities by June 30, 2009.[228] Finally, the SA explicitly stated that not only would all U.S. troops be required to leave Iraq by the end of 2011, but no permanent U.S. military bases would be permitted to remain.

Accompanying the SA was the SFA, which provided a road map for a long-term economic, political, and cultural relationship between Iraq and the United States.[229] The intent was to affirm "the genuine desire of the two countries to establish a long-term relationship of cooperation and friendship, based on the principles of equality in sovereignty and the rights that are enshrined in the United Nations Charter and their common interests."[230] It articulated seven broad areas for cooperation: political and diplomatic, defense and security, cultural, economic and energy, health and environmental, information technology and communications, and law enforcement and judicial. The SFA had no expiration date. If the time horizon for the SA was December 31, 2011, the SFA was to describe a relationship into the indefinite future, with the

---

[225] The significant limits of Iraqi jurisdiction over U.S. military personnel essentially provided immunities to U.S. soldiers who were always "on duty" and were not authorized to be off U.S. military bases except in the performance of their missions. The language of the SA allowed the government of Iraq to tell the Iraqi people that U.S. soldiers would be prosecuted under Iraqi law if they committed crimes in Iraq while, simultaneously, providing the legal protections to all military personnel to ensure that such prosecutions would never occur. Military lawyers argued that the wording of the SA ensured that crimes U.S. military personnel committed in Iraq would always be addressed under the Uniform Code of Military Justice and that under no circumstances would a U.S. service member be subject to Iraqi criminal jurisdiction. Brennan notes, 2009–2011.

[226] Mason, 2009, p. 8.

[227] SIGIR, 2009c, p. 7.

[228] Katzman, 2009, p. 40.

[229] Katzman, 2011b, p. 7.

[230] United States of America and the Republic of Iraq, 2008b.

ultimate goal of establishing an enduring strategic partnership between Iraq and the United States.

Figure 2.1 highlights the critical elements of both the SFA and the SA as MNF-I identified them. Together, the two agreements fixed the timetable for the withdrawal of U.S. military forces from Iraq and the transition of responsibilities for all missions, activities, functions, and relationships from the U.S. military to Embassy Baghdad, the embassy's yet-to-be formed OSC-I, USCENTCOM, or various elements within ISF and the government of Iraq. These two documents established the overarching political and strategic guidance that would govern the final transition and the end of U.S. military operations in the Iraq war.

Figure 2.1
Establish Enduring Strategic Partnership

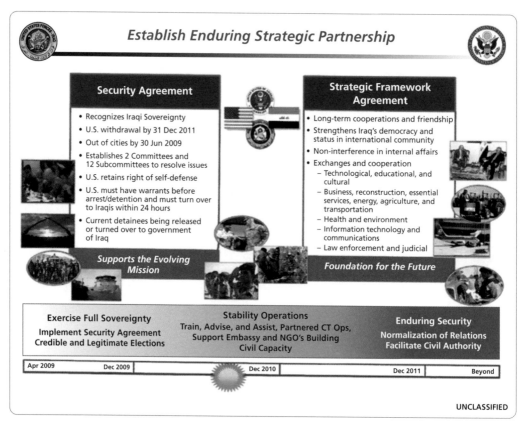

SOURCE: Embassy of the United States, Baghdad, and ISF-I, *Executive Core Group Briefing*, August 1, 2010, slide 15, Not available to the general public.
RAND RR232-2.1

CHAPTER THREE

Multi-National Force–Iraq Transition Planning and Execution, 2009–2010

Overview

In 2003, military activities in Iraq had begun with a relatively limited scope, including combat operations and relatively modest support for postconflict reconstruction and humanitarian activities. In time, that limited charter grew and changed to encompass a remarkable range of governance, institution building, economic development, and civil affairs activities. In compliance with the terms of the November 2008 SA, the expansive reach of U.S. military involvement in Iraq began to contract in early 2009, and the military began to develop processes and procedures for transitioning its myriad activities back to other U.S. and Iraqi institutions.

Going forward, MNF-I would operate under a deadline for the complete drawdown of U.S. forces. This meant that—barring a decision by either party to withdraw from or renegotiate the agreement—an end point was known for the presence of U.S. forces in Iraq. With that deadline established, U.S. forces would be required to develop and implement plans for the drawdown of forces and ultimate transition of all military activities.

If the SA equipped MNF-I with a deadline, it also bound MNF-I to new and stringent rules of engagement. This meant that military operations under the SA diverged significantly from those MNF-I had been authorized to undertake only months earlier. Instead of carrying out unilateral operations based on U.S. intelligence, MNF-I had to collaborate with Iraqi counterparts on targeting, secure warrants from Iraqi legal institutions, and partner with Iraqi forces to execute operations jointly. Moreover, the SA limited U.S. forces' ability to operate unilaterally in Iraqi cities, which changed the ways in which MNF-I approached civil-military affairs, governance, and other noncombat functions.

Yet the constraints the SA placed on U.S. combat operations were not definitive. The agreement noted that

> [t]he Government of Iraq requests the temporary assistance of the United States Forces for the purposes of supporting Iraq in its efforts to maintain security and

stability in Iraq, including cooperation in the conduct of operations against al-Qaeda and other terrorist groups, outlaw groups, and remnants of the former regime.

It required military operations by U.S. forces "be conducted with the agreement of the Government of Iraq . . . [and] fully coordinated with Iraqi authorities." However, it also provided that both parties would "retain the right to legitimate self-defense within Iraq, as defined in applicable international law."[1] This provision allowed continued unilateral U.S. action in limited circumstances.

Multi-National Force–Iraq Under the Security Agreement, Through Withdrawal from Cities (January 1, 2009, to June 30, 2009)

The SA took effect January 1, 2009, in the final weeks of the U.S. administration that had overseen its negotiation. Despite extensive discussion of Iraq during the presidential campaign, it remained to be seen exactly whether and how President-elect Obama would change U.S. policy in Iraq once he took office.

Camp Lejeune Speech

Five weeks after taking office, President Obama gave a major policy speech reaffirming the nation's commitment to meeting the deadline articulated in the SA and transitioning the military out of Iraq.[2] "Today, I have come to speak to you about how the war in Iraq will end," he explained. President Obama acknowledged ongoing challenges in Iraq but asserted that there must be realistic limits to U.S. goals.

> What we will not do is let the pursuit of the perfect stand in the way of achievable goals. We cannot rid Iraq of all who oppose America or sympathize with our adversaries. We cannot police Iraq's streets until they are completely safe, nor stay until Iraq's union is perfected. We cannot sustain indefinitely a commitment that has put a strain on our military, and will cost the American people nearly a trillion dollars.[3]

In essence, the President asserted, the United States would accept the progress and stability achieved to date, even in the face of continued challenges, as good enough.

President Obama outlined a transition that would proceed in two stages. In the first phase, he called for the removal of "combat brigades" from Iraq, with the target

---

[1] United States of America and the Republic of Iraq, 2008a, p. 13.

[2] Barack Obama, "Responsibly Ending the War in Iraq," remarks, Camp Lejeune, N.C., February 27, 2009.

[3] Obama, 2009.

of ending the U.S. combat mission by August 31, 2010.[4] The administration anticipated that this goal would require a drawdown to 120,000 troops by September 1, 2009, with all remaining "combat" forces removed after 12 additional months. In the second phase, the 50,000 troops who remained would have three discrete functions: training, equipping, and advising ISF; conducting targeted counterterrorism missions in partnership with Iraqi counterparts; and protecting ongoing civilian and military efforts. The forces charged with developing ISF capacity would be organized into newly formed advise-and-assist brigades (AABs) to continue training and mentoring Iraqi forces through the end of the mission.[5] All U.S. forces would be withdrawn from Iraq by the end of 2011.

President Obama also stressed his commitment to pursue a path leading toward a long-term strategic partnership with Iraq, as articulated in the SFA and SA. The final steps the military would take in Iraq would set what he called the "emerging foundation" for that long-term relationship. With the drawdown of the combat mission, the President explained, the U.S. mission in Iraq would transition toward a focus on sustained diplomacy in support of the goal of developing a more peaceful and prosperous Iraq. The military would support this goal by supporting the political process, reconciliation, and development of ministerial capacity to improve governance and transparency. Finally, the President established the goal of Iraq emerging as an important regional ally in support of U.S. security interests in the broader Middle East. Thus, while in one part of the speech the President clearly indicated his desire to end the war by 2012, other portions of the speech reaffirmed an expansive commitment to the future of Iraq. Military and civilian planners in Baghdad would discuss these twin goals as establishing a strategy that was "conditions based but time constrained."[6] It was clear to military planners that the strategic and operational objectives could not be met by 2012. The only question was whether or not U.S. forces would depart Iraq before achieving the goals established by the President and codified in the Iraq JCP.[7]

Out of the Cities

On June 30, 2009, U.S. forces withdrew from cities, in accordance with the timetable established by the SA.[8] The SA had required that

> [a]ll United States combat forces shall withdraw from Iraqi cities, villages, and
> localities no later than the time at which Iraqi Security Forces assume full responsi-

---

[4] Ending the U.S. combat mission in Iraq in August 2010 was not specified in the SA and represented a unilateral U.S. policy decision driven by then-Senator Obama's campaign promises in 2008.

[5] SIGIR, 2009c, p. 6.

[6] Interview with USF-I Directorate of Joint Strategy and Plans (J5) staff officer, December 3, 2011, Baghdad.

[7] Interview with USF-I J5 staff officer, December 3, 2011, Baghdad.

[8] Mason, 2009, p. 10.

bility for security in an Iraqi province, provided that such withdrawal is completed no later than June 30, 2009.[9]

By midnight on that day, all U.S. combat troops were required to be out of Iraqi cities and redeployed to bases in rural areas.[10] Thereafter, U.S. combat troops would require a request from the Iraqi government to reenter cities,[11] although General Odierno noted that U.S. forces would continue to conduct "significant operations outside of the cities and the belts around the major cities."[12]

In spring 2009, MNF-I worked with the government of Iraq to clarify the definition of "out of cities" and how the mandate would be implemented. While MNF-I felt some presence would be necessary to coordinate with Iraqi forces, some Iraqi leaders fought for a more categorical ban on U.S. troops operating in cities.[13] Some number of American noncombat forces would remain within city limits to conduct training and advising missions.

This was the first major milestone in the SA drawdown process.[14] Prime Minister Maliki declared June 30 "National Sovereignty Day," to henceforth be a national holiday.[15] "This day, which we consider a national celebration, is an achievement made by all Iraqis," Maliki proclaimed. "Those who think that Iraqis are unable to defend their country are committing a fatal mistake."[16]

General Odierno emphasized that the "out of cities" order would require a new mind-set if U.S. forces were to succeed. This meant that military forces could not interpret the order as meaning they should just do less; they needed to do things differently. In discussions with the MNF-I leaders and staff, General Odierno likened the order to the surge in that it was not just a change in number but a change in tactics, techniques, and procedures that had been the key to success. The CG emphasized that the "out of cities" order would require a similar degree of innovation and would require the collective attention of leaders at all levels of command to implement properly.[17]

Withdrawal from cities substantially reduced the amount of information available to U.S. forces. The intelligence gleaned from the presence of U.S. troops among

---

[9]   United States of America and the Republic of Iraq, 2008a, p. 15.

[10]   Tim Cocks and Muhanad Mohammed, "Iraq Regains Control of Cities as U.S. Pulls Back," Reuters, June 30, 2009.

[11]   Mark Thompson, "With U.S. Pullout, Iraq Takes Ownership of Its War," *Time*, June 30, 2009.

[12]   Thompson, 2009.

[13]   Interview with USF-I command historian, Baghdad, July 22, 2011.

[14]   Katzman, 2009, p. 25.

[15]   Katzman, 2011a, p. 7.

[16]   Cocks and Mohammed, 2009.

[17]   Interview with USF-I command historian, Baghdad, July 22, 2011.

the population suddenly disappeared, and the United States had to rely increasingly on Iraqis for information. As BG James Nixon explained, "[t]he major challenge that we will face is continuing to maintain situational awareness as we reduce presence inside the cities." Achieving this would require effective partnerships with the Iraqis and other allies.[18]

In April 2009, U.S. and Iraqi forces conducted a joint raid in Kut that left two Iraqis dead. The aftermath of the incident made clear how different the operating environment had become for U.S. forces. Maliki claimed the operation lacked requisite approvals and called it a violation of the SA. In response, General Odierno put in place a methodology for "transparent targeting" of terrorist and insurgent targets, in which U.S. forces would nominate targets to a combined targeting committee whose approval would be required to conduct operations. General Odierno explained that the new targeting procedure would foster greater cooperation between U.S. forces and their Iraqi counterparts. In addition, the process would demonstrate to both the government and the Iraqi people that U.S. forces were taking actions consistent with Iraqi laws and were not conducting unilateral actions.[19]

Multi-National Force–Iraq and U.S. Forces–Iraq Through the End of Combat Operations (June 30, 2009, to August 30, 2010)

On May 23, 2009, MNF-I released Operations Order (OPORD) 09-01, "Operations Order for the Responsible Drawdown of U.S. Forces from Iraq." The stated purpose of this plan was to integrate, synchronize, and prioritize MNF-I operations, plans, and emerging concepts to support the drawdown of forces in preparation for the August 31, 2010, end of combat operations. At the same time, MNF-I also began to plan and coordinate for the drawdown by setting up a drawdown fusion cell and a U.S. Army Central (ARCENT) support element to assist in planning.[20]

The drawdown between June 2009 and the end of combat operations in August 2010 led to the withdrawal of more than 35,000 pieces of equipment that were able to either move under their own power or be pulled behind another vehicle, what the military often calls rolling stock. The drawdown of forces would see the reduction in U.S. boots on the ground from more than 170,000 at the height of the surge in October 2007 to 86,000 by June 2009 and to 50,000 by the end of August 2010.[21] This reduction in U.S. combat power would be executed simultaneously with a hoped-for

---

[18] Tom A. Peter, "US Withdraws from Iraq Cities," *Global Post*, June 29, 2009.

[19] Interview with USF-I command historian, Baghdad, July 22, 2011.

[20] Interview with USF-I command historian, Baghdad, July 22, 2011.

[21] Rick Brennan, "USF-I Operations Thru December 31, 2011," J3, USF-I, July 5, 2010, Not available to the general public.

build-up in ISF capabilities. While it was understood that MNF-I would continue to provide key enablers such as command and control; intelligence, surveillance, and reconnaissance (ISR); aviation; fire support; maintenance; and logistics to the ISF, in June of 2009 it was unclear whether Iraqi forces were capable of conducting independent operations without direct U.S. assistance and/or backup.

Iraqi Security Forces Capacity

On August 19, 2009, Baghdad was rocked by two coordinated truck bombs near the Finance Ministry and Foreign Ministry, killing about 100 and wounding hundreds more.[22] The security breach was, by far, the most trying security challenge the ISF had faced since the U.S. withdrawal from cities two months before, and it raised questions about the readiness of the ISF to maintain internal security.

While the initial Iraqi response was characterized by a lack of coordination between the Iraqi Army and the Iraqi police, the government of Iraq made it clear that it did not want any assistance from MNF-I.[23] The incident highlighted the changing role of the U.S. military in Iraq mandated by the SA's prohibition on unilateral U.S. military actions except for self-defense. As a result, Americans at or near the scene of the August attacks were forced to stand aside and wait for an Iraqi request for assistance rather than contribute to the response.[24] While this transition from partnered operations to Iraqi operations had to overcome a number of early challenges, MNF-I soon concluded that the ISF had developed sufficient capacity to address its internal security concerns with limited U.S. military assistance.

Part of the reason the 2009 bombings did not destabilize the transition was the increasing capacity of the ISF to maintain internal security, even though they remained reliant on U.S. assistance for critical enablers, such as ISR, logistics, maintenance, and aviation support. In fact, much of the Iraqi Army took pride in protecting U.S. forces, bases, and facilities. While there was significant concern over the risk of a return to destabilizing levels of violence as the U.S. drew down force levels, such high levels of violence largely failed to materialize.

By the end of 2009, the general assessment was that the ISF had the operational capability to do much of what was needed for internal defense, although its leaders often lacked the political will to take action when doing so involved targeting individuals and groups with close ties to the ruling political coalition. What this meant in practice is that the ISF proved capable of targeting Sunni extremists, such as al-Qaeda in Iraq and other irreconcilable Sunni extremist groups, but remained reluctant to take

---

[22] Sam Dagher, "2 Blasts Expose Security Flaws in Heart of Iraq," *New York Times*, August 19, 2009.

[23] Interview with the USF-I command historian, Baghdad, December 13, 2012.

[24] Dagher, 2009.

necessary action to target Shi'a groups without the direct approval of the prime minister's office.[25]

2010 Joint Campaign Plan

On November 23, 2009, the U.S. Ambassador to Iraq, Christopher Hill, and the USF-I CG, Raymond Odierno, signed the 2010 JCP, which was to go into effect in January 2010.[26] This campaign plan, which was designed to reflect the SA and the SFA, provided the road map for the transition of U.S. military missions to U.S. civilian agencies and Iraqi government entities. "The path from today's transition to tomorrow's Iraq, a cornerstone of regional stability," it explained, "runs through this campaign plan."[27]

The JCP was distinct from past campaign plans because it was the first to have its primary focus on transition, as required by the SA and consistent with the thrust of the SFA. In light of recent security gains and improved demonstrations of Iraqi governance capacity, the JCP was intended to move the campaign forward by reflecting a "new era of transition" made possible by the efforts of civilian and military members of the U.S. government, who were working in concert with both Iraqi and coalition partners.[28]

As Figure 3.1 illustrates, the JCP identified four LOOs: political, economic and energy, rule of law, and security. The purpose of the plan was to coordinate and integrate the LOOs to ensure that all transition activities would contribute to the development of a long-term strategic partnership between Iraq and the United States.[29]

The JCP described the period of transition leading up to the final withdrawal of all troops as proceeding in three stages, defined by both conditions and time. Thus, the plan was often characterized as being "conditions based but time constrained."[30] If the objectives of a stage were met early, the campaign plan would proceed to the next phase. Otherwise, transitions would proceed in accordance with specified dates. Stage 1, "Transition to the New Security Environment," which was largely complete by the drafting of the 2010 JCP, would last from January 1, 2009, to December 31, 2009, the period under the SA leading up to the anticipated national elections. Stage 2, "Transition to a Stronger Bilateral Relationship," would run from January 1, 2010 to the

---

[25] Interview with USF-I command historian, Baghdad, July 22, 2011.

[26] Appendix B of this book supplies the JCP base document, USF-I and U.S. Embassy Baghdad, *2010 Joint Campaign Plan*, November 23, 2009. The document was originally unclassified but "for official use only" so that it could be shared with Iraqi officials when appropriate for the mission. The JCP contained both classified and unclassified (for official use only) annexes. In December 2011, USF-I removed the "for official use only" caveat for the main body of JCP and all unclassified annexes for the purposes of this book to create a historical record for wide dissemination. Annex F (Transition) of the JCP is included as Appendix C to this book. USF-I has also removed the "for official use only" caveat from the other citations from JCP annexes in this book.

[27] USF-I and U.S. Embassy Baghdad, 2009, p. 14.

[28] Interview with USF-I command historian, Baghdad, July 22, 2011.

[29] Interview with USF-I command historian, Baghdad, July 22, 2011.

[30] Brennan notes, 2009–2011.

Figure 3.1
2010 Joint Campaign Plan Lines of Operation

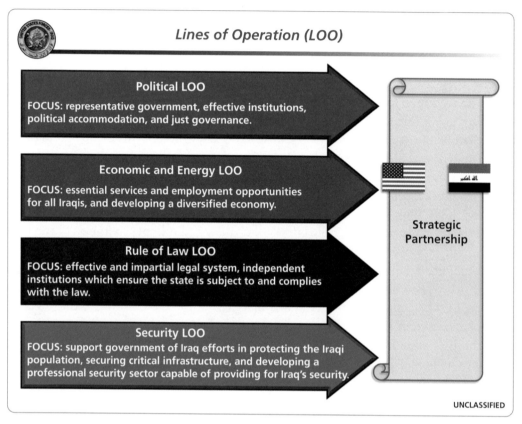

SOURCE: USF-I, "JCP Activity Transition," briefing to RAND Corporation, June 25, 2011g, side 4,
Not available to the general public.
RAND RR232-3.1

conclusion of U.S. combat operations on December 31, 2011. As the last stage before
the U.S. troop withdrawal, this phase placed a high priority on training, enabling, and
advising Iraqis to institute effective governance and conduct both internal and external
security missions. Stage 3—"Iraq, Strategic Partner"—would begin after the U.S. military mission in Iraq ended, at which point the U.S.-Iraqi bilateral relationship would
begin to normalize.[31] Figure 3.2 presents the elements of each stage of the campaign.

The JCP described the complex process of transition as the final campaign of the
long war in Iraq. The JCP outlined the military's "process of canvassing, categorizing,
and defining a 'handover' process of functions it now performs" that would be the core

---

[31] USF-I, 2011g.

Figure 3.2
2010 JCP Campaign Stages

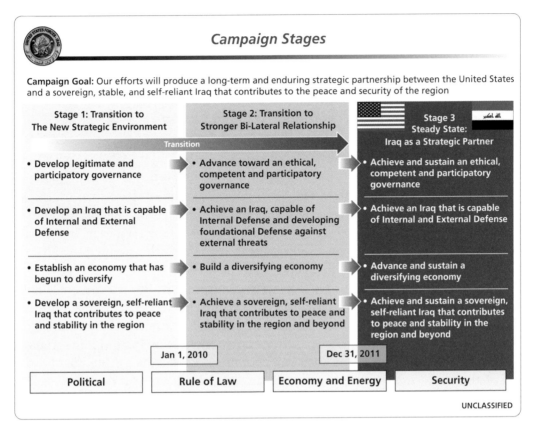

Campaign Stages

Campaign Goal: Our efforts will produce a long-term and enduring strategic partnership between the United States and a sovereign, stable, and self-reliant Iraq that contributes to the peace and security of the region

Stage 1: Transition to The New Strategic Environment	Stage 2: Transition to Stronger Bi-Lateral Relationship	Stage 3 Steady State: Iraq as a Strategic Partner
Transition		
• Develop legitimate and participatory governance	• Advance toward an ethical, competent and participatory governance	• Achieve and sustain an ethical, competent and participatory governance
• Develop an Iraq that is capable of Internal and External Defense	• Achieve an Iraq, capable of Internal Defense and developing foundational Defense against external threats	• Achieve an Iraq that is capable of Internal and External Defense
• Establish an economy that has begun to diversify	• Build a diversifying economy	• Advance and sustain a diversifying economy
• Develop a sovereign, self-reliant Iraq that contributes to peace and stability in the region	• Achieve a sovereign, self-reliant Iraq that contributes to peace and stability in the region and beyond	• Achieve and sustain a sovereign, self-reliant Iraq that contributes to peace and stability in the region and beyond

Jan 1, 2010 — Dec 31, 2011

Political	Rule of Law	Economy and Energy	Security

UNCLASSIFIED

SOURCE: USF-I, 2011g, slide 5.
RAND *RR232-3.2*

of the transition effort.[32] Under the transition process the JCP outlined, other U.S. government entities would take responsibility for hundreds of activities that the military previously performed. The majority of enduring U.S. functions would be transitioned to the U.S. embassy in Baghdad. For example, the new embassy OSC-I would assume the training and security cooperation functions that MNSTC-I had carried out, and it was anticipated USAID would take on some of the civil affairs roles USF-I had played with the support of DoD CERP funds.[33]

---

[32] Interview with USF-I command historian, Baghdad, July 22, 2011. USF-I and U.S. Embassy Baghdad, 2009, Annex F, "Transition" (included in this volume as Appendix C), explains this in greater detail. As with the main body of the JCP, the annex was provided to RAND in an effort to make this information widely available as a historical document.

[33] Interview with USF-I command historian, Baghdad, July 22, 2011. (CERP funds, however, would not be available for USAID programming.)

In essence, the transition plan established a process for gradually reducing the responsibilities the military had acquired for during the previous seven years of nation-building, including activities that fostered economic development, democratic governance, and support for the rule of law. In the final stages of the transition, USF-I would also phase out all security-related activities. All remaining activities would need to be transferred or transformed as responsibility shifted to Embassy Baghdad, USCENTCOM, other U.S. government agencies, or the Iraqi government.

Operations Order 10-01 and Fragmentary Order 10-01.4

On January 1, 2010, MNF-I became USF-I, and USF-I began executing OPORD 10-01, the first OPORD published pursuant to the 2010 JCP.[34] USF-I set out to prioritize efforts for U.S. forces in Iraq to ensure they accomplished all essential tasks by December 31, 2011.[35] This represented an important step toward the final transition. General Odierno's intent was to set the conditions for the establishment of an enduring strategic partnership between the United States and a "sovereign, stable and self-reliant Iraq that contributes to peace and security in the region."[36] This overarching goal for USF-I was consistent with the 2008 Camp Lejeune speech, in which President Obama announced his new strategy in Iraq. The strategy not only reaffirmed the aspirational goals for Iraq and strong U.S.-Iraqi partnership in regional security that the Bush administration had initially developed, but it established limits on what the United States would do to achieve these goals. The consequence of maintaining the strategic and aspirational goals President Bush had established while announcing the ultimate end of the USF-I mission in 2011 was the creation of policy uncertainty. What was clear to military and civilian planners was that the JCP Phase II goals could not be met by December 2011. This disconnect between stated ends and resources available (e.g., time) caused many to view the 2010 JCP as a conditions-based but time-constrained strategy.[37] This policy uncertainty led USF-I planners to pursue both options simultaneously, executing a plan that would ultimately take the force down to zero by the end of 2011 while anticipating that the mission might change to enable the achievement of mission objectives.

OPORD 10-01 marked a major shift in U.S. military operations in Iraq because it had to address a number of important tasks associated with the eventual redeployment of U.S. forces: the drawdown of forces to 50,000; the withdrawal of U.S. forces from the cities; the repositioning of forces and equipment; and many other key tasks and missions. In an effort to develop detailed plans addressing these missions, USF-I

---

[34] USF-I, 2011a.

[35] Interview with USF-I command historian, Baghdad, July 22, 2011.

[36] Brennan notes, 2009–2011.

[37] USF-I planners were not the only ones to make this observation; many outside analysts visiting USF-I over the final two years of the mission highlighted the same point. Brennan notes, 2009–2011.

published separate fragmentary orders (FRAGOs) covering major missions.[38] From the standpoint of the transition, OPORD 10-01.4 (Stability Operations) was the most important FRAGO because it directed USF-I to transition key functions and responsibilities, lay the foundation for a long-term bilateral relationship, and responsibly draw down forces and equipment.[39] OPORD 10-01.4 was published in August 2010 and provided guidance to execute the first portion of the final transition process. The purpose of the FRAGO was to provide a basic plan to guide U.S. forces during the last 15 months of the operation.[40] It directed USF-I to transition from "full spectrum" operations to stability operations while simultaneously shrinking the number of military personnel in Iraq to a 50,000-strong transition force by September 1, 2010.[41] From a historical perspective, OPORD 10-01.4 can be viewed as the first military planning document related to the transition process that would end the U.S. war in Iraq.

OPORD 10-01.4 was organized around three LOOs—strategic partnership, operations, and civil support and theater sustainment. This was in contrast to the four LOOs set out in the JCP that had been agreed one month earlier (see Figure 3.1)—political, economic, rule of law, and security. According to the USF-I command historian, the change reflected USF-I's pivot away from supporting Iraqi governance and capacity-building toward establishing a long-term strategic partnership and enabling the U.S. troop redeployment.[42]

Throughout the development of OPORD 10-01.4, General Odierno emphasized to his team that the transitions of responsibility should be executed in such a manner that USF-I would be able to redeploy "with honor and success" after having set the conditions for a long-term strategic partnership with Iraq.[43] As a planning document, OPORD 10-01.4 established the broad parameters of the final transition but left the detailed planning and execution, especially the retrograde and redeployment, to General Odierno's successor, GEN Lloyd J. Austin III. Consequently, USF-I planners involved in the development of OPORD 10.01-4 were instructed to plan through the end of 2011, informally and quietly, but not to publish detailed guidance beyond June 2011. It was clear that General Odierno's intent was to leave the details to General Austin and the team that would be charged with successfully completing the last phase

---

[38] A FRAGO provides changes to an existing order. Anything in the original OPORD that is not changed or rescinded by the FRAGO continues in force. A FRAGO is published as an addendum to the original order and is given an OPORD number that both identifies the base order and the specific FRAGO.

[39] Interview with USF-I command historian, Baghdad, July 22, 2011.

[40] Interview with USF-I command historian, Baghdad, July 22, 2011.

[41] Interview with USF-I J5 staff officer, Carlisle, Pa., January 6, 2013.

[42] Interview with USF-I command historian, Baghdad, July 22, 2011.

[43] Brennan notes, 2009–2011.

of the transition. Indeed, many portions of OPORD 10-01.4 were left intentionally vague to allow General Austin to define the last stage of the mission himself.[44]

National Elections, March 2010

The Iraqi Constitution required the passage of a law to govern the January 31, 2010, national elections. OPORD 10-01 was developed in such a way that USF-I could continue to support the parliamentary elections by providing a broad range of assistance and support to the ISF. The Iraqi Council of Representatives was extremely divided on critical issues the law would need to address, including voter eligibility, methods of allocating seats, whether and how to hold votes in disputed territories, and the size of the next Council of Representatives. As a result, the council repeatedly missed deadlines for passing the enabling law. Because the council did not adopt legislation until December 6, 2009, the election date was pushed back to March 7, 2010.[45]

In the first months of 2010, concerns swirled about the inclusivity and security of the coming elections. In January, a decision by the Maliki-appointed Justice and Accountability Commission (the successor to the de-Ba'athification Commission) to invalidate the candidacies of hundreds of individuals sparked fears of a Sunni boycott.[46] Critics argued that the commission's move threatened democratic rule and undermined ongoing efforts at national reconciliation.[47]

In such a superheated and sectarian atmosphere, election security was a high priority for USF-I. In particular, the U.S. government feared that al-Qaeda in Iraq would take advantage of the "seams" between Iraqi and Kurdish forces in disputed areas of northern Iraq to conduct attacks that could disrupt voting. To fill these security gaps, USF-I, Kurdish security forces, and ISF created the Combined Security Mechanism (CSM), in which all three entities would conduct joint patrols and staff joint checkpoints in and around disputed areas. As Chapter Six will discuss in more detail, the CSM also served for several years as an important confidence-building measure between Kurdish and Iraqi forces that helped build trust and defuse tensions.[48]

The Iraq Transition Conference

On July 22–23, 2010, a closed interagency conference was conducted at the National Defense University in Washington, D.C., to discuss the planned transition in Iraq. General Odierno and Ambassador Hill brought their most senior staff officers from Baghdad to meet with officials from the White House, DoS, the Office of the Secre-

---

[44] Interview with USF-I J5 staff officer, Carlisle, Pa., January 6, 2013.

[45] Katzman, 2011a, pp. 9–10.

[46] Brennan notes, 2009–2011.

[47] Kenneth M. Pollack and Michael E. O'Hanlon, "Iraq's Ban on Democracy," *New York Times*, January 17, 2010.

[48] Katzman, 2011a, pp. 17–18.

tary of Defense (OSD), and USCENTCOM. The stated purpose of the conference was to ensure interagency partners shared "situational awareness" on the status of transition planning, socialized planned transition activities, and identified issues, points of emphasis, potential resource gaps, tasks, and associated risks.[49] The conference was the occasion for a wide-ranging discussion on progress being made toward transition objectives, projections for future progress, and identification of challenges.

The conference made clear that the only substantive planning for the transition was taking place in Iraq.[50] The transition conference enabled USF-I and Embassy Baghdad to present their plans to the military and civilian leadership in Washington. However, it did little to motivate the interagency group to initiate the type of national-level planning required to support a transition of this magnitude. Neither DoS nor DoD would even designate Washington-based officials to coordinate transition planning efforts for more than six months after the conference. Moreover, as USF-I and Embassy Baghdad focused on the operational aspects of planning and executing the transition, no one in Washington was conducting a fundamental reassessment of the U.S. policy and strategy for Iraq and the region in light of the fact that many of the goals and objectives assigned to USF-I would likely be unachievable before the end of mission. Or, if such an assessment was done, it was not communicated to planners in Iraq.[51]

Change in Leadership

On August 31, 2010, Operation Iraqi Freedom concluded, and OND began the following day. The change of mission was accompanied by a change in leadership. General Austin, at the time serving as Director of the Joint Staff, replaced General Odierno as CG of USF-I. Ambassador James Jeffrey had taken over the embassy from departing Ambassador Hill a few weeks earlier in August.

Relations between Ambassador Hill and General Odierno had reportedly been strained, starting shortly after the beginning of the ambassador's 16-month appointment in April 2009.[52] According to USF-I staff officers, an icy relationship developed between the military and diplomatic staffs, in both USF-I and Embassy Baghdad. As planning for the transition started in the development of both the 2010 JCP and OPORD 10-1, the embassy staff avoided routine participation in working group meetings, frequently stating that senior embassy leadership did not see the value in such planning and noting that the amount of planning requirements USF-I had initiated

---

[49] Brennan notes, 2009–2011.

[50] Interview with USF-I command historian, Baghdad, July 22, 2011.

[51] Brennan notes, 2009–2011.

[52] Thomas E. Ricks, "Iraq, the Unraveling (XXIV): U.S. Embassy vs. U.S. Military, Again," *Foreign Policy*, September 28, 2009b.

prevented embassy officials from accomplishing their normal diplomatic duties.[53] Perhaps more important, this adversarial atmosphere also prevented the development of the types of close working relationships that had existed previously, during the tenure of Ambassador Crocker. Consequently, with some notable exceptions within USF-I J5 and Directorate of Strategic Effects (J9), who were either colocated at the embassy or had military staff officers embedded within the embassy staff, meaningful interaction between the embassy and USF-I was virtually nonexistent.[54]

When he was chosen to lead the diplomatic mission in Iraq, Ambassador Hill struck some as a surprising choice because he had had no experience in the Middle East or dealing with a large military operation being conducted in concert with a diplomatic mission. His successor, James Jeffrey, arrived in Baghdad with significantly more regional expertise. In addition to previous postings as Deputy Chief of Mission in Iraq (2004–2005) and Deputy National Security Advisor (2007–2008), Ambassador Jeffrey had most recently served as U.S. ambassador to neighboring Turkey.[55] Furthermore, prior to his career in DoS, Ambassador Jeffrey had served as a U.S. Army captain in Vietnam and, as a result, both understood and respected his military counterpart. In addition, because of his experience working the Iraq portfolio, Ambassador Jeffrey had also developed close working relationships with many senior military officers who had served in Iraq, including General Austin.

An effective working relationship between Ambassador Jeffrey and General Austin, and the institutions each led, would be especially critical during OND. The execution of the final transition, by means of the whole-of-government approach called for in the 2010 JCP, relied on effective interagency cooperation and coordination in Baghdad. Fortunately, both Ambassador Jeffrey and General Austin recognized the importance of working together and, under their combined leadership, Embassy Baghdad and USF-I would exhibit extraordinary cooperation and collaboration through the remainder of the transition planning process.[56]

---

[53] Conversations author Brennan held with embassy staff while serving as senior advisor to the MNF-I/USF-I J3, Baghdad, 2009–2011.

[54] Brennan notes, 2009–2011.

[55] Anthony Shadid, "Ambassador Leaves Iraq with Much Still Unsettled," *New York Times*, August 12, 2010.

[56] Brennan notes, 2009–2011. This conclusion was widely shared by military officers and diplomats within both Embassy Baghdad and USF-I.

June 30, 2011, Baghdad: The RAND team visited Iraq and interviewed the leadership and staff of USF-I and Embassy Baghdad in June and July 2011. Meeting with USF-I Chief of Staff MG William "Burke" Garrett were (left to right) Stephanie Young, Terry Kelly, Rick Brennan, Mike McNerney, General Garrett, Larry Hanauer, Scott McMahon, Jason Campbell, and Charles Ries. Photo courtesy of U.S. Department of Defense.

December 13, 2011, Baghdad: RAND conducted its final on-site interviews at USF-I; pictured are (left to right) MG Arthur Bartell (J3), Rick Brennan, General Garrett (CoS), and BG Jeff Snow (J5). General Garrett, as the Chief of Staff and the originating authority for classified and restricted material from the command, approved the release of details relating to the Joint Campaign Plan, operational plans and orders, internal staff processes and procedures, internal USF-I assessments of the Iraq operational environment, and a host of other details necessary to accurately document Operation New Dawn in an unclassified report. Photo courtesy of U.S. Department of Defense.

April 14, 2003, Baghdad: A convoy of U.S. Army armored vehicles kicks up dust as it crosses the flightline at Baghdad International Airport. The airport served as the primary base of operations for U.S. troops, cargo, and humanitarian airlift for Operation Iraqi Freedom. Photo courtesy of U.S. Department of Defense.

April 9, 2013, Baghdad: On the day coalition forces formally occupied the city, a group of celebrating Iraqis unsuccessfully tried to pull down this statue of Saddam Hussein in Firdos Square. U.S. Marines stepped in and used an M88 tank recovery vehicle to finish the job. Photo courtesy of U.S. Department of Defense.

June 28, 2004, Baghdad: L. Paul Bremer, U.S. presidential envoy to Iraq and administrator of the Coalition Provisional Authority, officially handed over sovereignty to the interim Iraqi government. Shown here are (left to right) Mouwafak al-Rabii, Ambassador Bremer, Ahmed Chalabi, Adnan Pachachi, and Adl Abdul Mahdi. Adl was not a member of the Governing Council but was likely representing Abdel-Aziz al-Hakim. Photo courtesy of U.S. Department of Defense.

2004, Mosul: Iraqi police officers practice breaching a building while training with International Police Liaison officers and the Civilian Police Assistance Training Team at Forward Operating Base Marez. Photo courtesy of U.S. Department of Defense.

March 22, 2007, Babil: Staff Sgt. Johnny Colon (USA), talks with Iraqi Lt. Gen. Hamza Qais, commander of Babil Province Police, about local crime statistics. U.S. Army photo by Sgt. Marcus Butler.

September 10, 2007, Washington, D.C.: GEN David H. Petraeus (USA), Multi-National Force–Iraq commander, told members of the House Foreign Affairs and Armed Services committees that "the military objectives of 'The Surge' are, in large measure, being met." Photo courtesy of U.S. Department of Defense.

February 18, 2008, Baghdad: Sgt. Jennifer Manzo (USA) and an Iraqi army medic treat a young child during a combined medical effort at the Ministry of Health. U.S. Army photo by Spc. Nicholas Hernandez.

September 29, 2007, Mahmudiyah: Spc. James Buron (USA) watches Iraqi soldiers maneuver through a rope obstacle course as part of their commando training at an Iraqi army compound. U.S. Air Force photo by Staff Sgt. Quinton Russ.

March 6, 2008, Baghdad: Corporal Ivan speaks with members of Sons of Iraq while establishing a new checkpoint in the Rashid district. U.S. Army photo by Pfc. Michael Hendrickson.

March 6, 2008, Baghdad: Iraqi National Police and members of the Sons of Iraq work together to establish a checkpoint in the Rashid district. U.S. Army photo by Pfc. Michael Hendrickson.

September 16, 2008, Camp Victory, Iraq: U.S. Secretary of Defense Robert M. Gates, General Petraeus, and GEN Raymond T. Odierno talk prior to the Multi-National Force–Iraq change of command ceremony at Camp Victory. Photo courtesy of U.S. Department of Defense.

November 4, 2008, Camp Victory, Iraq: One-hundred eighty-six service members became brand-new U.S. citizens during a naturalization ceremony at Al-Faw Palace. The palace served as headquarters for the U.S. military in Iraq. U.S. Army photo by Spc. Christopher M. Gaylord.

December 14, 2008, Baghdad: President George W. Bush and Iraqi Prime Minister Nouri al-Maliki shake hands after signing the strategic framework and security agreements at a joint news conference at the prime minister's palace. White House photo by Eric Draper.

December 14, 2008, Camp Victory, Iraq: President Bush reaches to shake as many hands as possible as he meets with U.S. military and diplomatic personnel at Al-Faw Palace. White House photo by Eric Draper.

December 14, 2008, Camp Victory, Iraq: President Bush stands on stage with General Odierno after Bush addressed U.S. military and diplomatic personnel at Al-Faw Palace. White House photo by Eric Draper.

January 23, 2009, Baghdad: Staff Sgt. Jesus Robles (USA) goes over a map in Istaqlal Qada with his Iraqi police counterpart during a combined patrol that focused on securing this northeast Baghdad neighborhood before the upcoming provincial elections. U.S. Army photo by Brad Willeford.

January 28, 2009, Baghdad: Outside the Zafraniyah polling site, an Iraqi Army colonel proves that he has voted in the provincial elections by raising his purple index finger. That vote took place before the general provincial election, scheduled for January 31, to accommodate Iraqi Security Forces, detainees, hospital patients, and other special-needs cases. U.S. Army photo by Spc. Douglas York.

January 31, 2009, Kadhimiya: Iraqi voters proudly display their ink-stained fingers after voting in the Iraqi provincial election. Photo courtesy of U.S. Department of Defense.

February 27, 2009, Camp Lejeune, N.C.: U.S. troops and civilians listen to President Barack Obama during his visit to the camp to discuss current policies and an exit strategy from Iraq. U.S. Marine Corps photo by Lance Cpl. Michael J. Ayotte.

March 26, 2009, Mahmudiyah: Staff Sgt. Jeffrey Eaken (USA) teaches Iraqi soldiers about the 120 mm mortar system, including hand and arm signals used to adjust the system. U.S. Army photo by Sgt. Kani Ronningen.

June 30, 2009, Baghdad: Iraqi government officials and U.S. military leaders look on as Iraqi security forces pass for review representing the Iraqi military, police, and civil leadership. The event marked the withdrawal of U.S. combat forces from Iraqi cities in accordance with the bilateral security agreement signed in 2008. U.S. Air Force photo by Capt. Tommy Avilucea.

June 30, 2009, Baghdad: General Odierno pays tribute at the Iraqi Monument of the Unknown Soldier during celebrations of Iraq's sovereignty day. U.S. Air Force photo by Capt. Tommy Avilucea.

PART II

Transition Management and Planning

CHAPTER FOUR

Transition Management

In July 2010, incoming USF-I Chief of Staff MG William B. "Burke" Garrett directed the USF-I staff to initiate a detailed planning process to identify opportunities, challenges, and constraints that would affect the final phase of military operations in Iraq.[1] The strategic question planners confronted was how the United States could withdraw military forces and capabilities in a manner that would enable follow-on organizations to advance U.S. national interests, goals, and objectives. Or, perhaps more succinctly: How can the military depart a country in a way that sets the conditions necessary for other organizations to be successful?

To understand how USF-I and Embassy Baghdad managed the transition of responsibilities during OND, it is necessary to examine the planning and decision-making structures the 2010 JCP created, explore how the new USF-I and embassy leadership employed and modified the structures, and review the concept for the operation that, as outlined in OPORD 11-01, would be used to direct USF-I activities from January 6, 2011, the date it was published, through the final departure of U.S. forces from Iraq on December 18, 2011.[2]

Team Building

In the months before and after General Austin's arrival, General Garrett developed a senior leader management plan to bring together the general officer team that would be necessary to plan and execute the transition while, simultaneously, continuing to conduct military operations in a war zone. This team would not only have to work closely and cooperatively with its Embassy Baghdad counterparts but would also have to work collaboratively with them. Over the course of their careers, especially since the invasion of Afghanistan in 2001, the general officers selected had not only worked

---

[1] Interview with USF-I J5 staff officer, Baghdad, December 14, 2011.

[2] The bulk of the planning for OPORD 11-01 took place between October and December 2010, immediately after the arrival of General Austin and the beginning of OND. While the OPORD was signed and published on January 6, 2011, it did not take effect until February 6, 2011.

together during combat but had also worked for General Austin in at least one previous assignment. According to one general officer on the USF-I staff, "all the generals within USF-I share common experiences, are friends with one another and absolutely trust one another." He went on to say that the assembled leadership team is not only "personally loyal to General Austin, but we also understand his goals, objectives and priorities."[3] This degree of camaraderie, forged by years of wartime service, was viewed crucial for managing the complexities and uncertainties associated with the last 16 months of U.S. military's involvement in Iraq.[4] This type of close teamwork also existed within the embassy, where the Deputy Chief of Mission, Chief of Staff, Political Counselor, Deputy Political Counselor, Chief and Deputy Chief Security Officers, Economic Chief, and Basra principal officer and General Austin's political advisor had all previously worked for Ambassador Jeffrey.[5]

Operations Order 11-01: Key Objectives and Guidance for Operation New Dawn Execution

The first step for this leadership team was to develop the detailed OPORD for the last 12 months of the mission. This would be the first—and only—OPORD USF-I published in 2011 and would be known as OPORD 11-01.[6] The OPORD directed USF-I to "conduct stability operations, support the U.S. Mission in Iraq,[7] and transition enduring activities to set conditions for an enduring strategic partnership that contributes to regional security."[8] Annex V to the order provided "military and interagency planners with coordination processes and policies for achieving the campaign goals as identified in the JCP, and amplifying guidance on how USF-I [would] conduct the transition of enduring activities" to the U.S. embassy, USCENTCOM, and the government of Iraq.

---

[3]   Interview with General Garrett, Baghdad, December 14, 2011.

[4]   By January 2012, the final leadership team for USF-I was in place: GEN Lloyd Austin III, Commander; LTG Michael Ferriter, DCG, Advising and Training (DCG [A&T]); LTG Robert Cone, Deputy Commanding General for Operations; MG Edward Cardon, Deputy Commanding General for Support; MG Nelson Cannon, Deputy Commanding General for Detention Operations and Provost Marshal Office; MG William "Burke" Garrett, Chief of Staff; COL Dick Kuhel, Directorate of Joint Personnel (J1); MG Mark Perrin, Directorate of Joint Intelligence (J2); MG Arthur Bartell, J3; MG Thomas Richardson, Directorate of Joint Logistics (J4); BG Jeffrey Snow, J5; RDML(S) Hank Bond, Directorate of Joint Communications and Information (J6); BG Rock Donahue, Directorate of Joint Engineering (J7); COL Paul Chamberlain, Directorate of Resource Management/Comptroller (J8); MG Jeffrey Buchanan, J9; MG Job Handy, Air Component Coordination Element (ACCE); and Ambassador Larry Butler, political advisor.

[5]   Email comment provided by Ambassador James Jeffrey to Charles Ries, January 11, 2011.

[6]   OPLAN 12-01 was prepared in the event that there would be a residual military presence in Iraq beyond 2011; however, that plan was never turned into an order to execute.

[7]   The term *U.S. Mission in Iraq* refers to the interagency team that works within Embassy Baghdad under the authority of the Chief of Mission, the U.S. ambassador to Iraq.

[8]   USF-I, 2011g, slide 9.

The approved mission specified that USF-I "conducts stability operations, supports the U.S. Mission in Iraq, and transitions enduring activities to set conditions for an enduring strategic partnership that contributes to regional stability beyond 2011."[9] The commander's intent further specified that USF-I would

> continue improving the ability of the Iraqi Security Forces to provide internal security, develop foundational external defense capabilities, and lead and manage their institutions. Simultaneously, USF-I must support the Chief of Mission–Iraq with military capabilities and engagement, while transitioning responsibility for operations in Iraq to USCENTCOM, the US Mission in Iraq, the GoI [government of Iraq] and others. It is essential throughout to ensure unity of effort and demonstrate continuity of US commitment to Iraq.[10]

To achieve this mission and commander's intent, OPORD 11-01 established ten key tasks as follows:

1. Protect the force.
2. Maintain situational awareness.
3. Advise, train, assist, and equip the ISF.
4. Conduct partnered counterterrorism operations and an enduring counterterrorism capability within ISOF.
5. Provide military capabilities in support of Embassy Baghdad.
6. Transfer responsibility to appropriate partners.
7. Establish and support OSC-I.
8. Support the establishment and mission of INL.
9. Support reconciliation efforts within Iraq to address grievances between Arabs and Kurds, and between the Sunni/SOI and the Shi'a extremist groups.
10. Redeploy the force.[11]

OPORD 11-01 also set forth three lines of effort (LOEs) and 13 supporting objectives for this mission (see Figure 4.1).[12] These LOEs depicted the three primary areas of focus during 2011:

---

9 Interview with USF-I Joint Future Operations (J35) staff officer, Baghdad, June 17, 2011.

10 Interview with USF-I J35 staff officer, Baghdad, June 17, 2011.

11 Interview with USF-I J35 staff officer, Baghdad, June 17, 2011.

12 OPORD 11-01 focused on activities associated with the Security LOO of the JCP. Theoretically, each of the four JCP LOOs would have an associated OPORD directing implementation. However, the Political LOO, Economic and Energy LOO, and Rule of Law LOO were the primary responsibilities of the embassy, with USF-I support as necessary. Since DoS does not prepare OPORDs, none were prepared for these three LOOs. That does not mean, however, that embassy officials were not conducting actions associated with these LOOs, which they were. It simply means that an OPORD was not prepared. Moreover, OPORD 11-01 did address issues associated with the other LOOs as it related to the transition. For example, while Embassy Baghdad had the lead for

Figure 4.1
Operations Order 11-01 Lines of Effort

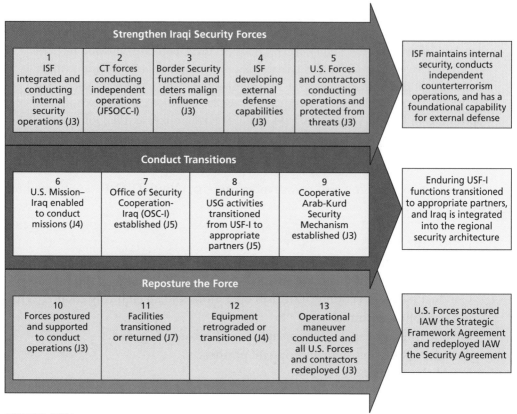

SOURCE: USF-I.
RAND *RR232-4.1*

1. continue to develop and strengthen ISF capacity to conduct internal and external defense missions
2. conduct transitions by handing over key functions to the embassy and other entities, establishing an embassy OSC-I, and ensuring that the U.S. embassy could accomplish its core missions
3. reposture the force, which would involve returning facilities to the Iraqi government, retrograding or disposing of equipment, and redeploying 50,000 troops and a like number of contractors—all while continuing to conduct military operations.

---

the rule of law, USF-I remained engaged with transitioning many of the activities it was performing, including helping to build the capacity of the Iraqi judicial system and transferring detainees from U.S. to Iraqi custody. All these types of activities were subsumed within the Conduct Transition LOE of OPORD 11-01. Brennan notes, 2009–2011.

ISF capacity-building would continue during the first phase of operation, from February 6, 2011, to August 31, 2011, and was designated as the decisive activity during Phase I of the OPORD (see Chapter Seven for more detail). At that point, USF-I's priority of effort would turn to reposturing the force out of Iraq, a task that would be complete by December 31, 2011 (see Chapter Nine for more detail). The large majority of the "conduct transitions" LOE was related to establishing OSC-I and enabling an expeditionary embassy (see Chapter Eight). Throughout the last year of the operation, USF-I, in close partnership with Embassy Baghdad, would spend an enormous amount of time and effort transferring the military's enduring activities to the U.S. Mission and other U.S. and Iraqi entities. This chapter focuses on the processes they developed to manage the transition.[13]

Transition Implementation Planning

The initial transition plan was outlined in Annex F (Transition) to the 2010 JCP, which established processes for managing the transition.[14] Annex F directed USF-I to identify all the programs, projects, and relationships that were being managed by any military entity in Iraq and to assess how each of these activities was aligned with the accomplishment of broader JCP tasks. Annex F also established a mechanism to transfer or transform enduring activities to a receiving organization no later than December 31, 2011.

It is important to highlight that the transition did not involve a reassessment of U.S. strategic goals and objectives in Iraq. The optimistic goals and objectives listed in the 2010 JCP remained unchanged from previous years, even though USF-I was scheduled to leave Iraq within two years. What Annex F did provide was a detailed process designed to hand over USF-I activities to other entities in a manner that helped set the conditions for the success of recipient organizations. Thus, USF-I defined the term *transition* as follows:

> The transfer, transformation, completion or termination of tasks, programs, projects or relationships that are owned, performed or managed by a military organization engaged in combat, COIN, SASO, or any other military operation to the host nation, U.S. Embassy and OSC, follow-on DoD organizations operating in country, or any other [U.S. government] or international entity.[15]

---

[13] Interview with USF-I J5 staff officer, Baghdad, March 13, 2011.

[14] USF-I, Annex F (Transition) to Joint Campaign Plan 2010, pp. F-1 through F-2-1. See Appendix C to this report.

[15] Interview with USF-I Chief of Staff General Garrett, USA, Baghdad, September 15, 2011.

The transition was to proceed through five phases: define, measure, analyze, implement, and assess. The first step, "define," consisted of defining transition objectives, assessing challenges, identifying key stakeholders, and outlining a timetable with major milestones to be met.

The second phase, "measure," was a countrywide data-gathering effort that sought to identify all tasks that would be transitioned. Annex F directed all USF-I elements to canvass subordinate elements to identify all USF-I activities and relationships.[16] As Figure 4.2 shows, the canvassing process identified more than 30,000 discrete military tasks, projects, programs, and relationships. By grouping these efforts, USF-I staff

Figure 4.2
Adjudication and Disposition of Activities

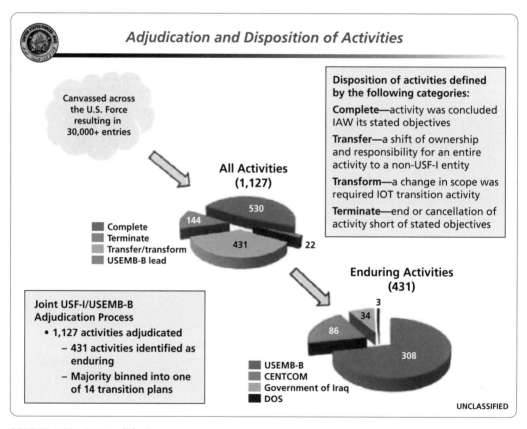

SOURCE: USF-I, 2011g, slide 6.
NOTES: IOT = in order to; USEMB-B = U.S. Embassy Baghdad.
RAND RR232-4.2

---

[16] USF-I, 2011e. This canvassing effort ran concurrently with the development of the 2010 JCP and was completed in October 2009

adjudicated them into 1,127 separate activities.[17] Staff identified 530 of them as completed; 144 as having already transitioned to the embassy; and 22 others as having been terminated short of their stated objectives, because it was judged either that no appropriate recipient organization existed or that the activity would not be feasible once military forces departed.

This left 431 activities considered to be "enduring," meaning that they needed to be transferred to a non-USF-I entity or modified before they could be transferred or ended.[18] Of these 431 enduring activities, 308 were slated for transfer to the U.S. embassy.[19] Other recipient institutions included DoS (three), USCENTCOM (86), and the government of Iraq (34).[20] The "define" and "measure" phases were completed by October 2009.[21]

In the third phase, "analyze," USF-I staff were to evaluate the effectiveness of each of the 431 enduring activities in achieving the JCP's objectives; determine whether to transfer, modify, or terminate each one; and identify the steps to be taken in what sequence. Initially, each of the activities was assigned to one of the four JCP LOOs, so the "owner" of each LOO could manage the transition process.[22] This proved difficult to manage, so the activities were instead binned into 14 transition plans whose topics mirrored the JCP annexes. Each plan defined responsibilities, resource requirements, milestones, policy guidance, and other material necessary to guide the transition process.[23] Table 4.1 provides the subjects of the transition plans, along with the USF-I staff lead and recipient organizations involved with each.

In the fourth phase, "implement," USF-I would take the steps needed to transition each of the enduring activities, including assessing whether the identified recipient organization had the capabilities to take on the activity, securing the recipient organization's agreement to accept the activity, identifying the manpower and financial resources (including training, if necessary) needed to enable the recipient to perform the activity, identifying a time line for the handover, and transferring data related to

---

[17] For example, interactions with five Kurdish security force commanders and 23 classes for Kurdish *peshmerga* would be grouped together as a single activity (build relationships with Kurdish security forces).

[18] USF-I, 2011g.

[19] Other transitions happened in compartmented and/or sensitive areas; these were far less structured and often did not involve the senior leadership of USF-I. Interview with former USF-I J5 staff officer, Carlisle, Pa., January 6, 2013.

[20] USF-I, 2011g.

[21] USF-I, "Strategic Transition Overview," in *Iraq Transition Senior Leader Conference*, briefing, Washington, D.C., July 22–23, 2010b, slide 36.

[22] USF-I J3 led the Security LOO, while the embassy's political counselor, economic counselor, and rule of law coordinator, respectively, led the Political, Economic, and Rule of Law LOOs. But the Political, Economic, and Rule of Law LOOs were subsequently dropped in OPORD 10-01, as we will discuss later.

[23] Drafting of the transition plans began in May 2010, as directed by USF-I FRAGO. This phase of the planning process ended in April 2010. See also USF-I, 2010b.

Table 4.1
Transition Plans

Transition Plan	USF-I Staff Lead	Recipient Organization
Intelligence	J2	USCENTCOM, embassy, OSC-I, and government of Iraq
Defense attaché office	J2	Defense Attaché Office
Arab-Kurd relations	J35	Embassy
Balance Iranian influence	J35	USCENTCOM, embassy, OSC-I, and government of Iraq
Counter–violent extremist organizations	J35	USCENTCOM, embassy, OSC-I, and government of Iraq
Turkey-Iraq-U.S. trilateral cooperation	J35	Embassy
Counterterrorism	JFSOCC-I	Government of Iraq, USCENTCOM
Police training	ITAM-Police	Embassy
Office of Security Cooperation Transition	PSG-I	USCENTCOM, embassy
Transition to police for internal security (police primacy)	J5 Plans	Embassy
Telecommunications	J6	Embassy
Transportation	ACCE	Embassy
Knowledge management	J6	Embassy
Rule of law	SJA	Embassy

NOTES: ITAM = Iraq Training and Advisory Mission; PSG-I = Partnership Strategy Group–Iraq.

the activity. If it was determined that an activity could not be transferred in its entirety, it would be rescoped into something that was within the capacity of the recipient organization. In reality, many activities could not be effectively transitioned without both increasing organizational capabilities and rescoping.[24]

The final phase, "assessment," called for macrolevel and microlevel assessments of the effectiveness of the transition process. This would require regular "state of the transition" reporting at all levels of USF-I and Embassy Baghdad.[25] This continuous evaluation process became integral to the management of the transition; it enabled USF-I and embassy staff to ensure that transition plans stayed on track and enabled senior USF-I and embassy officials to make informed decisions regarding the transition.

---

[24] Interview with USF-I J5 staff officer, Baghdad, June 25, 2011.

[25] Interview with USF-I J5 staff officer, June 15, 2011.

Transition Management Working Groups

Successful transition required close collaboration between USF-I and Embassy Baghdad. As discussed in the previous chapter, each organization had internal processes for identifying and managing their individual responsibilities, which led USF-I and the embassy to maintain some distance from each other. To implement the JCP, however, the civilian and military leadership agreed that it was necessary to establish a formal planning and decisionmaking structure. The concept was to develop a joint civilian and military process that would enable collaboration at all levels, from action officers to the USF-I commander and the ambassador.

The joint campaign management process was built around a quarterly assessment and decisionmaking cycle. Annex K to the 2010 JCP established a multilevel interagency framework for managing the transition of enduring activities that were seen as necessary to further U.S. goals in Iraq.[26] It created four staff and decisionmaking entities (see Figure 4.3) to monitor and guide LOO execution in accordance with the 2010 JCP, although as OND began, the four groups increasingly focused on the transition of USF-I–led activities to the embassy, USCENTCOM, other U.S. government agencies, and the Iraqi government.[27] Three of these met weekly:

- Joint Campaign Working Group (JCWG): This senior staff-level coordinating body met to discuss each of the four LOOs on a rotating basis, addressing each once a month. Chaired by the embassy's Political-Military Strategic Planning Team and the USF-I J5 Strategy staff, the JCWG assessed the implementation of each LOO, identified challenges affecting the transition, and developed strategies for keeping transitions on track. The JCWG would raise issues that needed higher-level review to the Executive Core Group. However, since the Political LOO and Diplomatic LOO had been merged during the creation of the 2010 JCP, neither the JCWG nor any other part of the Executive Core Group process addressed the effects of the transition and retrograde of U.S. forces from Iraq.
- Executive Core Group (ECG): This executive-level interagency group met to address critical issues affecting the implementation of the transition plans and the drawdown of U.S. troops. It issued guidance to the JCWG and also raised issues to the LOO Core group as needed.
- LOO Core ("Core"): This group, chaired by the USF-I CG and the U.S. ambassador, met to review each LOO on a rotating basis. The meetings were the venue in which the seniormost DoD and DoS officials in country reviewed the status of transition initiatives, assessed significant challenges, and decided on actions to be taken. The Core provided direction for the staff-level JCWGs to implement.

---

[26] Interview with USF-I J5 staff officer, June 15, 2011.

[27] USF-I and U.S. Embassy Baghdad, 2009, Annex K, Not available to the general public.

Figure 4.3
Campaign Management

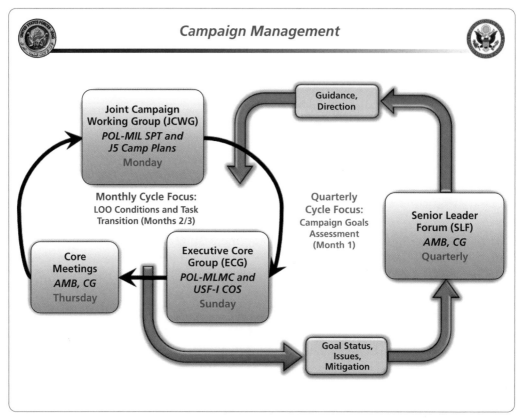

SOURCE: Embassy of the United States, Baghdad, Iraq, and SFF-I, *Executive Core Group Briefing*, August 1, 2010, Not available to the general public.
NOTES: AMB = ambassador; POL-MIL = political-military.
RAND RR232-4.3

In addition to these meetings, the USF-I CG and U.S. ambassador addressed strategic policy challenges on a quarterly basis at Senior Leader Forum (SLF) meetings. This was the primary venue for the ambassador and the CG to ensure that JCP goals were being advanced throughout the transition process. Decisions made at the SLF would then generate guidance to the ECG and JCWGs for further action.[28]

---

[28] USF-I and U.S. Embassy Baghdad, 2009, Annex K, Not available to the general public.

Changes to Transition Management Processes Under Operations Order 11-01

While the staff and decisionmaking entities worked to ensure that U.S. strategic objectives continued to be advanced throughout the transition process, OPORD 11-01 created additional interagency working groups to address the administrative and logistical side of the transition. These included the

- Embassy Support Group—an interagency team, cochaired by the USF-I J4 and the embassy's management section, charged with coordinating USF-I's logistics support to the embassy
- Base Transition Working Group—a forum, cochaired by the USF-I J7 and the embassy's management section, to ensure that the transfer of facilities from USF-I to the embassy—including property, buildings, support infrastructure, and security infrastructure—remained on track
- Operations Transition Working Group—a group, cochaired by the USF-I J3 and the embassy's regional security officer (RSO), to address the range of security, surveillance, and protective responsibilities that the embassy would need to assume after the withdrawal of U.S. troops
- USCENTCOM Iraq Transition Working Group—this forum for coordinating the transition of selected USF-I activities to USCENTCOM and discussing USF-I preparation for the redeployment phase was chaired by the USCENTCOM J5 and involved representatives from all USF-I staff directorates and the embassy's political-military section.[29]

To inform these working groups' discussions, multiple subgroups were created to examine such issues as medical support to enduring U.S. mission sites, transportation, and construction. ROC drills were conducted to test strategies for transitioning USF-I activities to the embassy.[30]

General Austin also established a new internal process for developing USF-I's positions on transition issues. This Joint Plans and Operations process served as the centerpiece of the commander's decisionmaking process for all aspects of military operations and the transition process (see Figure 4.4).

- The Joint Plans and Operations Group, a "council of colonels," chaired by the USF-I Deputy Chief of Staff, a one-star general officer, synchronized planning

---

[29] USF-I, OPORD 11-01, January 6, 2011b, pp. V-3 to V-4, Not available to the general public.

[30] Former senior official, U.S. Embassy Baghdad, email to authors, November 21, 2011; Jeffrey Stuart, USA, "Operations Transitions Working Group," briefing to RAND, Baghdad, June 26, 2011, slide 20; Rock Donahue, USF-I J7, "Transition of Engineer Activities, United States Forces—Iraq J7," briefing to RAND, Baghdad, June 26, 2011, slide 13.

Figure 4.4
USF-I Internal Decisionmaking Structures

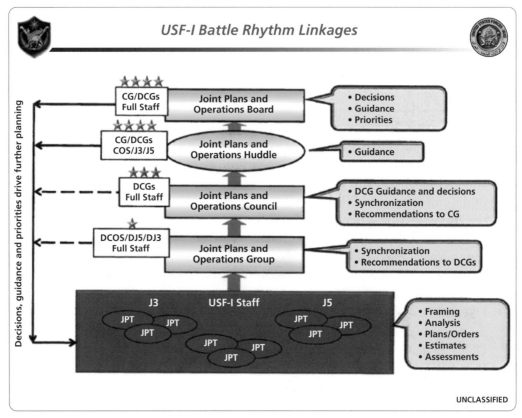

SOURCE: Slide courtesy of USF-I, June 16, 2011.
NOTES: DCOS = Deputy Chief of Staff; DJ5 = Deputy Director, Joint Strategy and Plans;
DJ3 = Deputy Director, Joint Operations, JPT = joint planning team.
RAND RR232-4.4

across the USF-I staff, reviewed transition plans and proposed changes, and identified transition issues that needed higher-level attention or resolution.

- The Joint Plans and Operations Council, chaired by the USF-I Deputy Commanding General (DCG) for Operations, included all DCGs, the Chief of Staff,
and senior staff.

- The Joint Plans and Operations Huddle, a meeting of select senior staff officers
chaired by the USF-I CG, discussed topics in a small-group setting that encouraged debate. This group could also be assembled quickly and served as a forum
for rapid decisionmaking when the USF-I commander's guidance was urgently
needed to support operations.[31]

---

[31]   Interview with former member of the USF-I command group, Washington, D.C., December 12, 2011.

- The Joint Plans and Operations Board, chaired by the USF-I Commander, General Austin, typically resolved issues escalated for adjudication and decision within a week, according to USF-I officials.[32]

Annex F of the JCP directed USF-I to manage and track the transition of enduring activities using a "transition dashboard."[33] The dashboard's purpose was to provide a numerical snapshot of the transition—to count how many projects had been transitioned and identify how many were still in process. However, the USF-I J5 Strategy Directorate's Joint Assessment Board used the dashboard more qualitatively to identify implementation sequencing inconsistencies and other problems, which were referred to the relevant coordination bodies for resolution (i.e., Joint Plans and Operations for military-to-military matters or the JCWG/ECG structure for civil-military issues).[34] Routinely, however, any action that was headed for the ECG would have previously gone through the Joint Plans and Operations process.[35]

Given the logistical complexity of the planned redeployment of tens of thousands of U.S. troops and their equipment—not to mention the associated security threats—the USF-I commander directed the establishment of a new coordination body to synchronize military units' redeployment maneuvers, sustainment, and transition activities across Iraq. The new forum, which met monthly, was named the Operations Synchronization Board (OSB).[36] The OSB enabled the principals and their staffs to identify upcoming decision points for OPORD 11-01 execution, execution risks, mitigation strategies, and the need for adjustments to the plan as conditions changed in Iraq.[37]

Force reposturing plans were evaluated using ROC drills that role-played "blue" (USF-I and other U.S.) activity to close a base and move forces, "red" (enemy) activity designed to disrupt the process, and "green" (Iraqi) decisions and activities that either facilitated or impeded the process. The exercises helped identify contingencies that might disrupt force movements, which enabled planners to develop alternative courses of action. They also allowed rehearsal of plans to ensure that all the steps needed to withdraw from a facility occurred in the right order.[38]

---

[32] USF-I, 2011g. See also USF-I J5 Strategy, "Joint Assessment Board," briefing to RAND, Baghdad, June 30, 2011.

[33] See Appendix C in this volume for JCP, Annex F, part of which describes the transition dashboard.

[34] USF-I, 2011g. See also USF-I J5 Strategy, 2011.

[35] Interview with former USF-I J5 staff officer, Carlisle, Pa., January 6, 2013.

[36] USF-I, "ExSum 11-01 Operations ROC Drill," memorandum, Baghdad, February 16, 2011c.

[37] USF-I, 2011c. Also former senior official, U.S. Embassy Baghdad, email to authors, November 21, 2011.

[38] Brennan notes, 2009–2011.

Iraqi Participation in Transition Planning and Execution

Although a significant number of activities were scheduled to transition to the government of Iraq, the JCP did not establish similar mechanisms to discuss strategic and policy issues related to the Iraqis' assumption of USF-I activities.[39] As Chapter Five will discuss, part of the reason for this was that Prime Minister Maliki, by refusing to appoint senior officials to lead MOD and MOI, effectively blocked these agencies from making decisions regarding the pending transition. As a result, strategic decisions on the transition of activities to the Iraqi government were largely made without formal U.S-Iraqi coordination.

Coordination between U.S. forces and their Iraqi counterparts was better at the tactical and operational levels. The U.S.-Iraq SA provided for a joint committee to support its implementation, including the transfer of property.[40] Under this rubric, USF-I and the Iraqi government created structures for managing the mechanics of returning bases and other facilities to Iraqi control, such as the establishment of a "receivership cell" in Prime Minister Maliki's office to serve as a central point of contact.

Even with coordination mechanisms in place, the process of base closure and transfer was complicated by the failure of the Iraqi government to target Shi'a extremist groups that would attack U.S. bases during the final days before departure. Base closures involved dismantling structures and protective measures that, if leaked to extremists, would put U.S. personnel and assets at risk during the drawdown process. Therefore, to ensure force protection, USF-I considered it unwise to share its closure schedule with the Iraqi government until the planned transfer was imminent.[41]

In theory, the joint committee was also intended to oversee the transfer of responsibility for detention operations and related facilities, management of radio frequencies, airspace control, and other topics, but no effective forums were ever created for senior U.S. embassy, USF-I, and Iraqi authorities to bring about the transfer of these activities to the Iraqis.[42] Instead, according to Ambassador Jeffrey, the United States typically decided what the Iraqis needed and informed them of the decision. The Iraqis acted in a passive, "receive" mode during the transition process.[43]

In some cases, the limited Iraqi participation in transition planning created risks. For example, the embassy's Transportation Attaché reminded her Iraqi counterparts of the pending termination of U.S. control over Iraqi airspace, which was managed through a combination of U.S. military and privately contracted air traffic controllers.

---

[39] Interview with USF-I J5 staff officer, June 15, 2011.

[40] United States of America and the Republic of Iraq, 2008a, Article 5, "Property Ownership."

[41] Donahue, 2011.

[42] USF-I, "Joint Campaign Plan Overall Goal Assessment," in *Iraq Transition Senior Leader Conference*, briefing, Washington, D.C., July 22–23, 2010a, slides 12–13, Not available to the general public.

[43] Interview with U.S. Ambassador to Iraq James Jeffrey, Arlington, Va., March 12, 2012.

A crisis ensued when the Iraqi government waited until 48 hours before airspace shutdown to take over the ongoing contracts.[44]

Summary

The structures, processes, and procedures that USF-I and the embassy established to manage the transition enabled officials at all levels to track developments, assess progress, and mitigate challenges. The process provided an effective tool for managing the transition of activities. Several lessons can be learned from the experience of the various transition management working groups. As the remainder of this book will discuss, both the transition of activities and the associated reposture and redeployment of U.S. military forces out of Iraq faced many challenges and, unsurprisingly, not everything went according to plan. However, the processes and procedures put in place to manage the operation served their purpose, facilitating close cooperation and collaboration within the USF-I staff; between USF-I and embassy Baghdad; and, to a lesser extent, between Baghdad and Washington. However, because of ISF's inability to keep information from falling into the hands of extremist organizations, as well as other factors that we will discuss in the next chapter, many aspects of the transition were not effectively shared with the government of Iraq.

As military planners often state, no plan ever survives first contact with the enemy. Thus, while the 2010 JCP and OPORD 11-01 provided a framework for how USF-I would conduct the last 11 months of the operation, there were a large number of influences on the mission over which neither USF-I nor Embassy Baghdad had any control. These external influences, emanating from both Washington and Baghdad, are the subject of Chapter Five.

---

[44] Interview with John P. Desrocher, Director, Office of Iraq Affairs, DoS, Washington, D.C., June 1, 2011; interview with Don Cooke, Director, Office of Iraq Economics and Assistance Affairs, DoS, Washington, D.C., June 1, 2011.

The Influence of Washington and Baghdad on the Transition

As USF-I and Embassy Baghdad were developing and executing the transition plan established by the 2010 JCP and refined in OPORD 11-01, senior government officials in Washington and Baghdad were making a range of policy decisions that affected U.S. strategy toward Iraq. This chapter will examine how civilian decisionmaking affected the transition process.

The planning and execution of the transition were shaped by a small number of critical decisions made by the White House (under both Presidents Bush and Obama) and by the Principals and Deputies Committees of the NSC.[1] Perhaps the most crucial of these was the Bush administration's decision to sign the U.S.-Iraq SA, which established the time line for the withdrawal of U.S. troops, and the U.S.-Iraq SFA, which established goals and consultation mechanisms for managing U.S.-Iraq relations after the end of the U.S. military presence. As noted earlier, transition initiatives were driven by the need to plan backward from the firm deadline established by the SA.

Also important was the Obama administration's decision to enter into negotiations for a possible post-2011 military presence while simultaneously directing comprehensive planning for only one scenario—the total withdrawal of all U.S. forces that the SA called for. Since the question of an enduring troop presence remained unresolved until the last possible moment, agencies were reluctant to make final decisions on infrastructure, equipment, contracts, personnel, and troop redeployment. This uncertainty also caused challenges for the planning efforts of USF-I and Embassy Baghdad because leaders were forced to keep options open for two alternative end states.

The executive branch of the U.S. government was not, however, the only important actor in this drama. Congressional opposition to key elements of the Bush and Obama administrations' Iraq policies affected the transition in significant ways. Most notably, Congress's refusal to fund key transition-related initiatives and the reliance

---

[1] The NSC Principals Committee comprises Cabinet-level officials from DoS and DoD, the Chairman of the Joint Chiefs of Staff, and the National Security Advisor. Other officials of similar rank, such as the Director of National Intelligence and the Director of the Central Intelligence Agency, play an advisory role in its discussions. The Deputies Committee comprises these officials' chief deputies (or their designees).

on continuing resolutions (CRs) limited the resources available to DoS for post-2011 embassy operations planning and for bilateral assistance programs.

For its part, the Iraqi government was notably absent from transition planning and decisionmaking. The long delay in forming a government after the March 2010 parliamentary elections was one factor. But even after a government was formed and seated in December 2010, the failure of Prime Minister Maliki and the Iraqiya party to agree on mutually acceptable nominees for ministers of defense and interior, as noted in the previous chapter, was cited as a reason that USF-I and Embassy Baghdad limited their coordination of transition activities with the Iraqi government.[2]

Without question, the transition was also complicated by the reluctance—on the part of Congress and of senior policymakers in both the Bush and Obama administrations—to make some decisions on goals, strategies, and resources early enough to enable meticulous planning. Delays in Iraqi and American decisions on terms and conditions that might apply to an enduring U.S. troop presence, indecision over the mission and size of a potential residual military force, repeated revisiting of the size and mission of OSC-I and other embassy staffs, and uncertainty regarding program funding—just to name a few examples—made it difficult or impossible for agencies to define achievable program objectives, construct facilities, let contracts, or implement other aspects of transition plans.[3] Assumptions had to be made, some on worst-case and others on best-case outcomes, while supporting multiple courses of action.[4] Agencies spent time and resources developing and executing multiple transition plans in the absence of clear guidance on the program and policy objectives they were working to advance; then, when decisions were made (or revisited), officials often had to revise plans already being implemented. Given the existence of a hard deadline of December 31, 2011, such starts and stops consumed the time of both commanders and staff and resulted in expenditure of scarce resources that could have been more profitably used elsewhere.[5]

Key Political Decisions Shaped Plans

Policy decisions as far back as 2007 shaped the eventual withdrawal of U.S. forces from Iraq. The Bush administration's surge of U.S. forces and resources into Iraq was

---

[2]   Interview with USF-I J35 staff officer, Baghdad, June 14, 2011.

[3]   Interviews with USF-I and Embassy Baghdad staff, Baghdad, June 7–18, 2011.

[4]   Interview with former J5 staff officer, Carlisle, Pa., January 6, 2013.

[5]   Interview with USF-I staff officer, Baghdad, November 15, 2011. For example, a significant portion of the USF-I staff moved to al-Asad Air Base in Western Iraq in July 2011 to establish this as a base U.S. aircraft could use post-2011. Not only did the staff move—requiring the commander and senior staff to travel between Baghdad and al-Asad every day—but funds were used to upgrade facilities. Shortly after base improvements were completed, the President decided to end the mission, and the base was turned over to the Iraqis.

intended to combat insurgent violence and to bridge gaps in Iraqi capabilities, thereby enabling ISF ultimately to operate more independently. The surge was also intended to create conditions that would permit the U.S. military to withdraw with honor and success. Throughout 2008, as the United States aggressively implemented its new strategy, Prime Minister Maliki and his political coalition began to demonstrate the government of Iraq's authority through military operations in Basra, Mosul, Sadr City, and cities along the Iranian border. The successful operations greatly strengthened the political position and reputation of Prime Minister Maliki, who continually pressed for the Iraqi government to take greater control over its own affairs.

Just one month after taking office, President Obama reiterated the U.S. commitment to these agreements in a February 27, 2009, speech at Camp Lejeune, North Carolina. The President made clear that subsequent U.S. engagement would be a "strong political, diplomatic, and civilian effort" led by DoS.[6] It would thus be axiomatic that the United States would need to be prepared to maintain a diplomatic mission in Iraq no matter how bad the security situation might get. Although the United States maintains embassies in many dangerous places, with the exception of present-day Afghanistan, DoS has operated diplomatic missions in few other countries where the U.S. military had been a participant in an ongoing conflict. However, both the Bush and Obama administrations were committed to "normalizing" U.S. relations with Iraq, so the commitment to maintain an embassy in Baghdad despite the security environment was never questioned. Highlighting this point, DoS Bureau of Diplomatic Security (DS) officials asserted that they knew of no discussions to define what security conditions (other than imminent threats) might trigger the closure of the U.S. mission.[7] The USF-I J5 did conduct noncombatant evacuation planning that included the Embassy Baghdad and USCENTCOM teams, updating that plan and transitioning it to USCENTCOM in late 2011.[8] From the standpoint of DoS, the continued existence of a U.S. embassy in Baghdad was determined to be a critical symbol of America's political commitment to Iraq.[9]

Deputies' Committee Decisions

Broad policy planning for the transition began in 2009 in the NSC's Deputies Committee and its Iraq Executive Steering Group. Several U.S. officials in both Washing-

---

[6]   Obama, 2009.

[7]   Interview with DoS DS officials, December 20, 2011.

[8]   Interview with former USF-I J5 staff officer, Carlisle, Pa., January 6, 2013.

[9]   Interview with DoS DS officials, December 20, 2011.

ton and Baghdad confirmed that key framing decisions were made in the summer of 2009 that affected the parameters for all that would follow.[10]

The Iraq Executive Steering Group reviewed MNSTC-I's proposal for sizing OSC-I, selecting the most robust option presented for core OSC-I staff (157 military and civilians), reflecting a decision to build into OSC-I the capability to do more than meet the minimal security assistance requirements MNSTC-I had identified. MNSTC-I originally developed widely varying estimates for required security assistance team (SAT) contractors—ranging from about 200 to over 1,200—to support implementation of Foreign Military Sales (FMS) cases. A more-rigorous assessment of requirements to carry out the technical aspects of FMS case implementation anticipated for 2012 and beyond led MNSTC-I to propose 763 SATs, which the Executive Steering Group also endorsed. Both the 157 OSC-I core staff ceiling and the 763 SAT ceiling were approved at the Deputies' Committee level in August 2009.

Establishing a planning number for OSC-I core staff was crucial because this staff would be the foundation for supporting future security cooperation activities with Iraq, ranging from U.S. equipment sales and training to educational exchanges to exercises. A substantial OSC-I staff would enable the office to manage a more-robust security cooperation agenda and to oversee security assistance programs, such as the provision of training and equipment through the FMS process.[11] Similarly, approval of the 763 SATs needed to execute the anticipated FMS cases enabled implementation of a comprehensive security assistance plan, although it left no headroom for additional trainers necessary for future sales of military equipment—most notably Iraq's expected acquisition of F-16 fighter jets.[12] Thus, before OSC-I's missions were even defined, the deputies decided for "political reasons" to keep the number of people associated with OSC-I (both OSC-I staff and security assistance trainers) under 1,000.[13] Moreover, they did not consider life-support requirements and security contractors in this original calculus.[14]

Several officials involved in transition planning commented that the Deputies' Committee's decisions in mid- to late-2009, which were designed to advance broad policy objectives and strategic considerations, were not accompanied by an objective "analysis of cost, feasibility, support requirements, or Iraqi input."[15] For example,

---

[10] Interview with Kathleen Austin-Ferguson, Executive Assistant to the Under Secretary of State for Management, Washington, D.C., October 25, 2011. Also interview with DoS officials, Washington, D.C., June 1, 2011.

[11] OSC-I, "OSC-I Transition Plan," briefing to RAND, June 27, 2011; interviews with DCG (A&T) officials, June 27–28, 2011, and DoD officials, November 18, 2011.

[12] In September 2011, Iraq signed a contract to purchase 18 F-16 jets at a cost of $3 billion.

[13] Interview with DoS officials, Washington, D.C., June 1, 2011.

[14] Interview with former USF-I J5 staff officer, Carlisle, Pa., January 6, 2013.

[15] Interview with Ambassador Jeanine Jackson, U.S. Embassy Baghdad Minister-Counselor for Management, Baghdad, June 28, 2011.

although the deputies approved a proposal to develop a consulate or embassy branch office (EBO) in Mosul, the costs of mitigating security risks proved so high that DoS later dropped these plans.

Withdrawal Time Lines

Perhaps the most important constraint on transition planning was the SA's deadline for the ultimate withdrawal of U.S. forces. On the ground, the December 31, 2011, deadline drove virtually all planning. As Chapter Seven will describe in detail, the time-constrained drawdown roadmap the SA required drove MNSTC-I to modify its incremental, open-ended ISF training program in favor of a training plan aimed at elevating the ISF to a defined "minimum essential capability" (MEC) by the end of 2011. The ISF's achievement of MEC, in turn, would allow an orderly withdrawal of U.S. forces.[16] However, the focus on achieving MEC came at the expense of creating the type of institutional capabilities that the Iraqis would need to sustain development of the ISF after the departure of U.S. forces, when the Iraqis would be responsible for training themselves.

Lack of Certainty Regarding an Enduring U.S. Troop Presence Complicated Planning

Although many officials in both the United States and Iraq could envisage an enduring U.S. force presence for training and selected counterterrorism missions after 2011, the President's remarks provided guidance to agencies that planning should assume no U.S. troops would remain in Iraq after 2011. Emphasizing the finality of the decision, a central USF-I planning assumption was commonly referred to as "zero means zero." However, although the White House took a clear public stance, U.S. and Iraqi officials in 2011 were indeed discussing behind the scenes whether and how to extend the presence of U.S. forces in Iraq, thus creating significant uncertainty about the size and character of a potential long-term U.S. military presence.[17]

From an interagency planning perspective, it would have been preferable to plan for multiple potential scenarios, but it might have weakened the U.S. hand in discussions with Iraq over immunities to do so.[18] Consequently, General Austin directed

---

[16] Interviews with DCG (A&T) staff, Baghdad, June 27–28, 2011, and with DoD officials, Washington, D.C., November 18, 2011.

[17] For a detailed discussion about the internal dialogue between Baghdad and Washington regarding the potential mission and size of a residual military presence beyond 2011, see Gordon and Trainor, 2012, pp. 651–671.

[18] Interview with DoS officials, Washington, D.C., June 10, 2011.

USF-I staff to build flexibility into all plans to provide the President with maximum options well into the final withdrawal of U.S. forces. The details of such planning were closely guarded to ensure that "military planning did not get ahead of policy decisions."[19]

Despite the publicly announced decision that all U.S. forces would depart Iraq by the end of 2011, USF-I planners developed flexible drawdown time lines with multiple branch options for a continued presence. The planning process assumed that this residual mission would entail limited ISF training and assistance efforts and undertake limited operational missions, such as continued involvement in the trilateral CSM, which served as an important confidence-building measure between the Iraqi Army and Kurdish *peshmerga* in disputed territories.[20] One former OSD official explained that DoD "had multiple contingency plans in case we were invited to stay," adding that, as early as mid-2010, military force planners scheduling unit rotations ensured that some units would have sufficient time left in their deployments to remain in Iraq for several months into 2012 if necessary.[21] Force rotation plans, sometimes referred to as "patch charts" that showed the emblems of units slated for upcoming rotations. These were maintained through 2012 to provide needed flexibility.[22] For many in USF-I and DoD, therefore, the question was not whether there would be a follow-on force but rather how large a force would be agreed on and what its missions would be.[23] Indeed, even the August 2011 decision by all of Iraq's political parties that U.S. troops could not be granted immunities after 2011 was largely viewed as just a hurdle to negotiations.[24]

To keep options open, as USF-I planned the logistical aspects of withdrawing tens of thousands of troops and their equipment, it developed a withdrawal timetable that delayed as much as possible the movement of assets that, once withdrawn, would be difficult, costly, or unlikely politically to redeploy to Iraq. Even after the mid-October 2011 decision to "go to zero," there were discussions about bringing significant forces back into Iraq as early as 2012 for exercises and training assistance, once the situation calmed down.

---

[19] Interview with USF-I J5 staff officer, Baghdad, December 22, 2011.

[20] Interview with senior DoS official, Washington, D.C., October 25, 2011. See also Tim Arango and Michael S. Schmidt, "Despite Difficult Talks, U.S. and Iraq Had Expected Some American Troops to Stay," *New York Times*, October 21, 2011. It is difficult to say for sure whether such planning assumptions were made because no one really believed that all troops would be withdrawn—that "zero means zero"—or because it was considered prudent military planning to have options available for potential policy shifts.

[21] Email from former OSD official, Washington, D.C., January 27, 2012.

[22] Interview with former USF-I J5 staff officer, Carlisle, Pa., January 6, 2013.

[23] Interview with USF-I J3 staff officer, Baghdad, August 13, 2011.

[24] From a military planner's perspective, USF-I had to be ready with a feasible option until it received the final decision. What planners thought was politically possible was irrelevant because the command had to be ready to execute any option the political leadership chose. Plans had to be on the shelf to maintain flexibility to execute either way. Interview with former USF-I Plans officer, Carlisle, Pa., January 6, 2013.

Early planning identified June 2011 as the point at which a final decision would need to be made regarding whether a possible follow-on force would be appropriate and acceptable to the Iraqis and, if so, what its size and composition would be. However, this deadline continued to be pushed back throughout the summer and well into fall 2011. Post-2011 DoD forces required for anticipated missions ranged from 24,000 to as low as 1,600.[25] Contingency planners focused on the feasibility and optimal solution to support these potential troop levels while simultaneously keeping the "go to zero" plan in the forefront. U.S. officials were impatient for a choice to be made so that the troop drawdown could proceed. Indeed, Secretary of Defense Leon Panetta famously expressed his frustration with the delays, telling American troops in Iraq in July 2011, "I'd like things to move a lot faster here, frankly, in terms of the decision-making process. Do they want us to stay? Don't they want us to stay? Dammit, make a decision."[26]

To preserve options, the best-case transition plan had to be modified to enable a much more rapid departure of forces and equipment in the final months of 2011,[27] which would affect the timetable for the transfer of control of U.S. bases, facilities, and equipment from USF-I to the Iraqi government or Embassy Baghdad. In May 2011, USF-I published Change 1 to OPORD 11-01 to update the directive. Information that had been blurry during planning in late 2010 was coming into focus, allowing the staff to provide better details in Change 1. Also, guidance in Change 1 needed refinement to direct ongoing operations, advance the effort to "go to zero," and improve flexibility for a follow-on force. Each delay required multiple critical decisions about how to best posture the force to meet the potential needs of a follow-on military force while simultaneously ensuring that USF-I would be able to exit Iraq by the end of 2011, as had been agreed to in the SA. In late September, General Austin established October 15, 2011, as the date by which USF-I needed a final decision, having determined that after that date, "the laws of physics" would make it impossible to withdraw fully by December 31.[28] President Obama announced his decision on the troop withdrawal on October 21, although USF-I was aware of the decision on October 16 and had begun executing its final redeployment plan.[29]

In spite of the SA's firm deadline, the President's public statements, and the requirement under OPORD 11-01 that all troops be out of Iraq by the end of the

---

[25] Gordon and Trainor, 2012, pp. 655, 669.

[26] Elisabeth Bumiller, "Panetta Presses Iraq for Decision on Troops," *New York Times*, July 11, 2011.

[27] It should be noted that the plan always envisioned the retention of 50,000 in Iraq as long as possible, not only to create options for policymakers but also to make as much progress as possible in training, advising, and equipping the ISF. The actual delay in troop movement was only a month, shifting from the end of August to the end of September. Interview with USF-I general officer, Washington, D.C., January 6, 2013.

[28] Interview with USF-I staff officer, Baghdad, November 15, 2011.

[29] Interview with USF-I staff officer, Baghdad, November 15, 2011.

year, the President's ultimate decision to withdraw all U.S. troops nevertheless came as a surprise to some in USF-I. In November 2011, when it was clear that "zero" really did mean "zero," the OSC-I chief, LTG Robert Caslen, told SIGIR that the office was reevaluating all its plans "in light of the withdrawal of U.S. troops."[30] This comment reflects the widely held conviction within USF-I, even at very senior levels, that the plan would be changed at the last minute to permit a follow-on U.S. force.[31]

Eleventh-hour decisions to end discussions of an enduring U.S. troop presence also affected transition planning at DoS, even though DoS had consistently followed the White House directive that agencies develop post-2011 plans that assumed no follow-on military force—the "worst case scenario" for the department, in that it would require the U.S. Mission to be entirely self-sufficient. Embassy Baghdad was slated to assume the lead for several activities that it would have managed very differently if a residual U.S. military presence remained in Iraq, particularly efforts to counter Iranian influence and violent extremist organizations and to manage Arab-Kurd tensions. As the embassy prepared to take over lead responsibility for these missions, it could not be sure whether or not to expect direct DoD involvement, so it sought to develop the greatest possible in-house capability.[32]

Furthermore, DoS dedicated staff time and scarce funding to create capabilities that would have proved unnecessary if a decision had been made to retain a force presence.[33] If U.S. troops were staying, for example, DoS would have been able to draw on medical assistance from the military instead of letting a $132 million contract to establish ten new medical facilities throughout Iraq. Similarly, as late as summer 2011, officials at DoS considered the possibility that, if DoD maintained a troop presence at bases that included DoS personnel, DoD might continue to be able to pay for security and life-support contracts. Although DoS planned as though it would need to be entirely self-sufficient, until the President's October 21, 2011 announcement, DoS officials still speculated that the U.S. military might continue to provide the embassy with some forms of direct support.[34] Overall, the pervasive mentality in DoS, USF-I, USCENTCOM, and other agencies through much of 2011 was on the potential for some undetermined number of DoD forces to remain post-2011. This distracted some

---

[30] SIGIR, *Quarterly Report and Semiannual Report to the United States Congress*, January 30, 2012a, p. 48.

[31] Given the Iraqi proclivity to wait until the very last second to make a decision during negotiations to give them leverage in bargaining, this was a perfectly plausible assumption. Prime Minister Maliki used this last-minute bargaining technique during the 2008 deliberations over the SFA/SA, and it was being used during formation of the Iraqi government.

[32] Interview with Deputy Assistant Secretary of State for Near Eastern Affairs Barbara Leaf, Washington, D.C., January 13, 2012; interview with DoS officials, Washington, D.C., June 10, 2011.

[33] Interview with senior DoS official, Washington, D.C., July 15, 2011.

[34] Interview with senior DoS official, Washington, D.C., July 15, 2011.

from vigorously planning for the significantly different and challenging transformation to be expected in U.S.-Iraq relations without a follow-on military presence.[35]

Executive Branch Planning Processes Complicated Transition-Related Decisions

Inefficiencies in the executive branch's transition planning caused delays in decision-making, which in turn caused difficulties in executing elements of the transition. First, DoS's initial planning efforts were somewhat limited, perhaps because resources to manage the transition were slow in coming. As early as July 2010, the Commission on Wartime Contracting (CWC) expressed concern

> that the ongoing planning for State's operations in Iraq during the drawdown and after the U.S. military exit has not been sufficiently detailed. It has lacked input on key decisions needed to resolve policy issues and identify requirements, and has not fully addressed the contract-management challenges ahead.[36]

The CWC emphasized that congressional support for DoS's new missions in Iraq would be critically important to a successful transition, writing, "particularly troubling is the fact that State has not persuaded congressional appropriators of the need for significant new resources to perform its mission in Iraq."[37]

Second, better DoD-DoS coordination on transition planning and decisionmaking early in the process might have identified more challenges earlier and given DoS more time to address them. Delays in developing an effective interagency coordination process, according to GAO, "have made the transition more challenging than it otherwise could have been, compounding State's relatively limited capacity to plan."[38] Both DoS and DoD should have had Executive and Legislative Branch parameters and/or limits to guide basic planning assumptions much earlier than they were provided.[39]

Despite barriers to collaboration in Washington, multiple officials at both USF-I and the embassy made clear that on-the-ground interagency collaboration in Iraq was excellent. One DoS official explained that, ironically, one of the reasons DoD and INL had "little contact" in Washington on law enforcement training was that good coordi-

---

[35] Interview with former USF-I staff officer, North Carolina, February 1, 2013.

[36] CWC, "Better Planning for Defense-to-State Transition in Iraq Needed to Avoid Mistakes and Waste," CWC Special Report 3, July 12, 2010, pp. 2–3.

[37] CWC, 2010, pp. 2–3.

[38] GAO, *Iraq Drawdown: Opportunities Exist to Improve Equipment Visibility, Contractor Demobilization, and Clarity of Post-2011 DOD Role*, Washington, D.C., GAO-11-774, September 2011, p. 36.

[39] Interview with former USF-I J5 staff officer, Carlisle, Pa., January 6, 2013.

nation in the field resulted in "less need for that kind of interaction" at home.[40] At the planning level, embassy personnel were invited and included in all USF-I J5 Plans and J5 Strategy planning efforts.[41]

Third, and one of the likely reasons DoS-DoD coordination got off to a slow start, neither DoS nor DoD dedicated senior officials to lead the transition effort until quite late in the process. Instead, the task of managing the complex transition was assigned to officials consumed primarily with managing daily developments in U.S.-Iraqi bilateral relations or with related administrative functions (contracting, security, etc.) on a global basis. Thus, no single official or office at either DoS or DoD headquarters "owned" the Iraq transition. On February 14, 2011, the Secretary of State appointed an Iraq Transition Coordinator, former U.S. Ambassador to Laos and Coordinator for Economic Transition at Embassy Baghdad Patricia Haslach—18 months after the Deputies' Committee's initial decisions on the transition in August 2009 and less than one year before the U.S. troop withdrawal.[42] Her office closed in February 2012, less than two months after the transition.[43] By the time of Haslach's appointment, however, the most critical elements of transition planning were already under way. From the perspective of the planning team in Baghdad, the DoS office had limited situational awareness as key decisions were being implemented, let alone in time to influence decisions in the first place.[44] One lesson learned, therefore, is that agencies should designate a Washington-based senior-level official to lead the operational aspects of the transition from the very beginning of the planning process, perhaps a full two years in advance of a planned transition. Moreover, such planning should be hosted by the responsible combatant command (in this case, USCENTCOM) because it is best situated with planners and processes to host the interagency team required to ensure success.

---

[40] Interview with DoS INL official, January 20, 2012.

[41] Interview with former USF-I J5 staff officer, Carlisle, Pa., January 6, 2013.

[42] OIG, "DOS Planning for the Transition to a Civilian-Led Mission in Iraq," Report No. MERO-I-11-08, May 2011, p. 5. See also Patrick Kennedy, "Statement of Ambassador Patrick Kennedy, Under Secretary of State for Management, U.S. Department of State," *U.S. Military Leaving Iraq: Is the State Department Ready?* hearing before the Subcommittee on National Security, Homeland Defense, and Foreign Operations, Committee on Oversight and Government Reform, U.S. House of Representatives, 112th Congress, 1st Sess., March 2, 2011a, p. 8.

[43] Interview with former DoS Coordinator for Iraq Transition Ambassador Patricia Haslach, Washington, D.C., July 15, 2011. In her subsequent assignment, as Principal Deputy Assistant Secretary in the Bureau of Conflict and Stabilization Operations, Ambassador Haslach led an effort to identify lessons learned from the transition in Iraq. This effort was not intended to resolve outstanding issues in Iraq but rather to prepare for similar transition efforts that might take place in the future. See Patricia M. Haslach, "Iraq Transition Lessons Offsite Followup," email to Larry Hanauer, RAND, May 4, 2012.

[44] Interview with former USF-I J5 staff officer, Carlisle, Pa., January 6, 2012.

Fourth, slow decisionmaking also caused DoS's scarce planners to waste time that could have been spent on more productive pursuits. As the DoS Office of the Inspector General (OIG) reported,

> key high-level decisions in Washington have been delayed, wasting scarce embassy time and effort. According to embassy officials, numerous staff and hundreds of hours were devoted to developing an alternative to the LOGCAP [Logistics Civil Augmentation Program] contract for provision of housing, food services, and other life support. After months, it was decided the embassy could continue to use LOGCAP under a shared cost arrangement with DOD.[45]

DoS and DoD cannot be faulted for all of the obstacles encountered in transition planning. Congress's failure to appropriate funds in a timely manner, for example, prevented DoS from finalizing the size and scope of its programs. Similarly, DoD could not determine the number of SATs needed to support security assistance cases,[46] in part because the U.S. and Iraqi governments had not yet decided on the full scope of FMS cases that would need to be supported.[47] There may be room for DoS to examine its planning approach in such cases, adopting a better-resourced and more-defined decisionmaking process. Although expensive in manpower, a process like the one the military used might help interagency planners avoid dead ends, such as the one made with LOGCAP, and ensure better integration in a "security heavy" operation, such as Iraq or Afghanistan.

Iraq's Failure to Form a Government Causes Delays

The drawn-out process of government formation in the wake of the March 2010 parliamentary elections caused extensive delays in the transition process and impeded the Iraqi government from having a voice in the transition at a critical time.

In 2009, U.S. officials expected that Iraqi parliamentary elections in January 2010 would be followed by the establishment of a fully functional Iraqi government by June 2010, which would enable the United States to either complete its "responsible" drawdown of forces by December 2011 or, alternatively, have an empowered Iraqi government partner with which to conclude a new SOFA and agreed-on mission for a continuing U.S. force presence. However, internal wrangling over the terms of the electoral law led to the postponement of elections from January 2010 to March 7, 2010, and the narrow outcome and subsequent political jockeying prevented the formation of a government for nine months after that.

---

[45] OIG, 2011, p. 13.

[46] Interview with DoS officials, Washington, D.C., June 10, 2011.

[47] Interview with senior U.S. Army official attached to DCG (A&T), Baghdad, June 27, 2011.

Throughout most of 2010, USF-I and the embassy were unable to engage the government of Iraq on any transition-related planning because the Iraqi government was essentially in a caretaker status. Iraqi political and military leaders were both unable and unwilling to make controversial decisions until a government was formed, and the U.S. government did not want to be perceived as interfering with the internal political process that was taking place in Iraq. From a planning perspective, this would have been the ideal time for Washington and Baghdad to reach agreements on the size and scope of a potential enduring U.S. military presence and the nature and size of the post-2011 U.S. diplomatic presence, but discussions on these issues had to be postponed until the Iraqis formed a government.[48]

The political stalemate was finally overcome when the major rival factions, Maliki's State of Law party and the Iraqiya coalition led by former Prime Minister Ayad al-Allawi, signed a power-sharing agreement in the Kurdish city of Erbil in November 2010. A new government, with Maliki as prime minister, was seated in December. But even this development was insufficient to enable transition planning with the Iraqis to proceed.

Prime Minister Maliki postponed appointing ministers to head the key security-related ministries of Defense, Interior, and National Security. Instead, he retained these responsibilities, while episodically exchanging possible names for the security ministries with Iraqiya and other political groups. In May 2011, Maliki selected interim candidates to lead the ministries of Defense (Sadun al-Dulaimi), National Security Affairs (Riyad Ghrayb), and Interior (Tawfiq al-Yasiri).[49] The first two began work in August 2011, but Maliki continued to serve as acting Interior Minister into 2012.[50]

While the political crisis had been attenuated, the prime minister's control over the bureaucracy further delayed transition planning. Without key ministers in place, the Iraqi agencies that were most critical to the transition lacked officials empowered to make significant or politically sensitive transition-related decisions.[51] Incumbent officials refused to take any steps that might be controversial, referring almost all decisions to the prime minister's office. This reluctance to act caused yet more delays in transition planning.[52]

---

[48] Interview with USF-I command historian, Baghdad, July 22, 2011.

[49] Reidar Visser, "Another Batch of Security Ministry Nominees: Turning the Clock Back to 2006?" *Iraq and Gulf Analysis*, May 6, 2011.

[50] Michael Schmidt, "Attacks in Iraq Heighten Political Tensions," *New York Times*, August 16, 2011; Central Intelligence Agency, "Chiefs of State and Cabinet Members of Foreign Governments: Iraq," September 1, 2011.

[51] Katzman, 2011b, p. 15. It could, of course, be argued that, even if Maliki had appointed key ministers much earlier, the fact that these officials would have come from parties other than Maliki's State of Law means that they would not have been empowered to make controversial decisions anyway.

[52] Interviews with DCG (A&T) officials, Baghdad, June 27, 2011, and senior DoS official, Washington, D.C., October 25, 2011. Engagement did continue at lower levels in the MOD and MOI on issues that would affect transition planning but that were not fraught with political implications. For example, Senior Deputy Minister

INL's evaluation of the Iraqi police's capabilities, for example, was delayed by several months because outside researchers hired to conduct a baseline assessment were unable to gain access to Iraqi officials; as a result, the implementation of INL's Police Development Program (PDP) was pushed back.[53] Similarly, the government of Iraq's inability to conclude agreements with Embassy Baghdad regarding land usage caused delays in facility construction necessary for the operation of the U.S. Mission after 2011.[54] Discussing DoS and DoD efforts to identify locations for enduring OSC-I and INL facilities, for example, USF-I wrote in October 2010 that "our ability to lock in those locations is dependent on formal agreement with the government of Iraq, something that cannot be obtained until a permanent government is in place."[55]

As noted earlier, because embassy and USF-I officials working on the transition lacked empowered counterparts in Iraqi ministries, the United States did not "formally consult" the Iraqi government on many key transition initiatives, according to a senior DoS official in Washington.[56] Ambassador Jeffrey stated that neither Prime Minister Maliki nor the U.S. government ever proposed a bilateral forum for handling transition issues. "We never really asked the Iraqis which activities they wanted," Jeffrey said. "We told them what we would give them."[57] For all practical purposes, therefore, the transition was an internal U.S. government effort, and Iraqi perspectives had only informal effects on U.S. transition planning.

Congressional Influence on the Transition

The U.S. Congress also significantly influenced the transition. Six major political and policy considerations affected critical funding decisions. First, many members simply lacked confidence that DoS would be able to execute its mission in Iraq even if adequately funded. Rep. Steve Chabot, Chair of the House Foreign Affairs Committee's Middle East South Asia Subcommittee, bluntly stated in June 2011 that it "requires a

---

of Interior Adnan al-Asadi worked with his MOD counterpart to divide roles and responsibilities between the MOI and MOD—an important step that would affect how the United States would transition related training programs and manage future training and assistance initiatives. See SIGIR, *Quarterly Report to the United States Congress,* October 30, 2011f, p. 4.

[53] Rebecca Santana, "'Bottomless Pit': Watchdog Slams US Training of Iraqi Police," Associated Press, October 24, 2011; interview with DoS INL official, Washington, D.C., January 20, 2012.

[54] OIG, 2011, pp. 6, 19.

[55] USF-I, *Quarterly Command Report, 4th Quarter, FY10 (1 July–30 September 2010),* October 31, 2010e, Not available to the general public.

[56] Interview with DoS officials, Washington, D.C., June 1, 2011.

[57] U.S. Ambassador to Iraq James Jeffrey, RAND Corporation Roundtable Discussion, Arlington, Va., March 14, 2012.

willful suspension of disbelief to believe that DoS alone—without the help of U.S. military forces on the ground—has the capability to satisfactorily execute this mission."[58]

Second, many members of Congress are far more comfortable working with DoD than with DoS and are more familiar with the former's capabilities. While the Armed Services Committees and Defense Appropriations Subcommittees in both the House and the Senate are intimately involved in the authorization and funding of programs related to DoD's operations, the authorizing Foreign Affairs Committees and (to a lesser extent) the appropriators on the Foreign Operations Subcommittees focus more on strategic foreign policy challenges and foreign aid than on DoS management and operations.[59]

Third, Congress approaches funding decisions for DoS and DoD very differently. For one thing, the Defense Appropriations Subcommittees are much more accustomed to working with high dollar figures than are the appropriations subcommittees that fund DoS. Moreover, for most of the past decade, Congress has funded military operations in Iraq and Afghanistan through "must pass" emergency supplemental appropriations, which are generally not subject to caps on government spending and need not be offset by cuts to other programs.[60] With a few exceptions, such as the funds for construction contained in the fiscal year (FY) 2012 supplemental, DoS's budgets have been accorded no similar sense of urgency.

Fourth, although the military's funding requirements for Iraq were large by any objective measure, the proposed Iraq operations budget comprised a higher percentage of DoS's overall budget than the Iraq war had represented to DoD's. As a result, DoS's Iraq request represented a particularly substantial financial commitment that merited close scrutiny, particularly because many skeptics questioned whether DoS could manage a program that, compared to its normal operations, was so complex and costly.

Fifth, Congress questioned the disproportionately high cost of supporting and securing a relatively small core diplomatic staff. As a Senate Foreign Relations Com-

---

[58] Steve Chabot, "Opening Statement of Rep. Steve Chabot," *Preserving Progress: Transitioning Authority and Implementing the Strategic Framework in Iraq, Part II*, Washington, D.C.: Subcommittee on the Middle East and South Asia, Committee on Foreign Affairs, U.S. House of Representatives, 112th Cong., 1st Sess., June 23, 2011.

[59] Interview with former staff member of the Senate Foreign Relations Committee, Washington, D.C., February 3, 2012.

[60] See GAO, *Supplemental Appropriations: Opportunities Exist to Increase Transparency and Provide Additional Controls*, GAO-08-314, January 2008a, p. 9. Between 2002 and 2008, the amount of supplemental appropriations that were offset by cuts to other programs shrank from 40 percent to less than 1 percent. Without the need to make difficult trade-offs to increase war spending, some critics claim, Congress gave DoD a virtual blank check for its activities in Iraq and Afghanistan—particularly given that the annual Defense Supplemental Appropriations bills were considered must-pass legislation for reasons related to both domestic politics and military operations. See also Veronique de Rugy, "What's the Emergency?" *Regulation*, Summer 2008, pp. 6–8, and Veronique de Rugy and Allison Kasic, The Never-Ending Emergency: Trends in Supplemental Spending," Arlington, Va.: George Mason University Mercatus Center, Working Paper No. 11-30, August 2011.

mittee staff report bluntly asserted, the "cost of the Mission may be its death knell."[61] Given that the overwhelming majority of DoS's Iraq budget is for nonprogrammatic functions, such as basic life support and security, members of Congress questioned whether the U.S. Mission in Iraq would generate sufficient foreign policy benefit to justify the enormous investment. The Senate Foreign Relations Committee, for example, recommended that "the State Department should reconsider whether the embassy branch offices will have sufficient freedom of movement to justify their considerable expense,"[62] a recommendation that would prove prescient. Since U.S. military units perform support and security missions for themselves, these functions do not appear as separate expenses in defense budgets; as a result, such costs did not receive the same level of attention during debates on defense expenditures.

Sixth, DoS enjoys far less popular support than the military. Even when the wisdom of the Iraq war was being fiercely debated, funding for deployed troops enjoyed broad, bipartisan support. In contrast, the most visible portion of DoS's budget, foreign aid, has little public support. Given that Congress is seeking to slash billions of dollars in government spending to address the nation's economic crisis, the absence of a domestic constituency for DoS missions made DoS's budget requests vulnerable to cuts.[63]

As a result of these political and budgetary realities, Congress habitually withheld small sums from DoS even as it funded similar capabilities at DoD. Sen. John Kerry castigated his colleagues for being unwilling "to provide the financial resources necessary for success by supporting our diplomatic efforts with the same vigor that we devote to our military mission."[64] Ambassador Peter Bodde, then Assistant Deputy Chief of Mission in Baghdad, noted that, when DoS proposed a contracted medical capability that would get patients to medical care within the "miracle hour"—a time

---

[61] U.S. Senate, Committee on Foreign Relations, *Iraq Report: Political Fragmentation and Corruption Stymie Economic Growth and Political Progress; A Minority Staff Trip Report*, S. Prt. 112-34, 112th Cong., 2nd Sess., April 30, 2012, p. 19.

[62] U.S. Senate, Committee on Foreign Relations, "Iraq: The Transition from a Military Mission to a Civilian-Led Effort," S. Prt. 112-3, January 31, 2011, p. 12.

[63] For example, in a hearing on DoS's FY 2012 budget, Rep. Harold (Hal) Rogers, the chairman of the House Appropriations Committee, bluntly told Secretary of State Clinton, that it was

> time that we get serious about reducing spending, putting a dent in our record-setting deficit. It's difficult to believe that the administration shares my goal to cut spending when the '12 State, Foreign Operations request of $59.5 billion is an increase of more than 22 percent above the '10 bill. Even if '10 supplementals are included, [the] budget still represents an 8 percent increase . . . . [W]e simply can't sustain the level of spending in this bill.

Harold Rogers, "Opening Statement of Rep. Harold Rogers," in Fiscal Year 2012 Budget Request for State Department and Foreign Assistance Programs, hearing before the Subcommittee on State and Foreign Operations, Committee on Appropriations, U.S. House of Representatives, March 10, 2011.

[64] John Kerry, "Opening Statement of Senator John Kerry," *Iraq: The Challenging Transition to a Civilian Mission*, hearing before the Committee on Foreign Relations, U.S. Senate, 112th Cong., 1st Sess., February 1, 2011.

frame that greatly increases survival rates—Congress initially balked at the cost even though it had never seriously considered scaling back U.S. troops' access to emergency medical care in Iraq.[65]

Similarly, although Congress funded large-scale growth in DoD's contracting and acquisition staff, it refused to provide DoS with funds to do the same—even as several members of Congress criticized DoS's weak contract management as a threat to the success of the transition. In the FY 2009 National Defense Authorization Act (NDAA), Congress established ten new flag-rank (general or admiral) contracting positions to reverse a decline in DoD's acquisition workforce that had contributed to contracting failures in Iraq and Afghanistan.[66] Congress also provided short-term funding to hire several hundred acquisition personnel as an interim measure until a larger number of permanent positions could be resourced.[67] Noting that, "in a military environment (especially in an expeditionary environment), the number and level of the Generals associated with a discipline reflects its importance,"[68] this budget authorization was a clear statement that Congress believed DoD contracting was a critically important mission that required additional resources.

After years of liberally funding military operations in Iraq, some key stakeholders argued that DoS's post-2011 budget was no place to be frugal. In a January 2011 report, the Senate Foreign Relations Committee pointed out that, even at $5 billion to 6 billion per year, DoS's Iraq operations "would constitute a small fraction of the $750 billion the war has cost to this point." The committee report also asserted that "Congress must provide the financial resources necessary to complete the diplomatic mission in Iraq."[69] The congressionally mandated CWC offered an even blunter assessment, writing that

> without a substantial increase in budgetary support from Congress, the post-2011 prospects for Iraq—and for U.S. interests in the region—will be bleak .... Given that Congress has appropriated more than $1 trillion for U.S. operations in Afghanistan and Iraq since 2001, the prospect of jeopardizing the gains in Iraq and U.S. interests in the region to save a small fraction of that sum looks like false

[65] Interview with Ambassador Peter Bodde, Assistant Chief of Mission for Assistance Transition in Iraq, Embassy Baghdad, June 27, 2011.

[66] Public Law 110-417, The Duncan Hunter National Defense Authorization Act for Fiscal Year 2009, October 14, 2008, Sec. 503. Also see Frank Kendall and Brooks L. Bash, "Joint Prepared Statement by Frank Kendall and Lt. Gen. Brooks L. Bash, USAF," *The Final Report of the Commission on Wartime Contracting in Iraq and Afghanistan*, hearing before the Subcommittee on Readiness and Management Support, Committee on Armed Services, U.S. Senate, 112th Cong., 1st Sess., October 19, 2011; U.S. Senate, Committee on Armed Services, "Report to Accompany S. 3001, the National Defense Authorization Act for Fiscal Year 2009," May 12, 2008, pp. 329–330.

[67] Kendall and Bash, 2011.

[68] U.S. Senate, Committee on Armed Services, 2008, pp. 329–330.

[69] U.S. Senate, Committee on Foreign Relations, 2011, pp. 4, 18.

economy indeed . . . . [F]alse economy in budgetary support can cause mission degradation or outright failure, with dire consequences for Americans and Iraqis alike, as well as risks to regional stability.[70]

U.S. Ambassador to Iraq James Jeffrey spelled out the potential consequences in further detail in testimony to the Senate Foreign Relations Committee in February 2011:

> To not finish the job now creates substantial risks of what some people call a "Charlie Wilson's War" moment in Iraq with both a resurgence of al-Qaida and the empowering of other problematic regional players . . . . Gutting our presence in Iraq could also provide Iran increased ability to create anxieties in the region that in turn could spiral out of control.[71]

Funding for the Transition

In addition to scrutinizing relatively small-ticket budget items, such as contract management and medical support, Congress withheld funding that DoS had requested for transition-related requirements. In the FY 2010 supplemental appropriation, Congress cut $540 million (34 percent) from the Department's $1.57 billion request for Iraq Diplomatic and Consular Program operations—funding requested for construction of permanent diplomatic and consular facilities outside Baghdad and for some security and operations functions. Congress explicitly stated that it was withholding the funds for construction of permanent facilities because such funding should be requested in the department's regular appropriations requests rather than in an "emergency" supplemental.[72]

The impact of this deduction, however, was significant. The cuts "resulted in no flexibility to accommodate increasing costs of security and interim facilities," according to a DoS budget summary.[73] In response to the funding uncertainty, DoS deferred construction of all permanent diplomatic facilities in Iraq, revised staffing levels to

---

[70] CWC, "Iraq—A Forgotten Mission?" CWC Special Report 4, March 1, 2011a, pp. 1, 3, 7.

[71] James F. Jeffrey, "Statement of Hon. James F. Jeffrey, Ambassador to Iraq, U.S. Department of State, Washington, DC," *Iraq: The Challenging Transition to a Civilian Mission*, hearing before the Committee on Foreign Relations, U.S. Senate, 112th Cong., 1st Sess., February 1, 2011a, pp. 6–8.

[72] U.S. Senate, Committee on Appropriations, "Making Emergency Supplemental Appropriations for Disaster Relief and Summer Jobs for the Fiscal Year Ending September 30, 2010, and for Other Purposes," Conf. Rpt. 111-188, 111th Cong., 2nd Sess., May 14, 2010. The argument that capital costs, such as construction, belong in base budgets rather than supplemental requests had been advanced (mostly unsuccessfully) for years by opponents of large supplemental appropriations; given that previous supplemental appropriations bills included funds for many similar routine and capital expenses, it is not clear why Congress decided to apply the principle to DoS's construction request.

[73] DoS, "11 02 01 State FY 2010–FY 12 Iraq Funding_FINAL.docx," undated a.

reflect only mission-critical personnel, reduced ground movements for embassy officials throughout Iraq, eliminated plans to open EBOs in Diyala and Mosul, and shifted construction and security costs to DoD by having OSC-I manage the proposed EBO in Kirkuk.[74] All these cost-cutting steps would hinder the embassy's ability to project a presence throughout the country when U.S. forces departed.

When congressional appropriations committees marked up FY 2011 budget legislation, they again slashed DoS's requests significantly. The administration requested $1.787 billion for the Iraq Diplomatic and Consular Program, but the Senate committee appropriated $1.65 billion (a 9-percent cut), and the House committee appropriated only $1.34 billion (a 25-percent cut). The final appropriation that passed Congress provided $1.524 billion, 15 percent below the administration's request.[75] Of course, the department "adjusted the mission to comport with the dollars available."[76] However, the delays and cuts had a significant effect on embassy operations and on the transition.[77] The lack of funds for key transition-related contracts in FY 2011, for example, forced DoS to delay some contracts while paying for others in installments using funds left over from prior years and committing portions of funds requested for the following year.[78]

Funding delays continued in FY 2012. After five CRs that extended funding at FY 2011 base levels, Congress provided DoS's FY 2012 funds in an omnibus appropriations act passed on December 23, 2011, that cut the administration's total base request from $50.8 billion to $42.5 billion, a reduction of 17 percent.[79] However, because the department needed time to allocate the appropriation—basically, to the extent Congress permitted flexibility to decide which programs would absorb the cuts

---

[74] DoS, Response to Question for the Record #2 Submitted by the Subcommittee on State, Foreign Relations, and Related Programs, Committee on Appropriations, U.S. House of Representatives, drafted August 20, 2010b. Also interview with Kathleen Austin-Ferguson, Executive Assistant to the Under Secretary of State for Management, Washington, D.C., October 25, 2011.

[75] DoS, Bureau of Management, Office of Resource Management, email to RAND researcher, October 27, 2011. See also U.S. Senate, Committee on Foreign Relations, 2011, p. 16.

[76] Patrick Kennedy, "Statement by Patrick F. Kennedy, Department of State, Under Secretary for Management Department of State," statement for the record, for *Department of State Contracting: CWC's Second Interim Report, the QDDR, and Iraq Transition*, hearing before the Commission on Wartime Contracting, June 6, 2011b, p. 92.

[77] Kennedy, 2011a, p. 2.

[78] According to Kennedy, 2011b, pp. 8–9:

> The Department has been able to meet the most critical operational needs in FY 2011 by delaying selected contract awards and through the judicious use of funds carried forward from prior appropriations. The most direct impact of the reduced funding level on contracts is that the Department can only fund several contracts for a portion of their period of performance. We will be relying on funds requested in FY 2012 to meet the remainder of their period of performance.

[79] Susan B. Epstein and Marian Leonardo Lawson, *State, Foreign Operations, and Related Programs: FY2012 Budget and Appropriations*, Congressional Research Service, R41905, January 6, 2012, p. 28.

in funding—bureaus still did not know a month later how much money they would have.[80] Thus, budgetary uncertainties continued for at least the first one-third of the fiscal year, just as U.S. forces completely departed.[81]

DoS's base request for FY 2012 Iraq Diplomatic and Consular Program operations was a mere $495.9 million to cover "enduring programs," defined as long-term operational needs for the embassy in Baghdad and the two consulates in Basra and Erbil.[82] The request was so low because, for the first time, the department submitted its wartime and contingency-related requirements in a separate OCO request[83]—a strategy that had helped DoD win political support for its requests.[84] Both agencies argued that DoS's $3.229 billion OCO request was necessary "for deploying, securing, and supplying the department's civilian presence in an extremely high-threat environment."[85] This figure included $1.39 billion for enhancements to existing security operations; $266.1 million for operations and basic life support (primarily logistical services and the LOGCAP contract); and $332.0 million for the Baghdad Master Plan,[86] which covered construction and upgrades to housing, office space, and the embassy's own powerplants and wastewater treatment facilities.[87]

DoS's request did not include funds for OSC-I, for which Congress provided $524 million in Title 10 OCO funds in the FY 2012 NDAA.[88] While operating funds for OSC-I personnel and equipment, as with all security cooperation programs, are provided by a surcharge imposed on Foreign Military Sales, OCO funds cover requirements that are not typical for a security cooperation mission, such as OSC-I's share of

---

[80] Interview with DoS INL official, January 20, 2012.

[81] This type of budgetary uncertainty had a much greater effect on Embassy Baghdad than it would have had on USF-I because funds for deployed forces are provided by Overseas Contingency Operations (OCO) Funding, which allows a great deal of latitude in how the monies are spent within the military theater of operations.

[82] Secretary of State, *Congressional Budget Justification*, Vol. 1: *Department of State Operations, Fiscal Year 2012*, Washington, D.C.: U.S. Department of State, February 14, 2011a, pp. 59–60.

[83] Secretary of State, 2011a, p. 779 (footnote).

[84] William Lynn, "Statement of William J. Lynn, III, Deputy Secretary of Defense," before the Senate Budget Committee, March 10, 2011, p. 3.

[85] DoS, *Executive Budget Summary: Function 150 and Other International Programs, Fiscal Year 2012*, p. 134.

[86] Secretary of State, 2011a, p. 766.

[87] Much of this work involved consolidating and upgrading preexisting support facilities from DoD bases to locations on or near the embassy compound. See Secretary of State, 2011a, p. 766. Upgrades to the embassy's own dedicated powerplant are among the most costly and critical elements of these expenditures, according to an official with DoS's Bureau of Overseas Building Operations (OBO); "without power," this official stated, "the embassy closes." Interview with DoS OBO, Washington, D.C., December 2, 2011.

[88] 10 U.S. Code, Armed Forces. See also Public Law 112-81, National Defense Authorization Act for Fiscal Year 2012, December 31, 2011, Sec. 1228.

embassy construction, security, and life support, particularly for the SATs who would be located at remote sites throughout the country.[89]

DoS's FY 2012 Iraq request appeared disproportionately costly in part because it represented a large portion of the department's worldwide budget and because it was a significant increase from the previous year's request. DoS's overall FY 2012 proposal for Iraq-related assistance and operations was $6.3 billion, a sobering 11.3 percent of the department's $55.7 billion total budget request and more than double its FY 2011 request of $2.7 billion.[90] Secretary of State Clinton, in testimony to DoS's Senate Appropriations subcommittee, pointed out that it would be far less expensive overall for DoS, rather than DoD, to lead the U.S. presence in Iraq. She told the committee that

> [s]hifting responsibilities from our soldiers to our civilians actually saves taxpay-
> ers a great deal of money. The military's total OCO request worldwide will drop
> by $45 billion from 2010 while our costs in State and USAID will increase by less
> than $4 billion. Every business owner I know would gladly invest four dollars to
> save forty-five.[91]

Nevertheless, many DoS officials assumed that Congress would slash the department's FY 2012 request for Iraq. In fact, DoS officials were so uncertain that they would receive the administration's full budget request that several officials asserted in July 2011 that they were planning to operate on as little as one-half the funds sought.[92] Expecting that Congress would either cut the request or fail to pass appropriation legislation by the beginning of the fiscal year, officials in Baghdad and Washington worked to scale back programs and develop contingency plans.[93]

Impact of Continuing Resolutions and Funding Delays

Regardless of the amounts ultimately appropriated, Congress's failure to pass appropriating legislation and its reliance on CRs to keep DoS functioning severely affected transition-related planning and programming. By continuing funding for the previous year's activities and at the previous year's levels, CRs have three primary drawbacks.[94] First, when requirements increase, as has been the case with DoS's transition-related

---

[89] Interview with senior DoS official, Washington, D.C., July 15, 2011.

[90] Hillary Rodham Clinton, "2012 State and USAID Budget Request" testimony before the House Appropriations Committee on Foreign Operations, March 10, 2011c. See also OIG, 2011, p. 5.

[91] Hillary Rodham Clinton, "Fiscal Year 2012 State Department Budget," testimony before the Subcommittee on the Department of State, Foreign Operations and Related Programs, Committee on Appropriations, U.S. Senate, March 2, 2011b.

[92] Interview with senior DoS official, Washington, D.C., July 15, 2011.

[93] Interview with senior DoS official, Washington, D.C., July 15, 2011.

[94] Lynn, 2011, p. 9.

activities in Iraq, CRs do not provide sufficient resources to implement new strategies. Second, CRs often allocate funding to the wrong programs; they can provide funds for activities agencies want to end, for example, while prohibiting funds for new initiatives that may be critical. Furthermore, CRs provide no money for activities originally funded in supplemental appropriations because they only continue appropriations for initiatives in an agency's base budget. All these factors affected DoS's transition planning in Iraq. As DoS OIG wrote in May 2011, before Congress passed DoS's FY 2011 budget: "Because in FY 2010, the transition was mainly funded through a supplemental appropriation that is not included in the CR, the amount of funding currently available for the transition is substantially less than the embassy estimated it needed in FY 2011."[95]

Several DoS officials stated that transition planning was severely hampered by the delays and uncertainty created by CRs and delays in passing appropriations bills. One senior official in the office of the Under Secretary for Management commented that, "[i]f Congress doesn't fund programs when it should, you get behind the eight ball from the beginning. You will never catch up."[96] Among the initiatives disrupted by the lack of a permanent appropriation were the following:

- Consulate Construction: Although the FY 2010 supplemental fully funded construction requirements for Baghdad and for nonpermanent facilities in Erbil, Basra, and Kirkuk,[97] DoS OBO could not identify construction requirements because the FY 2011 CR created uncertainty regarding the size and scope of many embassy programs. An INL official in Baghdad asserted that, because of the constant uncertainty regarding budgets, "the entire year has been spent redoing plans."[98] Similarly, OIG asserted bluntly in a May 2011 report that "the uncertain funding situation caused by FY 2011 continuing resolutions and the ongoing FY 2012 budgeting process have hindered the embassy in moving forward with planned construction."[99]
- Establishment of OSC-I: DoD did not have the authority to fund OSC-I until the day before it was scheduled to be operational. Because a CR provides no funding for new initiatives, OSD had to push Congress to include a special provision in a pending CR to authorize and appropriate funds for the establishment of the new OSC-I.[100] Congress finally gave DoD the authority to spend existing funds

---

[95] OIG, 2011, p. 12.

[96] Interview with Kathleen Austin-Ferguson, Executive Assistant to the Under Secretary of State for Management, Washington, D.C., October 25, 2011.

[97] Interview with DoS OBO, Washington, D.C., December 2, 2011.

[98] Interview with U.S. embassy official, Baghdad, June 27, 2011.

[99] OIG, 2011, p. 20.

[100] Interview with USF-I officer (Comptroller/J8), Baghdad, June 30, 2011.

on OSC-I's activities, operations, security, and facilities in its first FY 2012 CR, which was passed on September 30, 2011, but appropriated no additional funds for it.[101]

- Contracting: As stated earlier, DoS had to delay some contracts and juggle funding to pay for critical contracts. After passage of the fourth FY 2011 CR, which provided appropriations only through March 4, 2011, department officials believed they would be short $292 million for security and life-support contracts that needed to be executed in the second half of the fiscal year.[102]
- Interagency Reimbursements: The Economy Act requires agencies to reimburse each other for services provided, and the U.S. Mission in Iraq—like all embassies—developed a cost-sharing formula under which other agencies would reimburse DoS for overhead expenses and other costs. CRs did not provide these agencies with funds for this purpose, however, leaving DoS to cover these costs out of pocket until (and unless) subsequent legislation provided the funding.[103]

Perhaps the greatest effect of Congress's reliance on CRs, however, is that they drove DoS to scale back significantly the very assistance and training missions being transferred from the military. For example, INL's original 2009 PDP plan called for 350 police advisors. By December 2010, because of both funding uncertainties and increased cost estimates, INL decided to reduce the program to 190 advisors and requested $1 billion to enable the advisors to conduct training at 28 sites, seven of which they would reach by air from the program's hubs in Baghdad, Basra, and Erbil. But drawn-out debates over FY 2011 and FY 2012 appropriations compelled DoS to develop a strategy for phasing in a program that would need only one-half that amount of funding.[104] In the first phase, the PDP would deploy 115 advisors by June 2011—less than one-third the staffing levels originally envisioned.[105] Furthermore, to reduce costs, INL eliminated the air transportation resources necessary for trainers to visit the seven remote sites, thereby precluding INL from training police beyond the areas around its three hubs.[106] As of January 1, 2012, when the transition plan called for the PDP to be

---

[101] U.S. Congress, "Joint Resolution Making Continuing Appropriations for Fiscal Year 2012, and for Other Purposes," 111th Cong., 1st Sess., H.J. Res. 79, Sec. 116, September 14, 2011.

[102] DoS, "State FY 2010 and 2011 Iraq Funding," February 1, 2011a.

[103] DoS, 2011a.

[104] Interview with U.S. embassy official, Baghdad, June 27, 2011. Also interview with DoS INL officials, Washington, D.C., January 20, 2012.

[105] Interview with U.S. embassy official, Baghdad, June 27, 2011. Also Santana, 2011.

[106] Interview with U.S. embassy official, Baghdad, June 27, 2011.

at full operating capacity, only 89 trainers were in country,[107] a staffing level that PDP nevertheless called "mission capable."[108]

As a point of comparison, USF-I had approximately 400 trainers teaching basic police skills at more than 130 sites throughout Iraq in mid-2011.[109] Although INL's training strategy is different from the military's, funding constraints and delays prevented INL from even coming close to matching a police-training program on the scale that it had planned for or that the United States had previously undertaken. These challenges, combined with Iraqi doubts about the value of the program and opposition to the presence of a large number of American trainers, greatly limited the effectiveness of the initiative.[110] The experience of both the Coalition Police Assistance Training Team operating under the CPA in 2003–2004 and INL's PDP should call into question whether its reasonable to expect a small civilian organization, such as the CPA or an embassy, to assume a training mission as large as what was needed in Iraq.

Summary

Senior-level policymakers in Washington made a number of critical decisions early on, such as limiting the locations of provincial diplomatic posts and the number of OSC-I staff. Repeated changes to fundamental issues, such as location, staff size, and available resources, had a cascading effect, making it difficult for agencies to execute the plan by building housing, awarding contracts, and designing bilateral assistance programs. Because of the firm deadline for the U.S. military withdrawal, delays jeopardized the success of the transition.

Certainly, any complex process encounters its share of obstacles and new developments that require changing plans. However, the absence of a clear chain of command for transition-specific issues further complicated planning. For the most part, DoS and DoD headquarters staffs responsible for the transition were also focused on other Iraq policy matters and daily crises, and it was not until 2011 that either organization had a senior-level point person in Washington dedicated to the transition to ensure that planning remained on track, was politically realistic, and had congressional support. In retrospect, it would have been advantageous if agencies had appointed senior Washington-based transition coordinators at the very beginning of the transition planning process. Such an interagency coordination effort could have leveraged the planning capabilities at USCENTCOM to develop the types of detailed plans necessary for a successful transition—or transformation—to a new strategic relationship with Iraq.

---

[107] Interview with DoS INL officials, Washington, D.C., January 20, 2012.

[108] SIGIR, 2012a, pp. 33–34.

[109] Interview with USF-I staff officers, Baghdad, June 10, 2011. Also OIG, 2011, p. 18.

[110] Tim Arango, "U.S. May Scrap Costly Efforts to Train Iraqi Police," *New York Times*, May 13, 2012b.

Transition planning was also greatly hampered by Congress's reluctance to fund DoS's full budget requests or, at a minimum, make earlier decisions enabling DoS to adapt plans to budgetary constraints. Without knowing how much funding would be available—sometimes for months into a new fiscal year—DoS had difficulty developing effective plans for construction, program staffing, and other critical functions.

Political debates shaped the administration's ability to secure the resources needed for the transition. The domestic political discourse regarding Iraq was dominated not by the question of how to prepare DoS for its growing responsibilities but rather by public and often highly partisan debates over the basic wisdom of the military's withdrawal. Internally, the administration's emphasis was on the redeployment of U.S. military forces, rather than the effects of the withdrawal within Iraq or the transformational effects the withdrawal of forces would have on U.S.-Iraqi relations. This focus on the end of an eight-year military mission in Iraq, rather than the beginning of a new "strategic relationship," likely distracted from the administration's message that the United States sought to remain engaged in Iraq over the long term and undermined its efforts to secure the resources for DoS's post-2011 role.

PART III

Executing the Transition and Retrograde of Forces

CHAPTER SIX

Enduring Security Challenges

From 2003 forward, creating and maintaining security has been a preeminent U.S. goal in Iraq and a precondition for achievement of all other U.S. objectives. The sudden removal of Saddam Hussein's security apparatus began a chain of events that ultimately brought Iraq to the verge of uncontrolled civil war. By 2010, the level of violence had fallen to a level that the average Iraqi felt safe sending children to school, going to work, shopping, and conducting other activities associated with daily living.[1] However, this new peace was punctuated by periods of political crisis, car bombings, assassination, and other acts of violence that undermined the security gains both U.S. and Iraqi forces had made. This chapter will examine some of the major enduring security challenges in Iraq, describe the ways the U.S. military addressed these threats prior to its departure, and then highlight the steps taken to transition responsibility for these security related activities to Embassy Baghdad; other U.S. government agencies; and in many cases, the government of Iraq.

Drivers of Instability

From early 2009 until August 2010, General Odierno used the concept of the "drivers of instability" as a vehicle to discuss the enduring security threats within Iraq. The MNF-I staff concluded that four interrelated challenges threatened to undermine U.S. goals in Iraq: communal and factional struggle for power and resources; insufficient governmental capacity; the activities of violent extremist groups; and external interference from neighboring countries, such as Iran and Syria. These so-called drivers of instability served as the underlying cause of the enduring security challenges in Iraq.[2] While the leadership of USF-I did not use this terminology during the last year of tran-

---

[1] Brennan notes, 2009–2011.

[2] Brennan notes, 2009–2011. The following discussion of the drivers of instability was derived from multiple MNF-I sources made available to RAND for the purpose of this study. While these were internal staff assessments, both General Odierno and General Austin referenced these threats during congressional testimony and interviews given to the press.

sition, staff assessments of the Iraqi operating environment confirmed that the threats identified were likely to remain for the foreseeable future. The security environment remained fragile as USF-I and Embassy Baghdad began planning the Iraq transition in 2010. As part of the planning process, MNF-I conducted a strategic assessment to best understand the remaining challenges that could undermine U.S. efforts to assist Iraq develop into a sovereign, stable, and self-reliant country.

Communal and Factional Struggle for Power and Resources

MNF-I planners assessed that there was not yet an Iraqi consensus on the nature of the state accepted across ethnic, sectarian, and regional lines. At the core of the struggle was a competition for power and control over resources within Iraq. Unresolved debates over the nature of federalism and the power of the regions vis-à-vis Baghdad allowed conflicts over disputed internal boundaries, energy policy, and revenue sharing to fester. Ethnic and sectarian disputes—such as the failure of the Iraqi government to transition and reintegrate SOI and the assassination, intimidation, and detention of political opponents—created a political process characterized by mistrust and fear.

Insufficient Government of Iraq Capacity

MNF-I also concluded that two decisions the CPA made in 2003 plagued the government of Iraq with poor institutional capacity and a general lack of skilled technocrats: the decision to not recall the Iraqi military and de-Ba'athification—a law that was then codified in the Iraqi Constitution ratified in 2005. As a result, Iraq lacked public-sector expertise and remains challenged to provide essential services, medical care, access to education, employment, and security—shortcomings that, when compounded by endemic corruption, undermined the legitimacy of the government in the eyes of the populace. Other groups, such as the Sadrist movement and its JAM, emerged as alternative centers of power by meeting the population's needs in ways the government could not.

Violent Extremist Groups

Four categories of violent extremist groups exploited political fissures to gain power and influence in Iraq: internal Sunni, internal Shi'a, external Sunni, and external Shi'a. The first group consists of Sunni nationalist groups opposed to the government of Iraq and seeking to replace it with Sunni-led government. Many of these had historic ties to the Ba'ath regime of Saddam Hussein and conduct terrorist activities to discredit the government of Iraq. Such groups as 1920s Revolutionary Brigade, Jaysh al Islam, and Jaysh Rijal al-Tariq al-Naqshabandi (JRTN) had limited public support but are potential alternatives to the government of Iraq should ethnosectarian violence reemerge. The next category consists of internally focused Shi'a groups, such as Promised Day Brigade. The Promised Day Brigade is largely considered a Shi'a nationalist group, a militant organization associated with the Sadrist Trend. This group is the successor to JAM, which disbanded in 2008 after being defeated by the ISF in the battle "Charge of the Knights" in Basra. The Promised Day Brigade has publicly eschewed violence in the aftermath of the departure of U.S. forces but remains a militia with broad support in portions of southern Iraq and Sadr City. The third category consists of externally supported Sunni extremists, most notably al-Qaeda in Iraq and Ansar al-Islam. While these organizations lack extensive public support, they have been the most successful in conducting large-scale, simultaneous, and regionally dispersed operations within Iraq. As with the Sunni nationalist groups, the goal of these groups is to overthrow the government of Iraq and replace it with a government that enforces a strict interpretation of Islamic law. The fourth and final category of extremists includes Shi'a groups that work at the behest of Tehran: Kata'ib Hezbollah and Asa'ib al-Haq. While the members of these groups are Iraqi, both organizations reportedly swore allegiance to the Iranian Supreme Leader in Tehran and receive training, equipment, and funding directly from Iran. The U.S. Department of Treasury has formally designated Kata'ib Hezbollah, often viewed as a direct-action arm of the Islamic Revolutionary Guards Corps–Quds Force (IRGC-QF), as a terrorist organization. By the middle of 2011, USF-I staff members assessed that Asa'ib al-Haq was also evolving into a direct-action arm of the IRGC-QF.[3]

External Interference from Neighboring Countries

Both Iran and Syria sought to exacerbate tensions within Iraq for their own political purposes in the run-up to the U.S. transition. Ever since the Iran-Iraq war, Tehran had provided a safe haven for anti-Saddam Iraqi exiles and had armed, trained, and equipped militias willing to operate inside Iraq. Since 2003, Iran had been a major supporter of violent extremists in Iraq and, according to one USF-I estimate, is likely to have caused as many as one-half of the U.S. casualties in Iraq.[4] Through its proxies, Iran was also able to dial up or dial down the level of violence targeting Iraqi civilians and Iraqi government officials and the ISF. MNF-I assessed that Iran's goal in doing this was to keep Iraq weak and acquiescent to Tehran. For its part, Syria was believed to have turned a blind eye to the movement of al-Qaeda in Iraq foreign fighters and munitions through its territory in an effort to mollify extremists and forestall violence within its own borders. Moreover, Syria served as a safe haven for Iraqi Sunni extremist organizations and was the leading external supporter of the Sunni insurgency in Iraq. Thus, the Sunni insurgency and al-Qaeda in Iraq received support from Syria, and Shi'a militias received support from Iran.[5]

Addressing these security challenges was always a responsibility shared among Embassy Baghdad, MNF-I/USF-I, and the government of Iraq. The 2010 JCP con-

---

[3] Interview with USF-I J3 staff officer, Baghdad, August 1, 2011. This is further supported by the author's notes taken while assigned as senior advisor to the USF-I J3 from October 2009 to December 2011.

[4] Interview with USF-I J3 staff officer, Baghdad, August 1, 2011.

[5] Brennan notes, 2009–2011.

tained three specific annexes to address these threats: Balancing Iranian Influence (BII), Countering Violent Extremist Organizations, and Managing Arab-Kurd Tensions.[6] Because each of these was part of a joint embassy-military planning document, each plan addressed political, economic, diplomatic, and security related activities that the U.S. government as a whole was to pursue. From the standpoint of the transition, the challenge was how to effectively transfer responsibility for the security related activities while, simultaneously, assisting in the broader political, economic, and diplomatic goals each of the plans established. Planners anticipated the embassy and its OSC-I component would lead U.S. efforts to encourage the government of Iraq to address security issues in a balanced manner that did not aggravate ethnosectarian tensions. But it was expected that, at the end of the transition, most U.S. security activities would either be terminated because there was no longer a need (there would no longer be a need to protect U.S. bases, for example) or transferred to the ISF. As anticipated in the SA, the Iraqis would assume responsibility for their own security.

Enduring Threats and Responses

The United States and Prime Minister Maliki did not always agree on the nature, source, or degree of the threats various groups posed. Likewise, there was also a lack of consensus within Iraq concerning these issues because of the continuing communal and factional struggle for power and resources we discussed earlier, in the section on drivers of instability. Because of Prime Minister Maliki's ties to Iran and sectarian affiliation, he often downplayed the threats Iran and Shi'a extremists groups posed while, simultaneously, overplaying the danger of Sunni extremist groups, especially those with ties to the former Ba'ath party. Only on the issue of al-Qaeda in Iraq did the U.S. government and the government of Iraq share a common perception of the threat.[7] This chapter is based on a U.S. view of the enduring threats to Iraqi security viewed in terms of Iranian influence, including Iran's support of Shi'a extremist groups, violent extremist organizations operating within Iraq, and Arab-Kurd tensions. As will be discussed later, differences between this analysis and that of Prime Minister Maliki and his allies within the government of Iraq significantly affected the transition and operational maneuver of forces out of Iraq.

Iranian Influence in Iraq

The 2003 overthrow of Saddam Hussein improved Iran's position in Iraq and also in the wider Arab world. Saddam Hussein's regime stood as a bulwark against Iran's

---

[6]   Most of the specifics of these three annexes remain classified. As a result, many of the activities and tasks that had to be transitioned will not be addressed in this book.

[7]   Brennan notes, 2009–2011.

regional ambitions. Saddam not only helped prevent the "export" of Iran's revolution to the Arab world but also sapped the Islamic Republic's power and resources by waging war against Tehran for eight long years. Iran was quick to recognize the benefits of the U.S. invasion and overthrow of Saddam; it helped Shi'a and Kurdish parties with which it had close ties to play key roles in Iraqi politics. Iran also expanded its influence by supporting Shi'a armed militias, by organizing attacks against U.S. troops, and by building economic ties to key regions, including southern Iraq. Tehran's overall objectives in Iraq have been to ensure that Baghdad is governed by a friendly, Shi'a-dominated regime; that Iraqi political entities be sufficiently divided to empower Tehran to be a power broker in Iraqi politics; that Iraq remain stable and weak to prevent it from posing a significant security threat to Iran; and that U.S. troops depart Iraq without maintaining permanent or even temporary military bases in Iraq.[8]

While Iraq and Iran share Shi'a Islam as the faith of the majority of the population, the age-old enmity between the Persians (Iran) and Arabs (Iraq) prevented close relationships. However, since 2003, Iran and Iraq have built close political and economic relations based on a mix of Iranian intimidation, inducements, and shared interests between Tehran and the Shi'a-led government in Baghdad. While the U.S. government did not welcome such developments, policymakers realized that geography, economics, geopolitics, and a shared religion would drive the two countries to develop mutually beneficial bilateral relations. Unfortunately, Tehran has often exploited the lack of capacity within the government of Iraq to advance Iranian interests at a high cost to the Iraqis. As will be discussed, nowhere is this more evident than Iranian military support to Shi'a extremist groups and Tehran's intrusive interference in Iraqi internal political affairs.[9]

Iran's current activities in Iraq are rooted in the 1979 Iranian Revolution. The leader of Iran's revolution, Ayatollah Ruhollah Khomeini, viewed Shi'a-majority Iraq as a prime target for exporting his revolution to the Arab world. Khomeini's ambition was rooted in his own experience; his many years of exile in Iraq had allowed him to build ties with Iraq's myriad Shi'a groups and senior religious leaders. Moreover, the Iraqi Shi'a seethed under the rule of Saddam Hussein, who saw his Sunni brethren as the foundation of his tyrannical regime. Nevertheless, Khomeini's vision of *velayat-e faqih* [guardianship of the jurist][10] was not popular among Iraqi Shi'a clergy, and many

---

[8] Interview with MNF-I staff officer responsible for developing the BII strategy, Baghdad, January 5, 2009. Also, information regarding the BII strategy and Iranian influence in Iraq is drawn from the author's notes taken while assigned to the Joint Interagency Task Force–Iraq (JIATF-I), September 2008 to October 2009.

[9] Brennan notes, 2009–2011.

[10] The concept of *velayat-e faqih* as practiced in Iran holds that the Supreme Leader is entrusted by Allah to serve as guardian of the faith, who holds custodianship over the people. The power of the Supreme Leader in Iran is not absolute but is shared by other leading bodies, particularly the Assembly of Experts and the Council of Guardians.

Iraqi Shi'a fought on behalf of the Sunni-dominated Iraqi government against Iran during the eight-year-long conflict between the two countries.

The Iranian regime did manage to recruit many Iraqis to form the Supreme Council of the Islamic Revolution in Iraq—a political party that is now known as the Islamic Supreme Council of Iraq (ISCI). While the Supreme Council of the Islamic Revolution in Iraq served primarily as a political organization, a subordinate group named the Badr Corps fought against Saddam in the 1980s alongside Iran's Revolutionary Guards. The first Supreme Council chief was the Iraqi-born Ayatollah Mohammad Hashemi Shahroudi, who went on to serve as Iran's Minister of Justice and is now a close advisor to Iran's Supreme Leader, Ayatollah Ali Khamenei. The current president of ISCI, Ammar al-Hakim, is the nephew of the group's founder, Ayatollah Mohammed Baqir al-Hakim, and maintains close contact with the Iranian Supreme Leader. It is also important to note that the current leader of the Badr Corps, Hadi al-Amiri, served alongside the IRGC-QF chief Major General Qassem Soleimani during the Iran-Iraq war and, as the current Minister of Transportation, continues to maintain a close professional relationship with his former comrade.[11] Similarly, as a member of the outlawed Islamic Dawa party, Nouri al-Maliki was forced to flee Iraq in 1979 and lived in Tehran from 1982 to 1990, where he also developed close ties with Iranians who are now key leaders in Tehran. It would be a mistake, however, to suggest that Iran only has ties to the Shi'a leadership in Iraq. Over the years, Tehran has cultivated relationships with Jalal Talabani, President of Iraq and Secretary General of the Patriotic Union of Kurdistan; Massoud Barzani, President of Kurdistan and leader of the Kurdistan Democratic Party; and Ayad Allawi, former Interim Prime Minister of Iraq and current leader of the largely Sunni political bloc, Iraqiya. In fact, USF-I analysts concluded that Iran uses its money to influence most Iraqi political leaders. Using a combination of bribes, intimidation, and other forms of persuasion, Tehran has emerged as a critically important part of the Iraqi political process. Indeed, one senior U.S. military official privately stated that all large decisions in Iraq are made only after consultation with Tehran and, often, with Tehran's tacit approval.[12]

Iranian relations with Iraqi opposition parties, which also included the Patriotic Union of Kurdistan, are not devoid of tensions. To be clear, Iraqi Shi'a and Kurdish parties do not serve as Iranian proxies, and the government of Iraq will, at times, resist pressures from Tehran. However, some of the most senior leaders in the government of

---

[11] According to a senior USF-I officer, in 2011 then–Brigadier General Soleimani was promoted to major general, the highest military rank in Iranian military, as a result of his "success" in Iraq. Interview with a senior U.S. military officer, Baghdad, December 12, 2011; Brennan notes, 2009–2011.

[12] Interview with senior U.S. military officer, Baghdad, August 30, 2010. Also, information regarding the BII strategy and Iranian influence in Iraq is drawn from the author's notes taken while assigned to JIATF-I, September 2008 to October 2009.

Iraq often find their individual political interests aligned with gaining the support of Tehran, bringing together an expedient partnership.[13]

Tehran's support of the Iraqi opposition to Saddam Hussein ultimately paid a big dividend.[14] The U.S. removal of Saddam's regime enabled Iran to exercise significant influence in Baghdad through ISCI, al-Dawa, and the Patriotic Union of Kurdistan, based on personal relationships built on years of working together toward a common goal. But Tehran did not stop at that; it cultivated other Shi'a political and insurgent groups that could satisfy its multiple objectives.

The Islamic Republic, fearing U.S. efforts to overthrow the regime in Tehran, encouraged compliant proxies to tie down U.S. forces in Iraq. Iran was able to use these groups to wage a covert war with the United States and to intimidate Iraqi leaders.[15] For example, ISCI and the Badr Corps were valuable political allies, but their inclusion in the government of Iraq starting in 2004 prevented them from actively attacking U.S. forces. As disorder increased in Iraq after the U.S. invasion of 2003, Iran began to provide support to Muqtada al-Sadr's JAM. By the end of 2004, Iran was providing arms, ammunition, IEDs, and rockets to JAM.[16] To facilitate training, Iran employed the services of its closest proxy, the Lebanese Hezbollah. IRGC-QF and Lebanese Hezbollah conducted military training at military sites inside Iran. Fighters would then return to Iraq, where they attacked not only U.S. forces but also the ISF. During the height of the war, these militias were also responsible for much of the violence targeting Sunnis and Christians living in Baghdad and were largely responsible for the ethnic cleansing that took place in Baghdad and the surrounding security belts.[17] As the military operations became more complex, Sadr established a number of so-called Special Groups that specialized in use of rockets, IEDs, assassination, and kidnapping. For its part, Iran played an important role as a benefactor for these Special Groups, the most important of which were Kata'ib Hezbollah and Asa'ib ahl al-Haq.[18] These groups were frequently used to attack U.S. troops while providing plausible Iranian deniability. This ability to dial up or dial down the level of violence in Iraq through the use of proxy militia forces will remain an important tool in Tehran's arsenal.

Iran has attempted to cement its influence in Iraq through economic, religious, and ideological activities. Iranian companies, especially those suspected of being

---

[13] Brennan notes, 2009–2011.

[14] It should be noted that the United States also supported many opponents of Saddam Hussein, most notably Ahmed Chalabi, who would later become one of the most vocal opponents of the United States.

[15] Rick Brennan, "Iran's Covert War in Iraq," *Washington Times*, March 15, 2007.

[16] Interview with MNF-I staff officer responsible for developing the BII strategy, Baghdad, January 5, 2009.

[17] Brennan notes, 2009–2011; interview with military officer who served in G5 (Plans) of MND-B from October 2006 to December 2007, Washington, D.C., December 31, 2012.

[18] Michael Knights, "The Evolution of Iran's Special Groups in Iraq," Combating Terrorism Center, November 1, 2010.

affiliated with Iran's Revolutionary Guards, play an important role in southern Iraq's economy. The tourism industry in the holy city of Najaf, for example, is dominated by Iranian pilgrims and tour companies that not only own the buses that take pilgrims to Najaf but also own the hotels, restaurants, and many of the shops the tourists frequent.[19] Thus, while Iranian tourism is adding to the Iraqi economy, the tourism industry is also a cover for Iranian movement of arms and munitions throughout Iraq and, more recently, to Syria, Lebanon, and other countries in the Levant.[20]

Iran's influence in Iraq continues to have a negative influence on Iraq's relations with other Arab countries in the Gulf. Following the 2003 U.S. invasion of Iraq, Iran began a concerted effort to enhance its relationship with Baghdad to extend its influence throughout the wider Arab world. The Sunni-dominated Arab countries in the region viewed this growing Iran-Iraq partnership as a threat. Both Iraq and Iran shared common concerns over the policies of Iran's primary regional rival, Saudi Arabia. For his part, King Abdullah of Saudi Arabia has long viewed Prime Minister Maliki as a puppet of Tehran and, until recently, has resisted all efforts to establish diplomatic relationship with Iraq.[21] Moreover, the Arab countries in the Gulf Cooperation Council fear increased power in the region and, consequently, not only shun Iran but are increasingly isolating Iraq because of its friendly relations with Iran.[22]

There are clearly limits to Iran's influence in Iraq. The Sunnis and Kurds in northern Iraq are wary of Iran, and Shi'a religious leaders in Najaf will continue to resist Khamenei's (and Qom's) influence.[23] More important, the Iraqi Shi'a may once again come to view themselves as the region's natural power. There is no reason to believe that, after decades of being oppressed by Saddam's Sunni regime, Iraqi Shi'a will be willing to suffer Iranian-Persian dominance for long. However, much will depend on Iraq's internal stability. A fractious and relatively weak Iraq best serves Tehran's interests. Therefore, there is no reason to expect Iran to cease supporting Shi'a proxy groups that can cause instability and, if needed, pressure Baghdad to fall in line with Tehran.

---

[19] Gabriel Gatehouse, "Najaf's Return as a Religious Tourist Destination," *BBC News*, February 27, 2010.

[20] Interview with USF-I J3 staff officer, Baghdad, June 18, 2012.

[21] Interview with USF-I staff officer, Baghdad, August 1, 2011. On February 21, 2012, the Kingdom of Saudi Arabia named its first Ambassador to Iraq in 20 years. While the Saudi diplomat will not be resident in Iraq, this is an important step toward rebuilding relationships with neighboring Arab countries. See Jack Healy, "Saudis Pick First Envoy to Baghdad in 20 Years," *New York Times*, February 21, 2012.

[22] Meeting with senior Iraqi official, Washington, D.C., January 8, 2013. The Gulf Cooperation Council is a political and economic union of the Arab states bordering the Arabian Gulf that includes Bahrain, Kuwait, Oman, Qatar, Saudi Arabia, and the United Arab Emirates. Both Jordan and Morocco have been invited to join the council.

[23] The city of Qom is considered a Shi'a religious holy site and is the center of religious learning within Iran. Since the Islamic Revolution in 1979, Qom has emerged as the largest center of Shi'a scholarship in the world, largely espousing the teachings of the late Ayatollah Khomeini. The other major center of Shi'a religious scholarship, in Najaf, Iraq, espouses the traditional teachings of Shi'a Islam.

Finally, it is important to note that, while the Iranian Ministry of Foreign Affairs manages relations between Tehran and every other government in the world, the IRGC-QF maintains sole jurisdiction over Iran-Iraq relations. Consequently, Major General Qassem Soleimani, the commander of the Quds Force, is and has been the Iranian Supreme Leader's representative for all issues relating to Iraq. Not only is he responsible for many of the Shi'a extremist acts of violence in Iraq, but he also directed all "soft power," including the tourism, Iranian *bonyads* (which provide charitable services), and the disbursement of funds to a wide range of Iraqi political leaders and political parties.[24] In a 2008 letter to Secretary of Defense Gates, General Petraeus characterized Soleimani as a "truly evil person." The general further lamented that the "most sobering surprise of the week was probably the extent of direct Iranian involvement in Iraqi political intrigue."[25] As U.S. forces prepared to depart Iraq, Iranian influence permeated the political, economic, and internal security aspects of the country, posing an enduring threat to the development of a stable, independent, and sovereign Iraq.[26]

U.S. Military Efforts—Balancing Iranian Influence

One of the biggest security challenges for the U.S. military in Iraq was stemming the flow of arms, ammunition, and funding from the IRGC-QF in Iran to Shi'a militants. The porosity of the Iran-Iraq border, the corruption of Iraqi border guards, the historic smuggling routes, and the unwillingness of the government of Iraq to sustain border security all contributed to an ample flow of arms and munitions moving into Iraq. Compounding this challenge was the fact that the U.S. military in Iraq was precluded from taking any actions against the extremist training sites and munitions depots the IRGC-QF operates inside Iran.[27] In essence, the IRGC-QF and its proxies were able to stage operations from Iran with impunity, unless the U.S. government decided to respond to Iranian aggression by attacking training and logistics bases within Iran.[28] While there were those who favored such a response, neither the Bush nor the Obama administration desired to risk expanding the war.[29]

---

[24] Brennan notes, 2009–2011.

[25] As reported by Michael R. Gordon, "Iran's Master of Chaos Still Vexes U.S.," *New York Times*, October 2, 2012.

[26] Internal USF-I assessment developed in July 2011; Brennan notes, 2009–2011.

[27] The wartime authorities given to the commander of MNF-I and later USF-I authorized military actions only within the contiguous borders of Iraq, in accordance with policy and international law. Interview with USF-I SJA staff officer, Baghdad, January 21, 2011.

[28] Interview with USF-I J35 staff officer, Baghdad, December 10, 2011.

[29] Brennan notes, 2009–2011. It should be noted, however, that the decision not to take military action to stop Iranian-sponsored violence targeting U.S. military and diplomatic personnel and facilities was an action in and of itself. Indeed, some members of the USF-I staff argued that the failure to act was enabling Iran to sponsor increasingly more aggressive actions, knowing that the United States would not respond militarily.

In an attempt to counter the perceived growing influence of Iran within Iraq, MNF-I and Embassy Baghdad developed a joint strategy to balance Iranian influence in November 2008. This strategy sought to alter the power relationship between Baghdad and Tehran to help Iraq emerge as an equal partner with Tehran. The strategy accepted that Iran would have influence in Iraq, as any neighboring country would. However, the U.S. goal was that the relationship would be mutually beneficial and that Prime Minister Maliki would demand that Tehran cease nefarious activities within Iraq.[30] The goals of the BII strategy included the following:

- Iraq would establish a constructive bilateral trade relationship with Iran that would balance Iranian soft-power gains and would promote Iraqi equities as part of a mutually beneficial relationship.
- The competency, legitimacy, political accommodation, and public support of Iraq civil and governmental institutions would be increased, independent of Iranian influences.
- Iraq would provide its own internal security through competent and increasingly professional ISF and security ministries governed by the rule of law who have both the willingness and capacity to protect Iraqi sovereignty from extremist groups supported by Iran.
- In the middle to long term, Iraqi independence from outside influences, economic development, and movement toward a stable democracy would serve as an example of Arab modernization that will be promulgated throughout the region.[31]

In retrospect, these goals and objectives reflected a U.S. vision for how Iraq would interact with Iran. In private discussions, Prime Minister Maliki would sometimes agree with these goals, but he and some of the most senior ministry leaders became increasingly less interested in conducting any action that would alienate Tehran as the deadline for the departure of U.S. forces approached. While the government of Iraq was happy to accept U.S. assistance, it was rarely willing to take actions that would jeopardize its relations with Tehran.[32] In this area, perhaps more than any other, U.S. and Iraqi interests diverged.

---

[30] These observations and some others later are from contemporaneous notes author Richard Brennan took while working in JIATF-I, Baghdad, September 2008 to October 2009. Whenever the situation in Iraq started to threaten Prime Minister Maliki personally, he would be willing to stand up to Tehran. However, the prime minister was largely unwilling to expend any political capital with Iran to curtail Iranian sponsorship of Shi'a extremist organizations targeting U.S. forces and diplomats. This intransigence came to a head in January 3, 2011, when Kata'ib Hezbollah used an improvised rocket-assisted mortar to target a U.S. base located in southern Iraq about 3 km from the Iranian border. The threat of improvised rocket-assisted mortar attacks persisted until the end of the operation, forcing USF-I to take more aggressive unilateral action to ensure force protection.

[31] Notes taken by author while working within JIATF-I, Baghdad, September 2008 to October 2009.

[32] Brennan notes, 2009–2011.

A key component of the BII strategy was to strengthen the ISF's capacity to target extremist groups funded by Iran. In addition, a key goal of the strategy was to help develop the Iraqi political will to stand up to Iran and effectively use its security forces to protect its sovereignty as it relates to extremist groups. However, it is important to note that the BII strategy focused less on taking active steps to counter Iranian influence and more on strengthening Iraqi legitimacy, security, governance, and economic strength as ways of enhancing Iraq's ability to resist Iranian influence by itself. The logic was that, if Iraqi strength and self-sufficiency were the primary bulwarks against Iranian power projection in Iraq, the serial political crises, uneven economic development, and security shortcomings that Iraq faced opened the door to greater Iranian influence. Finally, an important aspect of the strategy was a realization that a free, democratic, and economically prosperous Iraq would provide a positive model of how Shi'as could govern in a manner that was not repressive. In essence, as Iraq grew stronger, it would exert its influence across the border into Iran.[33]

Finally, because the BII strategy was drafted at the same time that 2008 SA was being drafted, the emphasis shifted from U.S. unilateral actions targeting Shi'a extremists to partnered operations and then eventually to advise-and-assist operations. We will address the challenges associated with this transition later.

BII Transition: From U.S. Forces–Iraq to the U.S. Embassy

Annex L of the JCP outlined 43 tasks related to BII, each of which was tied to U.S. objectives in the political, economic and energy, rule of law, and security spheres. One-third (15) of the tasks were transferred to the embassy—primarily its political and economic sections—with the signing of the 2010 JCP, and two were terminated. USF-I worked to transfer or transform the remaining 26 BII tasks to the embassy, USCENTCOM, and the government of Iraq. In reality, a number of the tasks relating to targeting Shi'a extremist groups Iran supports would be terminated as part of the transition because of Prime Minister Maliki's reluctance to take actions that would both antagonize Iran and likely alienate elements of his Shi'a political base.[34]
Most BII tasks had been managed by the USF-I J9—the Directorate for Strategic Effects— which managed the command's strategic communication, public affairs,

---

[33] In 2006, MNF-I developed the "Counter Malign Iranian Influence Strategy," designing it to target the smuggling routes, funding, weapon caches, and Shi'a extremist organizations Iran was supporting. By the end of 2008, the Shi'a extremist threat had largely been eliminated, and MNF-I developed a "cross LOO strategy" designed to balance Iranian influence. At the macro level, the BII strategy mirrored the broader JCP, but specific activities undertaken at the tactical level focused directly on countering Iranian actions in Iraq. However, since the authorities of MNF-I/USF-I only extended to the Iran-Iraq border, the BII strategy did not address any action or activity that occurred within Iran itself. Interview with USF-I J5 staff officer, Baghdad, August 15, 2011.

[34] USF-I, J35, "J35 Transition Plan Presentation to RAND," Baghdad, July 1, 2011, slide 4, Not available to the general public.

and political and economic outreach.[35] Given that the BII strategy focused primarily on strengthening Iraq so it could push back against Iranian influence, the list of tasks included a number of initiatives to support Iraqi diplomacy, by promoting Iraqi involvement in regional affairs, by resolving bilateral Iraqi-Iranian disputes, and by promoting good governance. However, the list also included quite a few tasks by which the embassy would help Iraq counter Iranian activities directly. The following is a partial list of BII tasks that either were completed by USF-I or were transitioned to the government of Iraq, Embassy Baghdad, USCENTCOM, or to other U.S. agencies as specified in the 2010 JCP[36]:

- Governance tasks
  - The government of Iraq competently conducts legitimate and credible elections that Iraqis accept.
  - The government of Iraq maintains good relations with provincial and local governments to reinforce public legitimacy.
  - Advise and assist the government of Iraq to strengthen oversight of banks, financial institutions, charities, *hawalas*,[37] and other institutions conducting financial transactions in Iraq.
- Support Iraqi Diplomacy tasks
  - Iraq conducts and expands functioning diplomatic relations with regional states, including Sunni Arab states.
  - Iraq actively participates in regional and international organizations, institutions, and processes and with nongovernmental organizations.
  - Assist Iraq in negotiating mutually beneficial agreements for cross-border and gas fields with Iran.
  - Iraq and Iran establish procedures to peacefully resolve waterway and border disputes.
  - Water rights are negotiated with Iran, and agreements are established to ensure that Iran does not divert water and deprive Iraq of the water it needs for its people and economy.
- Counter Iran Directly tasks
  - Strengthen government of Iraq capacity to develop, secure, and oversee its customs and border measures to detect Iranian illegal smuggling of lethal and nonlethal material.

---

[35] See U.S. Department of the Army, "Biography of USF-I Spokesman and Director for Strategic Effects Major General Stephen R. Lanza," April 2010.

[36] USF-I, J35, 2011, slides 5–6.

[37] The Arabic *hawala* means *transfer* and is an informal value transfer system that involves a large network of money brokers located throughout the Middle East, North Africa, Horn of Africa, the Indian subcontinent, and other places around the world. It is widely used alternative banking and financial system that is not subject to governmental regulations or oversight.

- Secure government of Iraq enforcement of UN Security Council economic sanctions against Iranian banks and Iranian entities designated as terrorist organizations by the U.S. Treasury.
- Advise and assist the ISF to arrest and prosecute Iranian-sponsored violent actors and facilitators according to Iraqi law.
- The government of Iraq exerts diplomatic pressure on Iran to end lethal aid to Iraqi militias.
- The government of Iraq identifies and blocks Iran's attempts to unlawfully influence government officials.

Without doubt, the U.S. ability to pursue its BII goals diminished after the U.S. military's departure. During OND, the U.S. military transitioned many of its security activities from partnered U.S.-Iraqi operations to enabled operations, with the ISF in the lead as conditions evolved. However, it is important to note that USF-I continued to conduct a limited number of independent operations targeting Iranian-backed militias that posed an imminent threat to U.S. facilities and/or personnel through the end of the mission in December 2011.[38]

What is unclear at this point is the degree to which Iraq will be able to exert its sovereignty in relation to Iranian influence. Will Iraq be able play a major role in Arab world politics? Or will Iraq remain a weak state that is generally acquiescent to Tehran? Clearly, there are outcomes between these two opposite poles. However, how the Sunni-Arab states in the region view Tehran, it is unclear how Iraq can mend its relations with other neighboring states without first distancing itself from Tehran, an option that Prime Minister Maliki has been unwilling to take. Moreover, Iraq's relationship with Iran may well undermine U.S-Iraqi relations, threatening the loss of congressional support for continued funding of important security cooperation programs designed to strengthen the ISF.

Violent Extremist Organizations

Throughout the U.S. military presence in Iraq, U.S. troops, Iraqi forces, and Iraqi civilians were targeted by a diverse assortment of violent extremists. Despite repeated Iranian assurances that it would stop arming, funding, and training these groups, there is no evidence to suggest that this transformation occurred. According to DoS,

> [t]errorist groups such as al-Qaeda in Iraq committed attacks against a wide swath of society, including Sunnis, Shia, and members of other sects or ethnicities, security forces, places of worship, religious pilgrims, economic infrastructure, and government officials . . . . Certain militant organizations, such as those influenced by

---

[38] Brennan notes, 2009–2011.

Iran, also committed numerous terrorist attacks, primarily against foreign embassies and military forces.[39]

At the time of USF-I's departure in December 2011, the five most dangerous extremist groups operating in Iraq were the Promised Day Brigade, Asa'ib al-Haq, Kata'ib Hezbollah, JRTN, and al-Qaeda in Iraq. The following subsections discuss these groups.

Promised Day Brigade

The Promised Day Brigade was formed in November 2008 as an offshoot of Muqtada al-Sadr's JAM. The organization's primary purposes are disrupting U.S. security operations and destabilizing the political process in Iraq.[40] Iran provided the Promised Day Brigade with significant amounts of funding and training through 2011. However, the relationship between Sadr and Tehran has always been tenuous, and Tehran viewed him as an unreliable partner who would often seek his own political purposes rather than adhere to the desires of Tehran.[41] Nowhere was this more evident than during the 2010 government-formation process, when Sadr withheld support for Maliki and entered into dialogue with Ayad Allawi (Iraqiya), Ammar Al Hakim (ISCI), and Massoud Barzani (Kurdistan Democratic Party) in an effort to find an alternative to Maliki as the prime minister. Although this effort to find an acceptable replacement for Maliki failed, it demonstrated that Sadr and his political party were willing oppose the wishes of Tehran.[42] While Sadr, who sits at the top of the Promised Day Brigade's leadership, is more than willing to accept funding and material support from the IRGC-QF, he has also worked hard to ensure that his militia remains first and foremost loyal to him and his movement. Tehran, for its part, continues to provide funding at lower levels to retain some degree of influence.[43]

Although Sadr denounced sectarian violence in 2007, JAM and the Promised Day Brigade were nonetheless linked to attacks on U.S. forces. In June 2011, the group issued a statement claiming responsibility for a total of 52 attacks against U.S. troops.[44] In July 2011, after urging his followers to cease attacks against Americans,[45] Sadr charged the Promised Day Brigade with fighting U.S. troops if they remained in

---

[39] DoS, "Country Reports in Human Rights Practices for 2011," website, April 24, 2012a.

[40] Stanford University, "Promised Day Brigades," Mapping Militant Organizations website, August 27, 2012b.

[41] Interview with USF-I J35 staff, Baghdad, August 1, 2011.

[42] Brennan notes, 2009–2011.

[43] Interview with USF-I staff officer, Baghdad, December 10, 2011.

[44] Stanford University, 2012b. It should be noted that such claims are impossible to verify and that several Shi'a extremist groups would often claim responsibility for the same attack in an effort to enhance their reputation as an organization that was fighting to force the departure of U.S. forces.

[45] "Iraq's Sadr Says to Halt Attacks on U.S. Troops," Thomson/Reuters, September 11, 2011.

Iraq after their scheduled withdrawal on December 31, 2011,[46] even stating that U.S. embassy employees are "occupiers" who should be resisted.[47] While some members of the Promised Day Brigade appeared to collaborate with Kata'ib Hezbollah and Asa'ib al-Haq on attacks against U.S. forces, the brigade has also fought against these groups, such as when they came to blows over influence in and around Sadr City in 2009. As of June 6, 2011, the Promised Day Brigade was estimated to have about 5,000 members.[48]

Asa'ib al-Haq

From 2003 through 2005, the IRGC-QF focused its efforts on supporting Muqtada al-Sadr's JAM. However, because JAM grew so fast, it was disorganized, uncontrolled, and unreliable. Starting in 2004, the IRGC-QF enlisted the assistance of its strongest proxy—Lebanese Hezbollah—to train Iraqi Special Groups to use explosively formed penetrators and rockets and to conduct kidnapping.[49] Asa'ib al-Haq [League of the Righteous] was formed in July 2006 specifically to create a popular organization that would appeal to the masses as an alternative to JAM. The IRGC-QF handpicked one of Sadr's primary rivals, a protégé of Sadr's father, Qais al-Khazali, to lead this new group because he had consistently opposed Sadr's cease-fire agreements with both the U.S. and Iraqi military.[50] Tehran also selected an experienced member of Lebanese Hezbollah, Ali Mussa Daqduq, to facilitate the training of this new organization. In June 2006, IRGC-QF named Qais Khazali to head all Special Groups in Iraq and, simultaneously, named Daqduq as his chief advisor and primary liaison with the IRGC-QF.[51]

On January 20, 2007, Asa'ib al-Haq raided the Karbala Provincial Joint Coordination Center (PJCC). To break through security, Asa'ib al-Haq gunmen wore U.S. military uniforms, drove American sport utility vehicles and Humvees, and carried counterfeit identification cards.[52] This was the most complex operation ever executed by an Iranian-backed Special Group,[53] resulting in the kidnapping and murder of five U.S. soldiers. Intelligence collected after the attack indicated that Qais Khazali, his brother Laith Khazali, and Daqduq were all involved in the attack. On March 20,

---

[46] Stanford University, 2012b.

[47] "Iraq's Anti-US Cleric Considers US Embassy Employees Occupiers," *China Daily*, October 23, 2011.

[48] "June Deadliest Month for U.S. Troops in 2 Years," *USA Today*, June 30, 2011.

[49] Knights, 2010.

[50] Knights, 2010.

[51] Elizabeth O'Bagy and Stephen Wicken, "Fact Sheet: Ali Mussa Daqduq," Institute for the Study of War, May 14, 2012.

[52] O'Bagy and Wicken, 2012.

[53] Interview with USF-I staff officer, Baghdad, December 10, 2011.

2007, all three were captured.[54] However, both Khazali brothers were later released as part of a reconciliation effort between the government of Iraq and Shi'a militia groups. Although Qais Khazali assured the government of Iraq that he would remain in Iraq and enter the political process after his release, he fled to Lebanon immediately after securing his freedom. Since his release from custody, Qais revitalized Asa'ib al-Haq and, during 2011, was responsible for multiple attacks against U.S. forces.[55] By June 2011, USF-I assessed Asa'ib al-Haq to be a reliable proxy of the IRGC-QF, second only to Kata'ib Hezbollah in both loyalty to the Supreme Leader and technical capability.[56] Daqduq was the last prisoner released to the government of Iraq. While the United States sought to extradite Daqduq for the murder of American soldiers, Prime Minister Maliki—with pressure from Qassem Soleimani—refused to make this transfer.[57] Although Prime Minister Maliki continually assured the United States that Daqduq would remain in an Iraqi prison for his role in organizing the PJCC attack and subsequent murders, he was released from prison and returned to Lebanon on November 17, 2012.[58]

Kata'ib Hezbollah

The origin of Kata'ib Hezbollah [Brigades of the Party of God] was markedly different from other Shi'a groups. It is widely viewed that the leader of Kata'ib Hezbollah is Abu Mahdi al-Muhandis, whose real name is Jamal al-Ibrahimi.[59] Muhandis was born in Basra and was later forced to flee to Iran during the early 1980s for being a member of the outlawed Dawa party. While in exile, Muhandis became a member of the Badr Brigade, eventually rising to become its deputy commander in 2001, serving under its commander, Hadi al-Amiri. During his early years with the Badr Brigade, Muhandis participated in the bombings of the French and American embassies in Kuwait in

---

[54] O'Bagy and Wicken, 2012. Much of the intelligence collected at the scene of the PJCC attack has been declassified so that it could be used in Iraqi criminal proceedings. One item found was a computer disk containing the entire plan of attack that, when printed, was 31 typewritten pages long. On April 4, 2007, MNF-I produced a translation of the file for use in prosecution within the Iraqi Criminal Court system as Document Number NMEC-2007-624036, "Detailed Description of Preparations and Logistics and Support That Were Deployed in the Operation Against the Karbala PRCC Headquarters." A copy of that report was made available to RAND researchers for the purpose of this study.

[55] Brennan notes, 2009–2011.

[56] Brennan notes, 2009–2011.

[57] "Iraq Releases Suspected Hezbollah Operative Daqduk," Reuters, November 16, 2012.

[58] It should be noted that the United States declined to move Daqduq to Guantanamo or a similar facility when he was under the control of MNF-I in 2007. Having decided not to make hard political choices when Daqduq was under U.S. military control, the United States then asked the government of Iraq to make an extremely hard choice knowing that it would likely result in a strong response from Iran. Interview with a former member of the NSC Staff, Washington, D.C., December 26, 2012.

[59] "Abu Mahdi and Iran's Web in Iraq," United Press International, October 7, 2010.

1983. After these bombings, Muhandis fled to Tehran with the Badr Brigade. Then, in 1985, he returned to Kuwait, where he attempted to assassinate the emir. Muhandis was later convicted in absentia by a Kuwaiti court for these crimes.[60] After the fall of Saddam Hussein, he briefly returned to Iraq in 2003 but quickly fled back to Tehran to avoid U.S. capture.[61] Muhandis retains a close personal and professional relationship with Hadi al-Amiri, the current Iraqi Minister of Transportation, as well as with Prime Minister Maliki.[62] The successes of the Badr Brigade as a military organization under the control of Iran resulted in an expansion of the organization to a division and eventually to a corp size element. As the deputy commander of Badr Corps, Muhandis developed a close relationship with the IRCG-QF commander Qassem Soleimani, for whom he now serves as close advisor.[63] As a result of his position with the IRGC-QF and his relationship with multiple Iraqi political leaders, he served as the interlocutor between General Soleimani and Prime Minister Maliki from 2005 until the departure of U.S. forces in 2011.[64]

The IRGC-QF organized Kata'ib Hezbollah in early 2007 to employ its most experienced and trusted Iraqi operators, using the most sensitive Iranian equipment and weaponry.[65] Until 2011, Kata'ib Hezbollah was the only organization known to have been given improvised rocket-assisted mortars for use against U.S. forces.[66] Unlike other violent Iraqi Shi'a extremists, members of Kata'ib Hezbollah swear an oath of fealty to the Iranian Supreme Leader and accept the Iranian vision of *velayat-e faqih*. Consequently, Kata'ib Hezbollah, like Lebanese Hezbollah, is used as a tool to "export the Islamic revolution" as practiced in Tehran. Finally, while Kata'ib Hezbollah does not have a direct relationship with a specific Iraqi political party, USF-I staff officers asserted that members of the group are believed to have a familial relationship with members of Badr Organization, a political party in Iraq that was constituted after the

---

[60] Knights, 2010.

[61] Muhandis was elected to serve as a Member of Parliament in March 2006. By Iraqi law, Members enjoy immunity from prosecution. However, according to Muhandis, the U.S. military made it clear that it would not honor his claim of immunity, and he therefore decided to remain in Iran until the departure of U.S. forces. See Thomas Strouse, "Kata'ib Hezbollah and the Intricate Web of Iranian Military Involvement in Iraq," *Terrorism Monitor*, Vol. 8, No. 9, March 4, 2010.

[62] Interview with USF-I staff officer, Baghdad, December 10, 2011.

[63] Knights, 2010.

[64] Interview with USF-I staff officer, Baghdad, December 10, 2011.

[65] Knights, 2010.

[66] In June 2011, an internal USF-I assessment concluded that Asa'ib al-Haq was responsible for at least one improvised rocket-assisted mortar attack in 2011, confirming the assessments that Asa'ib al-Haq was becoming a trusted agent of the IRGC-QF. Brennan notes, 2009–2011.

U.S. invasion of Iraq.[67] This relationship made targeting the organization politically toxic for Prime Minister Maliki.[68]

The exact size of Kata'ib Hezbollah's membership is unknown, although a 2010 U.S. government estimate put it at between 400 and 1,000.[69] DoS has stated that the group is "almost entirely dependent on support from Iran."[70] USF-I staff officers took that one step further and asserted that Kata'ib Hezbollah, like Lebanese Hezbollah, "should be considered a direct action arm of the Quds Force."[71] On July 2, 2009, the U.S. government formally designated Kata'ib Hezbollah to be a foreign terrorist organization.[72]

Jaysh Rijal al-Tariq al-Naqshabandi

JRTN was founded by Izzat Ibrahim al-Duri, who had been Iraq's vice president under Saddam Hussein. Al-Duri announced the creation of the organization shortly after the execution of Saddam on December 30, 2006. The movement immediately tapped into Sunni fear of the Shi'a-led government and mistrust of the Kurdish population; however, JRTN vehemently rejected the violent tactics of al-Qaeda in Iraq. JRTN was a relatively exclusive organization that relied heavily on former elite members of the Iraqi military, such as the Special Republican Guard. Membership in JRTN was tightly controlled through a formal vetting process using tribal networks that the former regime had established to verify the credentials of applicants. Building on tribal networks, JRTN has significantly expanded its area of influence since its establishment, developing operational capability in Kirkuk, Hawijah, Ramadi, Fallujah, Northern Diyala, Lake Hamrin, Rashad Abu Ghurayb, and Salah al-Din.[73]

The publicly stated goal of JRTN was to expel the United States from Iraq. From the very beginning, JRTN tailored its message to provide the greatest appeal to the Sunni population it needed for support in its operational areas. For example, JRTN

---

[67] Interview with USF-I staff officer, Baghdad, December 10, 2011. After the collapse of the Iraqi Army in 2003, the Badr Corps dissolved, and its members returned to Iraq. Over time, many of its fighters became members of the Iraqi Army and Iraqi police. The leadership of the Badr Corps established a political party known as the Badr Organization. The political wing of Badr Corps, ISCI, registered in Iraq as a separate political party. From 2003 to 2009, ISCI/Badr was viewed as one party. However, since the 2010 national elections, there have been fissures within that organization, and ISCI and Badr often pursue different agendas. While ISCI has demonstrated a willingness to take positions contrary to the wishes of Tehran when its interests differ, Badr has consistently remained in line with goals and objectives established in Tehran.

[68] Brennan notes, 2009–2011.

[69] DoS, "Foreign Terrorist Organizations," in *Country Reports on Terrorism 2011*, July 31, 2012b.

[70] DoS, 2012b.

[71] Brennan notes, 2009–2011.

[72] DoS, 2012b.

[73] Michael Knights, "The JRTN Movement and Iraq's Next Insurgency," Combating Terrorism Center, July 1, 2011.

videos frequently focused on the concerns of the Sunni population, which feared Iranian influence, abuse of power by the Shi'a-led government of Iraq, and Kurdish actions in the "occupied territories" of the disputed areas.[74] JRTN's strong religious affiliation gave it credibility in the eyes of the population, and its policy of only attacking foreign forces enabled it to maintain popular support.

Al-Qaeda in Iraq

Al-Qaeda in Iraq, also known as the Islamic State of Iraq, was established in 2004 by Jordanian-born Abu Mus'ab al-Zarqawi, who immediately pledged the group's allegiance to Osama bin Laden. Although al-Qaeda in Iraq initially targeted coalition forces and foreign companies operating in Iraq, it soon began to use violence as a means of pressuring the Iraqi populace to stop supporting either the U.S. government or the nascent Iraqi government. From the organization's inception, there was a large debate concerning how many foreign fighters the organization had operating within Iraq. The Counterterrorism Center at West Point uncovered records that indicated that at least 700 foreign fighters joined al-Qaeda in Iraq during 2006 and 2007. According to the center, the surge of foreign fighters occurred at the same time that al-Zarqawi began to employ brutal tactics, such as recording the beheading of captives.[75] On June 7, 2006, a U.S. air strike killed al-Zarqawi near the city of Baqubah, just north of Baghdad.[76] By 2007, al-Qaeda in Iraq's violence and brutality targeting Sunni civilians not only weakened support for the organization but also led to a widespread backlash against it. The ensuing Sunni Awakening began at about the same time as the surge of coalition forces. The decreasing level of public support, opposition from Sunni tribal leaders, increase in coalition forces, and changed coalition tactics all helped significantly weaken al-Qaeda in Iraq.[77] By the end of 2008, al-Qaeda in Iraq was in disarray, and the level of violence it conducted was at historic lows.[78]

On June 30, 2009, U.S. forces fulfilled the SA commitment to withdraw forces from Iraqi cities. Throughout the remainder of 2009, al-Qaeda in Iraq conducted a number of high-profile attacks in an effort to demonstrate its continued relevance. On August 15, 2011, the organization conducted a coordinated attack targeting 17 Iraqi cities, killing more than 80 people and injuring hundreds more.[79] On September 20, 2011, however, MG David Perkins, the CG of U.S. Division–North, described the

---

[74] Knights, 2011.

[75] Joseph Felter and Brian Fishman, "Al Qaeda's Foreign Fighters in Iraq: A First Look at the Sinjar Records," West Point, N.Y.: Combating Terrorism Center, December 19, 2007.

[76] Ellen Knickmeyer and Jonathan Finer, "Insurgent Leader Al-Zarqawi Killed in Iraq," *Washington Post*, June 8, 2006.

[77] National Counterterrorism Center, "Al-Qa'ida in Iraq (AQI)," *Counterterrorism 2013 Calendar*, 2012.

[78] Interview with USF-I staff officer, Baghdad, December 10, 2011.

[79] National Counterterrorism Center, 2012.

decline of al-Qaeda in Iraq's capabilities as being "dramatic" and further characterized the terrorist network as a "system coming apart."[80] This assessment notwithstanding, al-Qaeda in Iraq continued to maintain the capability to conduct high-profile coordinated attacks throughout Iraq. The continuing threat from this terrorist organization was clearly demonstrated on July 23, 2012, when it initiated an offensive involving a sequence of 40 attacks in 16 Iraqi cities in one-third of the Iraqi provinces that left more than 100 dead.[81] In a June 2011 internal assessment, the USF-I J3 concluded that al-Qaeda in Iraq would remain an enduring threat to Iraq and would remain capable of "conducting coordinated high-profile attacks at multiple locations nearly simultaneously in an attempt to reignite ethnosectarian violence, undermine the Iraqi government and the ISF security efforts, and tarnish Iraqi public perceptions of stability in Iraq." The assessment concluded, "absent USF-I support and assistance to the Iraqi military and ISOF to target al-Qaeda in Iraq leadership, the organization is likely to increase in strength, although not to its 2006 level." Providing this continued assistance to the ISF and ISOF more specifically was a central mission envisioned for a post-2011 residual U.S. military presence in Iraq.[82]

U.S. Military Efforts to Defeat Violent Extremist Organizations

A key objective of the U.S. military was to help Iraq defeat violent extremist organizations. Indeed, a central element of the COIN strategy was to protect the Iraqi populace while building the capacity of the ISF and of the government of Iraq as a whole to help establish the legitimacy of that government. It is important to note that the role the U.S. military played changed dramatically over time. At the height of the Iraqi civil war, U.S. forces routinely conducted unilateral operations using both conventional forces assigned to MNF-I and U.S. Special Operations Forces (SOF) operating out of Balad Air Base, north of Baghdad. Two SOF task forces were employed in a manner designed not only to target individual extremists but also to destroy their networks and undermine the legitimacy of extremist organizations with the local populace. Task Force–16 focused on the Sunni extremist threat, while Task Force–17 targeted Iran-supported Shi'a extremist groups.[83] While U.S. forces conducted both partnered and unilateral operations through the end of 2008, the signing of the SA marked a turning point in how U.S. forces were employed, transitioning from partnered operations to advise-and-assist missions designed to enable the ISF to conduct these missions on its

---

[80] SIGIR, 2011f, p. 57.

[81] Yasir Ghazi and Rod Nordland, "Iraq Insurgents Kill at Least 100 After Declaring New Offensive," *New York Times*, July 23, 2012.

[82] Brennan notes, 2009–2011.

[83] Brennan notes, September 2008 to October 2009. For a detailed description of how Task Force–16 and Task Force–17 were used in conjunction with conventional forces to target extremist organizations, see McChrystal, 2013.

own. Throughout this period, the U.S. military worked with the government of Iraq in an effort to enhance border control, an effort that was largely ineffective because of corruption and lack of political will on the part of local-, provincial-, and national-level governments.[84] USF-I also worked closely with the embassy and the U.S. Department of Justice (DoJ) in an effort to target funding sources, both within Iraq and internationally. For example, in 2010, DoJ designated the IRGC and the IRGC-QF as terrorist organizations, freezing their international assets and making it more difficult to exchange funds using international banking institutions.[85] The same type of action was taken against Kata'ib Hezbollah in 2009.[86] Finally, USF-I helped the ISF collect and analyze releasable intelligence that could be used to target extremist organizations.[87]

Military Operations: From Lead Role to Supporting Role

As discussed earlier, the 2008 SA was a turning point in U.S. military operations against terrorists and insurgents; as a means of asserting Iraq's sovereignty, the document placed legal restrictions on what U.S. forces could do independently. The agreement stated that all military operations must be "fully coordinated with Iraqi authorities" and carried out with the agreement of the government of Iraq and in full compliance with Iraqi laws. Except in the case of combat operations, the agreement prevented U.S. forces from searching buildings or detaining people without an Iraqi judicial warrant and required U.S. forces to hand over any detainee within 24 hours.[88] As two U.S. Army attorneys wrote in the *Military Review*, "U.S. forces in Iraq have largely transitioned from intelligence-driven, unilateral combat operations to warrant-based operations led by Iraqi security forces."[89]

The Iraqi government's willingness to undertake counterterrorism operations, however, was greatly colored by political agendas and sectarianism. Freedom House's 2009 report on Iraq stated that "judges have come under immense political and sectarian pressure and have been largely unable to pursue cases involving organized crime, corruption, and militia activity, even when presented with overwhelming evidence."[90] The Shi'a-led Iraqi government was reluctant to take unilateral action against—or pursue warrants for—Shi'a militants, both because it did not want to alienate its own political base and because it did not want to antagonize Iran, which supported many of the militant groups. At the same time, it targeted Sunni militants aggressively, even

---

[84] Brennan notes, 2009–2011.

[85] DoS, "Treasury Targets Iran's Islamic Revolutionary Guard Corps," press release, February 10, 2010a.

[86] DoS, "Designation of Kata'ib Hizballah as a Foreign Terrorist Organization," press release, July 2, 2009a.

[87] SIGIR, 2011f, p. 57.

[88] Security Agreement, Articles 4 and 22.

[89] Mike Ryan and Jason Coats, "The U.S.-Iraqi Security Agreement and the Changing Nature of U.S. Military Operations in Iraq," *Military Review*, September–October 2009, p. 48.

[90] Freedom House, "Iraq," *Freedom in the World 2009*, 2009.

issuing arrest warrants for hundreds of Awakening members it accused of violence against Shi'a.[91] This selective pursuit of insurgents based on sectarian affiliation demonstrated that the Maliki government would use the security apparatus and judicial system to reinforce Shi'a dominance, a trend that highlighted for Sunnis the stakes of their exclusion from power.

Without authority to undertake unilateral counterterrorist or COIN operations after the implementation of the SA in January 2009 and without a troop presence in Iraqi cities after June 2009, the U.S. military increasingly acted in support of the ISF.[92] Beginning in April 2009, BCTs were designated as AABs and trained to support Iraqi forces. As then–USF-I commander General Odierno wrote in *Army Magazine*, "in line with the change of mission [to stability operations], our AABs are primarily focused on partnering with their ISF counterparts and building civil capacity, yet they retain the combat power necessary to defend themselves and their interagency partners."[93] When OND began on September 1, 2010, six AABs and one advise-and-assist task force (brigade headquarters) formed the core of the 50,000-strong force that remained in Iraq to continue the development and professionalization of the ISF.[94] One of the key roles of U.S. forces was to support Iraqi operations with critical enablers that included logistics and sustainment, ISR collection and fusion, and mission planning. Perhaps more important, U.S. forces served as an impartial "honest broker" to encourage their Iraqi counterparts to pursue their mission in a manner that did not exacerbate ethnic, sectarian, or regional disputes. This mediation role of the U.S. military was embraced by the sergeant and lieutenant at the proverbial "pointy end of the spear," as well as by General Austin, who often mediated high-level disputes that had the potential to flare into broader conflict.[95]

Elements of the AABs coordinated with local ISF counterparts—including Iraqi Army, police, and border security units—to identify requirements that could then be met through U.S. training, equipment, and support.[96] AABs provided assistance in a wide range of skills, including targeting, intelligence fusion, police forensics and inves-

[91] See also "Bad Blood Again," *The Economist*, April 2, 2009.

[92] Sydney J. Freedberg, Jr., "In Iraq, Combat Turns into Advise and Assist," *National Journal*, December 5, 2009.

[93] Raymond Odierno, "Operation New Dawn: Building a Long-Term Strategic Partnership Through Stability Operations," *Army Magazine*, October 2010, p. 98. See also Kate Brannen, "Combat Brigades in Iraq Under Different Name," *Army Times*, August 19, 2010.

[94] Ernesto Londoño, "Operation Iraqi Freedom Ends as Last Combat Soldiers Leave Baghdad," *Washington Post*, August 19, 2010. See also Ray Odierno, "Statement of Gen. Ray Odierno, USA, Commanding General, Multi-National Force–Iraq," in *Status of Ongoing U.S. Efforts in Iraq*, hearing before the Committee on Armed Services, U.S. House of Representatives, 111 Cong., 1st Sess., September 30, 2009, p. 9, and Odierno, 2010, p. 98.

[95] Interview with USF-I staff officer, Baghdad, December 10, 2011.

[96] Donna Miles, "Brigade Tests New Concept in Iraq," *American Forces Press Service*, November 4, 2009.

tigative techniques, operational planning, and command and control.[97] AABs also supported U.S. PRTs throughout Iraq by providing civil affairs experts, not to mention security, logistical, and transportation assistance that enabled the PRTs to undertake their missions more effectively.[98] While critics claimed that AABs were simply combat brigades under a different name, the reality was that they had a very different mission and focus. Furthermore, the symbolism of the name change was politically important because it reflected the changing character of the U.S.-Iraqi relationship that occurred throughout the transition. In addition to the AABs, joint service members provided assistance and advice to the Iraqi Air Force, Navy, and SOF.

Military Training and Countering Iran: From U.S. Forces–Iraq to the U.S. Embassy

Many of the military training activities AABs undertook transitioned from USF-I to OSC-I and USCENTCOM, which would continue to coordinate training through normal U.S. government security cooperation and exercise planning processes. USF-I police training efforts moved to six AAB-supported training centers by September 2010, which enabled the AABs to provide police trainers with security, transportation, and other support. Police training initiatives were then further consolidated by October 2011 into the three primary sites that INL's PDP was to use, at which point INL took over support responsibilities from the AABs.[99]

Interactions with the Iraqi government regarding U.S. concerns about the politicization of security increasingly transitioned from the military to the U.S. embassy. While the military worked to mitigate sectarianism on the ground and in the ISF units it was training, the embassy engaged senior Iraqi officials to advocate the integration of Sunnis into political institutions and security forces and to protect the judiciary's independence from political pressure or manipulation. The United States was concerned about the politicization of ISOF since December 2006, when Prime Minister Maliki issued an executive order removing ISOF from the MOD and placing it, without the approval of parliament, under the direct control of his office (as part of the quasi-ministerial Counterterrorism Service [CTS]). This step raised concerns that Iraq's most elite military forces would be used as a political tool to serve the prime minister's interests rather than those of the nation.[100] In the period leading up to and following the U.S. military withdrawal, U.S. embassy officials regularly urged their Iraqi interlocutors to return Iraqi counterterrorism forces to the MOD to ensure their independence

---

[97] Odierno, 2010, p. 99.

[98] Freedberg, 2009.

[99] SIGIR, "Iraqi Security Forces: Police Training Program Developed Sizable Force, but Capabilities Are Unknown," SIGIR 11-003, October 25, 2010a, p. 28.

[100] SIGIR, "Iraqi Security Forces: Special Operations Force Program Is Achieving Goals, but Iraqi Support Remains Critical to Success," SIGIR 11-004, October 25, 2010b, p. 15.

and to facilitate their integration with other Iraqi military units; however, the embassy had no more success than the military did in achieving this objective.

The Arab-Kurd Conflict

Importance of Arab-Kurdish Tensions

A wide range of political, economic, and emotionally laden historical disputes between Arabs and Kurds (as well as small numbers of Turkmen) in northern Iraq created a volatile situation that could undermine U.S. policy goals for Iraq and threaten its long-term peace and security. The lack of trust between the Kurds and Arabs resulted in each side attempting to leverage the period of transition to gain as much of the disputed land and control of resources as possible before the departure of U.S. forces. Although the United States played an active role in attempting to resolve disputes by working with Iraqi political leaders and community leaders at the grassroots level, the presence of over 200,000 Kurdish *peshmerga* and Kurdish regional forces,[101] and thousands of additional Iraqi Army troops in and around disputed areas, created a risk that tensions could escalate into armed conflict.[102]

Senior embassy officials involved in Arab-Kurd dynamics expressed concerns that political leaders on either side could take advantage of a crisis by attempting to extend their control in disputed territories and natural resources. These officials also warned that unintentional clashes between Kurdish and Iraqi Army forces could spiral out of control, leading to a broader conflict.[103] Such risks led then–MNF-I commander Odierno to state in late 2009 that Arab-Kurd tensions were "the greatest single driver of instability in Iraq."[104]

Even if deliberate or inadvertent violence between Arabs and Kurds could be contained, such an armed conflict between the ISF and Kurdish Security Forces could disrupt the delicate balance between the major Iraqi political blocs, which are based primarily on ethnic and sectarian identity. As RAND researchers wrote in 2010,

> Armed conflict between any of the mainstream parties in the Iraqi political system carries the most severe consequences for U.S. interests, because it could explode

---

[101] Telephone interview with former USF-I J3 staff officer who had also worked in U.S. Division–North, Fort Hood, Texas, January 6, 2013.

[102] For a detailed discussion of the tensions between Arabs, Kurds, and Turkmen in northern Iraq, see Larry Hanauer, Jeffrey Martini, and Omar al-Shahery, *Managing Arab-Kurd Tensions in Northern Iraq After the Withdrawal of U.S. Troops*, Santa Monica, Calif.: RAND Corporation, OP-339-USFI, 2011.

[103] Interview with U.S. embassy officials, Baghdad, June 28, 2011.

[104] David H. Gurney and Jeffrey D. Smotherman, "An Interview with Raymond T. Odierno," *Joint Force Quarterly*, No. 55, 4th quarter 2009, p. 123.

the entire political order. The greatest danger in this category is the possibility of an ethnic clash between Iraqi Kurds and the Iraqi state.[105]

Tensions in the north are exacerbated by the failure of Iraqi politicians to resolve a range of contentious national issues, including the nature of federalism, the legal and political status of disputed territories, and the allocation of budgets and natural resources. Until these underlying disputes are settled, it will be difficult to find a solution to local tensions over security, property, and minority cultural rights. The longer such disputes persist, the more likely it is that a local skirmish could escalate into a conflagration that could destabilize all of Iraq.

Political Background

Negotiations over the post-Saddam polity quickly took on an ethnic and sectarian dynamic. Kurdish leaders were in a very strong position as negotiations began over the 2004 TAL and the October 2005 constitution, which enabled them to advance a robust federalist agenda. Sunni Arabs, upset at the notion of being a junior partner in a government they had dominated for decades, had boycotted Iraqi politics since Saddam's overthrow. As a result, Sunnis held few seats in the transitional national assembly elected in January 2005 and played little role in drafting the constitution until the very end.[106]

The Kurds' strong position enabled them to advance a robust federalist agenda in negotiations over the TAL and subsequent constitution. They won recognition of the Kurdistan Region as a political entity with certain powers and rights vis-à-vis the central government. However, on many of the most controversial issues, they won only commitments for processes that would likely lead to greater autonomy—a census that would lead to a referendum that would determine the status of disputed territories, for example, and a joint federal-regional process for determining how to develop Iraq's oil and gas. Furthermore, language regarding several prerequisites for effective autonomy—such as how to calculate a "fair" distribution of oil revenues and an "equitable share" of national resources that should go to the regions and governorates—required further definition in legislation.[107]

As time went on, the Kurds' strength relative to other political blocs declined, particularly as Shi'a influence grew and after Sunni Arabs ended their boycott of Iraqi politics in May 2005. Newly strengthened Sunni and Shi'a parties delayed the imple-

---

[105] David C. Gompert, Terrence K. Kelly, and Jessica Watkins, "Security in Iraq: Emerging Threats as U.S. Forces Withdraw," Santa Monica, Calif.: RAND Corporation, RB-9481-OSD, 2010, p. 2.

[106] See especially Joost Hiltermann, "Elections and Constitution Writing in Iraq, 2005," *Mediterranean Yearbook*, 2006, pp. 38–41. See also International Crisis Group, *Unmaking Iraq: A Constitutional Process Gone Awry*, Middle East Briefing No. 19, September 26, 2005, and Kenneth Katzman, *Iraq: Politics, Governance, and Human Rights*, Washington, D.C.: Congressional Research Service, RS21968, August 21, 2012, pp. 2, 10.

[107] Iraqi Constitution, 2005, Arts. 112, 121.

mentation of the Kurdish-won constitutional provisions, including the census, the referendum on Kirkuk, and the passage of hydrocarbon legislation. As a result, Kurds found it difficult to solidify their gains, enable independent action, and reduce their dependence on the central government.

Unresolved Issues

The range of unresolved issues spanned constitutional, political, and legal questions that had to be settled at the national level. For the most part, these issues stem from the failure to settle debates over federalism—the division of power, territory, and resources between the central government and the regions—during the constitution-drafting process.[108]

Disputed Territories

Perhaps the most complex outstanding issue was the resolution of internal boundaries between the Kurdistan Regional Government (KRG) and the remainder of Iraq, most notably the status of the city of Kirkuk. The 2004 TAL called for Iraq's administrative and provincial borders to be redrawn to undo territorial modifications Saddam had for political reasons and to reverse changes to the region's demographics that the previous regime achieved through expulsions and forced resettlements. The TAL also detailed a process for determining the new boundaries, including authorizing a request that the UN Secretary-General appoint a neutral arbitrator if Iraqi leaders themselves proved unable to "remedy these unjust changes in the permanent constitution."[109] When Iraqi leaders did not meet this timetable, the 2005 Iraqi Constitution provided that the executive branch would implement the relevant terms of the TAL. The TAL, in provisions the constitution ratified, defined the Kurdistan Region as areas controlled by the Kurdish parties as of March 19, 2003. However, soon after the U.S. invasion, Kurdish leaders sent *peshmerga* across the boundary separating these areas from the rest of Iraq to take control of additional territory in which large numbers of Kurds lived. Baghdad did not accept the legitimacy of the KRG's de facto control over these disputed areas—although it agreed to steps to avoid clashes between the Iraqi Army and *peshmerga* forces along the disputed internal boundaries—leaving the borders of the Kurdistan Region unsettled. In the meantime, the Iraqi Army and Kurdish *peshmerga* squabbled over responsibility for security in the disputed areas,[110] and troop movements in and around these areas nearly led to armed clashes on a number of occasions.[111]

---

[108]Larry Hanauer and Laurel E. Miller, *Resolving Kirkuk: Lessons Learned from Settlements of Earlier Ethno-Territorial Conflicts*, Santa Monica, Calif.: RAND Corporation, MG-1198-USFI, 2012, p. 9. See also Raad Alkadiri, "Oil and the Question of Federalism in Iraq," *International Affairs*, Vol. 86, No. 6, 2010, p. 1328.

[109]CPA, 2004, Art. 58(B). See also the Iraqi Constitution, 2005, Art. 143.

[110]"We Are Ready to Die to Protect Kurds of Iraq's Disputed Areas: Kurdistan Interior Ministry Official," *AK News*, July 21, 2012.

[111] Hanauer, Martini, and al-Shahery, 2011, pp. 6–7.

One of the volatile disputed territories is the oil-rich city of Kirkuk. Each of the three largest ethnic groups in the city has dominated it at time or another—the Turkmen under the Ottomans, the Kurds during a long period when Kurds comprised the majority of residents, and Arabs under Saddam.[112] Kirkuk is an especially significant symbol for Kurdish nationalists; Iraqi President Talabani called the city "the Kurdish Jerusalem,"[113] and KRG President Barzani—referring to massacres and hardships Kurds suffered under Saddam's "Arabization" campaign—characterized Kirkuk as "the symbol of the suffering of the Kurdish people."[114]

While ethnic identity is certainly a critical element of the conflict over Kirkuk, the dispute took on particular importance because of the billions of dollars worth of oil and gas that lie underneath the city and the implications of this wealth for Kurdish autonomy and the viability of the Iraqi polity. If Kirkuk were under KRG jurisdiction, the revenues from the city's hydrocarbon resources could promote dramatic economic development in the Kurdistan Region and empower the KRG to pursue its political and economic priorities—including, potentially, secession—independently of Baghdad.[115]

The political dispute between the KRG and the government of Iraq is, at its base, a struggle for power; resources; and, for the Kurds, national identity. The disputed territory is a flash point because no solution to the dispute does not involve a significant loss to either the Sunni Arabs who live in the region; the Kurdish people who claim it is part of their historic homeland; or the Shi'a-led government in Iraq, which fears that giving concessions to the KRG could lead to increased demands for the creation of semiautonomous provinces in other areas of Iraq, undermining the power of the central government. As a result, when political leaders seek to advance the interests of their constituencies, it not only threatens a political crisis but also has the real possibility of causing violence that could lead to a military confrontation between the ISF and the Kurdish Security Forces, as will be discussed later.

---

[112] Liam Anderson, "Power-Sharing in Kirkuk: Conflict or Compromise?" paper, Globalization, Urbanization, and Ethnicity Conference, Ottawa, December 3–4, 2009, pp. 4–7.

[113] Rawya Rageh, "Iraq Tensions Rise over Kirkuk," Al-Jazeera, April 17, 2011. See also Raber Younis Aziz, "Talabani Criticized for Designating Kirkuk 'Jerusalem of Kurdistan,'" AKnews, March 9, 2011.

[114] Robin Wright, "Kurdish Eyes on Iraq's Future," *Los Angeles Times*, November 24, 2002.

[115] Under the Iraqi Constitution, all hydrocarbon resources, no matter where they are located, belong to all Iraqis. However, the KRG has signed agreements with international oil companies regarding oil and gas located in disputed territories that it controls but that remain outside the Green Line—despite the fact that Baghdad challenges the legitimacy of these agreements. The KRG could seize Kirkuk or take advantage of its de facto control of the city to exercise unilateral control over the subsurface oil fields. More likely, a negotiated settlement between Baghdad and Erbil regarding Kirkuk would almost certainly include agreed-on terms for exploiting the Kirkuk oil fields.

Security—Integration of Peshmerga *and Iraqi Army*

Although the KRG has the authority to maintain an independent militia under the constitution—a concession the Kurds negotiated as a means of preserving their autonomy—Baghdad and Erbil have been unable to agree on the appropriate size of the force, the extent to which Kurdish *peshmerga* fighters should be integrated into the ISF, and responsibility for funding the militia.

Baghdad and Erbil agreed in 2004 that the KRG would receive 17 percent of the federal budget.[116] However, the central government has argued that this allocation includes expenses related to the *peshmerga*, as the constitution states: "the regional government shall be responsible for all the administrative requirements of the region, particularly the establishment and organization of the internal security forces for the region such as police, security forces, and guards of the region."[117] The KRG, in contrast, maintains that the *peshmerga*, as an element of Iraq's armed forces with responsibility for defending Iraq's borders, should be paid for by the federal government, which the constitution charges with "establishing and managing armed forces to secure the protection and guarantee the security of Iraq's borders and to defend Iraq."[118]

Iraq's 2007 budget law appeared to resolve the problem by directing the federal government to pay the *peshmerga's* expenses but disputes then arose over the size of the force. The KRG proposed that Baghdad pay the salaries of 100,000 active-duty *peshmerga* in addition to pensions for 90,000 *peshmerga* veterans, although the Maliki government balked at a force of more than 30,000. In April 2010, Maliki agreed to fund, equip, and train a 100,000-strong *peshmerga* force, 70,000 of which would remain under KRG control, while the remaining 30,000 would be consolidated to form four Regional Guard Brigades (RGBs), which would be integrated into the Iraqi Army.[119] Since then, however, Baghdad has failed to provide the full funding that the KRG expected, which, coupled with oil production accounting disputes, led Erbil protest by stopping oil exports through the federal government's northern export pipeline in April 2012.[120]

---

[116] Bassam Francis, "KRG Accuses Baghdad of Reducing Its Budget Share," *al-Hayat*, trans. Sami-Joe Abboud, June 14, 2012.

[117] Iraqi Constitution, 2005, Art. 117, Sec. 5.

[118] Iraqi Constitution, 2005, Art. 107, Sec. 2.

[119] International Crisis Group, *Iraq and the Kurds: Confronting Withdrawal Fears*, Middle East Report No. 103, March 28, 2011, p. 24. See also Hanauer, Martini, and al-Shahery, 2011, p. 7.

[120] Francis, 2012. A KRG Ministry of *peshmerga* official claimed in September 2012 that Baghdad had not provided funds to the ministry for five years. See Nehro Muhammad, "Peshmerga Budget Next on Agenda in Baghdad-Erbil Talks," *Rudaw* (Erbil), September 26, 2012.

U.S. Military Efforts

Despite the intractability of political and economic differences between Baghdad and Erbil, interethnic violence in the north has been limited, largely avoiding armed conflict between Iraqi and Kurdish security forces, although there have been a few close calls. For example, in September 2008, Iraqi Army units conducted COIN operations near the city of Khanaqin—a Kurdish-occupied city outside the Green Line. The potential conflict between the ISF and *peshmerga* was avoided only because of the personal intervention of KRG President Barzani and Iraqi Prime Minister Maliki.[121]

U.S. military efforts to prevent Arab-Kurd violence played a significant role in keeping tensions below the boiling point, which, the leadership of USF-I reasoned, could help establish the stability and political space necessary for future negotiations on unresolved issues to occur. A key role for the U.S. military was helping prevent conflicts between the ISF and *peshmerga* from escalating into violence. For example, in June 2009, the 26th Brigade of the 7th Division of the Iraqi Army, an Arab unit, attempted to move through the Kurdish town of Makhmur en route to the primarily Sunni Arab city of Mosul.

Fearful that this might be a Baghdad government land grab for Makhmur, *peshmerga* forces took the high ground to prevent the advance. U.S. forces stationed in the area were used to diffuse the tactical situation, while senior U.S. military officials interacted with their government of Iraq and KRG counterparts to de-escalate the pending crisis.[122] In part to avoid future incidents like that at Makhmur,[123] USF-I established a trilateral CSM in late 2009 (see Figure 6.1). At the tactical level, the mechanism included Iraqi Army troops and police, *peshmerga*, and Iraqi police—along with U.S. troops—who conducted joint patrols and maintained joint checkpoints in selected disputed areas in Ninewa and Diyala, as well as on the outskirts of Kirkuk city. A small combined force also patrolled inside the city of Kirkuk. This critical confidence-building measure helped build trust between Iraqi and Kurdish forces, demonstrated to the populace that security forces patrolling the area were free from ethnic biases, and established procedures for de-escalating crises before they spiraled out of control.[124]

In addition, three tripartite combined coordination centers (CCCs) located throughout the disputed areas promoted transparency and communication between

---

[121] Hanauer, Martini, and al-Shahery, 2011, p. 6. See also Brian Katulis, "Standoff in Khanaquin: Trouble Brews Between Arabs and Kurds in a Volatile Corner of Iraq," Center for American Progress, August 29, 2008.

[122] Hanauer, Martini, and al-Shahery, 2011, pp. 6–7; Patrick Cockburn, "Arab-Kurd Tensions Rise as US Pulls Out," *New Zealand Herald*, August 12, 2009.

[123] The CSM's primary original purpose was to fill the security vacuum that existed in the "seams" between areas under Kurdish and central government control, which al-Qaeda in Iraq exploited to launch frequent attacks. However, it quickly became a valuable tool for managing ISF-*peshmerga* tensions across the same seams.

[124] International Crisis Group, 2011, pp. 16–17.

Figure 6.1
USF-I Combined Security Mechanism Disposition

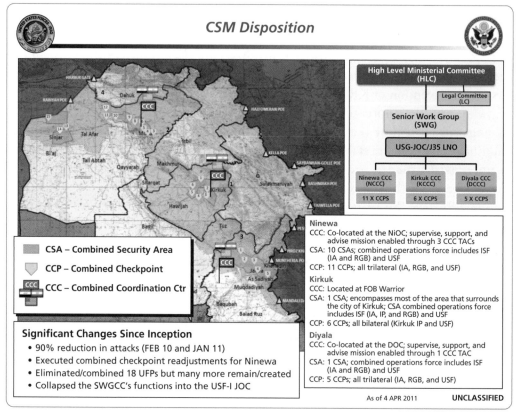

SOURCE: USF-I, "Arab-Kurd Transition Plan," update, April 4, 2011e.
NOTE: TAC = tactical command post.
RAND RR232-6.1

Iraqi Army and Kurdish forces and provided a mechanism for quickly de-escalating local-level conflicts.[125] In terms of resources, the CSM was a relatively modest effort. It involved a mere 1,000 U.S. troops; although in a contingency they could have drawn additional support from the division responsible for operations in northern Iraq.[126]

In August 2011, an internal USF-I J3 assessment bluntly warned that Arab-Kurd tensions would remain a potential flash point in northern Iraq that could result in a much broader conflict once USF-I stops mediating disputed over internal boundaries and access to natural resources. The key concern for USF-I planners was that, without

---

[125]Interview with USF-I J35 staff, Baghdad, August 1, 2012. In September 2010, USF-I issued press guidance explaining "the purpose of the CSM is to provide a mechanism for Iraqi Security Forces (ISF), Kurdish *peshmerga*, and U.S. Forces to work together to increase security in the disputed territories" (USF-I, "Combined Security Mechanism," press guidance, September 13, 2010d).

[126]Interview with USF-I J35 staff, Baghdad, August 1, 2012.

third-party mediation and overwatch, there was a strong likelihood that a political dispute could result in an armed clash that could extend throughout the disputed territory. This assessment not only drove the decision to keep forces in northern Iraq for as long as possible but was also one of the primary missions envisioned for a potential residual force to remain in Iraq beyond 2011.[127]

The Arab-Kurd Transition Plan

DoD and DoS agreed that the United States should be involved in the Arab-Kurd dynamic, both before and after the U.S. troop withdrawal. However, the plan for transitioning DoD's activities to the embassy posed significant challenges because DoS lacked the capability to continue the military's on-the-ground engagement at the tactical level. This aspect of the enduring security challenges received a significant amount of attention from USF-I and was identified as objective 9 in the "Conduct Transitions" LOE of OPORD 11-01.

Planners quickly pointed out that, while the embassy and OSC-I might be able to participate in trilateral command posts or otherwise serve as a diplomatic intermediary between Iraqi Army and Kurdish security forces after the U.S. military withdrawal, no forces would be available to conduct patrols or staff checkpoints. Without a continuing U.S. military presence of some sort, this important part of the U.S. military mission would simply have to be terminated. The big question USF-I and embassy officials confronted in planning the transition was how the embassy could acquire some capacity to limit potential conflict to give the political process sufficient time to work through the critical issues associated with Arab-Kurd tensions. Both embassy and USF-I officials believed no solution would be likely for at least five to ten years. In this view, the CSM was just one of many initiatives needed to manage Iraqi and Kurdish tensions in advance of a settlement.

As a result of the uncertainty about how the embassy would manage the CSM after the withdrawal of U.S. forces, the Arab-Kurd transition plan was the last to be signed, with the ECG approving it on April 5, 2011.[128] Although the plan called for six JCP activities to transition to the embassy, five of them involved primarily political initiatives—such as facilitating a peaceful resolution to internal boundary disputes—in which the embassy was already engaged.[129] The sixth activity was designed to encourage further incorporation of KRG security forces into the ISF as part of a broader goal to eliminate sectarianism from the ISF. This had also already been initiated by USF-I, with embassy officials reinforcing the message with the Iraqi government at policy

---

[127] Brennan notes, 2009–2011.

[128] Interview with Ambassador Larry Butler (USF-I political advisor), Baghdad, July 1, 2011.

[129] Interview with USF-I J35 staff, Baghdad, 25 June 2012. Also USF-I and U.S. Embassy Baghdad, "Arab-Kurd Relations Transition Plan," April 5, 2011, para. 4, Not available to the general public.

levels. The embassy continued to advocate for this policy objective after it formally assumed responsibility for this task on September 1, 2011.[130]

A key challenge associated with the transition was how to maintain the CSM—both the physical security elements of joint patrols and the de-escalation tools of the CCCs—as a sustainable security arbitration mechanism to dissuade either side from using security forces to expand territorial control and to prevent isolated security incidents from escalating into a broader Arab-Kurd crisis.[131] The senior leadership of both Embassy Baghdad and USF-I anticipated that there was an increased likelihood of Kurdish Regional Forces/ISF violence in the wake of U.S. troops' departure from Iraq.[132] However, the embassy had fewer personnel resources and far more limited capabilities, so there was little it could do to replicate a robust U.S. involvement in the CSM other than to take part in senior-level discussions.[133] Consequently, the Arab-Kurd transition plan called for USF-I to "systematically reduce" its participation in the CSM, after which key leader engagements and joint political and security meetings would be the only remaining CSM activities. The plan called for the embassy to participate in these meetings to the extent its capabilities and resources permitted.[134]

USF-I developed a methodical plan for gradually reducing its involvement in CSM checkpoints and patrols from active operational engagement to "overwatch."[135] The U.S. contingent pulled back, in stages, to local CSM hubs, then regional CSM hubs, and finally to U.S. bases. By keeping U.S. forces nearby in the immediate aftermath of their departure from CSM sites, this strategy enabled U.S. forces to intercede, if necessary, on short notice. The potential for Arab-Kurd violence was seen as so great that the "collapse" of forces into centralized locations that preceded units' final movement out of Iraq (described in detail in Chapter Nine) was delayed as long as possible to extend the presence of U.S. troops along the Arab-Kurd fault line. Moreover, as final negotiations on an enduring troop presence dragged into fall 2011, it was widely assumed that a primary mission of any continuing presence would be to participate in the CSM, so U.S. commanders made every effort to leave forces in place in the north until a final decision was announced.

---

[130]Brennan notes, 2009–2011.

[131]Brennan notes, 2009–2011.

[132]Brennan notes, 2009–2011.

[133]Interview with U.S. Embassy officials, Baghdad, June 28, 2011.

[134]USF-I and U.S. Embassy Baghdad, 2011, paras. 4, 6. By late June 2011, the embassy had not yet determined which embassy officials would take part in these discussions. Interview with U.S. embassy officials, Baghdad, June 28, 2011.

[135]Interview with USF-I J35 staff, Baghdad, August 1, 2012.

Summary

Ethnic and sectarian divisions have been primary forces shaping post-Saddam Iraq. Iraq's 2005 constitution—the result of drawn-out and contentious negotiations between ethnic and sectarian blocs—was a careful balance of political, economic, and security interests of Iraq's Shi'a, Sunni, and Kurdish constituencies. Although Iraqis of all stripes resisted the U.S. military presence and engaged in anti-American attacks, much of the violence during the U.S. military presence stemmed from conflict between Sunni and Shi'a, as well as among rival elements of each community. Although MNF-I developed the initial concept of drivers of instability in 2008, little progress was made to address the underlying causes of the continuing conflict in Iraq. As one senior military officer stated, "While the war in Iraq is over for the United States, it is not over for the Iraqis."[136]

U.S. military and diplomatic efforts between 2008 and 2011 helped calm these divisions in many ways, but enduring ethnic and sectarian tensions continue to pose substantial threats to Iraqi security and to the U.S. goal of developing a democratic, tolerant, prosperous, and secure Iraq. If Iraqi political leaders continue to engage in power plays, disregard the rule of law, and refuse to address outstanding issues in good faith, the outlook for Iraq is bleak. At best, the Iraqi government will vault from crisis to crisis, with each faction reluctant to make concessions until the last minute, if at all. These ethnic, sectarian, and ideological divides will contribute to greater political fracturing and the stalemate that prevents Iraq's most pressing issues—particularly federalism, the status of disputed territories, and hydrocarbons—from being resolved. However, such brinksmanship raises the stakes associated with each issue to be decided, increasing the incentives to win each battle by any means necessary. Furthermore, the radical rhetoric that such crises inspire provides justification for extremists to engage in violence against their adversaries. In a worst-case scenario, Iraq could disintegrate, once again, into civil war.

The persistence of sectarian divides in Iraq, particularly in the political arena, also allows Iran to influence Iraq through the use of political proxies and violent surrogates. Tehran worked to ensure that Iraqi politics was dominated by Iraqi Shi'a political figures with whom it had close ties. Iran and its allies resisted Sunni integration into the Iraqi government, particularly into positions of prominence, by undermining negotiations on Sunni political participation, enhancing support to Shi'a political parties, and fomenting greater sectarian violence. Although the U.S. strategy for balancing Iranian influence attempted to enhance Iraqi capacity in ways that would empower Baghdad to resist Iranian power projection efforts, Iraq's governance, security, and diplomatic capabilities were insufficient for Baghdad to withstand most pressures from Tehran. At least in the short term, the power vacuum that the rapid departure of U.S. forces and

---

[136]Interview with senior military officer, Baghdad, December 12, 2011.

the continued sectarian nature of Iraqi politics have presented an opportunity for Iran to increase its influence in Iraq, an opportunity Tehran will continue to exploit.

CHAPTER SEVEN

Enhancing the Security Sector Capacity of the Government of Iraq

CMATT, under the command of General Eaton, started the first concerted effort to rebuild the Iraqi military in 2003. During this period, DoS established separate efforts to build a new Iraqi police force through its Civilian Police Assistance Transition Team. Simultaneously, DoS established advisory missions to both MOD and MOI. As the CPA closed down in 2004, all these missions were consolidated into the newly established MNSTC-I, a subordinate command of MNF-I. The mission of MNSTC-I was to

> [a]ssist the Ministry of Interior, the Ministry of Defense, and the Counter-Terrorism Bureau; generate and replenish Iraqi Security Forces (ISF); and improve the quality of the ISF and institutional performance . . . to increase ISF capability to increasingly assume responsibility for population protection . . . with reduced coalition involvement.[1]

MNSTC-I's primary focus was on increasing the total number of Iraqi soldiers and police and providing them the capabilities necessary to perform their security functions. Sustaining this capacity would require ministries to manage the ISF after the departure of U.S. forces. Building ISF operational capabilities and improving ministerial performance were therefore viewed as critical requirements for transitioning security functions from the U.S. military to the government of Iraq.[2]

In January 2010, the colors of MNSTC-I were cased, and the responsibility for enhancing ISF capabilities transitioned to the DCG (A&T) under the newly established USF-I. The DCG (A&T) executed this responsibility through Army, Navy, Air Force, police, MOD, and MOI components of its ITAM and its weapon sale specialists in the Iraq Security Assistance Mission (ISAM). The DCG (A&T) was also dual-hatted as the commander of NTM-I, which had been established in 2004 to train Iraqi Special Police to serve as a national-level rapid-response force that could counter

---

[1] MNSTC-I, "Command Briefing," undated, slide 3, Not available to the general public.

[2] MNSTC-I, undated, slide 4.

terrorists and respond to large-scale civil disobedience.[3] Figure 7.1 diagrams the various U.S. security cooperation entities.

By the time General Austin assumed command of USF-I in September 2010, the United States assessed that the ISF could maintain internal security at an acceptable level but could not defend against external threats. The ISF was unable to conduct combined arms operations at any level of command, provide air sovereignty and an integrated air defense, sustain and maintain forces in the field, conduct counterterror-

Figure 7.1
U.S. Government Organizations Overseeing Security Sector Assistance to Iraq

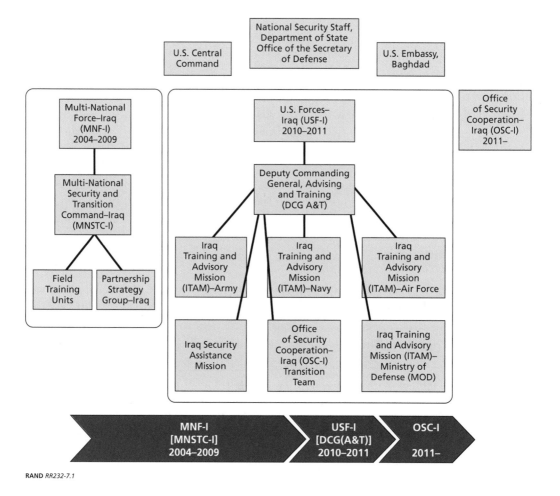

RAND RR232-7.1

---

3   The two dominant elements of law enforcement in Iraq are the local Iraqi police and the Federal Police. The NTM-I's primary focus was to assist in the leadership development of the Iraqi military and to enhance the capability of the national level police using Italian Carabinieri-led training. The national-level police were originally named Special Police when established in 2004. This organization was renamed National Police in 2006 and subsequently renamed as Federal Police in 2009. Brennan notes, 2009–2011.

ism operations without support from U.S. SOF, and conduct cross-ministerial intelligence and information sharing. Regardless of USF-I's level of effort, the ISF would not be able to learn how to execute these critical tasks independently in the 15 months remaining in the U.S. military mission.[4]

The transition of security-related functions from U.S. forces to the ISF required both improvements in ISF capacity and reduced U.S. force involvement. Thus, under MNSTC-I, Iraqi, rather than U.S., forces increasingly led joint operations. Under USF-I, U.S. forces took an increasingly hands-off approach; Iraqis undertook unilateral operations with U.S. forces providing overwatch and assistance in partnered and/ or enabled operations.

As discussed in Chapter Five, OPORD 11-01 identified "Strengthen Iraqi Security Forces" as the decisive LOE during phase one of the operation, losing its priority only when U.S. forces started the operational maneuver to exit Iraq. This chapter will review the policies and guidance the White House, DoS, and DoD developed to strengthen the ISF and the planning and tasks undertaken both at USCENTCOM headquarters and in the field. It will then provide an overview of the Iraqi security sector, followed by closer analysis of its components—the MOD, Army, Navy, Air Force and Army Air Corps, Special Operations Forces, and MOI and police—as well as Iraq's ability to manage its airspace. Throughout, it will assess the extent to which the transition succeeded in preparing the ISF to conduct its missions independently of U.S. forces.[5]

U.S. Policy, Strategy, and Plans for Security Sector Transition

In early 2009, MNSTC-I made three important assumptions in its planning for ISF development:

1. Iraq's principal threats for the foreseeable future would be the danger of intersectarian conflict and a continuing, albeit declining, insurgency.
2. No country in the region would have the capability to project forces into Iraq and sustain extended offensive operations during the subsequent five to ten years.

---

[4] Interview with USF-I J3 staff officer, Baghdad, October 10, 2011.

[5] While the focus of this chapter is on the transition from MNSTC-I and its successor headquarters to Iraqi forces, U.S. military companies, battalions, brigades, and divisions continued to provide advice and assistance to their Iraqi counterparts until the final retrograde of U.S. forces in first quarter of FY 2012. This tactical-level partnership was an important aspect of building partner capacity with the ISF. While senior USF-I leadership viewed this aspect of the operations as being vitally important for mission success, the results were largely observable at the Iraqi company and battalion levels. Interview with USF-I J3 staff officer, Baghdad, October 11, 2011.

3. It would be eight to 12 years before Iraq could develop a mature and fully self-reliant military establishment capable of defense against both external and internal threats.[6]

These assumptions suggested that, while Iraq's most pressing near-term requirement would be to maintain internal stability, it would need to begin developing the ability to defend itself from external aggression.

U.S. commanders widely accepted that the Iraqi Army would be unable to develop the capabilities needed to fight another army as long as it remained engaged in maintaining internal security within Iraq. MNSTC-I envisioned that police forces working under the MOI would assume the primary responsibility for counterterrorism and internal security, which would then free the Iraqi military to prepare for its external defense mission.

MNSTC-I expected that the global recession and resultant low oil prices would cause the Iraqi government to cut spending, including for the security sector. The pending U.S. troop withdrawal and likely reductions in Iraqi defense spending were thought to make it increasingly unlikely that the ISF would attain the robust capabilities the 2009 JCP and early MNSTC-I planning had envisioned.[7] MNSTC-I thus began to shift its long-term training vision from aspirational goals toward the development of the MEC necessary to allow Iraq to manage its own security when the United States would withdraw its forces at the end of 2011.[8]

Minimum Essential Capabilities
In October 2009, MNSTC-I defined Iraq's achievement of MEC as when "Iraq has adequate capability to secure the population, provide internal defense, and lay the foundation for basic external defense." MNSTC-I planners established five specific goals necessary to achieve MEC by December 2011:

- have the capacity to maintain internal security and stability through police primacy
- have the ability to maintain maritime security and defend key port and oil infrastructure against insurgents, terrorists, or limited external attacks
- present a credible, initial deterrent against external conventional threats

---

[6] MNSTC-I, "Shaping the Long-Term Security Partnership with Iraq," briefing to General Odierno, Commander, MNF-I, February 24, 2009a.

[7] The conditions-based 2009 JCP and early MNSTC-I planning envisioned an advanced Iraqi military force including 16 mostly mechanized Army divisions, an integrated air defense system, a regionally capable Air Force with modern helicopters and F-16s, a competent coastal Navy, and strong counterterrorism force that would include both ISOF and specialized Federal Police. Interview with DCG (A&T) staff officers, Baghdad, June 15, 2011.

[8] Interviews with former MNSTC-I and DCG (A&T) officials, November 2011.

- provide foundational capability to defend Iraq from external aggression[9]
- set conditions to achieve air sovereignty in the long term, which Iraq would do primarily through the acquisition and subsequent use of advanced fighter aircraft, such as the F-16 and an integrated air defense network.[10]

As depicted in Figure 7.2, the concept for MEC established desired capabilities within both the Iraqi military and police to prioritize resources and achieve specified strategic objectives. This concept was an integral component of the MNF-I and embassy JCP. The areas shaded green reflect MNF-I's assessment of Iraqi capabilities in 2009, while the areas shaded red reflect ISF shortfalls required for MEC. While U.S. forces could depart Iraq before the ISF achieved MEC, doing so would require changing JCP goals and objectives and accepting the long-term risk that Iraq would

Figure 7.2
Minimum Essential Capabilities for Iraqi Security Forces

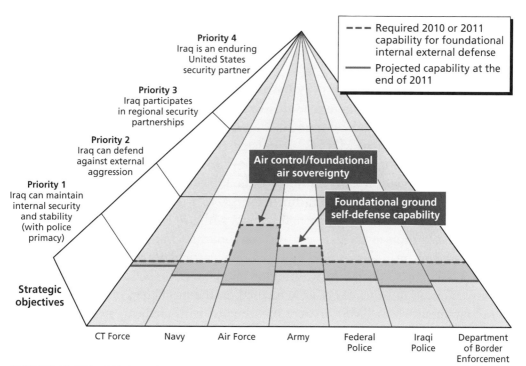

SOURCES: MNSTC-I, 2009b, and MNF-I, 2010.
RAND RR232-7.2

---

[9] Early planning envisioned this to be the capability to conduct battalion-level operations that could integrate both indirect fire and air support, demonstrating the basic elements of a combined arms operation. Interview with DCG (A&T) staff officers, Baghdad, June 15, 2011.

[10] MNSTC-I, "Terms of Reference, Iraqi Security Force (ISF) Capabilities," information paper, October 15, 2009b. Also MNF-I, "ISF Terms of Reference," briefing slide, October 26, 2010.

be unable to defend itself from external aggression and collaborate effectively with the United States and other partners. During 2009 and 2010, MNSTC-I focused primarily on the near-term challenge of enhancing the ISF's capacity to fight insurgents, protect the population against terrorist attacks, and incrementally move toward the achievement of MEC.

While the bulk of ISF training in 2009 and 2010 continued to be performed in the field by deployed units, including the Army's AABs, MNSTC-I also sought to enhance institutional training related to professionalism,[11] defense strategy, budgeting, planning, and doctrine, which were important to the legitimacy and success of the ISF over the long term. To ensure MNF-I and later USF-I would maintain a focus on these longer-term transition goals, MNSTC-I spent significant effort sorting through the array of lower-profile security cooperation tools, such as International Military Education and Training (IMET), Joint Combined Exchange Training, DoD Regional Centers for Security Studies, the Combating Terrorism Fellowship Program, and the National Guard State Partnership Program. While these programs were well understood in Washington, MNSTC-I—as a staff element of a field command—faced challenges developing requirements for these tools and incorporating the requirements into DoD's two-year programming, planning, and budgeting cycle.[12]

Moving Toward a "Normal" Security Assistance Model

In early 2009, MNSTC-I began planning for a bilateral security cooperation relationship after U.S. forces withdrew from Iraq. Security cooperation offices exist in countries around the world, so MNSTC-I initiated planning for the creation of OSC-I, which would operate under chief-of-mission authority. However, until late 2010, little thought had been given to how funding for the ISF training mission would transition from a wartime footing, with plentiful resources, to a more traditional U.S. security assistance plan funded through normal Foreign Military Financing (FMF) and FMS programs. According to civilian and military officials charged with this effort, the OSC-I transition planners in Baghdad were not particularly well positioned for the latter task. Moreover, officials in Washington and USCENTCOM who typically work security cooperation issues either were focused on the mechanics of security cooperation programs (such as FMS) or were preoccupied with challenges elsewhere in the world. Whatever the cause, the task of establishing OSC-I was left to a small cell that worked for the DCG (A&T), the successor organization to MNSTC-I.[13]

---

[11] Professionalism covered a host of subjects, including anticorruption, antisectarianism, interministerial cooperation, and respect for the rule of law.

[12] DoD IG, *Assessment of Planning for Transitioning the Security Assistance Mission in Iraq from Department of Defense to Department of State Authority*, Report No. SPO-2011-008, August 25, 2011, pp. 7–8 and 11; interviews with former MNSTC-I and DCG (A&T) officials, November 2011. See also MNSTC-I, 2009a.

[13] Interview with DCG (A&T) staff officers, Baghdad, June 15, 2011.

Switch to New Funding Mechanisms

USF-I helped the ISF move toward MEC through three programs: direct FMS; the Iraqi Security Forces Fund (ISFF); FMF; and the U.S. Equipment Transfer to Iraq (USETTI) program, which transferred used U.S. military equipment. (USETTI will be described in more detail in Chapter Nine.)

Since May 2005, the most important funding resource for moving the ISF toward MEC had been the ISFF. This Iraq-specific congressional appropriation ranged from $1 billion to $5.5 billion per year and was used to train, equip, and maintain all ISF elements.[14] ISFF was considered critical to the overall development of the ISF and was a lynchpin program for Iraqi combat training. For example, over $75 million of ISFF was invested at the Besmaya Combat Training Center (BCTC) to develop the heavy weapon ranges, support infrastructure for M1A1 tanks, and renovate physical infrastructure.[15]

The Iraqi government was required to fund 20 percent of all ISFF acquisitions, although exceptions were made in some cases.[16] As shown in Figure 7.3, over $18.5 billion in ISFF money was invested from 2005 through September 30, 2011. In its FY 2012 budget, Congress appropriated an additional $1.5 billion to ISFF.[17] Responsibility for executing these obligations fell on OSC-I as OND ended.

ISFF had a number of advantages over FMF, the traditional security cooperation program implemented around the world under DoS's Title 22 foreign operations authority.[18] First, it was a DoD (Title 10) wartime authority, thus giving MNF-I more control and flexibility over the allocation of program funds.[19] Second, ISFF could be used to support all ISF, whether military or police; in contrast, FMF could only be used to support the Iraqi military. Third, ISFF could fund a broader array of activities than FMF, including construction, force protection, training, equipping, life support, and sustainment. ISFF was, in effect, a one-stop shop for supporting all elements of the ISF, even if, over the long term, a transition to FMF—the lynchpin of security cooperation activities around the world—would be necessary to normalize the U.S.-Iraqi security relationship.

---

[14] U.S. Department of the Army, *Commander's Guide to Money as a Weapons System*, Center for Army Lessons Learned, April 2009.

[15] USF-I, "Commanders' Update," briefing, October 2011k, Not available to the general public.

[16] SIGIR, *Quarterly Report to the United States Congress*, October 30, 2010c, p. 53. Exceptions are for the items identified as belonging on the U.S. Munitions List (USML).

[17] SIGIR, 2012a, p. 20.

[18] 22 U.S. Code, Foreign Relations and Intercourse.

[19] The constraints on how FMF can be used are significantly more stringent than the standards applied to the utilization of ISFF. This flexibility allowed rapid reprogramming to address emerging needs and requirements. Brennan notes, 2009–2011.

Figure 7.3
Iraq Security Forces Fund Allocations, 2005–2011

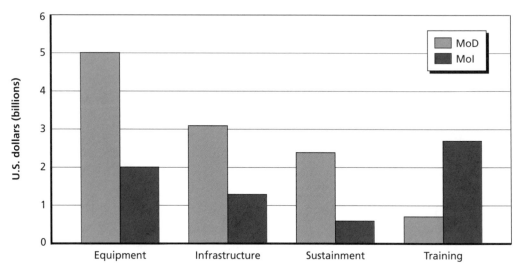

SOURCE: SIGIR, 2011f, pp. 20–22.
RAND *RR232-7.3*

Some Pentagon officials proposed transitioning U.S. capacity-building funding from ISFF to FMF as early as 2006. However, because MNF-I felt ISFF's flexibility gave it significant advantages over traditional FMF, planning to use FMF did not begin in earnest until late 2010.[20] The Under Secretary of Defense (Comptroller) directed the $1.5 billion in FY 2011 ISFF be managed in cooperation with the Defense Security Cooperation Agency, which manages the FMF program within DoD.[21] This action was an important step toward transitioning ISF support from a large-scale, operational effort to a more-traditional security-cooperation approach, paralleling the establishment of OSC-I.[22] The administration submitted its first request for FMF funds for Iraq (while requesting no further funding for ISFF) in DoS's FY 2012 OCO plan. (Although it requested $1 billion for Iraq alone, DoS ultimately allocated Iraq $850 million of the $1.1 billion in total OCO FMF funds that Congress appropriated.)[23] OSC-I had the authority to obligate remaining ISFF money through the end of FY 2012.

Starting in 2005, officials at DoS began to plan the development of an enduring funding mechanism to enhance MOI capabilities through its International Nar-

---

[20] Interview with OSD official, January 12, 2011.

[21] SIGIR, "Interim Report on Spend Plans for Fiscal Years 2011–2012 Iraq Security Forces Funds," SIGIR 12-015, April 26, 2012b, p. 4.

[22] Secretary of State, *Congressional Budget Justification*, Vol. 2: *Foreign Operations, Fiscal Year 2012*, Washington, D.C.: U.S. Department of State, April 2011b, p. 190.

[23] SIGIR, *Quarterly Report to the United States Congress*, April 30, 2012c, p. 23.

cotics, Counterterrorism, and Law Enforcement (INCLE). From FY 2006 through 2010, INL allocated a total of about $1.1 billion in INCLE for programs in Iraq. INL requested $1 billion in additional INCLE funding for FY 2012 alone.[24] Responsibility for police training transferred from DoD to DoS on October 1, 2011. The FY 2012 INCLE request was intended to fund the first year of operations for the PDP. INCLE funds would support the work of PDP advisors in mentoring and training programs; enhancing specialized policing skills, such as criminal investigations, forensics, and border security; and the development of programs at regional and national Iraq academies. The FY 2012 funds would also be allocated to support capacity building in the justice sector, as well as anticorruption, anti–money laundering, and antiterrorist financing programs.[25] Congress eventually appropriated $1.18 billion to INCLE in its FY 2012 budget.[26]

Planning for the Establishment of the Office of Security Cooperation in Iraq

After the President's Camp Lejeune speech, it became clear that the extensive military-led training program for the ISF would have to transition to a more "normal" security cooperation relationship. However, the planning for this transition was plagued by a lack of policy guidance, insufficient integration into strategic decisionmaking, and constant changes. OSC-I did not get sufficient focus in 2011, largely due to the uncertainty regarding the future mission and the planning workload caused by the potential for U.S. forces to remain beyond 2011.[27] In addition, the staff of the DCG (A&T), which was responsible for OSC-I transition planning during the last year of the U.S. military presence, was focused on improving the Iraqi military's capabilities as much as possible before the U.S. withdrawal; the question of how the United States would manage such training after 2011 was an important, but secondary, concern.

Perhaps the most critical shortcoming was the lack of policy guidance from Washington regarding security cooperation strategy after the U.S. withdrawal. The DoD Inspector General (IG) determined that senior OSD policy officials were unwilling "to limit strategic options before the nature of the relationship between the U.S. and the new government of Iraq could be better established."[28] USCENTCOM similarly failed to provide strategic direction until the planning process was well under way. For instance, it did not publish an Iraq Transition Plan until December 1, 2010. More important, USF-I developed the bulk of the transition plan and provided it to

---

[24] SIGIR, 2011f, pp. 6 and 110.

[25] Secretary of State, 2011b, pp. 188–189.

[26] SIGIR, 2012a, p. 20.

[27] Interview with former DoD official, January 17, 2012. It should be noted that, while OSC-I is part of the embassy team, DoD conducted the large majority of planning for the office's creation of OSC-I, mostly in Baghdad.

[28] DoD IG, 2011, p. 3.

USCENTCOM for coordination and approval.[29] Similarly, USCENTCOM did not update its 2007 Theater Campaign Plan until March 2011,[30] and it did not release a country plan for Iraq until October 26, 2011—after OSC-I became fully operational.[31]

While it was logical for Washington-based officials to want to set the parameters of U.S. strategy toward Iraq before developing institutions to manage it, officials on the ground at USF-I and the embassy who were responsible for establishing these institutions knew they would need to begin planning early. These officials could not draw on established procedures because "joint doctrine did not provide sufficiently detailed guidance about what was needed to transition from stability operations to robust security cooperation activities within a non-permissive environment," according to the DoD IG.[32]

In the absence of ISF-related transition guidance from Washington or USCENTCOM,[33] MNF-I, on its own initiative, established a small team, called the Partnership Strategy Group–Iraq (PSG-I), within MNSTC-I to identify the roles, missions, requirements, authorities, and organizational structure for an OSC-I.[34] PSG-I would later be renamed the OSC-I Transition Team and placed under the DCG (A&T).[35] As the only entity considering what a security cooperation office would need, the DoD IG wrote that PSG-I's "ad hoc group of strategic planners" were singularly responsible for the success of the OSC-I planning effort.[36]

PSG-I ultimately succeeded in spite of numerous obstacles. First, OSC-I planning was not its sole responsibility; it was also tasked with developing plans to cement a close, multifaceted bilateral relationship following the U.S. military withdrawal.[37] This prevented the office from focusing on OSC-I and ensuring that planning remained on track.

Second, as a part of an operational-level command staff, PSG-I was not prepared to engage in strategic planning with national-level implications. A higher-level organization might have considered how security cooperation could advance overall U.S.

---

[29] Interview with USF-I J5 staff officer, Baghdad, December 13, 2011.

[30] DoD IG, 2011, pp. 11–13. Also, interviews with former MNSTC-I and DCG (A&T) officials, November 2011.

[31] DoD IG, *Assessment of the DOD Establishment of the Office of Security Cooperation–Iraq*, Report No. DODIG-2012-063, March 16, 2012, pp. 5, 10–11.

[32] DoD IG, 2011, pp. 14–15.

[33] Interviews with former MNSTC-I and DCG (A&T) officials, Washington, D.C., November 2011.

[34] Interview with OSC-I Transition Team official, Baghdad, June 30, 2011.

[35] USF-I, OPORD 11-01, Change 1 to Appendix 4 to Annex V (Establishment of the Office of Security Cooperation–Iraq), May 5, 2011d, para. (3)(e)(3), p. V-4-20.

[36] DoD IG, 2011, p. i.

[37] Interview with OSC-I Transition Team official, Baghdad, June 30, 2011. Also DoD IG, 2011, p. 3.

goals and strategies in Iraq, but PSG-I focused on OSC-I's mechanics, such as how many positions it would have, where it would locate facilities, and how much funding would be necessary. As a result, instead of developing a resource plan that advanced a security assistance strategy, PSG-I designed the strategy around the level of resources that could be expected.[38]

Moreover, from its place on the organizational chart, PSG-I was not empowered to task anyone.[39] Although OSD and USCENTCOM allowed this USF-I staff element to lead the planning effort, PSG-I—buried within MNSTC-I and, later on, the DCG (A&T)—had to "lead from the bottom," according to the DoD IG.[40] One OSC-I planner said that PSG-I should have been housed at a higher-echelon entity, such as USCENTCOM, from which it could have directed in-country staffs to take actions in support of OSC-I's establishment.[41]

Furthermore, although PSG-I staff rose to the task, they were not particularly well suited to lead DoD planning for an enduring security cooperation institution. None of the initial PSG-I staff had a background in security assistance; even in mid-2011, when OSC-I's establishment was imminent, only three of the 18 staff on the OSC-I Transition Team—PSG-I's successor—had experience in security assistance positions, although the remainder had gone through an online training course administered by the Defense Institute of Security Assistance Management.[42]

Third, between March and September 2010, PSG-I staff were diverted to "special projects" for the DCG that focused on the post-2011 security environment, including police primacy, regional exercises, and potential NATO roles.[43] Thus, no one in Baghdad was focusing on OSC-I for about six months in the middle of 2010. This gap, driven by competing priorities, likely caused a senior DoS official to state that, despite PSG-I's earlier efforts, planning efforts for OSC-I "only got under way for real in late summer 2010."[44]

Fourth, OSC-I planning began too late for DoD to include the manpower resources in its FY 2012 base budget proposals, as had been anticipated. DoD officials thus decided in late 2010 to staff OSC-I with temporary positions and requested

---

[38] Interview with OSC-I Transition Team official, Baghdad, June 30, 2011.

[39] Interview with OSC-I Transition Team official, Baghdad, June 30, 2011. Also interview with DCG (A&T) official, Baghdad, June 27, 2012.

[40] DoD IG, 2011, pp. 8, 11–12.

[41] Interview with OSC-I Transition Team official, Baghdad, June 30, 2011.

[42] Interview with OSC-I Transition Team official, Baghdad, June 30, 2011.

[43] Interview with OSC-I Transition Team official, Baghdad, June 30, 2011.

[44] Interview with Kathleen Austin-Ferguson, Executive Assistant to the Under Secretary of State for Management, Washington, D.C., October 25, 2011.

funds for this purpose in the FY 2012 supplemental budget submission.[45] The failure to lock in the size of OSC-I's staff, however, contributed to frequent changes in the office's size and scope. The DoS IG reported that "planned full-time military and civilian U.S. government direct-hire staffing levels fluctuat[ed] from the low 100s to nearly 1,000" throughout the planning process and assessed that "the inability to decide on the OSC's size or its locations throughout Iraq" would likely prevent it from becoming fully operational by the target date of October 1, 2011.[46] USF-I expected that USCENTCOM would request DoS authorization in advance for permanent OSC-I staff to work under chief-of-mission authority, as required by National Security Decision Directive (NSDD) 38.[47] However, it was not until after OSC-I was supposed to be operational that DoD submitted an NSDD-38 request for any OSC-I staff except a few dozen contracting officers.[48] As will be discussed in the next chapter, this uncertainty made it extremely difficult for DoS to finalize plans for embassy facilities, ensure all construction could be completed by the end of 2011, and hire the appropriate number of guards for both static perimeter security and movement security teams.

Ultimately, it was decided to staff OSC-I within the 920 personnel ceiling—157 direct-hire U.S. government employees (118 military, 9 civilians, and 30 Iraqis) and 763 in-country contract SATs—that had originally been established in 2009. Such a direct-hire staff would be about 20 times the size of an average embassy-based security assistance office, which can have no more than six military staff members and a handful of civilians.[49] This staffing level was assessed to be the number of personnel needed to support the FMS cases that were currently in the pipeline. However, the number of cases was expected to grow significantly, and the Senate Foreign Relations Committee expressed concern that an increase in the number of contractors to support the sales—which would be accompanied by an increase in life support and security contractors—"could result in an even larger footprint likely to overwhelm DoS's already lean resources."[50]

Despite the SA's requirement that U.S. forces depart Iraq by December 31, 2011, the agreed-on OSC-I staffing level was based on an unstated assumption that some follow-on U.S. force would remain in Iraq after USF-I's departure to continue train-

---

[45] DoD IG, 2011, pp. 4–5.

[46] OIG, 2011, pp. 1, 19.

[47] NSDD 38, "Staffing at Diplomatic Missions and Their Overseas Constituent Posts," Office of Management Policy, Rightsizing, and Innovation, June 2, 1982.

[48] USF-I, PSG-I, "Office of Security Cooperation–Iraq (OSC-I)," July 9, 2010, para. 4. Also Interview with Kathleen Austin-Ferguson, Executive Assistant to the Under Secretary of State for Management, Washington, D.C., October 25, 2011.

[49] Defense Institute of Security Assistance Management, *The Management of Security Cooperation (Green Book)*, January 2013, p. 17-1.

[50] U.S. Senate, Committee on Foreign Relations, 2011, p. 17.

ing the ISF until it achieved the ability to conduct operations associated with external defense.[51] The failure to reach an agreement regarding a enduring troop presence post-2011 meant that such training would have to be conducted by U.S. units rotating through Iraq or not be conducted at all. As envisioned in late 2011, OSC-I would therefore have to coordinate a robust exercise and training program in addition to its traditional security assistance duties.[52]

USCENTCOM's *Iraq Country Plan* was built around an ambitious 2012–2017 capacity-building plan for the ISF that would require dozens, if not hundreds, of military-to-military interactions annually, utilizing the full range of security cooperation tools under the authorities of both DoS and DoD.[53] However, according to its *Iraq Country Plan*, USCENTCOM would not have an Iraqi Senior National Representative to help design security cooperation activities until 2013 and did not intend to have a detailed bilateral plan worked out with the ISF until 2014. Until it does, OSC-I will have to take an active role in developing and implementing security cooperation activities to execute USCENTCOM's country plan.[54]

With the arrival of LTG Robert L. Caslen, Jr., as the first OSC-I Chief on October 1, 2011, the newly established OSC-I began assessing its requirements for an expanded mission that was envisioned to assume the functions that would have been performed by a residual force. For example, one of the key roles General Caslen and his staff would perform was to mediate emerging conflicts between the Iraqi Army and Kurdish Security Forces within the disputed areas. To do this, OSC-I retained the Arab-Kurd cell that had formerly been part of the USF-I J3. It was also assumed that this new organization would undertake unit training that far exceeded the type of equipment training usually conducted by an OSC.[55] Consequently, a new organizational structure was soon established that looked less like a typical embassy security assistance office than a U.S. Army corps staff, with sections dedicated to personnel (J1), operations (J3), communications (J6), etc., as well as an in-house chief of staff, legal advisor, and force protection staff. The personnel requirements for this proposed expansion ballooned from 157 to 326. In late November 2011, DoD had approved a revised joint manning document for a direct-hire cadre of 326 and submitted it to DoS for submission to the government of Iraq for concurrence. The Iraqi government never approved this last-minute change, however, so the original staffing plan remained in effect, even though the mission had expanded significantly.[56]

---

[51] Interview with USF-I J5 staff officer, Baghdad, December 13, 2011.

[52] Brennan notes, 2009–2011.

[53] USCENTCOM, *Iraq Country Plan*, June 10, 2011a, pp. 5–6, Not available to the general public.

[54] USCENTCOM, 2011a, pp. 31–32.

[55] Interview with OSC-I staff officer, Baghdad, December 14, 2011.

[56] Interview with OSC-I staff officers, Baghdad, December 10, 2011.

The responsibility for training and equipping the ISF formally transitioned from USF-I to OSC-I on October 1, 2011, although OSC-I continued to receive support from USF-I until the latter's departure in December. However, OSC-I was initially ill prepared to assume responsibility for its expanded mission. It lacked the personnel, funding, and other resources necessary to conduct the type of expanded mission it was undertaking. Furthermore, along with other components of the U.S. embassy, OSC-I lacked formal agreements with the Iraqi government regarding critical issues, such as its use of Iraqi facilities and the size of its security footprint.

The difference between having a residual force in Iraq to continue the training mission and the actual withdrawal of all U.S. forces was jarring. Instead of having a three-star headquarters that was given responsibility for a training and exercise program and the number of personnel and availability of resources needed to accomplish the task, OSC-I emerged as the only DoD organization remaining in Iraq.[57] Instead of having a residual force present that could mitigate that challenge, the much smaller OSC-I fell victim to it. Rather than serving as a stabilizing force, OSC-I experienced the same rapidly paced operational tempo as the rest of USF-I, even revisiting training plans and seeking to increase its authorized core staff to 326 in the final days of 2011.

Lessons Learned from Office of Security Cooperation in Iraq Transition Planning

The process of establishing OSC-I offers a number of lessons that should be considered when transitioning security assistance programs from a wartime footing to a "normalized" bilateral relationship.

First, security cooperation planning needs to be informed by policy guidance issued by national-level policymaking entities—such as DoS, OSD, and/or the Joint Staff—and the combatant command headquarters. Furthermore, to encourage transition planners to take a strategic view, one or more of these agencies should lead the transition planning process itself; deferring to a transition cell in an operational command, as was done with USF-I's PSG-I, virtually ensures that plans will emphasize the operational aspects of establishing an office and handing over programs.

Without the benefit of such policy guidance, transition planners focused on operational decisions, such as the structure of OSC-I and the transitioning of specific tasks while giving insufficient consideration to strategic imperatives. Had planners first considered the role of security cooperation in the bilateral relationship and the ways in which OSC-I could advance U.S. strategic objectives in Iraq, they could have developed a budget, human resources plan, and other tools to make OSC-I successful. Moreover, OSC-I officials (like many DoD planners), fully expected that Iraq would accept an enduring U.S. troop presence and thus did not plan to take a leading role in the coordination of continued training for the ISF. This assumption rendered OSC-I

---

[57] Brennan notes, 2009–2011.

underprepared to support U.S. strategic objectives when the President announced that all U.S. troops would be withdrawn.

Second, such strategic questions should be resolved as far in advance as possible of a transition so operational decisions can be made early enough to be incorporated into budget requests, facility plans, and contracting requirements. In the absence of a security cooperation strategy, OSC-I planners could not determine how many staff members the office would need. By the time manpower levels were decided, it was too late to request funding in the normal budget process, and DoS had to adjust its construction plans repeatedly to accommodate changing numbers of OSC-I officers and the security and life-support contractors needed to support them (see Chapter Eight).

Third, early in the transition planning, it is essential to design a security cooperation organization focused on facilitating the full range of traditional cooperative activities, such as senior leader engagements, short-duration advise and assist events, joint exercises, and training and education exchanges, as well as the management of FMF and FMS. It could be argued that too much emphasis was put on OSC-I's FMS functions and not enough on the other aspects of security cooperation that are so important to the development of a professional and effective MOD and security sector. The focus on FMS created an impression that continued training and assistance initiatives would require the deployment of a large standing military force.[58]

Fourth, security cooperation transition plans should draw on personnel with experience in the field. Despite the truly impressive talents and dedication of PSG-I and OSC-I Transition Team staff, virtually none had any security cooperation expertise. Recruiting a core of subject-matter experts to participate in transition planning could help keep the focus on the security cooperation mission rather than on organizational mechanics.

Iraq Security Sector Overview

From 2003 onward, the lack of capacity within the ISF has served as the primary reason preventing the departure of U.S. forces. Thus, ensuring that all elements of the Iraqi security sector were functioning effectively was of critical importance to the transition effort. This section will examine the capabilities of the ISF by department and ministry both to assesses their capability and to highlight how USF-I transitioned responsibility for its remaining security-related activities to their Iraqi counterparts.

Iraq's security agencies are overseen by the MOD, CTS, and the MOI. The Iraqi National Security Council serves a coordinating role and provides oversight on behalf of the Council of Ministers, but the Office of the Commander in Chief retains the real power for overseeing the ISF. As of October 10, 2011, the ISF consisted of over 929,000

---

[58] Interviews with OSC-I official, June 27, 2011, and former MNSTC-I officials, November 10, 2011.

personnel. The MOD oversaw about 200,000 Army personnel, 68,000 training and support personnel, 5,100 Air Force personnel, 2,400 Army Air Corps personnel, and 3,700 Navy personnel. CTS, operating as a separate organization reporting directly to the prime minister, consisted of 4,200 personnel. The MOI oversaw over 645,000 personnel, with over half belonging to Iraqi police forces and the rest to various specialized departments. Table 7.1 is a detailed breakdown of ISF personnel by agency as of October 2011.[59]

In the first few years after it was disbanded and rebuilt from scratch in 2003, the Iraqi military grew rapidly. U.S. efforts initially focused on force generation, the recruiting and training of a large number of military personnel, particularly in the army. The 2007 Independent Commission on the Security Forces of Iraq found the MOD forces had greatly improved their internal security capabilities but still relied heavily on coalition combat support (e.g., air support, intelligence) and combat service support (e.g., logistics). MOI forces, on the other hand, still suffered from extreme dysfunction, including sectarianism, with limited capabilities and meager resources.[60]

Table 7.1
Iraqi Security Forces Personnel as of October 2011

Oversight Entity	Service	Personnel
MOD	Army	200,000
	Army Air Corps	2,400
	Air Force	5,053
	Navy	3,650
	Training and support	68,000
	Subtotal	279,103
MOI	Iraqi police	325,000
	Facilities protection	95,000
	Border enforcement	60,000
	Iraqi Federal Police	45,000
	Oil police	31,000
	Training and support	89,800
	Subtotal	645,800
CTS		4,200
Total		929,103

---

[59] SIGIR, 2011f, p. 54. See also SIGIR, *Quarterly Report and Semiannual Report to the United States Congress,* July 30, 2011d, p. 75.

[60] James L. Jones, chairman, *The Report of the Independent Commission on the Security Forces of Iraq,* September 6, 2007, pp. 8–9, 45–46.

Table 7.2
Funding for Iraqi Ministry of Interior and Ministry of Defense,
2006–2011

US$ (in billions)	2006	2007	2008	2009	2010	2011
MOI	1.9	3.2	5.7	5.5	6.14	6.31
MOD	3.4	4.1	5.3	4.1	4.90	5.84
Total Budget	5.3	7.3	11.0	9.6	11.00	12.15

SOURCES: DoD, *Measuring Stability and Security in Iraq*, June 2010, p. 13; USF-I,
"Responsible Redeployment of United States Armed Forces from Iraq," draft
report for Congress, October 2011f, pp. 8–9, Not available to general public.

At the same time, the resources available for security operations and enhancements declined. Table 7.2 shows MOD and MOI budgets for 2006–2011. Tumbling oil prices beginning in mid-2008 caused budget pressures in 2009, which led the Iraqi government to reduce the MOD's budget by 19 percent. The MOI budget, which was anemic early on but grew dramatically in 2008, absorbed a small reduction in 2009. Perhaps more problematic, systemic corruption exacerbated the budget shortfall for both ministries, with officials inflating numbers of new hires, embezzling pay, purchasing inferior equipment in exchange for kickbacks, and requiring officers to pay for promotions and training.[61]

Despite these challenges, both MOD and MOI forces continued to make progress, particularly in building capabilities useful for maintaining internal security. In its August 2010 report to Congress, DoD assessed that the MOD headquarters was on track to meet MEC objectives in all areas save logistics and sustainment. It assessed the army as making steady progress but falling short for equipping; training; and combined arms integration of the M1A1 tank, artillery, and mechanized enablers. The navy was on track, but the air force would not achieve MEC for fixed-wing airlift or "airspace control," the ability to detect and respond to airspace incursions. DoD assessed the MOI as having made extensive progress despite continued weaknesses in command and control, interoperability, resource and acquisition management, and operational sustainment. The report also warned that continued Iraqi progress would rely on continued U.S. funding through INL programs; ISFF; equipment transfers; and future FMF, IMET, and other traditional security cooperation appropriations.[62]

Iraq's security sector benefitted in 2010 and at least somewhat in 2011 from increases in oil production and a recovery in oil prices from about $50 per barrel in May 2009 to double that amount a little over a year later. But a classic guns-versus-butter debate also affected Iraqi budget decisions, with Iraqi protests driving the Iraqi government to divert $900 million in February 2011 planned for the first installment

---

[61] DoD, "2009–10 GOI Budget Shortfall Impact on Security," undated, p. 2, Not available to the general public.

[62] DoD, 2010, pp. viii–ix.

toward the purchase of 18 F-16 fighter jets—a $3 billion acquisition—to make up a shortfall in the national food ration program.[63] Iraq made its first payment toward the F-16 purchase in September 2011.[64]

Iraq's security sector had significantly improved by the end of 2011. Perhaps the most relevant statistic was the decrease in weekly security incidents from 1,500 in 2007 to fewer than 100 by fall 2011.[65] At midyear 2011, USF-I was still working with the ISF at over 70 sites across Iraq, including ministerial offices, national command and control centers, regional commands, institutional training and education centers, and forward operating bases.[66] In November 2011 testimony to Congress, Secretary of Defense Leon Panetta asserted that "Iraqis have some of the most capable counterterrorism forces in the region." He acknowledged gaps in Iraqi external defense capabilities, including logistics and air defense, but noted that these would be focus areas for OSC-I.[67] The U.S. government even promoted the ambitious goal for Iraq to contribute to stability across the Middle East by promoting regional cooperation on shared security interests, such as counterterrorism, maritime security, and joint exercises.

In contrast, many Iraqi officials were decidedly less upbeat about Iraq's security sector. For example, in August 2010, Iraqi Army General Babakir Al-Zibari, Chief of the Iraqi General Staff, stated that the Iraqi Army likely would not be ready to take over the full spectrum of its defense responsibilities until 2020.[68] Haidar Al-Mullah, a Shiite representative from Ayad Allawi's Iraqiyya party expressed concern about the politicization of the ISF, saying "Maliki considers the Iraqi security forces' ability to protect the government, and not the country, as the deciding factor."[69] While politics certainly play into such statements, they reflect concerns among Iraqis about the ISF's capabilities and legitimacy.

At a grassroots level, the Iraqi populace generally welcomed the notion of an independent and self-reliant ISF serving as the protector of a truly sovereign Iraq, with a majority of Iraqis expressing confidence in the Iraqi military and 74 percent of Iraqis nationwide supporting the withdrawal of U.S. forces from Iraq.[70] Iraqis' confidence in the ISF was not unwavering, however, as clear majorities expressed concern in Novem-

---

[63] Suadad al-Salhy, "Iraq Lawmakers Approve 2011 Budget of $82.6 Billion," Reuters, February 20, 2011.

[64] "Iraq: Procurement," *Jane's Sentinel Security Assessment—The Gulf States*, October 15, 2012.

[65] Leon E. Panetta, "Secretary of Defense Leon E. Panetta Submitted Testimony on Iraq," Committee on Armed Services, U.S. Senate, November 15, 2011, p. 1.

[66] USF-I, "Advising and Training and NATO Training Mission—Iraq," briefing, July 12, 2011j.

[67] Panetta, 2011, pp. 1–2.

[68] "Iraqi Army Not Ready to Take Over Until 2020, Says Country's Top General," *The Guardian*, August 12, 2010.

[69] Haidar Al-Mullah, Iraqiya spokesman media conference (in Arabic), Al-Iraqiya News, June 2011.

[70] "Economic Negativity Abounds in Iraq," Gallup poll, September 22, 2011; James Zogby, "Iraq: The War, Its Consequences & the Future," Zogby Research Services, LLC, November 18–20, 2011, p. 10.

ber 2011 that the ISF would be unable to prevent violence and terrorism in the wake of the U.S. withdrawal.[71] Sunnis, in particular, expressed concern that they would suffer after the U.S. military's departure, presumably out of fear that the Shi'a-dominated Maliki government would use the ISF as a tool of oppression.[72]

Great strides have been taken since 2003 to transform the ISF into a force that is both capable and seen as legitimate by the population, but much remains to be done before it attains the minimum capabilities necessary for both internal security and external defense. The following provides a brief assessment of the primary organizations within MOD and MOI as of the end of 2011.

Iraqi Ministry of Defense

The CPA established Iraq's MOD in March 2004, putting the New Iraqi Army under its authority, along with embryonic air and coastal defense forces. The CPA and U.S. military faced two basic challenges developing the Iraqi MOD. First, the MOD needed the basic capabilities to oversee the armed forces. The Iraqi Army, including the national guard it absorbed in January 2005, was a force created under fire, with quantity initially being its most important quality. Even as the focus shifted to improving capabilities, the primary criteria used to measure effectiveness and improvement evaluated combat skills rather than support functions or professional development. The navy, air force, and eventually army air corps each faced challenges common to relatively small, underfunded military forces evolving and establishing their roles in the shadow of a dominant army. The MOD would need to add value to each of these forces by providing strategic-level guidance for planning, budgeting, acquisition, logistics, personnel management, training, and education. In addition, the MOD would need to integrate these forces into an effective, self-sustaining national military by establishing clear roles and missions for each, developing doctrine that facilitated interoperability, coordinating command and control and planning processes, and prioritizing operational and force development requirements across the forces.

The second challenge was to ensure the MOD itself was a professional and efficient organization. Under Saddam Hussein, the MOD was a thoroughly politicized tool of the regime. If the post-Saddam MOD were to also become politicized, the ISF would likely follow the same eventual path. Moreover, if the MOD became a corrupt, bureaucratic maze doing more harm than good, it would become a drag on the military forces rather than an enabler. Over time, unfortunately, the MOD became hobbled by highly centralized decisionmaking processes, which hindered objective and

---

[71] Zogby, 2011, pp. 10–12. See also Bret H. McGurk, "Statement of Brett H. McGurk, Visiting Scholar, Columbia University School of Law," in *Security Issues Relating to Iraq*, hearing before the Committee on Armed Services, U.S. Senate, 112 Cong., 1st Sess., November 15, 2011, p. 4.

[72] Zogby, 2011, p. 11. Also interview with USF-I J3 staff officer, Baghdad, November 15, 2011.

prompt decisionmaking, undermined the development of a professional staff, and provided opportunities for corruption and politicization at senior levels. "Authorizations for even relatively low-level acquisitions [required] the involvement of Flag level officers and/or Ministry-level executives," according to DoD, while the Minister of Defense personally controlled "almost all procurement and maintenance funding decisions."[73]

These early challenges remained throughout the transition (and are characteristics of other Iraqi ministries as well). Effective oversight and institutional integrity take a generation or longer to develop. Thanks to intensive mentoring from U.S. and other forces, the MOD developed basic capabilities slowly but steadily. The primary weaknesses identified by 2007 were budgeting, contracting, personnel management, intelligence sharing, and logistics.[74] As noted earlier, DoD assessed the MOD in 2010 as making good progress except in the areas of logistics and sustainment, although planning and budgeting remained inadequate as well; DoD worried that the lack of a funding plan for sustaining the ground force, which was approaching full strength, inhibited MOD force improvements.[75]

Shifting the focus of the MOD and its military forces from force generation to force sustainment turned out to be a critical—and time-consuming—step in building a self-reliant military. The DoD IG reported in November 2010 that "USF-I did not have a comprehensive, integrated plan for developing the ISF logistics system." The IG explained that, "the more pressing need to generate ISF combat forces over the past several years precluded the generation and development of enabling forces and capabilities, including that of a logistics sustainment capability."[76] Long lead times are required to develop logistics management capabilities and foster a "culture" that prioritizes logistics and sustainment, and the MOD would likely have become far more self-supporting if sustainment had been a priority of U.S. training efforts early on.

Nevertheless, USF-I had many successes in developing the foundations for a capable, professional MOD. By mid-2011, the Iraqi MOD had made progress on a number of capabilities that were critical to strengthening and sustaining the defense sector over the long term, including materiel readiness measures; use of information technology; cross-ministerial cooperation; doctrine development; maintenance and supply infrastructure; and strategic planning, budgeting, and acquisition.[77] It also established a number of leadership development initiatives, including an English language training

---

[73] DoD, *Measuring Stability and Security in Iraq*, December 2009b, p. 56, and DoD, *Measuring Stability and Security in Iraq*, June 2010, p. 66.

[74] Center for Strategic and International Studies, 2007, pp. 47–54.

[75] DoD, 2010, p. viii.

[76] DoD IG, *Assessment of U.S. Government Efforts to Develop the Logistics Sustainment Capability of the Iraq Security Forces*, Report No. SPO-20 11-001, November 17, 2010, pp. 13, 17.

[77] USF-I, 2011j.

program; the Ministerial Training and Development Center; and the Center for Military Values, Principles, and Leadership Development.

NTM-I was also very active with the MOD and the Iraq armed forces. Despite having only 176 military personnel, NTM-I provided training support to several thousand Iraqi military personnel and police, with a particular focus on professionalization and institutional development.[78] NTM-I had hoped to retain a small retinue of military trainers in Iraq through 2013,[79] but NATO decided to end the mission on December 11, 2011, when negotiations on an extension of the program failed to secure legal immunities for NTM-I trainers.[80] The cancellation of the NTM-I mission reduced the number of foreign military trainers available to support Iraqi military and police training during the posttransition era.

Army and Army Air Corps

The Iraqi Army is the largest and most formidable military service under the MOD, due in large part to the intense focus placed on its development throughout the eight years of U.S. military presence. Its major subordinate commands include the Iraqi Ground Forces Command and the Baghdad Operations Center, both of which report directly to the National Operations Center. In addition to the Baghdad Operations Center, the Ground Forces Command operates in a number of area-centric operational commands (including Anbar, Basra, Diyala, and Ninewa), each of which has a number of divisions assigned to it. The Iraqi Army has a total of 14 divisions and an end strength of approximately 200,000, not counting the Iraqi Army Air Corps, which contains an additional 2,400 personnel.[81] In this subsection, as well as those on the other armed services, we primarily address the transitions that occurred during 2011 and note, where applicable, some major areas in which gaps continue to exist after the transition. The continued existence of such gaps does not necessarily imply a failure in transition planning or execution. Rather, many of the shortfalls simply reflect the fact that it takes a long time to build a military from scratch, and constructing a fully self-sustaining multiservice force that can develop the capacity to execute combined arms operations while battling an internal insurgency is a tall task for any military organization.

Prior to October 1, 2011, the DCG (A&T) was responsible for providing advice, assistance, and support to the Iraqi Army and to the other ISF services as well. For the most part, this took the form of capacity development and assistance at major training base locations, supply centers, and Iraq's version of a combined training center, and

---

[78] USF-I, 2011j.

[79] Serena Chaudhry, "NATO to Continue Iraq Training Mission to End: 2013," Reuters, September 12, 2011.

[80] "NATO to End Training Mission in Iraq," ABC News (Australia), December 13, 2011. See also "NATO Will Not Extend Iraq Training Mission Beyond 2011," *The National*, December 12, 2011.

[81] SIGIR, 2011f, p. 54.

the purchasing of and training on the operations and maintenance of new equipment. Training of units in the field was done primarily by U.S. forces under the DCG for Operations, although this function declined significantly toward the end of 2011 as U.S. tactical units prepared to shut down bases and leave Iraq. As of October 1, 2011, OSC-I assumed full responsibility for a reduced set of these functions. Plans for the OSC-I Army Section included personnel in billets focused primarily on supporting the Iraqi Army through DoS's Title 22 activities (FMS, IMET) and on personnel who would support DoD-funded Title 10 activities (military-to-military engagements, exercise support, and some training support).

The transition of responsibility for domestic security from MNF-I and later USF-I to the ISF was an ongoing task. Of particular note were several significant transitions of responsibility for operations and institutional development. On the operational side these included

- the PIC process, beginning in 2006, in which Iraqi forces became the "battlespace owners" in provinces, as conditions permitted (see Chapter Two)
- establishment of the Baghdad Operations Center to oversee the Battle for Baghdad (also in 2006) and, later, of similar operations centers in other locations (see Chapter Three)
- the "Charge of the Knights" operation, during which the Iraqi government deployed forces operating independently from MNF-I to take on Shi'a militias in Basra in 2008 (see Chapter Four)—although this required U.S. forces to provide significant amount of "invisible support," including ISR, fixed-wing attack aircraft, and special operations capabilities
- the departure of USF-I forces from Iraqi cities in accordance with the SA in 2009.

On the institutional side, major transitions included the Iraqi Training Brigade, with three subordinate battalions becoming fully operational in 2006 (many other elements of the Iraqi training base were developed and became operational around this time as well), and the completion and hand over of the national depot complex at Taji in 2009.

OPORD 11-01 directed ITAM-Army to transition eight key tasks and four enduring tasks to OSC-I, while ISAM tasks would remain with the security assistance element of OSC-I or be terminated.[82] The ITAM-Army key tasks to be transitioned to OSC-I, as outlined in OPORD 11-01, were as follows:

- Modernize the Iraqi Army, including fielding the M1A1, M109A5, M198, M113 family of vehicles, M88A2, heavy-equipment transport systems, and strategic

---

[82] USF-I, "Iraq Training and Assistance Mission–Army, Transition Plan," briefing to RAND, Baghdad, June 27, 2011h, Not available to the general public; USF-I, Iraq Training and Advisory Mission–Navy, "Transition Plan," briefing to RAND, June 28, 2011i, Not available to the general public.

bridging company, with priority to the 9th, 5th, 10th, 14th, 2nd, 7th, and 12th, Iraqi Army divisions.

- Support al-Tadreeb al-Shamil–partnered training by the U.S. divisions with the 3rd, 5th, 7th, and 10th Iraqi Army divisions.
- Complete the equipping and training of up to eight Kurdish RGBs and complete their integration into the Iraqi Army by end of mission, with the first four RGBs expected to begin equipping and training June–July 2011.
- Initiate the equipping and training of two joint operations commands (JOCs) to provide increased command and control capabilities for the Iraqi Army. Establish the first JOC and achieve initial operational capability in September–October 2011 with the arrival of the first sets of equipment.
- Prepare enduring partnerships to ensure future development of Iraqi Army professional competencies to manage and execute operational training and maintenance in accordance with Annex V (Interagency Coordination) of OPORD 11-01.
- Advise, train, assist, and equip the Iraqi Army to improve the ineffective Iraqi Army logistic systems to enhance the Iraqi Army's ability to sustain its own forces.
- Set the foundation for future recurring biannual joint exercises with support to Operation Lion's Leap.
- Coordinate extensively with OSC-I to set the conditions for the transfer of all enduring security assistance and security cooperation activities no later than October 1, 2011.
- Advise and assist Iraqi Ground Forces Command.
- Conduct security assistance and security cooperation activities to modernize and train the Iraqi Army with the ability to conduct combined arms and joint operations.
- Conduct security assistance and security cooperation activities to develop Iraqi Army sustainment capabilities
- Conduct key leader engagements.

Although these transitions appear to be simple transfers of responsibility from ITAM to OSC-I, the latter was expected to carry out these tasks far differently from the way USF-I had. OSC-I would operate under U.S. embassy management, have far more limited manpower, and be subject to greater Iraqi controls on its activities. It would also have far more limited mobility because Embassy Baghdad would base its rules of movement on more-limited intelligence and logistical support. Those delivering this support to the Iraqi armed forces would be either at OSC-I headquarters in Baghdad or colocated at ISF installations.

The Iraqi Army's efforts to modernize and mechanize its forces were facilitated by the delivery of a wide range of American materiel and weapon systems. Iraq agreed to purchase several important major pieces of equipment, such as M1A1 tanks, M113 personnel carriers, and M109A5 and M198 howitzers. Delivery of this equipment and training on their operations and maintenance continued throughout 2011, although units equipped with these items were not able to complete combined arms training by December 31, 2011.[83] DoD assessed that training shortcomings would prevent the Iraqi Army from reaching "combined arms proficiency above the company level (infantry with tanks, artillery, engineers, and army helicopters all conducting synchronized fire and maneuver training)" until FY 2014.[84] This kind of capability would be necessary for the Iraqis to defend their borders effectively against an external conventional threat, and its absence will hinder the army's ability to perform its primary mission in the interim.

Much progress was made in training, although gaps remained at the time of the U.S. military's departure. To manage a training program, ITAM-Army helped the Iraqi Army develop the Iraqi Army Training and Leader Development manual, similar to U.S. Army Regulation 350-1.[85] This document established a training program, guidelines, and requirements for the Iraqi Army. To address the shortcoming in combined arms training, USF-I developed and began implementation of a combined arms training program, al-Tadreeb al-Shamil, focused on small unit operations.[86] This training program was designed to enable Iraqi battalions to develop skills necessary for national defense, not just internal security. During the 25-day training rotation, training would include individual skills, such as marksmanship and land navigation. In addition, squads, platoons, and companies would conduct live-fire maneuvers, culminating with battalion combined arms maneuver training. By August 2011, this training program had been resourced, planned, and conducted completely by the Iraqis. The question for transition is whether or not these practices have been adequately institutionalized and are sustainable.

As USF-I transitioned bases to the Iraqi government, ITAM-Army facilitated the development or handover of nine fully functional training centers and 14 proponent schools (e.g., the Military Police school at Numaniyah).[87] BCTC—the largest base turned over to the Iraqis—hosts both the Armor School and the Combined Arms School.[88] These schools and the BCTC infrastructure will support the ongoing fielding and training of heavy equipment—such as 140 M-1A1 tanks and a number of M-113 armored personnel carriers, M-109A5 tracked artillery, M-88 tracked recovery vehicles, and M-198 towed howitzers—for the armored and artillery forces. BCTC has

---

[83] USF-I, 2011e.

[84] USF-I, 2011e.

[85] U.S. Army Regulation 350-1, Army Training and Leader Development, 2011.

[86] Brennan notes, 2009–2011.

[87] USF-I, 2011h, p. 10.

[88] USF-I, Public Affairs Office, "Iraqi Armor School Moves to Besmaya," December 12, 2010b.

approximately 220 km² of training area encompassing four maneuver training areas for large armored units; 44 separate live-fire training areas; and the capacity to house approximately 4,500 troops for rotational training.[89]

The Iraqi Army lacked a corpus of doctrine and training materials (similar to those of the U.S. Army Training and Evaluation Program) that would enable army officials to set training plans; identify training standards; set training conditions; and determine what materials, facilities, and equipment are needed for this training, particularly for higher echelon units.[90] OSC-I was not expected to have the manpower, authorities, or resources to work with large units, so unless the Iraqi Army developed its own training doctrine, with or without U.S. assistance, the Iraqi Army would not be able to sustain a self-directed training program.

Significant concern also existed about the Iraqi Army's ability to maintain the high-technology equipment it was receiving through FMS programs, such as M-1 tanks. Countless stories about the lack of a "maintenance culture," bolstered by the progressive abandonment of core capabilities of major maintenance facilities, such as the one MNF-I had built at Taji, were thought to indicate that the Iraq Army would need significant help if it were to keep its equipment functioning.[91] The impending departure of U.S. forces resulted in a "surge" toward building the ISF sustainment capacity along with a desire to provide a multitude of equipment. However, without an overarching long-term FMS strategy in place, the United States supplied equipment to the ISF without first developing life-cycle management plans, fully determining ISF logistics capacity and capability, or minimizing the number of equipment variants. For instance, the ISF received over 40 variants of night-vision goggles, as well as various models of cargo and other vehicles because they had become outdated for U.S. forces. The increase in FMS, ISFF, and USETTI equipment fielding quickly overwhelmed the ISF Taji National Depot, the MOD main repair parts and maintenance center, with stocks to support the equipment and its many variants. Although these issues were recognized in early 2011, and the flow was tailored, metered, and better coordinated

---

[89] USF-I, 2011k, pp. 38 and 39. Also see USF-I Public Affairs Office, "Besmaya Combat Training Center Transferred to Iraq Army," August 1, 2010a.

[90] Interview with BG Robin Maeler, June 2011. This combination of a body of training doctrine and facilities that units use to develop capabilities and a culture of unit training is, arguably, one of the significant strong points of the U.S. Army. One could argue that developing such a culture of training during an ongoing conflict is a tall order and that, in any event, eight years may not be enough time to do this completely. However, without doctrine that helps unit leaders plan and deliver training, it is hard to imagine how this culture would develop adequately.

[91] USF-I, 2011e, p. 10, and interview with Ginger Cruz, Deputy SIGIR, November 2011. Also, according to USF-I, "FMS Synchronization," briefing, October 2011m, maintenance and logistics military-to-military support will remain a key task in plans to build partner capacity.

with the ISF logistics personnel, it still left a logistics system that would need years to mature to support the ISF mission.[92]

The Iraqi Army Air Corps was a maturing force, capable of basic rotary-wing support of the Iraqi Army. By October 2011, it had successfully demonstrated the ability to conduct casualty evacuation, air movements, and very limited air-to-ground attack support. Although its capabilities were limited by the service's small size—it had just 2,400 personnel assigned at the end of September 2011—its ability to train these personnel as pilots and support staff was considered to be the greater obstacle.[93] As USCENTCOM's *Iraq Country Plan* states:

> The ability of the IqAAC [Iraqi Army Air Corps] to train its own pilots and pro- vide technical and professional development training for its personnel is extremely limited. The IqAAC will rely heavily on outsourced training for several years until its force generation capacity is developed.[94]

Navy and Marines

Iraq's naval capability was centered at Umm Qasr and consisted of riverine and border control elements operating a collection of small patrol craft. The head of the Iraqi Navy, a two-star admiral, oversaw the Iraqi Marine Brigade and the Navy Operational Command, both one-star commands, as well as the Basra Maritime Academy.

In the years prior to the transition, ITAM-Navy worked with the Iraqi Navy to develop a wide range of operational, sustainment, logistics, training, and command and control skills focused on both maritime security and maritime infrastructure protec- tion. It transferred some of these responsibilities to OSC-I at the end of 2011, although it handed off a roughly equal number of tasks directly to the Iraqi Navy. The follow- ing are the key tasks that USF-I transitioned to the Iraqi Navy and the Iraqi Marines, which also illustrate the range of topics addressed by ITAM-Navy training efforts:

- ITAM Navy tasks to be transitioned to OSC-I
  - Advise and assist Iraqi Navy in sustainment of force elements at readiness.
  - Advise and assist Iraqi Navy and Marines' ability to independently sustain force elements, through effective logistics, engineering, and training support.
  - Assist through oversight of Umm Qasr base expansion and related projects in quality assurance.
  - Continue key leader engagements with RADM Ali Hussain Ali—Iraqi Head of Navy and MOD—and his senior staff.

---

[92] Interview with USF-I J4 ISF logistics staff officer, North Carolina, February 1, 2013.

[93] SIGIR, 2011f.

[94] USCENTCOM, *Iraq Country Plan, Appendix 3 to Annex O (Security Cooperation), Iraqi Army Security Cooper- ation Roadmap*, Headquarters, October 12, 2011b, pp. O-4-1 through O-4-2, Not available to the general public.

- Assist in forming future ISR capabilities relative to threat levels of the area of responsibility.
- Train and assist Iraqi Navy joint operations to understand and execute procedures by organizing operations between Iraqi services essential in defense of their limited maritime access.
- Conduct assessments of FMS cases to prioritize with respect to finances.
- Maintain accountability over existing maintenance and warranty contracts for assets purchased through FMS/ISFF.
- ITAM Navy tasks to be transitioned to the Iraqi Navy or Iraqi Marines
  - Assess Iraqi Marines in point protection of oil platform and visit, board, search, and seizure procedures.
  - Advise and assist in command and control and protection operations of vital oil infrastructure and territorial waters.
  - Assess training of Iraqi Navy personnel in seamanship, ship handling, repair and maintenance, and logistics.
  - Assess competency of Iraqi Navy to patrol and defend the oil platforms and territorial waters.
  - Train and assist Iraqi Navy on logistical system and its proper usage to maintain Iraq's growing fleet.
  - Assist in advancement of mine countermeasure capabilities through training opportunities and relationships with neighboring countries.

Although the Iraqi Army is the largest and most capable element of the ISF by most standards, the Iraqi Navy was perhaps the most ready to perform its strategic missions—defending Iraqi oil platforms from harm. Not only does the Iraqi Navy have a narrow, well-defined mission, it has also been able to grow without the need to conduct simultaneous combat operations that the Iraqi Army has. The Iraqi Navy had been responsible for this mission since 2009, although initially with U.S. assistance.[95] According to the then-director of ITAM-Navy, RADM Kelvin N. Dixon, the Iraqi Navy was capable of accomplishing its mission without assistance as early as June 2011.[96] Furthermore, as far back as summer 2010, the Iraqi Navy showed significant proficiency in planning and executing complex combined training exercises with the Kuwaiti Navy.[97]

The Iraqi Navy had led its own operational training, with only limited U.S. assistance, for some time. However, its training was focused almost entirely on its coastal defense mission and was not yet capable of missions beyond Iraqi territorial waters.[98]

---

[95] USF-I, 2011e.

[96] Interview with ITAM-Navy Director RADM Kelvin N. Dixon, Baghdad, June 2012.

[97] DoD, 2010, pp. 77–78.

[98] Interview with ITAM-Navy Director RADM Kelvin N. Dixon, Baghdad, June 2012.

In the run-up to transition, ITAM-Navy highlighted maintenance deficiencies as an important area of concern. Admiral Dixon asserted in June 2011 that the Iraqi Navy has struggled to maintain its patrol boats and its more complex vessels, which he thought might hinder its ability to keep them seaworthy, a conclusion echoed by both SIGIR and the June 2010 DoD quarterly report to Congress, *Measuring Stability and Security in Iraq*.[99] Should this be the case, the Iraqi Navy's operational skills would be negated by its lack of maintenance and support capabilities, rendering it unable to secure Iraq's oil terminals or coast.[100]

To mitigate the maintenance problems, the Iraqis initiated an FMS case to cover the construction of a state of the art maintenance facility in Umm Qasr, which, when fully operational and staffed with trained mechanics, would permit the Iraqi Navy to maintain its fleet. However, developing the cadre of trained personnel needed to operate this facility will take some time; training is taking place in waves, with personnel being trained on one skill at a time. If the maintenance culture in the Iraqi Army is any guide, it will take some time for the Iraqi Navy to develop and institutionalize this capability. OSC-I will administer this FMS case and oversee the continued provision of training on maintenance capabilities.[101]

Air Force

At the end of the transition, the Iraqi Air Force was not as mature a force as the Iraqi Army and was not as prepared for its most important missions as either the Iraqi Army or the Iraqi Navy. This was largely because of two facts. First, the development of the Iraqi Air Force was not an early high priority because coalition air forces addressed immediate needs, so capacity-building efforts started much later than for the other services. Second, facilities to support modern air forces, whether they be operational units, maintenance facilities, or command and control headquarters, required sophisticated equipment and highly skilled personnel, all of which take a long time to put in place. Ideally, air force pilots, mechanics, and other skilled personnel should receive their training—which can take several years—before aircraft are delivered, both to get maximum use out of the capital investment and to prevent new planes from deteriorating needlessly.[102] As a result, most functions previously performed by U.S. air assets could not be transitioned adequately to the Iraqi Air Force.

---

[99] Iraq currently has an assortment of small vessels, which includes five of 12 U.S. *Swift*-class patrol boats and four Italian *Saettia*-class boats. Interview with ITAM-Navy Director RADM Kelvin N. Dixon, Baghdad, June 2012. Also SIGIR, 2011f, and DoD, 2010.

[100] Interview with ITAM-Navy Director RADM Kelvin N. Dixon, Baghdad, June 2012.

[101] USF-I, 2011i.

[102] Iraqi pilots are currently in the United States training to fly F-16 aircraft, despite the fact that the first F-16 is not scheduled for delivery to Iraq until 2014. Interview with former Deputy Assistant Secretary of Defense, November 2011.

The Iraqi Air Force has several missions. Among the most important are air mobility, support to ground forces, air sovereignty, and ISR. Air mobility was one of the first capabilities developed in the Iraqi Air Force, and Iraqi pilots had been flying C-130s for several years. Moreover, the Iraqi purchase (through FMS) of additional C-130Js, scheduled for delivery in 2013, would increase these capabilities.[103]

The capability for the Iraqi Air Force to provide ISR and armed support to ground forces lagged significantly. Although it was able to provide limited ground support capabilities using light turboprop aircraft, it will take several more years to utilize the F-16 multirole fighters it purchased to perform this mission.[104] Additionally, the technical skills necessary to synchronize ground and air forces are difficult to develop and were not present in the ISF. According to Chief of the Iraqi General Staff, General Babakir Zibari, neither the Iraqi Air Force nor the Iraqi Army Air Corps were capable of supporting ground forces in contact.[105] While USF-I remained in Iraq, U.S. Army attack helicopters or U.S. Air Force fighter aircraft frequently supported Iraqi ground forces engaged in conflict.[106] With the departure of USF-I, Iraqi ground forces lost all but very limited attack aviation support.

Air sovereignty is a product of three major functions—the ability to "watch, warn, and respond."[107] Establishing air sovereignty for Iraq required a combination of appropriate hardware and facilities, the ability to operate and maintain them, and command and control capabilities to tie it all together. With respect to hardware, the Iraqi Air Force opened a $9 million air operations center in April 2011 and soon afterward put into operation long-range radars capable of covering most of Iraq's airspace.[108] However, Iraq did not have the command and control capabilities to make effective use of the "watch" output of these radars, either to warn air and ground forces or to respond.[109] The Iraqi Air Force was able to conduct limited ISR using its exist-

---

[103] USF-I, 2011m.

[104] Iraq contracted to purchase 18 F-16 fighters from the United States, with the first deliveries expected in 2014. See SIGIR, 2011f, p. 54

[105] SIGIR, 2011f, p. 5.

[106] Brennan notes, 2009–2011.

[107] BG Russell Handy, 9th Air and Space Expeditionary Task Force Commander, briefing to RAND, Baghdad, June 26, 2011; USF-I, 2011f, p. 9.

[108] Jason Lake, "Iraqi Air Force Leaders Thank U.S. Advisors, Celebrate 80th Anniversary," 321st Air Expeditionary Wing Public Affairs, April 22, 2011. See also Andrew Slovensky, "New Radar Paints Iraq's Air Defense Picture," 362nd Mobile Public Affairs Detachment, November 3, 2011; Jim Loney, "Iraqi Air Defense: A Work in Progress," Reuters, September 27, 2011; and DoD, "Contracts," press release 391-11, May 9, 2011c.

[109] Handy, 2011.

ing aircraft,[110] primarily King Air 350ERs,[111] but information from U.S. ISR missions was largely lost when USF-I departed. Furthermore, Iraq lacked the ground-based air defense capabilities and the air superiority platforms to respond to threats, even if they did have adequate "watch" capabilities to "warn" the force. For example, in September 2011, the ITAM-Air Director Brig Gen Tony Rock stated that Iraq would not have an air intercept capability until it received and was able to use its F-16 fighter jets.[112] Then, in November 2011, ACCE-Iraq director Maj Gen Russell Handy forecast that it would be at least two years from the transition before Iraq was able to do so.[113] So, while Iraq had the first element of the watch, warn, and respond trifecta in place and while it had programs under way to create response capabilities (such as the F-16 FMS purchase and related pilot training), it would need to create the missing elements or ask for further assistance after 2011.

By the end of the transition, the Iraqi Air Force did not have adequate training capabilities of its own. USCENTCOM assessed in October 2011 that the Iraqi Air Force was not capable of "recruiting, training, and equipping enough personnel across all career fields; maintenance suffers significantly."[114] It had, however, made some progress. Since May 2011, it had run the Air Force Technical School without U.S. assistance, and it also ran the Air Force College, with a mission of producing air force officers. While the Iraqi Air Force had the ability to train basic airmen and while it had a small cadre of instructor pilots, it was expected to need assistance from abroad to develop more advanced skills.

Finally, the Iraqi Air Force's challenges were exacerbated by the fact that it was undermanned. Given the long lead times to develop highly skilled airmen and pilots, Iraq looked likely to be without adequate air force for years to come at the transition and will not be able to defend its own airspace until 2020, according to General Zibari.[115]

Iraqi Special Operations Forces

Prime Minister Maliki established CTS to function as a quasi-ministerial agency over which he exercised effective control. CTS comprised the two ISOF brigades and a

---

[110] Stacia Zachary, "Iraqi Air Force Builds ISR Foundation with Help from Airmen," Armed Forces News Service, June 16, 2009. See also Stacia Zachary, "Iraqi Air Force Brings ISR Capabilities Online," Armed Forces News Service, August 24, 2009.

[111] "Standing Up the IqAF: King Air 350s," *Defense Industry Daily*, August 25, 2009.

[112] Loney, 2011.

[113] Geoff Ziezulewica, "USAF General: Iraqi Air Defenses to Have Two-Year 'Gap,'" *Stars and Stripes*, November 7, 2011.

[114] USCENTCOM, *Iraq Country Plan, Appendix 5 to Annex O (Security Cooperation), Iraqi Air Force (IqAF) Security Cooperation Roadmap*, October 12, 2011c, p. O-5-1, Not available to the general public.

[115] SIGIR, 2011f, p. 5.

dedicated intelligence organization. Two Federal Police brigades primarily tasked with a counterterrorism function also operated as part of CTS. In addition, while not officially part of CTS, the Iraqi Army's 54th Brigade and its 56th Brigade (also known as the Baghdad Brigade) were often given CTS-related missions because of their level of training, extensive resourcing, and close relationship with the prime minister. Like CTS, both the 56th and 54th Brigades received their orders directly from General Farouk Al-Araji, Director of the Office of the Commander-in-Chief.[116]

The ISOF mission was to conduct COIN operations and deny safe haven to terrorists. As discussed earlier, Prime Minister Maliki moved the ISOF out from under MOD control and placed if directly under the prime minister's office through three executive orders in late 2006 and early 2007. U.S. SOF had trained ISOF and accompanied them on missions since the end of 2003. In addition to providing training in Iraq, some ISOF attended more-advanced training in the United States and in Jordan, where they partnered with Jordanian Special Forces.[117]

By 2007, according to the *Report of the Independent Commission on Security Forces in Iraq*, Iraqi SOF were not only the most capable force within the ISF but also one of the most capable special forces in the Middle East, well trained in both individual and collective skills. They were assessed as being capable of leading counterterrorism missions but still in need of coalition support, especially military airlift, close air support, and targeting intelligence.[118] Because it was outside the MOD/MOI structure, CTS was able to build and draw on more robust support capabilities than were available to general purpose forces, which depended on the MOD/MOI's flawed maintenance and sustainment structure.

USF-I's JFSOCC-I and its predecessor organizations steadily progressed from 2008 through 2011, from direct training of ISOF, to a train-the-trainer approach, to a mentoring and advising role, all the while continuing to accompany ISOF on operations when requested. By the end of 2009, DoD had spent over $237 million in ISFF money for sustainment, infrastructure, equipment, training, and operations for ISOF development.[119]

OPORD 11-01's "Strengthen the ISF" LOO included a requirement to help the ISF develop forces capable of counterterrorism operations. OPORD 11-01, Annex C,

---

[116] The 54th and 56th Brigades are both entirely Shi'a organizations with unquestioned loyalty to the prime minister, serving more as a palace guard than as an institutionalized counterterrorism unit. During the height of the civil war in 2006 and 2007, both organizations were linked to sectarian violence targeting Sunni leaders, political violence targeting opponents of Prime Minister Maliki, and the maintenance of secret prisons in and around Baghdad. Interview with USF-I J3 staff, Baghdad, August 1, 2011.

[117] SIGIR, 2010b, pp. 1–4, 15. Also, interview with USF-I staff, December 18, 2011.

[118] Jones, 2007, pp. 14, 55. The Senate Foreign Relations Committee expressed concern in 2011 that ISOF might become less capable on the departure of U.S. forces, because it still relied on the United States to integrate intelligence. See U.S. Senate, Committee on Foreign Relations, 2011, p. 6.

[119] SIGIR, 2010b, p. 8.

Appendix 4, created an integrated operational framework to guide continued ISOF training under JFSOCC-I in 2011. JFSOCC-I developed its own plan based on an assessment that U.S. SOF would continue "training a force in contact" through the departure of U.S. forces from Iraq in 2011 and beyond.[120] No other Iraqi force faced a more challenging combination of intense training requirements while simultaneously conducting intense operational missions. As characterized by members of JFSOCC-I, Iraqi SOF units were learning to build a plane while flying it—and being shot at.[121]

The Independent Commission on Security Forces in Iraq was concerned about the politicization of ISOF units, given their direct reporting chain to Prime Minister Maliki, and there was ample evidence that Maliki had used ISOF for missions that may have been politically motivated.[122] Officials in charge of ISOF at the regional level as of the end of 2011 had personal connections to Prime Minister Maliki and were entirely loyal to him; several were linked to judicial abuses and alleged to have directed raids that led to the arrests of Sunni political rivals.[123] In March 2011, the government reportedly deployed ISOF to break up Arab Spring–inspired protesters and arrest critics of the prime minister.[124] Furthermore, although ISOF had been relatively nonsectarian, the CTS experienced a gradual "Shi'afication" beginning in 2011. Despite U.S. advisors' consistent recommendation that the demographics of CTS units reflect those of the population as a whole (roughly 70 percent Shi'a, 15 to 20 percent Sunni, and 10 to 15 percent Kurd), the percentage of Shi'a in CTS classes became disproportionately high.[125]

It is common for Middle Eastern leaders to select elite security units with personal loyalty to the leader not only to provide protection but also to prevent senior military officials from conducting a coup. Such actions are often called "coup-proofing"—the development of highly trained and equipped praetorian guards that can also serve as tools of political repression.[126] U.S. military officials consistently emphasized to Iraqi leaders the need to develop an apolitical counterterrorism capacity that is integrated

---

[120] USF-I, OPORD 11-01, Annex C, Appendix 4, 2011, Not available to the general public.

[121] Interview with JFSOCC-I J3 staff, Baghdad, November 15, 2011.

[122] Jones, 2007, pp. 14 and 55.

[123] Interview with USF-I staff, December 18, 2011; Shane Bauer, "Iraq's New Death Squad," *The Nation*, June 3, 2009; Jack Healy, Tim Arango, and Michael Schmidt, "Premier's Acts in Iraq Raise U.S. Concerns," *New York Times*, December 13, 2011; and Roy Gutman, Sahar Issa, and Laith Hammoudi, "Iraq's Maliki Accused of Detaining Hundreds of Political Opponents," McClatchy Newspapers, January 19, 2012.

[124] Stephanie McCrummen, "Protesters Say Maliki Is Using Special Security Forces to Shut Down Demonstrations in Iraq," *Washington Post*, March 3, 2011.

[125] The graduates of the eight selection classes preceding November 1, 2011, were 85 percent Shi'a, 13 percent Sunni, and only 2 percent Kurd. Interview with JFSOCC-I J3 staff, Baghdad, November 15, 2011.

[126] For a detailed assessment of the practice of coup-proofing in the Middle East, see James T. Quinlivan, "Coup-Proofing: Its Practice and Consequences in the Middle East," *International Security*, Vol. 24, No. 2, Fall 1999, pp. 131–165.

into the military chain of command, but the prime minister's office disregarded this counsel.[127] By the end of 2011, Prime Minister Maliki had successfully surrounded himself with loyal forces assigned to the 54th and 56th Brigades and the 6th Iraqi Army Division, located in Baghdad. Prime Minister Maliki has also used his powers to personally promote loyal officers and retiring officers he suspected would be independent. The debate within USF-I was whether Prime Minister Maliki was simply consolidating power and coup-proofing the government or whether he was taking systematic actions that would enable him to serve as an authoritarian leader with military forces at his disposal to buttress the power of a nascent and corruptible legal system to target political adversaries.[128] That debate continues today.

Ministry of Interior

Although the CPA disbanded the old Iraqi Army, it did not dissolve the MOI because it believed that continued police functions would be critical and because it wanted to put Iraqi entities in the lead for internal security as quickly as possible.[129] However, MOI reform was slow and difficult.[130] For instance, in December 2006, the Iraq Study Group reported that the Iraqi MOI was "confronted by corruption and militia infiltration and lacks control over police in the provinces."[131] In 2007, the Independent Commission on Security Forces in Iraq found the MOI to be "dysfunctional and sectarian" and "struggling to be even partially effective as a government institution."[132]

These problems endured. As the transition loomed, the challenges of reforming the ministry were made even more difficult by the lack of a permanent minister to provide guidance and direction and Prime Minister Maliki's tenure as the acting minister.[133] Iraq's MOI was responsible for supporting Iraq's police forces, protection of government facilities, border control, tribal affairs, and immigration and passport regulation. With about 645,000 personnel, the MOI was Iraq's largest security agency,

---

[127] DoD, 2010, pp. 79–80.

[128] Brennan notes, 2009–2011.

[129] Dobbins et al., 2009, pp. 71–74. See also Andrew Rathmell, Olga Oliker, Terrence K. Kelly, David Brannan, and Keith Crane, *Developing Iraq's Security Sector: The Coalition Provisional Authority's Experience*, Santa Monica, Calif.: RAND Corporation, MG-365-OSD, 2005.

[130] See Andrew Rathmell, *Fixing Iraq's Internal Security Forces: Why Is Reform of the Ministry of Interior So Hard?* Washington, D.C.: Center for Strategic and International Studies, November 2007. Also see Robert Perito, *The Interior Ministry's Role in Security Sector Reform*, Washington, D.C.: United States Institute of Peace, May 2009.

[131] Iraq Study Group, 2006, p. 13.

[132] Jones, 2007, p. 86.

[133] SIGIR, *Quarterly Report to the United States Congress,* April 30, 2011c, p. 9.

dwarfing the MOD, which was about 43 percent as large (with a force strength of approximately 280,000).[134]

Police

The Iraqi MOI is responsible for federal police forces in 15 of Iraq's 18 provinces. However, the MOI does not oversee the police forces in the Kurdistan region, which are directed by the Kurdistan MOI.[135] The MOI also manages several Iraqi police entities:

- Iraqi Police Service (325,000): Composed of local patrol and station police, as well as specialists, such as forensic specialists, who are assigned throughout 15 of Iraq's 18 provinces. Its mission is to enforce the law, safeguard the public, and provide internal security at the local level.
- Federal Police (45,000): Serves as a bridging force between the Iraqi police Service and the Iraqi Army and could be called on if the capabilities of the local police are unable to control an incident.
- Border Police (60,000): Provides law enforcement at border crossings and ports of entry to protect Iraq from unlawful entry.
- Oil Police (31,000): Provides security for Iraq's oil infrastructure.
- Facilities Protection Services (95,000): Provides protection to the personnel and facilities of Iraqi government ministries.[136]

In addition to expanding the size of the police force tenfold from the Saddam era, DoD police training provided all officers with basic skills and human rights instruction, and many received specialized instruction in criminal investigations, forensics, and other advanced skills. In 2006, the MOI assumed responsibility for all police training centers, although the United States and other nations continued to provide specialized training courses.[137]

In the long run, the Iraqi government intended to disengage the Iraqi Army from routine internal security responsibilities and to transfer primary responsibility for this mission to police forces through the "Transition of Responsibility for Internal Security to the Police," a plan ITAM-Police and the MOI developed jointly in 2009. The objective, as described by then-USF-I commander General Odierno, was to

> assist the ISF in establishing goals and plans to begin the transition to police primacy, which will entail the Ministry of Interior and Iraqi police assuming full

---

[134] SIGIR, 2011c, p. 4.

[135] SIGIR, 2010a, p. 3.

[136] These numbers reflect payroll data, not present-for-duty totals (SIGIR, 2011f, p. 54).

[137] SIGIR, 2010a, pp. 11–18.

responsibility for internal security. This will allow the Ministry of Defense and the Iraqi Army to focus on training to deter or defeat external threats.[138]

While Iraq's national security advisor established a high-level committee to implement this transition plan, the initiative was stalled throughout the period in which Prime Minister Maliki served as interim Minister of Defense and Interior. Regardless, as long as sectarian and extremist organizations continued to engage in violence, USF-I expected that the Iraqi Army would continue to focus primarily on defeating these insurgents.

Senior officials in the MOI and MOD informed SIGIR during 2011 interviews that the Iraqi police forces were unable to secure all of Iraq's urban areas without assistance from the Iraqi Army as of October 2011. The government of Iraq was conducting a province-by-province assessment to determine when the MOI could assume complete responsibility for security in all of Iraq's major cities.[139]

Transitioning U.S. Responsibility for Training Iraqi Police

Before the Senate Committee on Foreign Relations on February 1, 2011, Ambassador Jeffrey and General Austin testified, "we need to help the Iraqis to professionalize their police, an absolutely critical component to the country's long-term stability."[140] Between 2003 and the end of 2011, the United States spent approximately $8 billion to train, staff, and equip Iraqi police forces.[141] However, police reform in Iraq was slow and difficult to implement due to sectarianism and endemic corruption.[142]

As noted, INL was responsible for the initial Iraqi police training program that began soon after major combat operations ended. However, due to the deteriorating security situation and magnitude of the police training challenge, on May 11, 2004, NSPD 36 transferred the mission of organizing, training, and equipping Iraq's security forces, including the police, to USCENTCOM until the secretaries of State and Defense agreed that DoS should reassume these responsibilities.[143] Even when police training was run by MNSTC-I or ITAM, however, INL recruited and paid for international civilian police advisors for the training centers.

---

[138] Odierno, 2010, p. 100.

[139] SIGIR, 2011f, p. 8.

[140] Jeffrey, 2011a, and Lloyd James Austin III, "Statement of GEN Lloyd James Austin III, Commanding General, United States Forces–Iraq," *Iraq: The Challenging Transition to a Civilian Mission*, hearing before the Committee on Foreign Relations, U.S. Senate, 112th Cong., 1st Sess., February 1, 2011a, pp. 12–13.

[141] SIGIR, *Iraqi Police Development Program: Opportunities for Improved Program Accountability and Budget Transparency*, SIGIR 12-006, October 24, 2011e, p. 1.

[142] For instance, Jones, 2007, pp. 112, 114, found that "in its current form the National Police is not a viable organization" and that "the National Police should be disbanded and reorganized under the MOI."

[143] NSPD 36, 2004; SIGIR, 2010a, p. 1.

In preparing to transfer its responsibilities, DoD drew down and then closed its large-scale ITAM police advisory and training efforts and transferred lead responsibilities back to INL. Before the transition, DoD oversaw more than 600 civilian police advisors at 12 locations around the country.[144] A 90-day transition from DoD to INL began July 1, 2011, and coordination in the field went very well. As was the case with other aspects of transition planning, Washington-based officials from DoD and DoS had very little contact regarding the handover. While the transition of responsibilities from ITAM to INL appeared to be a success, as will be discussed, the long-term viability of the program should have been carefully scrutinized from a strategic and policy perspective in Washington early in the transition planning process.

In planning for post-2011 MOI-related training, DoS changed the focus of police training activities in Iraq, suffered from budgeting uncertainties, and failed to achieve Iraqi buy-in for the type of training and assistance the United States wanted to provide (see Chapter Eight).

The Bureau of International Narcotics and Law Enforcement Affairs Police Development Program

On October 1, 2011, INL reassumed responsibility for the police training program from DoD, launching its billion-dollar-per-year PDP. However, the program had been troubled since the planning stages. It lacked Iraqi government buy-in, and INL had little understanding of the Iraqi police's baseline capabilities. INL had trouble recruiting qualified trainers, underestimated the difficulty of traveling in a hostile security environment, and spent hundreds of millions of dollars on facilities despite having no formal permission to use the sites.[145]

Since ITAM had been responsible for building the Iraqi police forces, PDP was designed to "professionalize" these forces, an objective seen as critical to long-term stability in Iraq.[146] It had been envisioned as an intensive five-year program, with a gradual transition in years four and five to a program similar in scale to other INL training initiatives around the world.[147] The program was intended to be very different

---

[144] SIGIR, 2011e, p. 3.

[145] Part of why INL had little understanding of the Iraqi police capabilities is that ITAM-Police—to the extent it evaluated police capabilities at all—assessed only quantitative measures, such as the number of officers trained, rather than the actual skill levels and capabilities of the force (see SIGIR, 2010a, pp. 18–19). By the time DoD provided its insights to INL, the latter was unable to develop its own baseline assessment before the program launched on October 1, 2011. That said, a senior INL official asserted that DoD's information was not particularly helpful for INL's baseline study. Such a view is unsurprising, given that DoD focused its training on beat cops, while INL was seeking to mentor senior police leaders. Interview with INL official, Washington, D.C., January 20, 2012.

[146] OIG, 2011, p. 18.

[147] U.S. Department State, Bureau of Near Eastern Affairs, Iraq, "Policy and Mission Overview—Iraq," briefing, April 2011, slide 21.

from the police training DoD had previously carried out, which offered basic police training to officers who would be on the streets in their communities. PDP, in contrast, was intended to provide advice and mentoring to senior-level Iraqi police officials, help with the design of police standards and policies, and aid development of a cadre of Iraqi instructors (a train-the-trainer approach).

The PDP plan, which the Deputies Committee approved in August 2009,[148] called for regularly deploying advisors to disparate training locations, which in turn required planning for extensive security, transportation, and logistics operations. The plan called for advisors to be located at hubs in Baghdad, Basra, and Erbil. From there, they would travel by land and air to 28 advisory sites (21 by land and seven by air) in ten of 18 provinces. INL intended to have 12 helicopters in Iraq to transport PDP advisors to training locations.[149] Figure 7.4 provides a December 2010 list of planned PDP training sites.

The PDP did not proceed as planned, even before the transition. Mitigating the risks of traveling between sites caused anticipated security costs to balloon. Security concerns, congressional skepticism, and other challenges caused INL to scale the program back repeatedly.[150] Table 7.3 shows PDP staffing levels at various times. In early 2011, DoS reduced its staffing plans for the PDP from 350 advisors (52 government employees and 298 contractors) to 190 (50 mostly direct-hire manager-supervisors and 140 mostly contract advisors).[151] However, continued expectations that Congress would not fully fund the program led to further reductions,[152] and in October 2011, INL chose to pursue a phased-in program that would begin by deploying 115 advisors (one-third the original plan), reduce the number of training locations from 28 to 21, and eliminate dedicated air transportation.[153] Yet DoS was unable to meet even this limited objective by the time it was to achieve initial operating capability; as of October 1, 2011, only 90 advisors were on board.[154]

From the outset of planning, the PDP was expected to be extremely costly—as a five-year program, the entire effort was expected to cost approximately $5 billion—

---

[148] SIGIR, *Iraq Police Development Program: Lack of Iraqi Support and Security Problems Raise Questions About the Continued Viability of the Program*, SIGIR 12-020, July 30, 2012d, p. 4

[149] SIGIR, *Quarterly Report and Semiannual Report to the United States Congress*, January 30, 2011b, p. 41.

[150] OIG, 2011, p. 18.

[151] SIGIR, 2011c, p. 47.

[152] SIGIR, 2011e, p. 10. See also William R. Brownfield, Assistant Secretary of State for INL, "Subject: INL Comments on the SIGIR Draft Report 'Iraqi Police Development Program: Opportunities for Improved Program Accountability and Budget Transparency' (SIGIR 12-006, September 30, 2011)," letter to Glenn D. Furbish, Assistant Inspector General for Audits, SIGIR, October 14, 2011. Printed in SIGIR, 2011e, p. 43.

[153] SIGIR, 2011e, pp. 9–10.

[154] SIGIR, 2011e, p. 14.

Figure 7.4
Police Development Program Site List as of December 2010

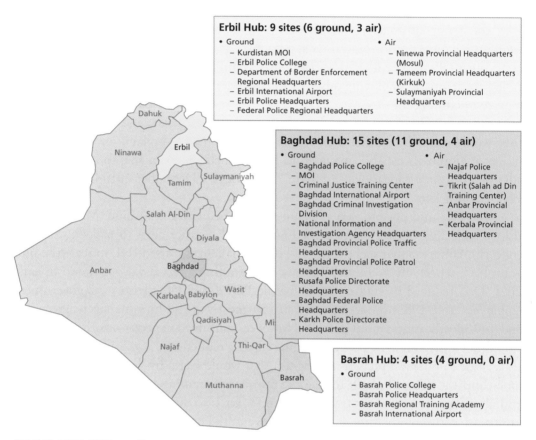

Erbil Hub: 9 sites (6 ground, 3 air)
- Ground
  - Kurdistan MOI
  - Erbil Police College
  - Department of Border Enforcement Regional Headquarters
  - Erbil International Airport
  - Erbil Police Headquarters
  - Federal Police Regional Headquarters
- Air
  - Ninewa Provincial Headquarters (Mosul)
  - Tameem Provincial Headquarters (Kirkuk)
  - Sulaymaniyah Provincial Headquarters

Baghdad Hub: 15 sites (11 ground, 4 air)
- Ground
  - Baghdad Police College
  - MOI
  - Criminal Justice Training Center
  - Baghdad International Airport
  - Baghdad Criminal Investigation Division
  - National Information and Investigation Agency Headquarters
  - Baghdad Provincial Police Traffic Headquarters
  - Baghdad Provincial Police Patrol Headquarters
  - Rusafa Police Directorate Headquarters
  - Baghdad Federal Police Headquarters
  - Karkh Police Directorate Headquarters
- Air
  - Najaf Police Headquarters
  - Tikrit (Salah ad Din Training Center)
  - Anbar Provincial Headquarters
  - Kerbala Provincial Headquarters

Basrah Hub: 4 sites (4 ground, 0 air)
- Ground
  - Basrah Police College
  - Basrah Police Headquarters
  - Basrah Regional Training Academy
  - Basrah International Airport

SOURCE: SIGIR, 2011e, p. 10.
RAND *RR232-7.4*

and the vast majority of the funds would pay for support functions.[155] DoS requested $1 billion in its FY 2012 budget to operate the program. Only 12 percent of this amount was for hiring, training, and deploying police advisors; the remaining 88 percent was for life support, security, aviation, and other support functions. Each police advisor, SIGIR later calculated, would have cost $2.1 million per year.[156] Personnel ratios were roughly the same, with each advisor requiring nine support personnel.[157] The reason for this high cost was that DoS would now be funding activities that DoD had previously performed at no cost to DoS.

---

[155] SIGIR, 2011e, p. 12.

[156] SIGIR, 2012d, executive summary and p. 19.

[157] SIGIR, 2011c, p. 47.

Table 7.3
INL Footprint

Date	Plan	Advisors (number)	Sites (number)	Dedicated Aircraft
August 2009	Original plan	350	28	Yes
Early 2011	Scaled-back	190	28	Yes
October 2011	Phased implementation	115	21	No

SOURCE: SIGIR, 2012d, p. 37.
NOTE: In July 2012, the MOI had recommended that PDP training take place at two sites in Baghdad.

The program was plagued from the start by management and planning problems. SIGIR Stuart Bowen testified to a congressional subcommittee on November 30, 2011, that although

> INL spent more than two years preparing to take over police training from DOD. . . . it did not produce (a) a detailed plan identifying what it planned to accomplish, (b) a comprehensive set of intermediate and longer-term milestones to judge progress, or (c) sufficient goals and metrics to assess program outcomes.[158]

SIGIR had earlier warned that, "without specific goals, objectives, and performance measures, the PDP could become a 'bottomless pit' for U.S. dollars intended for mentoring, advising, and training the Iraqi police forces."[159] A question never asked was why DoS thought it could manage something this large without any relevant previous experience.

The PDP also suffered from a lack of Iraqi government buy-in, a problem identified in many areas of the transition. Senior Deputy Iraqi Minister of Interior Adnan al-Asadi complained to Deputy SIGIR Ginger Cruz in late 2011 that, "with most of the money spent on lodging, security, support, all the MOI gets is a little expertise." She encouraged the United States to "take the program money and the overhead money and use it for something that can benefit the people of the United States, because there will be very little benefit to the MOI from the $1 billion."[160]

---

[158] Stuart W. Bowen, Jr., "Statement of Mr. Stuart W. Bowen, Jr., Inspector General, Office of the Special Inspector General for Iraq Reconstruction," in *Preserving Progress in Iraq, Part III: Iraq's Police Development Program*, hearing before the Subcommittee on the Middle East and South Asia, Committee on Foreign Affairs, U.S. House of Representatives, 112th Cong., 1st Sess., November 30, 2011b.

[159] SIGIR, 2011e, p. 8.

[160] Quoted in SIGIR, 2011f, p. 4.

Rule of Law

If law enforcement initiatives are to be effective, police must operate within a broader rule-of-law framework that includes, in the words of DoS IG, "the entire legal complex of a modern state, from a constitution and a legislature to courts, judges, police, prisons, due process procedures, a commercial code, and anti-corruption mechanisms."[161] Without a clear legal framework, the police can easily abuse their power; without a functioning court system that ensures due process, suspects cannot get a fair trial; without effective prisons, convicts can escape or suffer abuses at the hands of guards; and without effective anticorruption measures, justice is easily subverted by wealth.

The U.S. government's focus on the development of rule of law in Iraq began soon after the government of Saddam Hussein fell when, in May 2003, a team of federal judges and prosecutors visited Iraq to assess the situation.[162] The DoS, DoJ, and DoD all played a large role in promoting the rule of law in Iraq, although many other agencies were involved in related efforts. There has been much consistency across DoD and INL programs. Both have focused on bringing Iraqi prisons up to international standards, improving pretrial detention facilities,[163] and fighting corruption.[164] As of September 30, 2011, the United States had obligated $2.5 billion and spent $2.21 billion to improve the rule of law in Iraq.[165]

Early Rule-of-Law Initiatives

MNF-I and a variety of U.S. civilian agencies (chiefly DoJ components) quickly began conducting programs to improve the rule of law in Iraq. Early DoD efforts focused on training for police and corrections officers, as well as construction of detention facilities capable of housing suspected insurgents rounded up by coalition forces.[166] DoD embarked on several large-scale construction initiatives, including the establishment of several secure rule-of-law complexes in which Iraqi judges—who were frequent targets of assassination—could both live and work.[167] At the local level, military civil affairs units engaged Iraqis regionally in ad hoc efforts to reestablish judicial func-

---

[161] U.S. Department of State and the Broadcasting Board of Governors, Office of Inspector General, *Inspection of Rule-of-Law Programs*, report of inspection, Report Number ISP-IQO-06-01, October 2005, p. 5.

[162] OIG, "Inspection of Rule-of-Law Programs, Embassy Baghdad," OIG Report No. ISP-IQO-06-01, October 2005, p. 5.

[163] INL will pursue a targeted model facility approach, whereas DoD tried to address all local detention facilities.

[164] Interview with DoS officials, January 20, 2012.

[165] SIGIR, 2011f, p. 62. Also, email correspondence with senior SIGIR official, November 1, 2012.

[166] Office of Management and Budget, *Section 2207 Report on Iraq Relief and Reconstruction, 3rd Quarter, Fiscal Year 2004*, Appendix 1, "Sectoral Descriptions," July 2, 2004 July 2, 2004, pp. 1-15 to 1-17.

[167] SIGIR, *Quarterly and Semiannual Report to the United States Congress*, July 30, 2007b, p. 45.

tions, including financing the construction of courts and police stations.[168] Within two years, MNF-I had spent $26 million in CERP funds alone on almost 6,000 small-scale projects.[169]

On the civilian side, USAID, DoS, and DoJ began training a wide range of Iraqi officials. DoJ's International Criminal Investigative Training Assistance Program began training police officers and border security personnel in 2003. The U.S. Marshals Service established a witness protection program; DoJ's Office of Overseas Prosecutorial Development Assistance and Training trained prosecutors and judges; and USAID funded the Iraqi Special Tribunal charged with prosecuting the former regime's crimes against humanity. DoJ components also funded construction and renovation projects, primarily courthouses and judicial training facilities, and protective details for Iraqi judges and prosecutors.[170]

For the most part, the military and civilian agencies implemented their rule-of-law programs unilaterally, and interagency coordination was limited. The DoS IG found that, as of August 2005, "a fully integrated approach to justice-sector reform in Iraq is essential and does not exist at present."[171] A significant amount of overlap existed between the projects that DoD and DoJ had undertaken, with both conducting multiple programs related to witness protection and training for prosecutors, judges, and corrections officers and both financing construction of detention facilities, courthouses, and training facilities.[172]

Transitioning U.S. Rule-of-Law Efforts

In 2008, MNF-I and the embassy agreed to make rule-of-law matters a LOO in the JCP, which would be implemented under the joint auspices of the embassy's rule-of-law coordinator and MNF-I's SJA. These officials established the Interagency Rule-of-Law Coordinating Center to deconflict the myriad rule-of-law activities being that the military and various U.S. civilian agencies had undertaken.[173]

The 2010 JCP contained a rule-of-law annex that outlined U.S. goals, conditions, and objectives with regard to rule-of-law issues in Iraq. The annex specified the following U.S. rule-of-law goals:

---

[168] U.S. Department of the Army, Judge Advocate General's Legal Center and School, Center for Law and Military Operations, *Rule of Law Handbook: A Practitioner's Guide for Judge Advocates*, 2010, pp. 212–214.

[169] U.S. Department of State and the Broadcasting Board of Governors, Office of Inspector General, 2005, p. 1.

[170] Office of Management and Budget, 2004, pp. 1-22 to 1-26. Also DoS, *Section 2207 Report on Iraq Relief and Reconstruction*, January 5, 2005a, pp. 1-26 to 1-44.

[171] OIG, 2005, p. 30.

[172] DoS, *Section 2207 Report on Iraq Relief and Reconstruction*, April 6, 2005b, pp. 1-16 to 1-33.

[173] Richard Pregent, "Rule of Law Capacity Building in Iraq," *International Law Studies*, Vol. 86, 2010, p. 331. Also U.S. Department of the Army, Judge Advocate General's Legal Center and School, 2010, pp. 212–214.

1. promote a transparent, independent, efficient, and secure judiciary
2. improve law enforcement practices and capacities
3. improve pretrial detention and prison practices and capacities
4. enact civil justice reforms, including commercial and property laws
5. combat corruption
6. increase awareness of rights and equal access to justice through enhanced professionalism in the legal community.[174]

The annex contained 63 tasks that would all be transferred to Embassy Baghdad. These tasks relate to USF-I support for law enforcement training, judicial reform, judicial security, increasing the capacity to process scientific and technical evidence, legal reform, anticorruption efforts, prison reform, and increasing ministerial capacity. A handful of tasks, such as management of detention facilities, would be transitioned to the Iraqi government. While transitioning many initiatives simply involved shifting oversight of training programs, some programs, such as those related to detainees, were quite significant in scale. At the height of the insurgency, more than 25,000 detainees were in U.S. custody. In mid-July 2011, USF-I, working closely with advisors from the International Criminal Investigative Training Assistance Program, transferred the few remaining Iraqi detainees in U.S. custody to the Iraq Corrections Service.[175]

By the time U.S. forces departed Iraq in December 2011, the country had made significant improvements to its judicial system. U.S. assistance had transformed the corrections system, for example, "from near non-existence into a functional institution requiring minimal technical assistance and mentoring" by late 2012, according to SIGIR.[176] However, many shortcomings remained. The court system remains overtaxed and underresourced. Security threats to judicial officials continue to undermine the effectiveness of the courts, with attacks against judges rising in the last quarter of 2011. On December 22, 2011, a car bomb killed 32 anticorruption officials working at the Commission of Integrity—a clear attack on the rule of law just days after U.S. troops left the country.[177]

SIGIR reported in April 2011 that, despite the wide variety of efforts to stem corruption, "Iraq's anticorruption institutions continue to be hampered by myriad factors, including political interference, judicial inaction, inexperienced staff, and legal uncertainty."[178] Iraq's chief justice asserted in November 2011 that corruption is "pervasive," particularly at the most senior levels, and "has seemingly become the norm

---

[174] USF-I, *Joint Campaign Plan*, Annex C, "Rule of Law," November 23, 2009.

[175] SIGIR, 2011f, p. 69.

[176] SIGIR, *Sustaining the Progress Achieved by U.S. Rule of Law Programs in Iraq Remains Questionable*, SIGIR 13-001, October 25, 2012e, executive summary and p. 19.

[177] SIGIR, 2012a, p. 12.

[178] SIGIR, 2011c, p. 5.

at many ministries."[179] Indeed, Transparency International's 2012 index of corruption perception ranks Iraq as 169th out of the 176 ranked countries for its level of corruption.[180]

Airspace Management

Although management of air traffic control is not a security function per se, the ability to monitor what is in the air effectively plays a critical role in air defense, air support to ground forces, and border control. The U.S. military managed Iraq's airspace from soon after the invasion until beginning to transfer responsibility gradually to Iraqi civilian authorities in 2007. Even then, USF-I's air component remained closely involved in efforts that Iraqis led in principle but that expatriate air traffic controllers hired under U.S.-funded contracts actually executed.

After the conclusion of major combat operations, the United States and its allies operated all aspects of Iraqi airports and air traffic control systems. Responsibility for air traffic control is typically divided both geographically and by altitude. In August 2007, the process of transitioning control of Iraq's airspace began by transferring all civilian traffic in Iraqi airspace over 29,000 feet to the Baghdad Area Control Center, a modern air traffic control center built with Iraq Relief and Reconstruction Funds and operated by technical trainer-advisors recruited by the U.S. Federal Aviation Administration.[181] During this period, Iraq depended heavily on contract air traffic controllers. Although it has begun a nascent training program to develop Iraqi controllers, Iraq is not expected to have a full complement of trained staff until at least 2014.[182] It takes two to four years to fully train and certify an air traffic controller in the United States.[183] By the end of 2009, however, Iraq had only ten certified Iraqi air traffic controllers and 39 recently hired trainees in the pipeline against a requirement for 208 fully trained staff.[184] The shortage of qualified Iraqi air traffic controllers slowed the transition of airspace control during 2010 and 2011.

The U.S.-Iraqi SA specified that "surveillance and control over Iraqi airspace shall transfer to Iraqi authority immediately upon entry into force of this Agreement" but allowed that "Iraq may request from the United States Forces temporary support for

---

[179] SIGIR, 2012a, p. 12.

[180] Transparency International, "Corruption Perception Index 2012," 2012.

[181] Stacie Shafran, "Air Force Assists in Historic Final Transfer of Iraq's Airspace," U.S. Air Forces Central, Baghdad Media Outreach Team, 2011.

[182] DoD, *Measuring Stability and Security in Iraq*, September 2008, p. 19.

[183] U.S. Department of Labor, Bureau of Labor Statistics, *Occupational Outlook Handbook*, 2010–2011.

[184] DoD, 2009b, p. 64.

the Iraqi authorities in the mission of surveillance and control of Iraqi air space."[185] On January 1, 2009—the day the SA entered into force—the Iraqis assumed operational responsibility for control of airspace 24,000 feet and above. The Transportation Attaché Office in the U.S. embassy prepared a plan to return control of all airspace to the Iraqis in accordance with the SA, although the plan acknowledged that the United States would continue to help manage Iraqi airspace.[186]

In 2010, USF-I created the Airspace Transition Team, which reported to ACCE, the senior air advisor to the USF-I CG. The team consisted of three personnel within ACCE who, working in conjunction with Transportation Attaché Office, assisted the Iraqi Ministry of Transportation and the Iraq Civil Aviation Authority to develop the capacity to take control of Iraqi airspace incrementally. As USF-I drew down, the plan was to transfer this team's responsibilities completely to the embassy's Transportation Attaché Office.

While there were some delays in the transfer of airspace because of "technical and controller resource issues," the overall process proceeded generally on schedule with support from contract controllers, who both provided on-the-job training and air traffic control services in Iraq.[187] In mid-2010, additional airspace in the northern one-third of Iraq at altitudes from 15,000 feet and above transitioned to Iraq Civil Aviation Authority control.[188] As of July 2011, more than 100 air traffic controllers were in various stages of hiring, training, or certification, with an unknown number qualified.[189] As of October 1, 2011, the airspace transfer was complete, with all airspace under Iraqi control.[190] Each of Iraq's civil airports had radar systems able to track aircraft for air traffic control, and aircraft equipped with transponders could be tracked over 90 percent of Iraqi airspace. However, the ability of these radars to track aircraft without transponders (i.e., a hostile or malfunctioning aircraft) was far more limited, covering only about one-third of Iraq's airspace and very little of its border area.[191] Unfortunately, this level of coverage creates vulnerabilities in Iraq's ability to detect and track aircraft transiting its airspace.

The airspace transfer process was particularly challenging because of the long time required to create a certified air traffic controller force. Iraq faced many challenges in creating a certified controller force, given that the trainees first needed to learn Eng-

[185] SA, Art. 9, paras. 3–4.

[186] DoD, *Measuring Stability and Security in Iraq*, March 2009a, p. 17.

[187] SIGIR, 2011b.

[188] SIGIR, 2010c, p. 107.

[189] SIGIR, 2011d.

[190] Shafran, 2011.

[191] Iraq Civil Aviation Authority, "Iraq Aeronautical Information Publication (AIP)," November 17, 2011, p. ENR 1.6-3.

lish (the official language of air traffic control internationally), create an indigenous training system from scratch, and manage security risks associated with the convergence of civilian aircraft and large numbers of both fixed- and rotary-wing military aircraft in the same airspace.

Nevertheless, the transition associated with the loss of U.S. military and air traffic control assets proceeded reasonably well because it maintained effective coverage by using contract controllers and by transferring the airspace in segments from higher to lower altitudes. The Iraqis first had control of the traffic at higher altitudes and gradually increased their responsibility down to the more-complex approach and tower environments. This worked well for two reasons. First, when the transfer started, the United States was still conducting a significant number of tactical air operations at low altitudes. By retaining control of low-altitude airspace, the U.S. and coalition controllers were able to provide safe and effective air traffic control for their own air operations. Second, as the Iraqis' capabilities improved, they were better equipped to control the more-complex air traffic environments found at the lower altitudes.

Summary

Overall, USF-I and Embassy Baghdad made great strides enhancing ISF and related capabilities, particularly from 2009 through 2011. The MOI, in particular, evolved from what U.S. forces considered to be a relatively ineffective organization prior to 2010 to a ministry that USF-I assessed as having achieved almost all of its MEC goals by 2011. As many USF-I staff said in the final months of 2011: "It's time."

Once the SA made clear that U.S. training would come to a close (or at least change significantly) in 2011, the United States radically changed its aspirations for the ISF. Instead of striving to develop requirements-based competencies, U.S. officials' focused on the development of MEC that should be met before the end of the U.S. military mission in Iraq. MEC was thus envisioned to be a foundational capability that would serve as the baseline for the development of future capabilities. However, achieving MEC was never viewed as anything more than an interim step along the path of development capabilities necessary to fulfill critical security missions. Thus, while ISF may have been able to undertake some core security functions on the departure of U.S. forces, Iraq has a long way to go before the ISF and police can be viewed as self-sustaining forces that can guarantee both internal security and external defense.

Of greater concern than achieving the operational skills inherent in MEC, however, was whether the ISF would continue on the path toward professionalization. While the various ITAM staffs worked with the ISF components to put Iraqi training institutions in place and to train a cadre of qualified instructors, all elements of the ISF had serious deficiencies in their training capabilities at the time of the transition and thus in their abilities to sustain the process of recruiting, training, and fielding professional military and police forces.

Planning for the transition of ISF capacity development programs was marked by delays in decisionmaking and insufficiently detailed planning guidance from senior levels of the U.S. government. While the senior officials in Washington laid out broad and somewhat unrealistic goals for ISF capabilities, they appeared to have little interest in detailed, realistic discussions of how the United States would manage the full range of security cooperation programs if USF-I redeployed in 2011.[192] Staff at all levels of the U.S. government highlighted the need for more-specific policy guidance to enable execution of OSC-I's transition plan.[193] OSC-I would likely have been more prepared to oversee these programs if policymakers made decisions on the nature of the post-2011 security relationship earlier, thus providing more time for detailed planning. Furthermore, OSC-I would have been better able to address strategic U.S. objectives, rather than just the management of security cooperation operations, if a policy agency or higher-echelon command had led planning from the top down. Such an entity would have given more consideration to the role OSC-I played in U.S.-Iraqi relations than the staff-level PSG-I could.

Earlier planning (and execution) for the transition of security cooperation initiatives would have been beneficial. First drafts of detailed plans for security cooperation should be formulated at least two years prior to transition, despite inevitable U.S. and host-nation political uncertainties, competing priorities, changing guidance, and leadership turnover. Security cooperation plans may be executed more effectively if the transition from an operational force (e.g., USF-I) to an enduring organization (e.g., OSC-I) occurs at least six months prior to the deadline for full troop withdrawal. An earlier transition may provide political benefits that outweigh the reduced time operational units have to assist their host-nation counterparts. For example, an earlier security cooperation transition sends clear political signals about U.S. resolve to continue an enduring partnership and puts the new organization in a leadership role early enough to serve as a stabilizing influence during the final transition period.

---

[192] Interview with USF-I official, December 12, 2011.

[193] USF-I/DoS, "Iraq Transition Senior Leader Conference," briefing, National Defense University, July 23, 2010.

CHAPTER EIGHT

Enabling an Expeditionary Embassy

A large part of the transition effort in Iraq related to the establishment of OSC-I and creating within the embassy the capacity to function securely and effectively in what was in essence a war zone once U.S. military forces redeployed and USF-I was disestablished. These efforts, associated with the "Conduct Transitions" LOE of OPORD 11-01, were designed to enable what planners referred to as an "expeditionary" embassy.[1]

Even before the transition, the U.S. embassies in Baghdad and Kabul were unique among nearly 200 bilateral diplomatic posts in that they were established in the context of an ongoing conflict and, consequently, received significant support from the military to allow them to perform their official duties.[2] In Iraq, the extremely dangerous security environment and the severe limitations in public services meant the embassy, its support sites, and its constituent posts in Basra, Erbil, and Kirkuk had to develop an independent capability to operate their own power plants, water treatment facilities, and other critical infrastructure; run their own fleet of airplanes and helicopters to facilitate secure transportation; manage a network of both elementary and advanced medical facilities; and import virtually all food, fuel, and other critical supplies from neighboring countries. Embassy officers therefore did not have the opportunities for the frequent, routine, and unofficial contact with Iraqis of all walks of life that are normal in most other countries.

Embassy Baghdad depends heavily on contractors for security and life support. To operate multiple self-sufficient compounds, DoS hired thousands of private contractors to provide life-support services, such as food, laundry, logistics, and other critical functions. To manage such complex logistical and contracting matters on its own after the transition, the embassy planned to continue using a number of DoD contracting vehicles for at least several years, aided by more than 50 in-country DoD contract management specialists.

To enable the mission to function independently in a hostile environment, Congress appropriated hundreds of millions of dollars to build physical security at embassy,

---

[1] Interview with USF-I J3 and J5 staff officers, Baghdad, August 1, 2011.

[2] Unless specified otherwise, the terms *U.S. mission* and *U.S. embassy* refer collectively to Embassy Baghdad; its support sites; and its constituent posts in Basra, Erbil, and Kirkuk.

consulate, and branch office facilities, including the construction of protective cover over living spaces, offices, and other locations where staff congregate. Acknowledging that it would have closed its diplomatic mission in virtually any other country with similar threats,[3] DoS put in place a security operation unparalleled in the history of the department. The embassy's RSO, comprising DS agents and security specialists, oversees a multilayered network of intelligence, surveillance, protective, and response capabilities involving some 8,000 employees of private security firms under contract to the U.S. government. Even so, the embassy's organic security capabilities are far more limited than those USF-I provided, which is appropriate given its very different mission and legal standing.

With roughly nine support staff for every programmatic staff member, DoS would need to spend a great deal of money to support its diplomatic engagement, outreach, and reporting effort. Should the security situation deteriorate, the continuation of the diplomatic mission may become politically and financially unfeasible.

To operate in Iraq, DoS has had to develop a new paradigm and change the way it does business.[4] Working out of multiple interim facilities without U.S. military forces authorized to protect them, managing a hostile environment with field-expedient protective measures and extensive movement security, and mobilizing an array of support activities to ensure its self-sufficiency, the U.S. Mission in Iraq was intended to operate, unlike any other U.S. embassy in the world, as an "expeditionary" embassy.[5]

Not surprisingly, therefore, some aspects of the transition did not go entirely as planned. USF-I's provision of equipment and supplies—most of it used and already in Iraq—helped in some areas and complicated transition efforts in others. Delays in decisionmaking, insufficient congressional funding, a lack of Iraqi government support, and other hindrances prevented DoS from implementing 100 percent of the capabilities it hoped to have in place by January 1, 2012. By the time the last American soldier left the country, however, the U.S. diplomatic mission was able to operate multiple self-sufficient facilities and keep its personnel safe, enabling it to engage senior Iraqi government officials on leading U.S. policy priorities.

That said, DoS was unable to secure the funding needed to open diplomatic posts in several cities that it had initially deemed important, potentially limiting the insight that could be gained into developments in these areas. Furthermore, the cost and extent of required security measures limited U.S. diplomats' ability to travel around

---

[3]   Eric J. Boswell, "Statement of the Honorable Eric J. Boswell, Assistant Secretary for Diplomatic Security, U.S. Department of State," in *The Diplomat's Shield: Diplomatic Security and Its Implications for U.S. Diplomacy*, Washington, D.C.: Subcommittee on Oversight of Government Management, the Federal Workforce, and the District of Columbia, Committee on Homeland Security and Governmental Affairs, U.S. Senate, June 29, 2011b.

[4]   Interview with Ambassador Peter Bodde, Assistant Chief of Mission for Assistance Transition in Iraq, Embassy Baghdad, June 27, 2011.

[5]   Interview with Ambassador Peter Bodde, Assistant Chief of Mission for Assistance Transition in Iraq, Embassy Baghdad, June 27, 2011. Also, interview with DoS OBO officials, Washington, D.C., December 2, 2011.

the country and engage working-level Iraqi counterparts (see Chapter Ten). As a result, the embassy had great difficulty reporting on local and provincial developments and managing continued capacity-building efforts in such areas as the rule of law and civilian policing.

Scale of Effort

A number of U.S. embassies around the world are quite large, because they either are important for U.S. interests or serve as regional hubs that support multiple other embassies in the area. Not counting security and support contractors, the number of core diplomatic staff planned for the posttransition U.S. Mission in Iraq was similar to that for other large embassies, such as those in Paris, Tokyo, and Bangkok.[6]

DoS has also long operated embassies in extremely dangerous places. For example, for many years, diplomats at the U.S. embassies in Beirut and Algiers have left embassy compounds only in heavily armored convoys with large security details. Similarly, the U.S. embassy in Yemen continues to operate despite widespread violence and multiple armed attacks on the chancery.

Despite such extensive experience, Deputy Secretary of State Lew observed in August 2010 that "the scope and scale of the transition [in Iraq] is virtually unprecedented" for the department.[7] In addition to being the largest diplomatic mission in the world, the U.S. Mission in Iraq would be operating facilities in an especially dangerous environment, managing a complex web of contract staff that would far outnumber U.S. government employees, and operating in near-total isolation from the local economy. Ambassador Jeffrey observed that, with its own multimode fleet of secure transportation (e.g., mine-resistant ambush protected vehicles [MRAPs], helicopters, and fixed-wing aircraft) and its independent supply chains, the U.S. mission would be run like a military rear-area support command, something DoS has never done.[8]

With such a large percentage of its human and financial resources devoted to safety and life support in a generally hostile environment, the U.S. Mission in Iraq would be breaking new ground for DoS. As the department's IG wrote in May 2011, "the Department's transformation into an expeditionary organization working in an overseas contingency operational environment" is "unprecedented."[9] The mission's third-highest-ranking official echoed this characterization a month later, stating that

---

[6]   Michele Kelemen, "Huge Embassy Keeps U.S. Presence in Iraq," National Public Radio, December 18, 2012.

[7]   Jacob Lew, "The Next Phase in America's Relationship with Iraq," Statesmen's Forum video, Center for Strategic and International Studies, August 5, 2010.

[8]   Interview with Ambassador to Iraq James Jeffrey, Baghdad, June 27, 2011.

[9]   OIG, 2011, p. 27.

the post's ability to operate and sustain itself independently in a hostile area would make it "an expeditionary embassy."[10]

Embassy Staffing Numbers

DoS's initial plan called for a full complement of 17,000 staff, which decreased to 16,000 by October 2011.[11] However, only one-sixth of these staff members were planned to be full-time U.S. government employees, with the rest (approximately 84 percent) consisting of contractors. A mere one-tenth of the total staff would be engaged in "core" diplomatic and training functions.[12] As Secretary of State Clinton testified to the House Appropriations Committee in March 2011, mission staff would be "50 percent security, 30 percent life support contractors . . . 10 percent management and aviation security . . . and then 10 percent programmatic staff."[13] OSC-I alone, with a projected footprint of 4,000 personnel, would comprise about 25 percent of the mission's total staff.[14]

Even at these levels, embassy operations in Iraq would involve far fewer personnel than what had been typical during the previous eight years. For example, in 2008 the U.S.-coalition presence in Iraq included more than 170,000 troops plus more than 154,000 contractors, or a total of nearly 325,000 personnel.[15] In comparison, the total planned post-2011 U.S. mission presence was just 10 percent of the peak number of military personnel alone. Thus, although many observers emphasize how large the embassy would be compared to other diplomatic missions around the world, Ambassador Jeffrey reminded the Senate Armed Services Committee that "the overall U.S. Government footprint in the country will be a dramatic decrease of way more than 90 percent from its highest point."[16] Nevertheless, although the decrease of the U.S.

---

[10] Interview with Ambassador Peter Bodde, Assistant Chief of Mission for Assistance Transition in Iraq, Embassy Baghdad, June 27, 2011.

[11] Interview with Kathleen Austin-Ferguson, Executive Assistant to the Under Secretary of State for Management, Washington, D.C., October 25, 2011.

[12] Clinton, 2011c.

[13] Clinton, 2011c. A senior DoS official told RAND researchers seven months later that only 1,500 of 16,000 (9.4 percent) would be U.S. government direct-hire employees. Interview with Kathleen Austin-Ferguson, Executive Assistant to the Under Secretary of State for Management, Washington, D.C., October 25, 2011.

[14] Clinton, 2011c.

[15] Office of the Chairman, Joint Chiefs of Staff, "Contractor Personnel vs. Military Personnel in Iraq," cited in Austin B. Smith, "Experts: State Dept. Unprepared to Oversee Contracting in Iraq," *Medill News Service for Military Times*, November 9, 2011.

[16] James F. Jeffrey, "Statement of Hon. James F. Jeffrey, U.S. Ambassador to Iraq," *Hearing to Receive Testimony on United States Policy Toward Iraq*, Washington, D.C.: Committee on Armed Services, U.S. Senate, February 3, 2011b, p. 31.

footprint from some 300,000 troops and support personnel to a civilian-led presence of 16,000 seems like an enormous reduction to Americans, that number still represents a significant presence that some Iraqi political leaders find hard to accept.[17] No other country has a similarly sized diplomatic mission in Iraq. Indeed, the embassy of Turkey, Iraq's northwestern neighbor, employs a mere 55 people.[18]

Iraq Operating Environment

In an environment as complex, dynamic, and dangerous as Iraq, the mere functioning of the U.S. mission is a success for U.S. diplomacy. While the security situation in Iraq has steadily improved in recent years,[19] the security gains are reversible. Many of the root causes of the civil war of 2005–2007 remain unresolved. Consequently, escalating Sunni-Shi'a tension, an outbreak of Arab-Kurd violence in northern Iraq, a resurgence of al-Qaeda in Iraq, or a politicized ISF repressing political opponents of Prime Minister Maliki could destabilize the country.[20] Al-Qaeda in Iraq, Sunnis resisting perceived government discrimination, or Iranian-backed Shi'a militia groups could directly undermine Iraqi government institutions.[21] Most important, once U.S. troops left, extremists wishing to strike at the United States would likely target U.S. diplomatic personnel and facilities.

A worsening security environment would further constrain embassy activities, dilute important U.S. diplomatic initiatives, and weaken already tepid congressional support for a continued large-scale U.S. diplomatic presence in Iraq. By contrast, sustained improvements in security would allow the embassy to increase its engagement with the Iraqi government while requiring less security and support, thus enabling the mission to operate at lower cost over the long term.

Significant improvements in Iraqi security and stability could enable the mission to rely more on supplies and services from the local economy and thus eliminate the need to provide extensive (and costly) organic life support. According to Under Secretary of State for Management Patrick Kennedy, diplomatic "missions and their staff can normally obtain food and fuel on the local economy. However, this is not the case

---

[17] Interview with Deputy Special Inspector General for Iraqi Reconstruction Ginger Cruz, Washington, D.C., November 28, 2011.

[18] Tim Arango, "U.S. Planning to Slash Iraq Embassy Staff by as Much as Half," *New York Times*, February 7, 2012a.

[19] Lloyd J. Austin, III, "Statement of Gen Lloyd J. Austin III, USA, Commander, U.S. Forces–Iraq," *Hearing to Receive Testimony on United States Policy Toward Iraq*, Washington, D.C.: Committee on Armed Services, U.S. Senate, February 3, 2011b, pp. 6–7. According to Panetta, 2011, "the number of weekly security incidents across Iraq has decreased from 1,500 in 2007 to fewer than 100 in recent weeks."

[20] SIGIR, 2011f, p. 56.

[21] Austin, 2011b, p. 6.

in Iraq."[22] Embassy Baghdad planned to continue to rely on the Defense Logistics Agency (DLA) through the end of calendar year 2013,[23] when it expected to be able to arrange its own contract vehicle for such support. The embassy initially continued to import food, fuel, and other critical supplies from Kuwait and other neighboring countries. This reliance on imported essentials was driven in part by the need for more reliable supply chains than Iraq can provide but also by the need for safety. An Army Contracting Command document stated that local fuel trucks could not be permitted on U.S. facilities because of the "catastrophic risk" that they might be rigged with explosives.[24]

Acquiring more goods locally would save costs, in large part by eliminating security and transportation costs. Ambassador Jeffrey told a visiting congressional staff delegation in December 2011 that the cost of shipping food from Kuwait was 50 times the cost of procuring it locally.[25] With the redeployment of U.S. Army security escorts for DLA convoys coming from Kuwait in mid-2011, the Army Contracting Command increased the number of contract convoy escort teams from 10 to 45. Although the cost of this eight-month contract expansion was not made public, the *Washington Post* reported that the effort would require around 385 additional security personnel and between 105 and 175 additional armored vehicles.[26]

Despite plans to enable Embassy Baghdad to operate in a hostile environment, the mission would nonetheless require some support from the Iraqi government consistent with Iraq's Vienna Convention obligations to foreign diplomatic missions on its territory. While a more-benign security environment might enable the mission to reduce its security posture, improvements in the Iraqi government's ability to provide security support would also enable the mission to expand its engagement and consider further reducing resources spent on security measures.[27]

---

[22] Patrick F. Kennedy, Under Secretary of State for Management, letter to the Department of Defense, April 7, 2010a (quoted in CWC, 2010, p. 50).

[23] Peter F. Verga, Chief of Staff to the Under Secretary of Defense for Policy, "Statement of the Department of Defense," in *Assessment of the Transition from a Military- to a Civilian-Led Mission in Iraq*, Subcommittee on National Security, Homeland Defense, and Foreign Operations, Committee on Oversight and Government Reform, U.S. House of Representatives, 112th Cong., 2nd Sess., June 28, 2012.

[24] U.S. Department of the Army, "Limited Source Justification for Exception to Fair Opportunity Ordering Process Under Multiple Award Indefinite Delivery Contracts," Control Number 12-003, October 27, 2011b, p. 4.

[25] U.S. Senate, Committee on Foreign Relations, 2012, p. 20.

[26] Walter Pincus, "State Department Could Buy Local Food in Iraq," *Washington Post*, November 15, 2011. See also U.S. Department of the Army, 2011, p. 4.

[27] CWC, 2011a, p. 1.

U.S. Forces–Iraq and Embassy Baghdad's Transition Planning

As discussed in Chapter Four, USF-I and embassy leadership jointly managed and directed all aspects of transition planning.[28] While all aspects of the transition received high-level attention, no aspect of the Iraq transition was of greater importance than the creation of an expeditionary embassy that could safely and effectively carry on its mission on the departure of U.S. forces. Consequently, a large number of planning teams, working groups, and other leadership forums worked tirelessly to facilitate the transition of activities the military had previously performed. A senior DoS official involved in transition planning testified to Congress that "the planning for the transition in Iraq was probably the most complex planning effort ever undertaken by DoS and perhaps one of the most complicated civilian planning efforts ever taken by the U.S. government."[29] Indeed, this was the largest and most complex military to civilian transition that has occurred since World War II, requiring an extensive and continuous planning effort. Both military and civilian officials shared the same goal: the long-term success of Embassy Baghdad operating independent of direct U.S. military support.

Planning for the transition took place in both Washington and Baghdad. Washington managed the policy issues associated with U.S.-Iraqi bilateral relations, including the scope and scale of the U.S. military presence in Iraq after 2011. However, most programmatic and operational decisions associated with the transition were delegated to the field for resolution. This atypical delegation of authority to the field proved to be a boon; interagency collaboration in Iraq, particularly between the embassy and USF-I, was widely reported to be far more collegial and productive than that in Washington.[30]

One of the most significant transition planning challenges originated in Washington. INL and OSC-I planners frequently changed the scale of police training and security assistance programs, which repeatedly caused planners to revisit estimates for contract support, construction, and other characteristics of the mission (see Chapter Three). Estimates for the total mission population ranged from 13,000 in October 2010 to 17,000 in March 2011.[31] Personnel estimates for each site varied greatly throughout the planning process. As a result of these fluctuations, the embassy could

---

[28] While USF-I and Embassy Baghdad made most of the operational decisions, some required approval and/or authorization from higher headquarters within DoD and DoS. On the military side, MNF-I set many of the early plans in motion, and USF-I leadership, in close collaboration with its embassy counterparts, did a large majority of detailed planning and made a multitude of decisions. However, for the purposes of this chapter, *DoD* refers to any organization within DoD, recognizing that senior military officials in Iraq applied most of the effort and made most of the decisions about the transition to the embassy.

[29] Boswell, 2011b.

[30] Interviews with U.S. embassy and USF-I officials, Baghdad, June 25–30, 2011; interviews with DoS officials, Washington, D.C., June 10, 2011, and January 20, 2012.

[31] SIGIR, 2011c, p. 43; interview with Kathleen Austin-Ferguson, Executive Assistant to the Under Secretary of State for Management, October 25, 2011.

not finalize plans and initiate "the actions needed to construct facilities and provide for life support, security, and other logistical requirements" in a timely manner.[32]

In Baghdad, embassy leadership was concerned primarily with the challenges of assuming USF-I's missions and securing the resources and permissions to execute them. Operationally, the embassy concentrated on seven critical components of the transition:[33]

- *Property.* One of the embassy's greatest challenges was securing permission to acquire property for mission operations. Despite continual efforts to obtain land-use agreements from the Iraqi government, few such agreements were in place for DoS to begin construction early enough to meet the January 1, 2012, transition deadline. Between January and June 2011, Under Secretary of State for Management Kennedy authorized construction to begin in the absence of land-use agreements, despite the risk involved.[34] The department did not have a "Plan B." As the DoS IG reported in May 2011,

   the embassy is negotiating with the GOI to obtain more property currently occupied by the U.S. military, but there are no contingency plans if these property leases are delayed or denied. Further, the Office of the Inspector General (OIG) finds NEA's [Bureau of Near Eastern Affairs'] proposed accommodation solutions neither optimal nor sustainable in the long term.[35]

- *Facilities.* Although the embassy would need some additional facilities at its planned sites, senior DoS officials told congressional overseers that "to the maximum extent possible we will repurpose existing Defense Department infrastructure and property for each of the Consulate General and EBO sites."[36]

- *Security.* The embassy's RSO prepared to take on the responsibility to coordinate all aspects of the embassy's security, which entailed many functions RSOs around the world did not typically performed. "In addition to static and movement secu-

---

[32] OIG, 2011, p. 13.

[33] Jeffrey D. Feltman, "Statement for the Record of Assistant Secretary of State for Near Eastern Affairs Jeffrey D. Feltman," in *Transition to a Civilian-Led U.S. Presence in Iraq: Issues and Challenges*, hearing before the Committee on Foreign Affairs, U.S. House of Representatives, 111th Cong., 2nd Sess., November 18, 2010a, p. 15.

[34] Interview with DoS OBO officials, Washington, D.C., December 2, 2011. Six months after the transition, GAO reported that the DoS had secured Iraqi government permission (title, leases, or land-use agreements) for only five of the mission's 14 sites. See Michael J. Courts, "State and DOD Face Challenges in Finalizing Support and Security Capabilities," written testimony for *Assessment of the Transition from a Military- to a Civilian-Led Mission in Iraq*, hearing before the Subcommittee on National Security, Homeland Defense, and Foreign Operations, Committee on Oversight and Government Reform, U.S. House of Representatives, 112th Cong., 2nd Sess., June 28, 2012, p. 4.

[35] OIG, 2011, p. 7.

[36] Feltman, 2010a, p. 15.

rity," Assistant Secretary of State for Near East Affairs Jeffrey Feltman told a House committee, DoS DS

will include operating and maintaining 60 MRAPs, explosive ordnance disposal teams, Unmanned Aerial Vehicle [UAV] reconnaissance/surveillance, tactical communications, advance warning capabilities for indirect fire, tactical operations centers at each site, and tactical intelligence.[37]

- *Life support.* DoS prepared to assume management of DoD's LOGCAP life-support contract vehicle (with extensive DoD support) for at least one year after the transition.[38] Embassy and constituent post officials in country would take on responsibility for ensuring that LOGCAP contractors were properly vetted, housed, protected, and transported.
- *Aviation.* Embassy operations and INL's PDP would depend on the functioning of the embassy's air wing, with a hub in Baghdad and supporting hubs in Erbil and Basra.[39] While INL has experience operating aviation assets in other countries and while DS operated air assets in Baghdad for a number of years, air operations are especially challenging without host-nation support.
- *Medical.* Because Iraqi medical facilities were deemed inadequate (and insufficiently safe) to treat U.S. mission personnel, the embassy employed contract support to provide comprehensive medical care, including emergency evacuation, at multiple locations in Iraq.
- *Contractor oversight.* The U.S. mission would have dozens of contract managers in country, approximately 48 on loan from DoD agencies, to oversee the implementation of the myriad contract vehicles in place to support the mission.

The embassy was hindered to some extent by the lack of preexisting contacts with Iraqi officials necessary to facilitate government approval of routine items necessary for the mission to function. For example, the embassy needed the Ministry of Customs to approve the import of food and fuel, and the MOI needed to approve residency permits and visas for contractors. However, because USF-I did not need to acquire the same permissions, the embassy had to develop procedures and relationships from scratch.[40]

---

[37] Feltman, 2010a, p. 16.

[38] Feltman, 2010a, p. 28.

[39] Feltman, 2010a, p. 16.

[40] Interview with DoS Iraq Transition Coordinator Ambassador Patricia Haslach, Washington, D.C., February 17, 2012.

Constituent Posts

The embassy determined that it was critical to maintain outreach throughout all the provinces to preserve the security gains made in Iraq and that the best way to do this was to build strong relationships with local leaders. To enable effective regional outreach, the administration planned to open diplomatic establishments in multiple locations throughout Iraq, including consulates general in Erbil and Basra and an EBO in Kirkuk.[41] These were intended to enable the mission both to follow and to influence regional developments in Iraq, particularly along "key faultlines."[42] (Although the facility in Kirkuk had been planned as an EBO, DoS upgraded the facility to a consulate sometime around the U.S. military's departure.[43]) These posts would help fill the roles the 31 PRTs had played previously.[44] Closure of the PRTs had, according to SIGIR, resulted in "a significant loss of provincial presence and the political outreach, development assistance, and situational awareness that their presence facilitates."[45]

Specific missions for the consulates and EBO were to include the following:

- mitigating and mediating Arab-Kurd, Sunni-Shi'a, and provincial-national tensions
- strengthening the capacity of provincial institutions in key flashpoint locations
- balancing foreign interference
- providing a platform for the UN and other organizations

---

[41] Deputy Assistant Secretary of State for Iraq Michael Corbin described an EBO as "a way diplomats can have presence, but these are going to be [a] temporary presence." Speaking specifically of planned EBOs in Iraq, Corbin added: "These are a three to five-year presence where we, again, will use all the tools we can in the diplomatic toolkit and in the development toolkit to reach out to Iraqis." See Michael Corbin and Colin Kahl, "U.S. Transition in Iraq," press briefing, August 16, 2010. DoS has (or has had) other EBOs in commercial centers other than the capitals of several countries where it is impractical—either because of the difficulty of providing full consular services or because of political sensitivities—to establish a U.S. consulate. It is not, however, clear that EBOs are always temporary. Currently, for example, DoS has an EBO in Douala, Cameroon, which handles cargo shipments for diplomatic posts in central Africa; see Embassy of the United States, Yaounde, Cameroon, "Embassy Branch Office (EBO) Douala: Mission Statement," undated. In Bosnia, DoS has EBOs in both Banja Luka (the seat of government of the Republika Srpska) and Mostar (the center of the ethnic Croat population); see Embassy of the United States, Bosnia & Herzegovina, "Contact Us," undated. See also OIG, "Report of Inspection: Embassy Sarajevo, Bosnia and Herzegovina," Report No. ISP-I-09-55A, September 2009. DoS also had an EBO in Almaty, Kazakstan, the country's commercial center and former capital, from 2006 to 2009, at which point the EBO was upgraded to a consulate; see DoS, "U.S. Official Visits Kazakstan, Opens U.S. Consulate General in Almaty," December 11, 2009b.

[42] SIGIR, 2011c, p. 48. See also Feltman, 2010a, p. 13.

[43] Embassy of the United States, Baghdad, "U.S. Embassy Announces New Websites for Consulates in Basrah, Erbil, and Kirkuk," press release, January 17, 2012a.

[44] At the height of the PRT program, in summer 2008, a total of 31 PRTs, ePRTs, and provincial support teams were in place throughout Iraq. See GAO, *Provincial Reconstruction Teams in Afghanistan and Iraq*, Washington, D.C., GAO-09-86R, October 1, 2008d, pp. 2–3.

[45] SIGIR, 2011c, p. 48.

- promoting the safe return and resettlement of displaced persons
- encouraging foreign investment and economic development
- reporting on strategic trends, events, and drivers of Iraqi instability
- presenting U.S. policy and promoting mutual understanding and respect for American values
- providing limited services to U.S. citizens.[46]

Each of the posts would have unique missions as well. The consulate in Erbil, the capital of the KRG, would be the primary interlocutor with KRG officials; the EBO in Kirkuk would help monitor "the most highly charged internal boundary dispute in the country"; and the consulate in Basra, a key city in southern Iraq, would report on trade and infrastructure developments at Iraq's principal port (located at nearby Umm Qasr) and on Western investment, particularly in the hydrocarbon sector.[47] A planned EBO in Mosul, ultimately scrapped on budgeting grounds, was intended to follow developments in a city that served as the center of the Sunni insurgency and continued breeding ground for al-Qaeda in Iraq.[48] A lack of funds also led DoS to eliminate a third EBO in Diyala, an eastern Iraqi province along the Iran-Iraq border that had experienced extensive sectarian violence.[49]

As Figure 8.1 shows, a range of other facilities would fall under the U.S. mission. Three INL hubs, which would support INL training and advising activities at multiple dispersed facilities, were to be located in Erbil, Baghdad, and Basra.[50] The mission would also operate support sites in the Baghdad area at what used to be referred to as Sather Air Base and Forward Operating Base (FOB) Prosperity while they were under USF-I and DoD control. Finally, OSC-I planned to be present at ten separate locations. Table 8.1 lists DoS and DoD-managed sites in Iraq; Figure 8.2 and Table 8.2 do the same for DoD–managed sites and provides more detail about them.

Embassy facilities were built on preexisting USF-I sites and made extensive use of existing USF-I infrastructure.[51] Among the facilities on the bases that USF-I transferred was critical infrastructure necessary for the mission to operate independently of host-nation support, including seven power plants, four water treatment plans, three incinerators, and one wastewater treatment plant.[52] USF-I and embassy staff coordi-

---

[46] U.S. Department State, Bureau of Near Eastern Affairs, Iraq, 2011, slide 17.

[47] OIG, "Iraq Transition," briefing, October 6, 2010c, slide 8.

[48] Interview with USF-I J35 staff officer, Baghdad, December 10, 2011.

[49] CWC, 2011a, p. 3.

[50] U.S. Department State, Bureau of Near Eastern Affairs, Iraq, 2011, slide 22.

[51] USF-I, J7, "Base Transition to Enduring U.S. Entities," September 29, 2010.

[52] Donahue, 2011, slide 7. These power plants are located at Taji (three), Kirkuk (two), and Tikrit (one); another, at Victory Base, did not transfer to DoS. Water treatment plants are located at Taji, Besmaya, Prosperity, Shield

Figure 8.1
Department of State– and Department of Defense–Managed Sites in Iraq

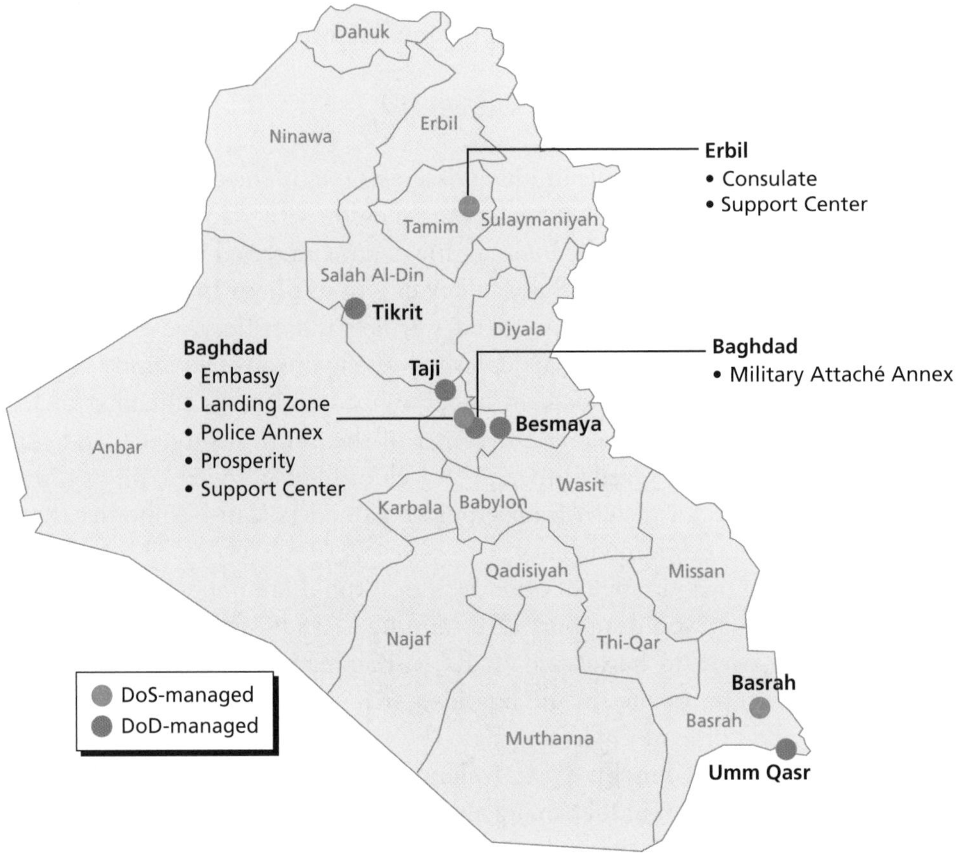

SOURCE: SIGIR, 2011d, p. 31. Based on graphic in Courts, 2012, p. 3.
RAND *RR232-8.1*

nated closely on the transfer of these sites and equipment through monthly meetings of the embassy Support and Enduring Base Transition Board, which involved the
embassy, USF-I J4 and J7, and OSC-I Transition Team.[53]

Construction Planning Hindered by Many Factors
A number of factors made construction planning for embassy sites difficult. First, DoS
had to plan for a worst-case security scenario of the embassy being cut off from outside supplies of power or water. Each site therefore had to be completely self-sufficient,
with each having all the critical infrastructure needed to operate (power, water, sewage,

---

(one each). The incinerators are located at Besmaya (two) and Taji (one). The wastewater treatment plant is located
at Besmaya.

[53] Donahue, 2011.

Table 8.1
Department of State Facilities in Iraq

Site Manager	Embassy Facility	Former USF-I Facility	Post-Transition Purpose	Planned Return to Iraqi Control
DoS	New embassy compound (Baghdad)	N/A	Embassy	—
	Conulate General Basra	Contingency Operating Base (COB) Basra	Consulate aite INL PDP hub	—
	Conulate General Erbil	Contingency Operating Station (COS) Erbil	Consulate	—
	Erbil Diplomatic Support Center	COS Erbil	INL PDP hub	—
	Baghdad Logistics Support Site	Camp Prosperity	Embassy logistics support	End of 2013
	Baghdad Aviation Support Site	Sather Air Base	Logistics and aviation	—
	Baghdad Police College Annex (BPAX)	COS Shield	INL PDP hub	End of 2012
OSC-I	Consulate Kirkuk	COS Warrior	Consulate	July 2012

SOURCES: Courts, 2012, p. 3; Patrick F. Kennedy, "Statement of the Honorable Patrick F. Kennedy," in *Assessment of the Transition from a Military- to a Civilian-Led Mission in Iraq*, hearing before the Subcommittee on National Security, Homeland Defense, and Foreign Operations, Committee on Oversight and Government Reform, U.S. House of Representatives, 112th Cong., 2nd Sess., June 28, 2012.

etc.). This took a great deal of planning and coordination with USF-I, and in the end, the required facilities were either transferred or built (at substantial expense) at each embassy site. The finite capabilities of this infrastructure, however, limited the embassy's ability to accommodate personnel growth far more than if the embassy had been able to connect to the host nation's power grid, water network, and wastewater treatment system.

Second, despite extensive regulations governing operations in hazardous countries, DoS lacked codified construction standards for "expeditionary" diplomatic posts, requiring OBO to develop ad hoc solutions along the way. For example, in the wake of the 1998 bombings of the U.S. embassies in Kenya and Tanzania, DoS developed regulations to protect chancery buildings and staff from truck bombs and other ground-based threats; these include the use of shatterproof glass and the placement of buildings some distance from the perimeters of their compounds. No security guidelines existed, however, on how to construct a facility designed to protect embassy personnel from incoming mortar fire. Similarly, while OBO has previously constructed housing

Figure 8.2
Department of Defense–Managed Sites in Iraq

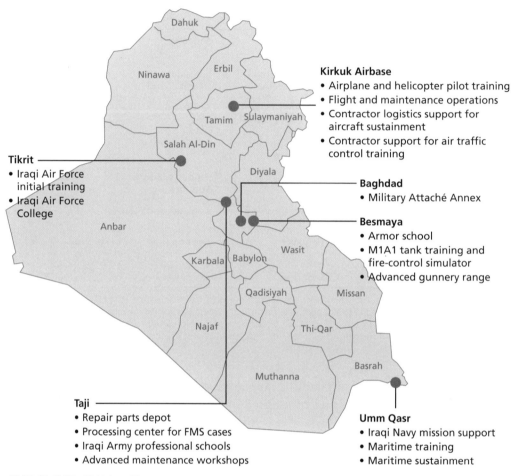

SOURCE: SIGIR, 2011d, p. 53.
RAND *RR232-8.2*

in secure embassy compounds, it had not previously had to determine security standards for the construction of costly protective cover over large numbers of temporary containerized housing units that typically served as billeting and office space on sites where permanent facilities were limited or nonexistent.

Third, DoS had to contend with a range of nonsecurity rules affecting construction that were not easy to follow in a contingency environment. For example, DoS facilities abroad must adhere to all U.S. laws and regulations, including those regarding the handling of waste. Embassy facilities in Iraq would not be connected to host-nation wastewater infrastructure (as they are in most countries), and the construction of multiple dedicated wastewater treatment plants was cost prohibitive. Embassy facili-

Table 8.2
OSC-I Facilities and Activities in Iraq

Site Manager	Current Facility	Former USF-I Facility	Post-Transition Purpose	Planned Return to Iraqi Control
OSC-I	Military Attaché and Security Assistance Annex	FOB Union III	OSC-I headquarters	Mid-2013
	EBO Kirkuk	COS Warrior	Pilot training, aircraft maintenance and support	July 2012
	Taji	COS Taji	Logistics and maintenance training	—
	Umm Qasr	COB Umm Qasr	Iraqi navy training	—
	Besmaya	FOB Hammer	Iraqi army training (armor, artillery)	—
	Tikrit	COB Speicher	Iraqi air force training	—
DoS	Conulate General Erbil	COS Erbil	OSC-I liaison to KRG	—
	Conulate General Basra	COB Basra	Support to air sovereignty	—
	Baghdad Aviation Support Site	Sather Air Base	Support to iraqi air operations center	—
	BPAX	COS Shield	OSC-I liaison to MOI	End of 2012

SOURCE: DoD IG, 2012, p. 22.

ties would therefore have to transport their effluent by truck to host-nation treatment facilities.[54]

Fourth, DoS's Baghdad Master Plan, finalized in June 2010, allowed little flexibility.[55] To facilitate the protection of each location, the plan called for building compounds on the smallest possible footprint within former USF-I bases. As a result, site plans had little "swing space" to accommodate additional personnel or functions. Agencies proposed larger staffs, added new types of facilities, and requested redundancy of some functions (e.g., vehicle maintenance facilities in multiple locations), so the master plan had to be revised frequently, and the area required for each site grew. In August 2011—four months before the transition—Deputy Secretary of State Nides directed a return to the master plan's principle of using the smallest possible footprint.[56]

Nevertheless, the lack of flexibility caused problems later on. By May 2011, DoS's IG found the main embassy compound had already nearly reached full capacity because of the addition of civilian staff and contractors and the relocation of others from sites being returned to the Iraqi government. The IG rejected the "creative solutions" DoS's

---

[54] Interview with DoS OBO officials, Washington, D.C., December 2, 2011.

[55] Interview with DoS OBO officials, Washington, D.C., December 2, 2011.

[56] Interview with DoS OBO officials, Washington, D.C., December 2, 2011.

Bureau of Near Eastern Affairs had offered, which included having staff sleep in shifts and requiring contractors to live outside the compound despite the dangers of doing so.[57]

Fifth, although USF-I provided DoS with millions of dollars worth of security-related equipment and miscellaneous supplies without reported incident, problems with the construction equipment USF-I provided caused multiple delays. The reality is that this equipment was still essential for supporting the USF-I mission and bases, so DoD could not release the equipment until it was no longer necessary. Thus, many of the delays were tied to the much larger delay in DoD actually executing the "go to zero" plan as it sought to offer political flexibility as long as possible for a potential contingency force in 2012.[58] DoS OBO had evaluated USF-I equipment in January and June 2010, tagging many of the specific items it sought to receive in support of post-transition operations. In December 2010, DoS developed a formal equipment request and, in February 2011, issued a request for proposals for construction based in part on the specifications of the equipment USF-I was to provide. Up to this point, planning for the transfer of equipment to DoS appeared to be going well. As Ambassador Jeffrey testified to the Senate Foreign Relations Committee on February 1, 2011,

> we're getting an extraordinary amount of effort by the U.S. military on all of the locations where we will be taking over . . . to do engineering, do joint planning, provide equipment, provide, for example, the containerized trailers, if you will, that people are living in. We're getting extraordinary support . . . . [E]verything that we've needed other than the [Black Hawk] helicopters, which we have another fix for that I'm perfectly happy with, has gone forward.[59]

However, in May 2011, DoS began to receive different types of the items requested, including some in unusable condition. For example, a shipment of containerized housing units from Balad to the consulate in Basra was looted en route, with wires, plumbing, and other equipment stripped out.[60] In addition, when containerized housing units of different sizes arrived, DoS OBO had to redesign housing areas to accommodate different foundations, overhead cover, fire safety systems, and other requirements. Given DoS's procurement and contracting requirements, these changes required either issuing new contracts or revising multiple construction contracts, the negotiation of

---

[57] Harold W. Geisel, "Testimony of Harold W. Geisel, Deputy Inspector General Office of Inspector General U.S. Department of State and the Broadcasting Board of Governors," in *Assessment of the Transition from a Military- to a Civilian-Led Mission in Iraq*, hearing before the Subcommittee on National Security, Homeland Defense, and Foreign Operations, Committee on Oversight and Government Reform, U.S. House of Representatives, 112th Cong., 2nd Sess., June 28, 2012.

[58] Interview with former USF-I J4 staff member, North Carolina, January 30, 2013.

[59] Jeffrey, 2011a.

[60] Interview with DoS OBO officials, Washington, D.C., December 2, 2011.

which caused delays, consumed staff planning time, and incurred additional costs.[61] As of December 2011, contract changes resulted in $40 million in additional costs, and sites that were supposed to have been completed by October 1, 2011, were expected to take an additional eight months to finish. These delays had significant implications for personnel safety because many housing units were not under overhead cover, and some infrastructure may have been inadequate for the number of personnel present. DoS OBO officials suggested, based on this experience, that a critical lesson learned is that DoS should "never again depend on another agency's secondhand goods" to build an embassy compound, particularly under tight time constraints.[62] While procuring new items might preferable, alternatives should also be examined in an effort to determine how to build flexibility into the OBO contracting system to enable it to respond more quickly should last-minute changes occur.

Funding Shortfalls Caused Scaling Back of Embassy's Provincial Presence

By mid-2010, funding shortfalls had caused DoS to scale back its plans for a permanent provincial diplomatic presence in Iraq. In April 2010, the department estimated that construction of permanent consulates in Basra and northern Iraq (either Mosul or Kirkuk)—to be planned within three to five years—would cost $526.8 million and requested these funds in FY 2010 supplemental appropriations legislation.[63]

Congress denied DoS's request, specifically withholding the entire amount ($526.841 million) requested for permanent consulate construction—a 34-percent cut to the administration's $1.57 billion request for diplomatic and consular programs in Iraq.[64] Congressional conferees for the FY 2010 supplemental funding legislation wrote that the appropriated funds "will support interim facilities for two consular operations and three provincial diplomatic team posts" and directed that funding requirements for permanent facilities "should be prioritized within the amounts made available for embassy Security, Construction, and Maintenance in regular appropriations acts."[65] In essence, Congress scolded the administration for seeking construction-related funds in its supplemental request rather than through normal budget mechanisms.[66]

As a result, DoS was forced to defer all permanent construction plans and use temporary facilities to meet mission and security needs. The cuts in funding, which also included a $10 million cut to contracted security operations in Iraq, also led DoS

---

[61] Interview with DoS OBO officials, Washington, D.C., December 2, 2011. See also Courts, 2012, p. 5.

[62] Interview with DoS OBO officials, Washington, D.C., December 2, 2011.

[63] U.S. Department of State and U.S. Agency for International Development, "Supplemental Budget Justification, Fiscal Year 2010," 2010b, pp. 13–14.

[64] U.S. Senate, Committee on Appropriations, 2010, p. 55.

[65] U.S. Senate, Committee on Appropriations, 2010, p. 55.

[66] Interview with DoS OBO officials, Washington, D.C., December 2, 2011.

to reduce ground movements throughout the country. Furthermore, anticipating that Congress would reduce future funding requests for Iraq operations, DoS reacted to the supplemental by greatly reducing its plans for a provincial presence in Iraq. Expecting cuts to its FY 2012 operations (diplomatic and consular programs) request as well,[67] the department was concerned it would not be able to fund sustained operations for all of the proposed provincial posts.[68]

To cut operating costs, DoS planners decided to eliminate the EBO in Diyala and debated which northern Iraqi EBO—Mosul or Kirkuk—should be dropped. Mosul was seen as less urgent, primarily because of concerns that Arab-Kurd tensions in Kirkuk made that city a likely source of future conflict.[69] Practical reasons were also a factor; given that no one could leave the Mosul PRT compound at the time for security reasons, "security tipped the scale" on the decision.[70] DoS therefore decided in late spring or early summer 2011 to postpone the opening of EBO Mosul indefinitely, which saved the department $204 million to $97 million in FY 2010 construction costs and $107 million in FY 2012 operations and security costs.[71] The decision was made a mere two weeks after construction contracts had been signed, potentially resulting in cancellation penalties.[72]

Anticipated shortfalls in DoS's budget also led DoS and DoD to decide to shift responsibility for management of the enduring presence in Kirkuk from DoS to OSC-I.[73] DoD was planning on a large OSC-I presence in Kirkuk, and the costs of security modifications to make the entire site comply with DoS standards were extremely prohibitive. By transferring responsibility for upgrading and operating the Kirkuk site to DoD, DoS was able to free $133 million for other construction priorities and to reduce anticipated operating costs by $87 million in FY 2012.[74]

---

[67] Interview with Kathleen Austin-Ferguson, Executive Assistant to the Under Secretary of State for Management, Washington, D.C., October 25, 2011.

[68] Telephone interview with DoS resource management official, October 27, 2011.

[69] Telephone interview with DoS resource management official, October 27, 2011. Also, Congress expressed a clear preference for Kirkuk, asserting that it "supports the placement of posts along the volatile Arab-Kurdish fault lines" (U.S. Senate, Committee on Appropriations, 2010, p. 55).

[70] Interview with Kathleen Austin-Ferguson, Executive Assistant to the Under Secretary of State for Management, Washington, D.C., October 25, 2011.

[71] Interview with Kathleen Austin-Ferguson, Executive Assistant to the Under Secretary of State for Management, Washington, D.C., October 25, 2011.

[72] Interview with senior DoS official, Washington, D.C., February 17, 2012.

[73] Interview with Kathleen Austin-Ferguson, Executive Assistant to the Under Secretary of State for Management, Washington, D.C., October 25, 2011.

[74] Telephone interview with DoS resource management official, October 27, 2011. As DoD's "tenant" in Kirkuk, DoS paid an agreed-on share of life support, medical, and communications costs, while DoD assumed the rest of the operating costs, most of which covered static security guards and other security measures. Interview with

The enormity of DoS's funding shortfalls raises questions about whether the department should have done more than just scale back its planned initiatives. Not only did it eliminate new facilities because of cuts in the FY 2010 supplemental, but DoS officials said that they had contingency plans in place to pursue programs in FY 2012 with as little as one-half the requested funding if necessary.[75] The DoS IG suggested in October 2010 that funding cuts might have merited a reevaluation of what the department could hope to achieve in Iraq:

> It appears that provincial staffing is now being driven by budget constraints, rather than an appraisal of what is needed to accomplish the mission; certainly there is no indication that the missions have been redefined or reduced as funding and staffing projections shrink. In June 2010, the Embassy provided the Department with its staffing requirements to achieve the defined provincial missions. After Congress approved the supplemental funding levels in August, the Embassy was told to cut staffing levels for the provinces in half, but was given no amended policy guidance.[76]

Slashing programs back was certainly simpler than reconsidering U.S. policy in Iraq. However, by not at least evaluating whether post-2011 U.S. strategy in Iraq should be modified to reflect drastically reduced funding, DoS ran the risk of maintaining objectives that could not reasonably be accomplished within resource constraints.

Personnel Challenges

Given DoS's relatively small size, it has been remarkably successful in filling thousands of positions throughout Iraq since 2003. Several aspects of the U.S. government's personnel system have complicated the department's task. First, the department has had a hard time institutionalizing expertise and experience on the ground. Since the scale of the Iraq commitment was so large, it required mobilizing the entire Foreign Service. This meant that many Foreign Service officers doing tours in Iraq had little or no experience in the Middle East before arriving. Since Foreign Service officers typically change jobs every two to three years, however, they are accustomed to taking assignments in regions or on topics outside their normal areas of expertise.

That said, Embassy Baghdad was disadvantaged by the fact that the standard assignments for such a high-threat environment as Baghdad would last only one

Kathleen Austin-Ferguson, Executive Assistant to the Under Secretary of State for Management, Washington, D.C., October 25, 2011.

[75] Interview with DoS Iraq Transition Coordinator, Ambassador Patricia Haslach, Washington, D.C., February 17, 2012.

[76] OIG, "Compliance Follow-Up Review of Embassy Baghdad," Report ISP-C-11-0SA, October 2010b, p. 12. Quoted in SIGIR, 2011c, p. 25.

year, although some officers do extend for second tours. As a result, mission staff offi-cers are constantly rotating out, and every section is staffed by officers learning new portfo-lios.[77] The military also suffered from frequent rotations, but many officers and noncommissioned officers served multiple tours and therefore required less time to "refresh" on their return.[78]

Furthermore, the security threats and hardships that embassy staff experienced in Baghdad entitled them, for a yearlong assignment, to as many as 66 days of rest and relaxation (R&R) trips and/or regional rest breaks, in addition to their 13–26 work days of leave (depending on length of service) and as much as 20 work days of spe-cially authorized administrative leave. As a result, officers who took all their available vacation and R&R or regional rest break days could be entitled to be away from post for as much as 18 weeks, or roughly one-third of their tours. Those who also received authorization for administrative leave could have been away for as much as 22 weeks, or 42 percent of their tours.[79]

Second, while Foreign Service officers rotate to other assignments, other DoS personnel in Iraq were hired for Iraq-specific assignments of limited duration. Under the law, "temporary organizations" may hire appropriately skilled people for three years with an extension of up to two additional years.[80] Such staff members are treated as regular civil service employees in all ways except for their limited-term appointments.[81] Because of DoS's critical need to surge personnel to Iraq and Afghanistan, it has used this "3161" authority to hire staff to work specifically on Iraq- and Afghanistan-related issues, both in Washington and in the field. As a result, "many of the State employees with the most experience on the ground are not career employees, but 3161s who are temporary hires."[82] Once they have served five years, however, they are not eligible for extension in the same position; while some may take other Iraq-related jobs, many leave, taking their expertise with them.

Third, a complex interaction of U.S. government personnel regulations and U.S.-Iraqi negotiations made it difficult for INL to hire retired federal employees with law enforcement and judicial experience as police trainers or rule-of-law advisors. When

---

[77] Interview with Ambassador Larry Butler (USF-I political advisor), Baghdad, July 1, 2011.

[78] Interview with Ambassador Larry Butler (USF-I political advisor), Baghdad, July 1, 2011.

[79] See DoS, "2012 Iraq Service Recognition Package (ISRP)," cable to all diplomatic and consular posts, SecState #052041, May 26, 2011, paras. 10–12; DoS, "Iraq Jobs: Benefits," undated c; interview with Ambassador Larry Butler (USF-I political advisor), Baghdad, July 1, 2011.

[80] 5 U.S. Code (USC) 3161, Employment and Compensation of Employees, January 3, 2012.

[81] 5 USC 3161. Office of Personnel Management regulations regarding 3161s appear in 5 Code of Federal Regu-lations Part 534, "Basic Pay for Employees of Temporary Organizations," *Federal Register*, Vol. 67, No. 17, Janu-ary 25, 2002.

[82] Michael Eisenstadt, testimony before the House Committee on Foreign Affairs, Subcommittee on the Middle East and South Asia, June 23, 2011.

the Iraqi government refused to extend diplomatic privileges and immunities to U.S. government contractors, INL decided to fill police and rule-of-law positions with 3161s, who, as full-time U.S. government employees, receive administrative and technical status under the Vienna Convention on Diplomatic Relations and thus have limited diplomatic privileges and immunities.[83] However, whereas retired federal employees can take government-funded contractor positions without penalty, U.S. law prevents them from "double dipping" by earning a government salary in addition to their government pensions. While these rules do not apply to former state and local government officials or to retired military personnel,[84] INL officials in Baghdad asserted that the statute deterred people with extensive federal law enforcement and judicial experience from accepting positions as 3161 advisors in Iraq. Over the longer term, INL officials in Washington asserted, DoS was able to hire police trainers despite these restrictions,[85] but the "double dipping" rules clearly complicated efforts to get the program off the ground.

Fourth, for a variety of reasons related to both suitability and security, Embassy Baghdad hired virtually no Iraqi nationals for a long time.[86] As of October 6, 2011, DoS had 4,410 contractors and grantees in Iraq, only 34 of whom were Iraqi nationals. The need to utilize American citizens and third-country nationals even for support and security roles sharply raises personnel and contracting costs,[87] deprives the mission of the cultural expertise and local knowledge that Iraqi employees could offer, and creates enormous logistical challenges (such as the need to transport, house, and feed thousands of employees).[88] While the "insider threat" Iraqi employees pose is real, a senior embassy official stated in June 2011 that it was not clear whether anyone ever weighed the benefits against the costs or whether anyone had assessed whether Iraqis could fill certain types of jobs in which they might pose less of a security risk.[89] That said, local security conditions affected the mission's hiring practices in each location. For example, because the Kurdistan Region is more secure and has a labor force more skilled than that of southern Iraq, the mission expected to hire a greater percentage of Iraqis in Erbil than in Basra.[90]

---

[83] Interview with INL-Baghdad official, Baghdad, June 27, 2011.

[84] Interview with ITAM Police officials, Baghdad, June 30, 2011.

[85] Interview with INL officials, Washington, D.C., January 20, 2012.

[86] SIGIR, 2011f, pp. 32–33.

[87] Brownfield, 2011, p. 43.

[88] Interview with Ambassador Jeanine Jackson, U.S. Embassy Baghdad Minister-Counselor for Management, Baghdad, June 28, 2011.

[89] Interview with Ambassador Jeanine Jackson, U.S. Embassy Baghdad Minister-Counselor for Management, Baghdad, June 28, 2011.

[90] Interview with Ambassador Jeanine Jackson, U.S. Embassy Baghdad Minister-Counselor for Management, Baghdad, June 28, 2011.

224 Ending the U.S. War in Iraq

Fifth, the Iraqi bureaucracy was neither prepared nor willing to handle the planned influx of thousands of U.S. officials and embassy contractors, who previously would have entered the country under procedures established by the U.S. military. The prime minister's office questioned the embassy's explanation of why it needed so many foreign contractors, particularly in security roles, and the Foreign Ministry was unable and/or unwilling to handle the large number of visas required. Many contractor positions were vacant on January 1, 2012, and could not be filled for months.

Ratio of Support to Programmatic Staff

Given the embassy's challenging operating environment, the vast majority of personnel working for the mission in 2012 were security and life-support contractors. Ratios of support staff to programmatic staff vary at each constituent post in Iraq, but the ratio for the entire U.S. mission is approximately 9 to 1—only 10 percent of the staff perform core diplomatic functions. As a point of comparison, the ratio of support staff to program staff at the U.S. embassies in Beijing, Cairo, and New Delhi is approximately 3 to 4, meaning roughly 57 percent perform core diplomatic functions.[91] Thus, the percentage of staff in Iraq performing support functions was more than twice as high as in other large posts. While the size of the embassy may get smaller, the ratio of support to programmatic staff is likely to remain fixed as long as the situation in Iraq remains unchanged.

Security

The U.S. mission operates in an extraordinarily dangerous environment. In 2010 alone, "the Embassy and other U.S. [diplomatic] interests in Iraq were targeted 50 times," and "multiple attempts" were made on the lives of DoS personnel through the use of IEDs.[92] It was anticipated that continued tensions could escalate and lead to political instability or armed conflict. Both Iranian-backed Shi'a extremists and Sunni extremists, including al-Qaeda in Iraq, were expected to continue to target Iraqi government institutions and U.S. government staff in country through violent attacks.[93] Further, planners concluded that Iraqi security institutions would remain less than fully capable

---

[91] SIGIR, 2011c, p. 44.

[92] DoS DS, "Vigilant in an Uncertain World: Diplomatic Security 2010 Year in Review," March 2011, p. 4. The context of the statement and its inclusion in a report on DoS activities make clear that the number cited does not include attacks that targeted U.S. military forces and/or facilities.

[93] Austin, 2011b, p. 6. See also James F. Jeffrey and Lloyd J. Austin, III, "Prepared Joint Statement of Ambassador James F. Jeffrey and Gen Lloyd J. Austin," *Iraq: The Challenging Transition to a Civilian Mission*, hearing before the Committee on Foreign Relations, U.S. Senate, 112th Cong., 1st Sess., February 1, 2011, pp. 8–12.

of maintaining internal security and mitigating the threat to mission personnel and facilities. Under Secretary of State for Management Kennedy delicately described the security situation as "less than permissive in nearly all areas of the country."[94]

DoS DS is responsible for the security of diplomatic missions in many dangerous countries around the world, including Pakistan and Yemen, and DoS officials argued that Embassy Baghdad would be more than capable of protecting itself after the U.S. military withdrawal. DS provided security for the embassy since mid-2010 "without any assistance from the military beyond certain very specialized functions,"[95] such as an incident-response quick-reaction force; downed aircraft and vehicle recovery capabilities; route clearance; explosive ordnance disposal support; and counter rocket, artillery, and mortar warning. After the military withdrawal, DS asserted, "the greatest challenge will be replacing capabilities currently unique to the military."[96]

DoS had no illusions that the U.S. Mission in Iraq would be able to address the full range of security threats in the same way the U.S. military had. Indeed, rather than replace some of these functions with in-house capabilities, the embassy planned to rely on the ISF for such tasks as route security. Thus, the DoS IG reported in May 2011 that, despite the presence of 7,000 security officers, "Department and U.S. military personnel acknowledge that the overall U.S. security capability will be reduced" after the U.S. troop withdrawal.[97]

While the retrograde of U.S. forces significantly reduced security-related capabilities, the proposed scale of the embassy-led security operations planned for Iraq would be unlike what DoS has managed anywhere else. On January 1, 2012, Embassy Baghdad had 5,100 contract security personnel (1,500 for movement and 3,600 static guards),[98] almost double the 2,700 it had managed a year earlier.[99] In June 2011, embassy security programs were managed by a cadre of 81 DS special agents in Baghdad, Erbil, and Talil, a number that was to be augmented by 25 additional special agents and 68 security protective specialists by the end of 2011.[100] Table 8.3 shows the five-year cost of security contracts for DoS-led sites throughout Iraq.

These enormous budgets were to fund robust security capabilities throughout Iraq to ensure the protection of embassy facilities, personnel, and information. The embassy

---

[94] Patrick F. Kennedy, Under Secretary of State for Management, letter to Ashton Carter, Under Secretary of Defense (Acquisition, Technology, and Logistics), December 20, 2010b.

[95] Boswell, 2011b.

[96] DoS DS, 2011, p. 4. See also Assistant Secretary of State for Diplomatic Security Eric J. Boswell, "Security Boost: DS Meets Challenge Posed by Iraq Drawdown," *State Magazine*, May 2011a, pp. 30–31.

[97] OIG, 2011, p. 22.

[98] Interview with Kathleen Austin-Ferguson, Executive Assistant to the Under Secretary of State for Management, Washington, D.C., October 25, 2011.

[99] Jeffrey, 2011b, pp. 20, 22.

[100] Boswell, 2011b.

Table 8.3
Worldwide Protective Services Contract Costs

City	Embassy Sites	Contractor	Five-Year Cost ($)	Average Annual Cost ($)
Baghdad (static)	New embassy compound JSS Shield (INL Baghdad hub)[a] Baghdad Diplomatic Support Center	SOC	764,852,577	152,970,515
Baghdad (movement)	New embassy compound JSS Shield[a]	Triple Canopy	1,501,915,639	300,383,127
Basra	Consulate-General Basra compound INL Basra hub Basra Aviation hub	Global	387,107,414	77,421,482
Erbil and Kirkuk	EBO Kirkuk Consulate-General Erbil INL Erbil hub Erbil Diplomatic Support Center (aviation)	DynCorp	654,391,925	130,878,385
Total			3,308,267,555	661,653,509

SOURCE: DoS DS.

[a] JSS was a term coined during the OIF "surge" for a site manned simultaneously by both U.S. and government of Iraq forces.

expected to have "a large security force at each location, consisting of armed Regional Security Officers/DS Agents, contracted protective support details, quick reaction and static guard forces."[101]

While DoS has relied on its own contract static guards and private personal security details (PSDs) in Baghdad for several years, the U.S. military provided both static and movement security at all PRT posts outside Baghdad, except Tallil and Erbil, through FY 2011.[102] To provide security for provincial posts after the military withdrawal, DoS had no choice but to continue to rely extensively on contractors after January 1, 2012.[103] Under Secretary Kennedy told a House subcommittee that "it makes no sense" to add 7,000 permanent hires to the worldwide DS force of 1,800 just to cover Iraq, where the requirement for such heavy security is expected to be short lived. The need for a short-term, large-scale surge capacity, Kennedy asserted, would best be addressed through the use of contractors.[104]

Since the infamous Nisour Square incident in September 2007, in which Blackwater contractors protecting an embassy convoy killed 17 Iraqi civilians, DoS has put

---

[101] Kennedy, 2010b.

[102] Brennan notes, 2009–2011.

[103] CWC, 2010, p. l.

[104] Kennedy, 2011a.

in place a number of steps to ensure that private security contractors operate within clear rules of engagement and under stricter government supervision. One requirement is that every personal security detachment include at least one DS special agent or security officer. Operational oversight of contract security staff had been ensured by what Under Secretary Kennedy called the "very, very good" ratio of one U.S. government direct-hire security officer to 35 security contractors.[105] In fact, DS created a new Security Protective Specialist employment category to "serve primarily in an oversight role in protective operations manned by contracted security personnel" in Afghanistan, Pakistan, and Iraq.[106]

Yet the U.S. mission's reliance on contracted security forces generated concerns in some quarters regarding contractor performance, reliability, cost, and oversight. Several members of Congress considered fighting terrorists and engaging in armed combat as an "inherently governmental" function that should be undertaken by either the military or government security officers accountable to the taxpayer. While these concerns and criticisms may be legitimate, DoS's reliance on contract security forces in Iraq is long-standing and is thus not a new issue arising from the transition.

Embassy Security Programs

Embassy Baghdad security programs fall into three categories:[107]

- the security of facilities, including personnel living and working on them
- the security of personnel traveling outside a diplomatic compound, which DoS has termed *assured mobility*
- the ability to respond quickly to emergency situations, for which a quick-reaction force could be utilized.

DoS's congressional overseers expressed doubts about its ability to manage the mission's security after 2011: "It is unclear whether the State Department has the capacity to maintain and protect the currently planned diplomatic presence without U.S. military support."[108] DoS officials, however, maintained that multilayered plans and security capabilities would enable it to address all aspects of potential threats.

To save funds and make use of materiel already in country, the embassy worked closely with USF-I to acquire security-related equipment from the U.S. military at little or no cost. USF-I provided DoS with MRAPs, sense and warn technologies, ammunition for U.S. government and contract security forces, and biometric equipment.[109]

---

[105] Kennedy, 2011a; SIGIR, 2011c, p. 53.

[106] DoS DS, 2011, p. 5. See also Boswell, 2011b.

[107] Courts, 2012, p. 6.

[108] U.S. Senate, Committee on Foreign Relations, 2011, p. 2.

[109] Kennedy, 2010b.

Just as OBO lacked guidelines for constructing facilities in a contingency environment, DoS DS had no policies in place for managing movement security in an "expeditionary" area. For example, while guidelines existed to determine whether local security threats merited the use of armored vehicles and, if so, what level of armor should be used, these guidelines were written to protect diplomatic personnel from kidnapping and small-scale attack rather than from powerful IEDs buried along vehicle paths. DoS never had to develop its own armored vehicle standards for conditions in which regular armored vehicles were inadequate; for years, it had simply relied on the military's MRAPs and up-armored High Mobility Multipurpose Wheeled Vehicles (HMMWVs).

Assured Mobility

Prior to the departure of military forces, USF-I provided route clearance—e.g., clearing roads of IEDs and other hazards—and other types of support to ensure the safety of mission-related travel, as required. The military's standard for route clearance entailed eliminating hazards as much as possible—an effort that required large numbers of personnel and specialized equipment that the embassy would not be able to match.

Starting in October 2010, an operations support working group jointly managed by RSO and the USF-I J3 explored how the embassy could provide security during personnel movements once it could no longer rely on this extensive U.S. military support.[110] The working group developed the concept of *assured mobility*—the ability to travel despite the existence of risks, even if that means postponing or otherwise altering travel plans. The shift from the military's risk-elimination strategy to the embassy's risk-mitigation approach illustrates the compromises and adjustments that needed to be made to transition from a military-led to a civilian-led presence in Iraq.

Assured mobility was and remains critical to the embassy's success in Iraq. U.S. diplomats trapped at their desks in Iraq, relying solely on phones, email, or other indirect means to communicate with Iraqis, are of little more value than DoS employees with access to the same communications tools in Washington. As Deputy Secretary of State Lew stated in an August 2010 speech at a Washington think tank, "[i]f we don't put people in a place where they have mobility, where they can go out and meet with the people and implement their programs, there's very little argument for being in the place we send them."[111] In Baghdad, many Iraqi government offices are in close proximity to the U.S. embassy compound. Outside the capital, however, mobility is especially critical because embassy personnel are required to travel extensively to visit Iraqi provincial and city government officials in an effort to maintain relationships that had been established over the years. As the Senate Foreign Relations Committee noted

---

[110] Interview with USF-I J35 staff officer, Baghdad, June 14, 2011.

[111] Lew, 2010. Karen DeYoung and Ernesto Londoño, "State Dept. Faces Skyrocketing Costs as It Prepares to Expand Role in Iraq," *Washington Post*, August 11, 2010, p. A1.

in its January 2011 report, "the satellite sites will only be as effective as their inhabitants' ability to get off their compounds . . . . The State Department should reconsider whether the EBOs will have sufficient freedom of movement to justify their considerable expense."[112]

In a hostile security environment, freedom of movement requires the ability to identify and avoid threats, as well as a significant degree of flexibility in route selection and the timing of movements. According to an embassy security officer, if the mission's RSO could not determine with confidence that a proposed movement could be conducted safely, the movement simply would not go that day.[113] This approach is very different from the U.S. military's goal of total route clearance, which was predicated on identifying and *eliminating* threats, so that a mission could take place as scheduled.[114]

To maximize security on proposed embassy movements, the RSO's assured mobility plan had multiple elements that included planning; information gathering; movement; and, if necessary, incident response:[115]

- *Coordination mechanisms.* The RSO created a staff specifically charged with overseeing assured mobility plans and activities. The embassy, consulates, and EBOs would each have a tactical operations center to integrate threat information, assess threats to embassy convoys, track movements, and coordinate movement security missions.[116]
- *Route planning.* The U.S. military had the capability to clear a route of IEDs or other hazards ahead of a convoy, thus raising confidence that any given route would be safe. In contrast, the embassy would not have an organic route clearance capability and could not rely on the Iraqi government to provide one. To provide maximum flexibility to adjust to changing security conditions, the RSO would have to plan multiple routes for every proposed movement.[117]
- *Consultation with the Iraqi government.* With USF-I's assistance, the RSO developed close ties to Iraq's Baghdad Operations Command and the Iraqi National Intelligence Center. Through this cooperative relationship with the Iraqi government, the RSO would be able to develop travel routes based on both Iraqi information and U.S. threat reporting. This type of information sharing would be consistent with what is done in embassies around the world.[118]

---

[112] U.S. Senate, Committee on Foreign Relations, 2011, pp. 11–12. It should be noted that the U.S. military did not frequently move around in Iraqi cities after 2009 because of the need for an ISF escort.

[113] Interview with Embassy Baghdad RSOs, Baghdad, March 15, 2011.

[114] Interview with USF-I J3 staff, March 15, 2011.

[115] Interview with DoS DS officials, Washington, D.C., December 20, 2011.

[116] DoS DS, 2011, p. 26.

[117] Interview with Embassy Baghdad RSOs, Baghdad, March 15, 2011.

[118] Interview with Embassy Baghdad RSOs, Baghdad, March 15, 2011.

- *Intelligence analysis.* DoS sought continued access to USF-I intelligence databases so that the embassy RSO could draw on historical and current information on security incidents to prepare for personnel movements and other security operations.[119] "Prior to the departure of this critical USF-I capability," Under Secretary Kennedy told Under Secretary of Defense for Acquisition, Technology, and Logistics Ashton Carter that "a solution must be developed that allows continued use of these applications by all Chief of Mission . . . locations to query, update, and report critical intelligence and operational data."[120]
- *Route reconnaissance.* The embassy's tactical operations center staff would include contractors to conduct advance overhead surveillance of travel routes using long-endurance UAVs.[121] Contractor teams were planned to be present in Baghdad, Basra, and Kirkuk.[122] According to the *New York Times*, the information these UAVs gathered would be similar to what the U.S. military's ground-tethered blimps used to provide. The embassy began using UAVs in 2011 and deployed them more frequently after the U.S. military withdrawal.[123]
- *Ground movement.* Private security contractors, under the close supervision of DS agents and security officers,[124] would provide security when embassy employees move outside the mission's compounds. Although the embassy has long relied on private security contractors to provide secure ground movement in Baghdad, DoS has less recent experience using private security contractors in the provinces, where the U.S. military provided security for ground movements at all PRT locations except Tallil and Erbil through FY 2010.[125] Embassy convoys thus use small numbers of up-armored civilian vehicles for personnel movements.
- *Overhead surveillance.* RSO staff and contractors would have at their disposal a fleet of UAVs and "little bird" helicopters to conduct real-time overhead surveillance of embassy convoys as they travel.[126]

---

[119] OIG, 2011, p. 22.

[120] Kennedy, 2010b.

[121] DoS DS, 2011, p. 27. See also Boswell, 2011a, pp. 30–31.

[122] Interview with Embassy Baghdad RSOs, Baghdad, June 27, 2011.

[123] Responsibility for physical and movement security in the provinces transitioned from the military to DoS DS over the course of FY 2011, with the final transition scheduled for October 2011. Eric Schmitt and Michael S. Schmidt, "U.S. Drones Patrolling Its Skies Provoke Outrage in Iraq," *New York Times*, January 29, 2012.

[124] Schmitt and Schmidt, 2012.

[125] After the notorious Nisour Square incident, in which embassy security contractors working for Blackwater used excessive force, resulting in the deaths of 17 Iraqis, State has required that every contractor security team have at least one DS special agent or security officer embedded in the team.

[126] Interview with Embassy Baghdad RSOs, Baghdad, March 15, 2011. See also Boswell, 2011a, pp. 30–31, and Schmitt and Schmidt, 2012.

- *Incident response.* Were embassy vehicles and personnel to find themselves in trouble, the RSO would dispatch both ground- and air-inserted quick-reaction forces to extricate them.[127] Such quick-reaction forces would be specially equipped for the evacuation of casualties.

Perhaps not surprisingly, given their experience in Iraq, DoS and military officials had different perspectives on the manner in which DoS planned to ensure secure travel. Members of the USF-I J3 staff indicated that diplomats would likely face greater risks while traveling after the U.S. military withdrawal because the embassy would lack a route clearance capability of its own.[128] That said, senior DoS officials never planned to match the "gold standard" capabilities of the U.S. military and expressed confidence that careful route planning, close coordination with the ISF, and a lower profile—not to mention the flexibility to cancel a mission if it is deemed unsafe to travel—would enable embassy staff to accomplish their missions with an acceptable margin of safety. Furthermore, although ISF route clearance skills might not match those of their U.S. military counterparts, a senior DS official reported that "Embassy Baghdad was particularly impressed with the level of expertise the Iraqis demonstrated in checking for IEDs—and otherwise providing route security and clearance—during Vice President Biden's visit to Baghdad in November 2011."[129]

Facility and Static Security

As with its plan for assured mobility, the RSO developed a multifaceted strategy for ensuring the security of embassy facilities and the personnel who live and work on them:

- *Static and perimeter security guards.* U.S. mission facilities have large numbers of contractor static guards around their perimeters and at access control points.[130] OSC-I training sites would be located inside Iraqi military bases, which would enable the facilities to be set back from the compounds' perimeters and to benefit from outer rings of Iraqi security in addition to embassy static guards.[131] The U.S. military provided force protection training to selected ISF personnel to enhance their ability to protect U.S. facilities.[132]

---

[127] Interview with Embassy Baghdad RSOs, Baghdad, March 15, 2011.

[128] Interview with Embassy Baghdad RSOs, Baghdad, March 15, 2011.

[129] Email from DoS DS officials, December 30, 2011.

[130] Boswell, 2011a, pp. 30–31.

[131] Oral briefing by USF-I J35 Protection, Baghdad, June 26, 2011.

[132] Oral briefing by USF-I J35 Protection, Baghdad, June 26, 2011.

- *U.S. Marine security guard detachment.* As at many U.S. embassies, Embassy Baghdad would have a detachment of U.S. Marine security guards charged with protecting U.S. diplomatic facilities, personnel, and information. Typically, Marines are stationed at the last access control point before entering the chancery or other U.S. mission office buildings.
- *Physical security measures.* Facilities on embassy compounds have many security features common to U.S. diplomatic missions around the world. However, because the consulates in Basra and Erbil and the EBO in Kirkuk were planned as interim facilities, DoS did not plan to install permanent and more costly security measures.[133] Security surveys undertaken collaboratively by USF-I and multiple elements of DoS in February 2010 and June 2010 recommended the creation of tailored "field expedient" physical security measures that would use or improve on locally available materials, such as existing structures and movable concrete blast walls, as much as possible.[134] However, even these temporary security measures would not be finalized until February 2013 for DoS-led facilities and September 2013 for DoD-led sites, according to GAO.[135]
- *Sense-and-warn radars.* In a letter to DoD to request equipment for the embassy, Under Secretary Kennedy wrote that the embassy has a "requirement for a flexible, moveable, and adjustable compound defense system for persistent ground observation and surveillance"[136]—radars and other equipment to detect incoming mortar and rocket fire with enough advance notice to allow people to duck and cover.[137] DoD agreed to provide a certain number of radars on nonexcess loan,[138] and the embassy planned to use USF-I's existing contract for sense-and-warn capabilities until DoS could develop and award its own contract. Although the embassy would not have the capacity to shoot down incoming rockets or to attack the location from which the rockets were fired, the sense-and-warn system would be able to identify the point of origin of the attack, which embassy officials could then provide to the ISF.[139]

---

[133] Interview with DoS OBO officials, Washington, D.C., December 2, 2011; interview with DoS DS officials, Washington, D.C., December 20, 2011.

[134] Interview with USF-I J3 staff officer who participated in the Operations Support Working Group, Baghdad, June 15, 2011.

[135] Courts, 2012, p. 6.

[136] Kennedy, 2010b.

[137] Interview with USF-I J3 staff, Baghdad, March 15, 2011.

[138] USF-I, J4 Joint Logistics Operation Center (JLOC), "J-4 Transition to DoS Plan Presentation to RAND Corporation," June 25, 2011a, slide 8. See also Kennedy, 2011b, pp. 11–12. See also GAO, 2011, p. 39.

[139] Interview with USF-I J3 staff, Baghdad, March 15, 2011.

- *Biometrics.* The embassy sought to transfer USF-I's badging system to enduring sites so Iraqi staff, contractors, and others who might need access to embassy facilities could be vetted and issued biometrically enabled badges. USF-I's system, DoS officials claimed, had been "instrumental in identifying individuals with derogatory information, some of whom have attempted to apply for positions at the U.S. Embassy."[140] DoD agreed to provide DoS with Biometric Authentication Toolset equipment and connectivity to its Biometric Identification System Access database.[141] In late June 2011, the RSO anticipated receiving $7 million under LOGCAP to create an off-site badging unit.
- *Overhead cover.* To protect against indirect fire, DoS planned to construct hard cover with sidewall protection over areas where DoS personnel work, sleep, and otherwise congregate in large numbers.[142] However, overhead cover has been the single largest construction line item for every embassy location.[143] Having already spent $146 million on overhead cover in FYs 2009 through 2011, the construction of overhead cover at yet more facilities after 2011 was too costly for DoS to implement.[144]

Differences in DoS and DoD security standards had important implications for the construction of embassy facilities. DoD security standards, according to an official with DoS OBO, are based on two assumptions that do not apply to DoS: Everyone is armed, and casualties are inevitable. In contrast, diplomats are unarmed, and DoS does not view any level of casualties as being acceptable. OBO's security standards are thus defensive only and are based on the needs of a downtown chancery site located in a relatively secure capital city rather than in an "expeditionary" environment.[145]

Consequently, neither the Foreign Affairs Manual (FAM) or the Foreign Affairs Handbook established a requirement for overhead cover; likewise, DoS's centralized Overseas Security Protective Board, which sets physical security protection standards for diplomatic missions abroad, also did not impose this requirement. Rather, the requirement came from the U.S. Mission in Iraq.[146] Although hardened cover cannot

---

[140] Kennedy, 2010b.

[141] Kennedy, 2011b, p. 12.

[142] Interview with USF-I J3 staff officer who participated in the Operations Support Working Group, Baghdad, June 15, 2011.

[143] Interview with DoS OBO officials, Washington, D.C., December 2, 2011.

[144] Interview with USF-I J3 staff officer who participated in the Operations Support Working Group, Baghdad, June 15, 2011.

[145] Interview with DoS OBO officials, Washington, D.C., December 2, 2011; interview with DoS DS officials, Washington, D.C., December 20, 2011.

[146] Interview with Kathleen Austin-Ferguson, Executive Assistant to the Under Secretary of State for Management, Washington, D.C., October 25, 2011. Also, interview with DoS OBO officials, Washington, D.C.,

protect against all incoming threats, it has proven to be highly effective against the most common forms of indirect fire, including the ubiquitous 107 mm rocket that has been extensively used throughout Iraq since 2004.[147]

One year before the transition, DoS and DoD had yet to agree on security standards or identify sources of funding for hundreds of millions of dollars' worth of security-related construction.[148] The inability to reconcile DoD and DoS security standards, and thus to identify the necessary the funds, impeded the transition in a number of important ways. First, debates over security standards led to lengthy inter-agency and intraagency negotiations that consumed a significant amount of staff planning time and delayed the transition of key embassy programs. One notable example was the preparation to transfer Joint Security Station (JSS) Shield from USF-I to INL. JSS Shield, which was to serve as one of INL's three logistics hubs, has been described as "the center of gravity for the PDP."[149] It is located next to the MOI and Baghdad's police college, and approximately two-thirds of all police advisory and mentoring engagements were planned to take place within 5 km of JSS Shield.[150] The site needed extensive physical security upgrades to meet DoS's requirements. A senior INL official in Baghdad stated that "the transfer of JSS Shield should have been a turnkey operation."[151] Instead, debates between INL and OBO over what security measures would be required delayed the beginning of PDP training and prevented construction from being completed by January 1, 2012.[152]

Second, Congress had not appropriated the $750 million needed to implement DoS's more onerous physical security requirements at provincial sites.[153] Thus, rather than scale back the mission's regional presence, DoS and DoD agreed that management of the Kirkuk compound would fall to DoD rather than to DoS. This transfer of responsibility enabled DoD to take on the $133 million in construction costs

December 2, 2011, and interview with DoS DS officials, Washington, D.C., December 20, 2011.

[147] Interview with USF-I J3 staff officer who participated in the Operations Support Working Group, Baghdad, June 15, 2011.

[148] OIG, 2011, p. 20.

[149] Interview with INL-Baghdad official, Baghdad, June 27, 2011.

[150] Interview with INL-Baghdad official, Baghdad, June 27, 2011.

[151] Interview with INL-Baghdad official, Baghdad, June 27, 2011.

[152] Interview with INL-Baghdad official, Baghdad, June 27, 2011. Also Paco Palmieri, "Update to Core Group: MOI/Police Advise and Train Mission Transition to INL," Baghdad, June 27, 2011, slide 7. Also interview with DoS OBO officials, Washington, D.C., December 2, 2011.

[153] U.S. Senate, Committee on Foreign Relations, 2011, p. 17.

and $87 million in annual operating costs.[154] Thus, the 100 DoS personnel located at Kirkuk operated under the security standards of their DoD landlord.[155]

Third, because DoS was not funded to enhance security to its own standards at all U.S. mission sites, management of security would be complicated by an arrangement in which "security will be a shared responsibility" of both DoS and DoD.[156] OSC-I planned to take advantage of statutory provisions that allow a DoD organization under chief-of-mission authority to opt out of DoS's more onerous security standards.[157] Therefore, at OSC-I–led sites, DoD would be responsible for the overall security of all facilities, while DoS and DoD personnel would each manage security for their own movements using different contract security companies.[158] At DoS-led sites, the embassy RSO would be responsible for security of the facilities and for DoS personnel, while DoD would remain responsible for the movement of its personnel using DoD security details.

Instead of having a unified security architecture operating under common guidelines, embassy security operations would be managed by two separate embassy staff sections (RSO and OSC-I), which would be guided by the security standards of two different agencies (DoS and DoD) and would report to different headquarters (DoS

---

[154] Telephone interview with DoS resource management official, October 27, 2011. As DoD's "tenant" in Kirkuk, DoS will pay for an agreed-on share of life support, medical, and communications costs, while DoD will assume the rest of the operating costs, most of which cover static security guards and other security measures. Interview with Kathleen Austin-Ferguson, Executive Assistant to the Under Secretary of State for Management, Washington, D.C., October 25, 2011.

[155] Interview with DoS OBO officials, Washington, D.C., December 2, 2011; interview with Kathleen Austin-Ferguson, Executive Assistant to the Under Secretary of State for Management, Washington, D.C., October 25, 2011.

[156] Secretary of State Clinton informed the Senate Foreign Relations Committee that "DOD (through USCENT-COM) has accepted responsibility for securing all OSC-I personnel, and for securing the field sites that will not be co-located with our other embassy constituent posts, including our OSC-I headquarters at FOB Union III in Baghdad" (Secretary of State Hillary Clinton, Response to Question for the Record #6 submitted by Sen. Richard Lugar, Committee on Foreign Relations, U.S. Senate, March 10, 2011). Kennedy, 2011b, p. 8, noted that DoS DS is "responsible for all State Department sites and DOD responsible for OSC-I personnel. As such, DOD will be staffing and funding its security operations."

[157] 22 USC 4802(a) gives the Secretary of State responsibility for protecting all U.S. government personnel and facilities abroad other than those under a combatant command (22 USC 4802, Responsibility of Secretary of State, January 7, 2011). (OSC-I staff are under chief-of-mission authority and thus not covered by this exemption; however, DoD is responsible for security for the 753 security assistance trainers, who are also in Iraq under chief-of-mission authority.) However, 22 USC 4805 allows the Secretary of State to delegate operational control of security functions protecting other agencies' personnel and facilities back to these agencies (22 USC 4805, Cooperation of Other Federal Agencies, January 7, 2011). Interview with DoS DS officials, Washington, D.C., December 20, 2011. Also, interview with Kathleen Austin-Ferguson, Executive Assistant to the Under Secretary of State for Management, Washington, D.C., October 25, 2011.

[158] DoD has engaged different companies than DoS did for static and movement security at OSC-I sites. Whereas DoS DS has contracted to SOC, Triple Canopy, Global, and DynCorp, OSC-I uses Triple Canopy, Aegis, and Olive Group. Interview with DoS DS officials, Washington, D.C., December 20, 2011.

and USCENTCOM).[159] Compared to most embassies, where all security responsibilities are centralized under the RSO, such a disjointed management and decisionmaking structure has the potential to undermine the critical goal of ensuring the safety of all U.S. mission personnel.

Future Security Challenges

DoS planned to take a broad range of dramatic steps to ensure the security of embassy personnel and facilities in Iraq. More important, DoS officials expressed confidence that the department would have the resources and capabilities necessary to address anticipated security threats. That said, however, a number of security-related factors affected the embassy's plans for future operations.

The embassy's security plans depended on the Iraqi government permitting it to undertake such initiatives as operating surveillance drones and helicopters, driving MRAPs and other military-style vehicles, and acting in many ways like the departed U.S. military. Given that a sizable portion of the Iraqi public was pleased to be freed from what many saw as a U.S. military occupation, the prospect of U.S. diplomats adopting the same profile was poorly received. Ambassador Jeffrey observed that the Iraqi government was eager to erase every sign of the U.S. military presence,[160] and DoS use of militarylike tactics and equipment blurs the distinctions between diplomatic activities and the military activities that Iraqis disliked. Iraqis viewed the continuation of such visible and proactive security measures as an affront to their country's sovereignty.

Perhaps the greatest transition challenge for DoS was the inability to predict what the security situation would be like in 2012 and beyond. U.S. officials assumed that the country would continue to experience some degree of violence and political turmoil and that U.S. personnel and facilities would continue to be targeted. An internal USF-I staff assessment indicated that, while current trends suggested that Iraq's security situation would likely remain stable—or gradually improve—through 2015, violent extremist organizations would continue to conduct operations designed to undermine U.S. presence, influence, and interests in Iraq. The assessment bluntly warned that, while the overall level of violence in Iraq would likely remain stable, embassy facilities and personnel would face a high degree of threat once U.S. forces departed Iraq.[161]

Nevertheless, security planning at the embassy was governed by an optimistic assumption that security would likely improve after the departure of U.S. forces.

---

[159] The U.S. ambassador is ultimately in charge of all U.S. personnel attached to the diplomatic mission, but day-to-day security decisions are made by personnel in the two staff sections in accordance with standards and directives from the two agencies.

[160] U.S. Ambassador to Iraq James Jeffrey, RAND Corporation Roundtable Discussion, Arlington, Va., March 14, 2012.

[161] Brennan notes, 2009–2011.

Indeed, officials in the political-military section of the embassy were convinced that the departure of USF-I would "remove the symbol of American power" from Iraq and thereby lessen Iraqi opposition to the United States and remove the rationale for attacking Chief of Mission personnel.[162] This is a stark contrast to OBO's facility planning, which assumed that embassy facilities would be totally cut off from host-nation infrastructure because of security threats and thus need to be self-sustaining. DoS DS officials explained that when they began to make security decisions in 2010, they planned as though security conditions would remain the same.[163] However, Secretary of State Clinton was more optimistic in a January 2011 letter to then-Senator John Kerry, in which she stated that operating costs for the diplomatic posts outside Baghdad would decline "as security improves"—not "if" or "when" it does.[164]

Deputy Assistant Secretary of Defense for the Middle East Colin Kahl offered an even more upbeat assessment of the future security environment in testimony to a congressional committee in November 2010:

> I feel confident that the State Department is planning against the security conditions as they are now, and since we expect the security conditions to continue to improve I think that the State Department will be well positioned beyond 2011 to continue to operate and contribute to Iraq's progress beyond that point. If circumstances were to deteriorate substantially then we would have to re-evaluate, but we don't anticipate that circumstances are going to deteriorate substantially.[165]

It is not clear how—or whether—the embassy would be able to operate in Iraq if the security situation were to get worse. While many U.S. officials believed that insurgents and extremists would not target Americans after the U.S. military withdrawal, it was plausible that the embassy and its constituent posts would be attacked once they become the highest-profile U.S. targets in the country; in fact, Shi'a cleric Muqtada al-Sadr asserted in October 2011 that, after U.S. troops depart, embassy staff should be considered "occupiers" and must be "resisted."[166] While he subsequently dialed down his rhetoric regarding U.S. diplomatic presence, other Iranian-backed Shi'a militants did not.

---

[162] Brennan notes, 2009–2011.

[163] Interview with DoS DS officials, Washington, D.C., December 20, 2011.

[164] Hillary Clinton, Secretary of State, letter to Senator John Kerry, January 12, 2011a.

[165] Colin Kahl, "Statement of Colin Kahl, Ph.D., Deputy Assistant Secretary of Defense for the Middle East, U.S. Department of Defense," and testimony, in *The Transition to a Civilian-Led U.S. Presence in Iraq: Issues and Challenges*, hearing before the Committee on Foreign Affairs, U.S. House of Representatives, 111th Cong., 2nd Sess., November 18, 2010, p. 37, in response to Representative Jim Costa.

[166] Reidar Visser, "Sadr Demands Resistance Against the US Embassy in Baghdad," *Iraq and Gulf Analysis*, October 22, 2011.

If the security environment were to worsen, the mission's operating costs would almost certainly increase dramatically, or the scale of the effort would be downsized. The mission might be forced to cease hiring Iraqis, increasing the costs associated with hiring third-country nationals for security, life support, housing, and other support services. Security costs would almost certainly increase, making embassy programs more expensive as well.[167]

Given both reduced government spending across the board and Congress's previously expressed reservations regarding DoS's operations in Iraq, the Department did not expect an influx of funding to cover increased security costs. Thus, any increase in security requirements would almost certainly require the embassy to shift funds from its outreach, training, and assistance programs or to downsize quickly as it does at other diplomatic posts that face deteriorating security situations.

A Low Tolerance for Civilian Casualties

DoS and the U.S. public have a low tolerance for civilian casualties.[168] Civilian U.S. casualties would likely erode political support for embassy outreach and assistance activities. Referring to the U.S. public, a senior DoS official stated in July 2011 that "our stomach for accepting casualties is extremely low," predicting that it would only take one big attack to undermine support for the mission at DoS and in Congress.[169]

Despite RSO Baghdad's zero-risk policy,[170] senior embassy officials recognized that they, in fact, would have to accept some degree of risk to accomplish their objectives in Iraq.[171] As sound a strategy as this may seem given the security environment, it remains to be seen how the U.S. mission will be affected if its personnel are subject to a high-profile attack or suffer civilian casualties.

---

[167] Citing GAO, Special Inspector-General for Afghanistan Reconstruction, and World Bank audits of reconstruction activities in Iraq and Afghanistan, a CWC document stated that "numerous audits estimate that unanticipated security costs increased project expenses by 25 percent." See CWC, *Transforming Wartime Contracting: Controlling Costs, Reducing Risks, Final Report to Congress*, August 2011b, p. 74.

[168] Interview with INL-Baghdad official, Baghdad, June 27, 2011.

[169] Interview with senior DoS official, Washington, D.C., July 15, 2011. The September 2012 attack on the U.S. consulate in Benghazi, Libya, which caused the death of U.S. Ambassador J. Christopher Stevens and three other DoS employees, demonstrated the degree to which casualties among DoS civilians abroad can indeed generate political controversy back home; the incident received extensive media coverage, sparked a heated exchange between the presidential campaigns of President Obama and Republican nominee Mitt Romney, and led commentators to question U.S. strategy in the Middle East in the wake of the Arab Spring. See Josh Rogin, "Romney Campaign Links Embassy Attacks to Obama's Failed Middle East Policies," *Foreign Policy* (The Cable), September 12, 2012. See also Ross Douthat, "In Search of a Middle Eastern Strategy," Evaluations Blog, *New York Times*, September 14, 2012.

[170] Interview with INL-Baghdad official, Baghdad, June 27, 2011.

[171] Interview with Ambassador Peter Bodde, Assistant Chief of Mission for Assistance Transition in Iraq, Embassy Baghdad, June 27, 2011.

Reliance on the Iraqi Government for Embassy Security

The embassy will rely on the Iraqi government for security far more than the U.S. military did. DoS officials knew long before the transition that the embassy would lose valuable force protection, intelligence, and firepower after the U.S. military redeployment. The mission's more-limited organic security capabilities would require it to "rely on the Iraqi forces and the Iraqi police for these functions."[172] In the provinces, however, the ISF could not be counted on to match the support and security that USF-I had previously provided the DoS PRTs.

DoS recognized early on that the ISF would need extensive training to augment U.S. mission security effectively. In FY 2009 and FY 2010, DS trained ISF officers in incident response, postblast investigation techniques, and other counterterrorism skills using $5.6 million in nonproliferation, antiterrorism, demining, and related programs funds. Using $6 million in FY 2008 economic support funds, which are available for five years, DS also trained 249 ISF officers to support movements of U.S. mission personnel, and all these Iraqi officers were integrated into U.S. mission PSDs.[173] Similarly, the embassy developed plans to train ISF in route clearance skills so they could support secure travel of mission staff,[174] although USF-I assessed in mid-2011 that the ISF's route clearance capabilities would be limited even with the envisioned training.[175]

Contracting

DoS planned to depend on contractor support for a wide range of fundamental functions in addition to security, including life support, facility operations and maintenance,[176] aviation transportation, medical care, and many others. Although DoS improved its in-house contract oversight practices, contracting challenges undermined the department's readiness to operate on its own in Iraq, thereby placing U.S. programs and funding at risk.

The number and value of contracts DoS oversees increased dramatically since the invasions of Iraq and Afghanistan. SIGIR notes that DoS administered $1.2 billion in

---

[172] Boswell, 2011b.

[173] Interview with DoS DS officials, Washington, D.C., December 20, 2011.

[174] Interview with BG Rock Donahue, USF-I J7, June 26, 2011.

[175] Interview with USF-I J3 staff, Baghdad, July 7, 2011.

[176] The maintenance contract includes everything from operating power plants and water purification systems to shampooing office carpets. See DoS, Office of Logistics Management, "Baghdad Operation and Maintenance Support" presolicitation notice, Solicitation Number SAQMMA-12-R-0012, November 3, 2011; see also DoS, Office of Acquisitions, "Baghdad Operation and Maint Support," Solicitation Number SAQMMA-12-R-0012," January 17, 2012.

contracts worldwide in 2003, but two years later it managed $5.3 billion—an almost 450 percent increase.[177] By 2007, the value of DoS contracts had risen to $7 billion.[178]

DoS recognized the need to improve its contracting capabilities to manage the surge in requirements, even highlighting the significance of contracting capabilities in its 2010 Quadrennial Diplomacy and Development Review.[179] Nevertheless, although the department has improved its contract management capacity, it was not resourced the necessary level of oversight. As two CWC commissioners testified to a congressional committee,

> This hearing poses the question, "US Military Leaving Iraq: Is the State Department Ready?" The short answer is "no," and the short reason for that answer is that establishing and sustaining an expanded US diplomatic presence in Iraq will require DoS to take on thousands of additional contractor employees that it has neither the funds to pay nor resources to manage.[180]

CWC Commissioner Dov Zakheim went so far as to say that the potential for waste, fraud, and abuse due to DoS's weak contract management abilities is not only "a huge risk," but is also "a likelihood."[181]

Embassy Baghdad's Use of Contractors
DoS's two biggest contracting expenses in Iraq after the transition would be for security and life support, although DoD has been managing life-support contracting for DoS ever since 2004 and planned to continue doing so for two years after the transition.[182]

Security Contract—Worldwide Protective Services
Security contractors under the DS-managed Worldwide Protective Services (WPS) contract vehicle comprised almost half of the 6,500 contractors under chief-of-mission

---

[177] SIGIR, 2009a, p. 18. Cited in Angel Rabasa, John Gordon IV, Peter Chalk, Christopher S. Chivvis, Audra K. Grant, K. Scott McMahon, Laurel E. Miller, Marco Overhaus, and Stephanie Pezard, *From Insurgency to Stability*, Vol. 1: *Key Capabilities and Practices*, Santa Monica, Calif.: RAND Corporation, MG-1111/1-OSD, 2011, p. 48.

[178] Kennedy, 2011b, p. 22.

[179] DoS and USAID, *Leading Through Civilian Power: The First Quadrennial Diplomacy and Development Review*, Washington, D.C., 2010a.

[180] Grant S. Green, "Statement of Grant S. Green, Commissioner, Commission on Wartime Contracting," *U.S. Military Leaving Iraq: Is the State Department Ready?* hearing before the Subcommittee on National Security, Homeland Defense, and Foreign Operations, Committee on Oversight and Government Reform, U.S. House of Representatives, 112th Congress, 1st Sess., March 2, 2011, p. 2.

[181] Smith, 2011.

[182] Jeffrey D. Feltman, Responses to Questions for the Record from the Honorable Russ Carnahan, in *Transition to a Civilian-Led U.S. Presence in Iraq: Issues and Challenges*, hearing before the Committee on Foreign Affairs, U.S. House of Representatives, 111th Cong., 2nd Sess., November 18, 2010b, pp. 53–63b, p. 58.

authority in Iraq as of October 5, 2011,[183] and the number of security contractors was expected to increase. Early in FY 2011, DoS's WPS contract employed 1,800 static guards and 900 PSDs in three locations in Iraq (Baghdad, Erbil, and Tallil). These security contractors were supervised by 81 DS special agents.[184] The FY 2012 WPS plan anticipated 5,500 security contractors (4,000 static guards and 1,500 PSDs) in five locations (Baghdad, Erbil, Mosul, Kirkuk, and Basra). By early FY 2012, after plans for a facility in Mosul had been dropped, the numbers declined to a total of 5,000 security contractors (3,800 static guards and 1,200 PSDs). These contractors were to be managed by 101 DS agents.[185] Thus, between FY 2011 and FY 2012, the WPS contracting vehicle was used to hire 110 percent more static guards and 33 percent more PSDs, requiring a doubling of DS staff to oversee them (see Table 8.4).

Life-Support Contract—Logistics Civil Augmentation Program

Since 2003, DoS relied on the military's LOGCAP contract to provide life-support services for the embassy in Baghdad, the PRTs, and other civilian facilities. DoS determined early on that it would not be able to develop and award its own life-support contract in time for the transition and, on April 7, 2010, formally requested that DoD continue LOGCAP (as well as DLA support).[186] The Deputy Secretary of Defense approved the request on September 27, 2010.[187]

The LOGCAP IV contract supporting U.S. embassy facilities would manage the embassy's food service; fuel; water; solid waste; power generation; firefighting; 3,910 mobile containerized housing units; and material handling equipment, such as fork-

Table 8.4
Increase in WPS Contract Security Personnel from FY 2011 to FY 2012

Time/Plan	Static Guards	PSDs	U.S. Government Supervision (DS special agents)
FY 2011 (as of November 2010)	1,800	900	81 DS agents
FY 2012 (as of October 2011)	3,800	1,200	101 DS agents
Increase (percent)	110	33	25

---

[183] SIGIR, 2011f, p. 52.

[184] Feltman, 2010b, p. 58.

[185] Feltman, 2010b, p. 58. Modified numbers were provided by DoS as a written reply to RAND questions, October 12, 2012.

[186] Interview with Kathleen Austin-Ferguson, Executive Assistant to the Under Secretary of State for Management, Washington, D.C., October 25, 2011.

[187] Interview with USF-I J3 staff, Baghdad, June 26, 2011.

lifts, cranes, and flatbed trucks.[188] It would also support embassy weather and fuel supplies at Baghdad and Basra and taxiway and runway maintenance at Sather.[189]

Under Secretary of State Kennedy told the CWC that the "use of LOGCAP, on an interim basis, is giving the department sufficient time to solicit for and award a life support contract of its own."[190] DoS planned to use LOGCAP only for one year, believing it would have a mechanism of its own in place by then.[191] DoD hoped for the same timetable and planned to train DoS personnel to take over the management of the contracts by the beginning of FY 2013.[192]

A senior DoS official stated in October 2011 that the embassy would depend on contracted life support "for the next several years," whether through LOGCAP or some other mechanism, because "the security environment is not likely to improve enough to allow staff to run to the store."[193] While necessary, DoS's extended reliance on LOGCAP would increase costs for embassy support. Noting in congressional testimony that "LOGCAP is a contingency contract and thus is considered 'a contract of last resort' for customers because of the potential additional costs arising from its noncompetitive aspects," SIGIR Stuart Bowen stated that a competitive bidding process would likely reduce costs and generate savings for the taxpayer.[194]

The Department of State Has Improved Its Contract Oversight Capability

DoS, GAO, DoD, SIGIR, and the CWC all agreed that DoS had a limited ability to oversee contracts,[195] both because it had a small number of contracting officers and because the officers were not adequately trained to manage contracts on as large a scale

---

[188] USF-I, J4, June 25, 2011a.

[189] Interview with USF-I J3 staff, Baghdad, June 26, 2011.

[190] Kennedy, 2011b, p. 11.

[191] Interview with senior DoS official, Washington, D.C., July 15, 2011.

[192] Interview with USF-I J5 staff, Baghdad, December 1, 2011.

[193] Interview with Kathleen Austin-Ferguson, Executive Assistant to the Under Secretary of State for Management, Washington, D.C., October 25, 2011.

[194] Stuart W. Bowen, Jr., "Testimony of Stuart W. Bowen, Jr.," in *Transition in Iraq: Is the State Department Prepared to Take the Lead?* hearing before the Committee on Oversight and Government Reform, U.S. House of Representatives, September 23, 2010, p. 5. Emphasis in the original.

[195] DoS and USAID, 2010a; GAO, 2011, p. 40; Frank Kendall, "Statement of Hon. Frank Kendall, Principal Deputy Under Secretary of Defense for Acquisition, Technology, and Logistics," in *The Final Report of the Commission on Wartime Contracting in Iraq and Afghanistan*, hearing before the Subcommittee on Readiness and Management Support, Committee on Armed Services, U.S. Senate, 112th Cong., 1st Sess., October 19, 2011b, pp. 24–25; Stuart W. Bowen, Jr., "Statement of Stuart W. Bowen, Jr., Inspector General Office of the Special Inspector General for Iraq Reconstruction," in *U.S. Military Leaving Iraq: Is the State Department Ready?* hearing before the Subcommittee on National Security, Homeland Defense, and Foreign Operations, Committee on Oversight and Government Reform, U.S. House of Representatives, 112th Cong., 1st Sess., March 2, 2011a, p. 2.

as Embassy Baghdad required.[196] Unfortunately, DoS was slow to rectify this situation. Although Under Secretary Kennedy's April 7, 2010, letter to DoD acknowledged that DoS "does not have within its Foreign Service cadre sufficient experience and expertise to perform necessary contract oversight,"[197] CWC Commissioner Katherine Schinasi told Congress 18 months later that "the State Department has not trained up its contracting officials sufficiently to be able to make good use of that LOGCAP contract."[198] Without improvements to its contracting capabilities, SIGIR expressed great "concern about whether State's current structure and resources provide a sufficient basis for managing very large continuing contracts and programs."[199]

As an interim measure, DoS planned to rely on DoD for both contract vehicles and support in Iraq. The Defense Contract Management Agency (DCMA) planned to detail 52 contract managers to DoS to work at embassy facilities and DLA, the Defense Contract Audit Agency, and Army Contracting Command to provide additional support.[200] As a senior DoD official told Congress, "DOD is basically providing the contracting support to State Department for all of its essential functions."[201] In June, Under Secretary of State for Management Patrick Kennedy informed Congress that DoS would require assistance from DCMA and the Defense Contract Audit Agency "through calendar year 2014 or until the Department develops the capacity to perform similar in-depth duties."[202]

DoS had, in fact, been working to improve contract management and oversight capabilities, particularly regarding Iraq and Afghanistan. After a 2005 review that found problems with INL contract management and oversight, DoS created distinct offices to manage contracts for both Iraq and Afghanistan,[203] with 102 contract management staff added to the Washington-based Iraq contracting team alone.[204] DoS elevated the status of contract oversight personnel through both recognition and awards. In addition, DoS expanded training opportunities available to contracting officers,[205] including classes at both DoS's Foreign Service Institute and DoD's Defense Acquisi-

---

[196] Bowen, 2010, p. 4.

[197] Kennedy, 2010a.

[198] Katherine Schinasi, testimony, in *The Final Report of the Commission on Wartime Contracting in Iraq and Afghanistan*, hearing before the Subcommittee on Readiness and Management Support, Committee on Armed Services, U.S. Senate, 112th Cong., 1st Sess., October 19, 2011, p. 23.

[199] Bowen, 2011a, p. 2.

[200] Kendall, 2011b, p. 41.

[201] Kendall, 2011b, pp. 24–25.

[202] Kennedy, 2012.

[203] Feltman, 2010b, p. 60.

[204] Kennedy, 2011b, p. 4.

[205] Kennedy, 2011b, p. 4.

tion University.[206] DoS also planned to deploy more contracting officers to ensure on-site oversight, with two or three contracting officer representatives at each DoS-led site in Iraq.[207]

In part because of the scale of the WPS contract, DS had taken significant steps since 2010 to improve its contract oversight.[208] DS hired more than 200 contract administrators to ensure compliance for both Iraq and Afghanistan,[209] including 39 to oversee the Iraq WPS contracts alone. As Assistant Secretary of State for Diplomatic Security Eric Boswell told a congressional committee in June 2011, contract oversight is now "a major function of our agents overseas."[210]

Although DoS made some missteps, the primary obstacle preventing the department from developing the capability to oversee its contracts in Iraq was a lack of funding from Congress. DoS requested a mere $3 million in its FY 2011 and FY 2012 budgets to improve contract oversight capabilities,[211] but Congress did not appropriate any additional resources—money or personnel—for this purpose.[212] SIGIR Bowen gently prodded Congress in March 2011 to give DoS more resources, testifying that Congress might wish to provide more support to DoS to bolster its overall contract management capacity, arguing that such funding would, in all likelihood, pay for itself within a short time.[213] In a previous House hearing in September 2010, SIGIR noted that resources to improve DoS's acquisition and contracting "are likely to prove to be a bargain compared with waste that may occur if State's program management and acquisition efforts continue to be underresourced."[214]

Embassy Air

Ever since 2003, U.S. military aviation assets had played an important part in providing mobility for U.S. military and civilian personnel throughout the country. Planning for a post-2011 U.S. civilian-led presence, therefore, included a requirement for avia-

---

[206]Kennedy, 2011b, p. 4.

[207]Interview with Ambassador Jeanine Jackson, U.S. Embassy Baghdad Minister-Counselor for Management, Baghdad, June 28, 2011.

[208]Interview with DoS DS officials, Washington, D.C., December 20, 2011.

[209]Kennedy, 2011a, p. 4.

[210]Boswell, 2011b.

[211]Bowen, 2011a, pp. 6, 8.

[212]Interview with Kathleen Austin-Ferguson, Executive Assistant to the Under Secretary of State for Management, Washington, D.C., October 25, 2011.

[213]Bowen, 2011a, p. 8.

[214]Bowen, 2010, p. 7.

tion assets, given the risks associated with intercity road travel and the lack of viable commercial air links between major Iraqi cities. Since INL had managed air wings as part of counternarcotics initiatives in a number of countries around the world since the 1980s,[215] planners envisioned having INL manage the 46-aircraft fleet that would constitute "Embassy Air."[216] Although most Embassy Air personnel are contractors, INL had a long history of providing oversight for the contract air crews that make up more than 95 percent of its global air wing staff.[217] However, as with many other aspects of the Iraq transition, the scale of what INL planned to undertake in Iraq would be far greater than any previous related undertaking.

Although Embassy Air would provide a wide range of services, including "quick reaction force movement, search and rescue, medical and casualty evacuation, and route reconnaissance and convoy escort," transportation for Chief of Mission personnel who are moving to, from, and within Iraq would be one of its most critical responsibilities.[218] In June 2011, the embassy management counselor said that the mission's ability to manage this transportation responsibility was one of her greatest concerns, both because of the scale of the operation and the still-unresolved questions of funding and authorities.[219] She stated that embassy contractors would be responsible for getting themselves to Iraq and that Embassy Air would transport them to mission facilities in country,[220] although it was not clear whether DoS would have the ability to do so for all mission contractors. USF-I officials expected that a C-130 "ring route" would need to be flown twice per week just to transport OSC-I contractors to work sites, as the embassy's DASH-8s would not offer sufficient capacity. However, it was not clear at the time whether DoS or DoD would have the authority or the funding to operate such a "milk run" service.[221] By January 2012, such milk runs were indeed operating, with passengers' employing agencies reimbursing DoS $800 per person per flight.[222]

---

[215] Robert B. Charles, "State Department's Air Wing and Plan Colombia," press briefing, Washington, D.C.: U.S. Department of State, October 29, 2003.

[216] Kahl, 2010, p. 57. See also OIG, 2011, p. 23.

[217] Rob Neil, "Targeting Terror's Roots—Part 1," *Pacific Wings*, August 1, 2007. Note, however, that the State IG found extensive shortcomings in INL's management of its air wing program in Afghanistan and Pakistan. See OIG, "The Bureau of International Narcotics and Law Enforcement Affairs Air Wing Program in Afghanistan and Pakistan: Performance Audit," Report Number MERO-A-10-03, March 2010a, p. 28.

[218] OIG, 2011, p. 23.

[219] Interview with Ambassador Jeanine Jackson, U.S. Embassy Baghdad Minister-Counselor for Management, Baghdad, June 28, 2011.

[220] Interview with Ambassador Jeanine Jackson, U.S. Embassy Baghdad Minister-Counselor for Management, Baghdad, June 28, 2011.

[221] Interview with USF-I J33 staff, Baghdad, June 26, 2011.

[222] Interview with senior DoS official, Washington, D.C., February 17, 2012.

The embassy's aviation transition plan had six phases, with initial operating capacity planned for July 2011 and full operating capacity planned for the fleet of 46 fixed-wing and rotary aircraft by December 2011.[223] The department faced many challenges to launching a fully operational air service by January 1, 2012. First, it sought land use agreements with the Iraqi government that would permit the embassy to house and operate its aircraft, which were not forthcoming. As a result, necessary construction was not completed or, in the case of Basra and Erbil, even begun by summer 2011. (Nevertheless, by February 2012, the facility at Erbil was able to accommodate Embassy Air flights.)[224]

Second, the embassy had to develop the ability to maintain, refuel, and support its aircraft without outside assistance because "[m]aintenance hangars with cranes are not available and Iraqi commercial aviation fuel delivery capability and dependability is poor."[225]

Third, the department had to conclude agreements with the governments of Jordan and Kuwait regarding the use of facilities there and the operation of flights in and out of Iraq. (These agreements were in place by February 2012.)[226]

Medical Support

One of the most critical components of the self-sufficient "expeditionary" embassy was the ability to provide both routine and emergency medical care for its personnel. Whereas embassies in most countries rely to at least some degree on local medical facilities, the U.S. Mission in Iraq could not do so at all. As Under Secretary Kennedy wrote in December 2010: "Other than two hospitals in Erbil, all other hospitals in Iraq are unavailable for our use because of security, inadequate medical care, or practical reasons."[227]

The embassy established two levels of medical facilities in Iraq.[228] It created seven health units to provide primary, urgent, and initial emergency care and four diplomatic support hospitals, which offer all health unit services; more-advanced medical, trauma, and diagnostic capabilities; and the ability to perform emergency surgery.[229] Figure 8.3 is a map of medical facilities. While embassy medical facilities cannot provide com-

---

[223]Kahl, 2010, p. 57. See also OIG, 2011, p. 23.

[224]Interview with senior DoS official, Washington, D.C., February 17, 2012.

[225]OIG, 2011, p. 24. See also Kennedy, 2010b.

[226]Interview with senior DoS official, Washington, D.C., February 17, 2012.

[227]Kennedy, 2010b.

[228]Kennedy, 2010b.

[229]Interview with USF-I officer, Baghdad, June 30, 2011.

prehensive treatment of serious injuries, patients in need of such services will be stabilized and evacuated to Jordan or Kuwait.[230] Although U.S. Air Force assets based in neighboring countries will, in principle, be able to evacuate a patient from Iraq within 24 hours, USF-I officials expressed concerns that the Iraqi government could pose obstacles to Air Force medical evacuation flights.[231]

In December 2010, DoS requested a wide range of medical equipment from DoD to save funds and eliminate the long lead time needed to purchase, transport, and

Figure 8.3
Map of U.S. Mission Health Care Facilities

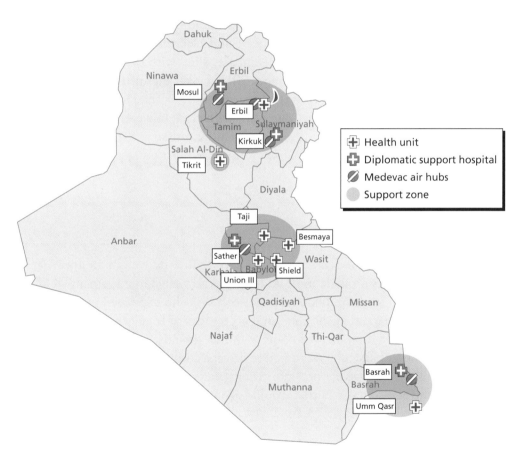

SOURCE: DoS, "Solicitation SAQMMA11-R-0010: Department of State Health Care Service Support in Iraq," February 9, 2001, p. 111.
RAND RR232-8.3

---

[230]Kennedy, 2011a, pp. 4–5.

[231]Interview with USF-I officer, Baghdad, June 30, 2011.

install critical equipment.[232] The most difficult element of medical transition planning was fulfilling DoS's decision that embassy personnel be treated within the "golden hour"; studies have shown that medical treatment within one hour of the injury greatly increases the chance of survival. Since this capability had been provided to U.S. military personnel (and DoS personnel serving alongside the military in Iraq over the previous eight years), DoS made a commitment to keep the capability available to embassy personnel in 2012 and thereafter. DoS medical planners achieved the golden hour, according to Assistant Deputy Chief of Mission Peter Bodde, "but at tremendous cost."[233] The medical contract that was awarded to Medical Support Services–Iraq on May 15, 2011, will cost $132 million over five years.[234]

The embassy's medical facilities were scheduled to reach initial operating capability by the end of June 2011; however, construction and contracting issues delayed the transfer of responsibility. Between September 1 and December 31, 2011, USF-I established a 104-person "bridge" to support all embassy sites, with the facilities planned to reach full operating capability by December 2011. USF-I planned to provide standby 24/7 casualty evacuation services through January 2012.[235] The DoS IG had expressed concern that the transition deadline might not be met;[236] however, a senior DoS official asserted that it was, in fact, fully operational by January 2012, adding that the establishment of the medical capability was one of the most successful elements of the transition.[237]

Knowledge Management

Transition planning seems to have succeeded extraordinarily well in an area that was not initially identified as an issue to be addressed: knowledge management (KM). An embassy staff member in the public diplomacy section, who happened to have a degree in library science, realized that she could not access lists of Iraqi officials who had participated in U.S. public diplomacy programs. When she raised this issue up the chain of command in the embassy, she was asked to direct a comprehensive effort to integrate all U.S. government knowledge on Iraqi reconstruction efforts before it disappeared into agency archives or databases that would become inaccessible.

---

[232]Kennedy, 2010b.

[233]Interview with Ambassador Peter Bodde, Assistant Chief of Mission for Assistance Transition in Iraq, Embassy Baghdad, June 27, 2011.

[234]Kennedy, 2011b, p. 11.

[235]Interview with USF-I officer, Baghdad, June 30, 2011.

[236]OIG, 2011, p. 2.

[237]Interview with DoS Iraq Transition Coordinator Ambassador Patricia Haslach, Washington, D.C., February 17, 2012.

The leadership of USF-I and embassy initiated a process to ensure that "data, information and knowledge about operations, projects, programs and Iraqi contacts at USF-I & PRTs will remain accessible to U.S. Mission–Iraq after the PRTs close and USF-I departs."[238] By preserving this information and making it easily accessible, the KM effort was intended to help make future projects more effective and to prevent repetition of past mistakes.

It is important to note, before 2011, that USF-I did not have an effective "KM" program of its own to transfer.[239] While both MNF-I and later USF-I had made efforts to categorize information from the various staff sections, the pace of operations and the continued addition of new officers to the staff made the task all but impossible. For example, by the end of Operation Iraqi Freedom on August 31, 2010, USCENTCOM had taken 500 terabytes of data from Iraq to establish a historical archive.[240] However, because this information was not well organized, it was not expected to be readily accessible for years to come.[241] As part of the KM transition plan that USF-I and embassy developed, USCENTCOM agreed to index and catalog USF-I's data, a process that was expected to take 18 to 24 months.[242] Similarly, DoS did not have an organized KM strategy until the embassy directed the preservation of historical data.[243]

Once begun, KM transition planning proceeded quickly. In October 2010, the ECG reviewed a KM plan[244] and put Ambassador Bodde in charge of the effort.[245] Soon afterward, the embassy's Office of Provincial Affairs and the KM Transition Office began providing guidance to PRTs on how to preserve records, and someone in each PRT was designated KM lead.[246] In December 2010, the USF-I CG and the embassy Deputy Chief of Mission agreed to add KM to the 12 transition LOOs.[247]

In addition to the planning process, the transition consisted of four overlapping phases.[248] First, an audit was conducted to identify information likely to be valuable

---

[238] Embassy of the United States, Baghdad, KM Coordinator, "Knowledge Management Transition Plan," briefing to ECG, March 14, 2011a.

[239] Interview with U.S. Embassy KM official, Baghdad, June 28, 2011.

[240] Interview with Ambassador Peter Bodde, Assistant Chief of Mission for Assistance Transition in Iraq, Embassy Baghdad, June 27, 2011.

[241] Interview with USF-I historian, Baghdad, December 13, 2011.

[242] Interview with U.S. embassy KM official, Baghdad, June 28, 2011.

[243] Interview with U.S. embassy KM official, Baghdad, June 28, 2011.

[244] Interview with U.S. embassy KM official, Baghdad, June 28, 2011.

[245] Embassy of the United States, Baghdad, KM Coordinator, "USF-I/USM-I Knowledge Management Transition," briefing, June 28, 2011b.

[246] Email from U.S. embassy KM official, June 29, 2011.

[247] Interview with U.S. embassy KM official, Baghdad, June 28, 2011.

[248] Embassy of the United States, Baghdad, KM Coordinator, 2011b.

in the future. After consulting with more than 80 offices in USF-I, the embassy, and PRTs, the combined USF-I and embassy KM team identified 477 information sources to transition,[249] more than 300 of which were not hosted on any network at all and would thus likely have been lost in the absence of a KM effort.[250] Data to be preserved included contact lists, biographical data, PRT data, program histories, situation reports, and data on significant security-related activities in the Combined Information Database Network Exchange.[251] Second, information was made widely accessible on networks or through other means. Portals were to be placed over data to facilitate widespread access,[252] and a directory was to be established to enable users to access data no matter where it resided.[253] Third, applications were transitioned to the embassy to ensure that remaining entities could access data in critical USF-I systems. Finally, the embassy established a permanent KM position to enable continued information access.

Embassy and USF-I officials cited three critical lessons learned from the KM effort. First, to avoid the eventual loss of information, an interagency KM team should be established at the beginning of a complex operation.[254] The early institution of sound KM practices would have facilitated the eventual KM transition, but by the time the KM effort started, a great deal of information had already been archived or lost and was thus no longer accessible.[255] Second, KM structures should be designed at the very beginning to serve the entire mission, not just individual agencies participating in it.[256] Without a comprehensive KM strategy, information becomes stovepiped and inaccessible during the mission, and it is harder to make such data widely accessible as part of a mission transition. Third, transition planning must be flexible. Although transition plans were established in the 2010 JCP, neither USF-I nor the embassy recognized the need for a KM transition plan until months later.[257] By not holding rigidly to the JCP transition annex, they were able to add a KM plan in December 2010, more than a year after the JCP was drafted.[258]

---

[249] Interview with U.S. embassy KM official, Baghdad, June 28, 2011.

[250] Embassy of the United States, Baghdad, KM Coordinator, 2011a.

[251] The Combined Information Database Network Exchange is a USCENTCOM data collection, correlation, analysis, and reporting tool used to assist theater decisionmaking and ease information flow, sharing, and reporting. Interview with U.S. embassy KM official, Baghdad, June 28, 2011.

[252] Interview with Ambassador Peter Bodde, Assistant Chief of Mission for Assistance Transition in Iraq, Embassy Baghdad, June 27, 2011. Also, interview with U.S. Embassy KM official, Baghdad, June 28, 2011.

[253] Interview with U.S. embassy KM official, Baghdad, June 30, 2011.

[254] Interview with U.S. embassy KM official, Baghdad, June 30, 2011.

[255] Embassy of the United States, Baghdad, KM Coordinator, 2011a, slide 5.

[256] Interview with U.S. embassy KM official, Baghdad, June 30, 2011.

[257] USF-I, 2011g, slide 12.

[258] Interview with U.S. embassy KM official, Baghdad, June 28, 2011.

Summary

There is no question that DoS broke new ground by planning a diplomatic outpost that would be completely self-sufficient and that would assume responsibility for functions a U.S. military force ten times its size had previously undertaken. Many embassies have operated in contingency environments in which they depended on armored vehicles and security guards, but never before had an embassy managed support functions of this size and scale—everything from field hospitals and a small airline, to military-style life support mechanisms and a small army of security guards who would both protect facilities and ensure secure movements of personnel. Because this is not a typical DoS mission, DoD in general, and USF-I more specifically, were intimately involved in helping the embassy prepare to conduct diplomatic operations in a hostile environment without the direct support of military forces. As one senior military officer noted, "this is the first time the United States [has] left an embassy in a war zone since Vietnam."[259] Understanding the perils that Embassy Baghdad would likely confront after the departure of forces in December 2011, the leadership of USF-I committed an unprecedented amount of manpower, resources, and funding, consistent with laws and regulations, to ensure the embassy was as prepared as possible to assume its new expeditionary mission.[260]

At a tactical and operational level, many aspects of the transition went well. The department established comprehensive medical and air transportation capabilities on schedule, working through obstacles created by contracts, unclear legal authorities and funding, and the need to secure aviation and support agreements with multiple countries in the region. Despite doubts as to whether DoS would be able to manage large-scale contracts for security, life support, medical operations, and other necessary functions, the department improved its contract oversight capabilities, trained all deployed DS officers to ensure government oversight of contract security personnel, and made effective use of DoD contract managers to fill gaps. Transition planners—particularly those on the ground in Iraq—demonstrated considerable flexibility by recognizing midstream that the embassy would benefit from a robust KM initiative that would ensure it could benefit from USF-I's considerable collections of information.

Transition planners encountered tactical challenges as well. Although DoS and USF-I had agreed at senior levels to transfer excess USF-I equipment to the embassy (such as housing units and generators), the unanticipated challenges that occurred during execution of these agreements created problems for DoS OBO that led to construction delays and short-term operational and security shortcomings. DoS was unable to adjust quickly, suggesting that the department should institute more flexible contingency zone contracting procedures.

---

[259] Interview with a USF-I senior military officer, Baghdad, December 14, 2011.

[260] Interview with former USF-I general officer, Washington, D.C., February 11, 2013.

However, the biggest obstacles to standing up the expeditionary embassy as planned were strategic, policy, and legislative failures and disappointments. For example, the lack of congressional funding for outposts in Mosul and Diyala hindered the embassy's original plans for diplomatic outreach in the Iraqi provinces. In addition, the lack of congressional funding for the construction of permanent facilities in Erbil, Kirkuk, and Basra made it much more difficult to plan and budget for long-term construction, security, and life support requirements.

Perhaps equally as important, transition planners in Washington made optimistic assumptions (1) that the security situation in Iraq would only get better in 2012 and beyond and (2) that, despite years of ups and downs in the U.S.-Iraqi relationship, Iraqi political leaders would welcome a large-scale, highly visible, proactive American civilian presence after the departure of U.S. troops. As will be discussed in Chapter Ten, neither assumption proved to be correct.

DoS successfully assumed the lead U.S. role in Iraq from DoD after an extensive transition planning effort, and it appeared to be well positioned for success as long as the security environment were to become increasingly permissive and if U.S.-Iraqi political ties were to gradually improve. However, it remains to be seen how effectively the expeditionary embassy this transition process created will be able to advance U.S. interests if political and security dynamics either remain the same or deteriorate. Most likely, to adjust to such changes, the embassy will have to revisit a number of central tenets of the transition effort and reassess the viability of establishing and maintaining large expeditionary embassies in the immediate aftermath of a large U.S. military operation.

CHAPTER NINE

Reposture the Force

Throughout 2011, USF-I executed OPORD 11-01 and OPORD 11-01 Change 1. The third LOE established in these OPORDs was defined as "Reposture the Force." It would be convenient to view this LOE simply as that portion of the OPORD that postured U.S. forces so USF-I could redeploy by the end of December 2011 in accordance with the SA.[1] However, the reality is much more complex. All actions taken in the pursuit of this LOE were conducted first and foremost to accomplish the numerous missions assigned to USF-I during the last year of the operation. For example, the sequence of base transitions to the Iraqis was determined by the operational needs of the force. Moreover, the tremendous logistical effort was designed to meet operational and transition requirements first while, simultaneously, taking necessary actions to retrograde equipment and supplies. Thus, while it is true that the redeployment of tens of thousands of troops, an equal or larger number of contractors, and millions of pieces of equipment from Iraq during 2011 was the largest movement of troops and military materiel since the buildup to World War II,[2] that was only one portion of the logistical challenge. An examination of how redeployment planning was conducted and executed during this last phase of the USF-I mission will help identify best practices and lessons that could be useful for policymakers and military staffs planning the impending transition in Afghanistan and similar transitions at the end of future large-scale military operations.

The Reposture the Force LOE depicted in Figure 4.1 had four components:

1. Forces are postured and supported to conduct operations.
2. Facilities [and bases] are transitioned or returned.
3. Equipment is retrograded or transitioned.
4. Operational maneuver is conducted and all U.S. forces and contractors are redeployed.

---

[1] Interview with USF-I J3 staff officer, Baghdad, November 6, 2011.

[2] LTG William G. Webster, Commander of the U.S. Third Army, in Brandon Babbit, "Webster: Drawdown and Buildup Dubbed 'Nickel II,'" *Desert Voice*, April 14, 2010, p. 3. See also William Branigin, "U.S. Shift from Iraq to Afghanistan Presents Massive Logistical Operation for Army," *Washington Post*, April 2, 2010.

The last three activities will be the focus of this chapter. However, it must be remembered that units were repostured throughout the operation in an effort to accomplish as much as possible before redeployment.[3] Further, it is critical to recognize that "operational maneuver" related to equipment and supplies as much as it did to forces. It included the synchronization of efforts to maintain security; sustain key aviation, intelligence, communications, and logistic nodes; support the transition to DoS; continue to advance ISF development; and continue the establishment of OSC-I, all while meeting base closure and unit and materiel retrograde requirements.[4] OPORD 11-01 called for the retrograde of forces to start on September 1, 2011. However, Change 1 to OPORD 11-01 delayed that initial movement by well over a month. Once the retrograde began, units would begin an operational maneuver that required the synchronization of base closures, retrograde of equipment, and the tactical movement of subordinate units conducting the redeployment.[5] The plan also specified that redeployment of tactical units (battalions) would be supported by other forces that would provide security along the route of movement and that units conducting the movement would receive priority for ISR, aviation assets, medical evacuation, route clearance, and communications assets.[6] And although the reposturing of USF-I forces was viewed from the very beginning as a tactical military operation, not an administrative movement of forces, it was not until the final months that the USF-I staff also included the movement of 1st Theater Support Command (TSC) supporting convoys in the same operational maneuver focus. This highlights the fact that, while combat operations had officially ended with the beginning of OND, military units participating in OND remained prepared to conduct combat operations when necessary to ensure force protection. It also highlighted the fact that the transfer and ending of the combat unit missions were the catalysts for the operational maneuver, base transfers, and supporting logistics operations. J3 directed the operational maneuver and the subsequent tasks of base closure, while the J7, J4, and other J-staffs supported the redeployment.[7] Finally, OPORD 11-01 specified that all activities not transitioned prior to the beginning of the operational maneuver would take place during the last phase of the operation, culminating when theater specific responsibilities for Iraq were returned to USCENTCOM and the simultaneous inactivation of the Iraq joint operations area and the disestablishment of USF-I.[8]

---

[3]   Interview with former USF-I general officer, Washington, D.C., January 8, 2013.

[4]   Interview with a member of the USF-I command element, Washington, D.C., December 31, 2012.

[5]   Brennan notes, 2009–2011.

[6]   Brennan notes, 2009–2011.

[7]   Interview with former USF-I J4 staff officer, North Carolina, January 31, 2013.

[8]   Brennan notes, 2009–2011.

This chapter focuses on how USF-I actually executed the final phase of the mission by closing and/or transitioning responsibility for 90 bases and facilities during 2011; moving millions of pieces of equipment out of Iraq and delivering millions of dollars of equipment to Embassy Baghdad and/or the Iraqi military; conducting an operational maneuver as the mechanism for moving military forces out of Iraq; and completing the transition by returning responsibility for theater-specific responsibility relating to Iraq to USCENTCOM.

Closing and Transitioning Bases and Facilities to Iraqi Control

Between January 2008 and December 2011, USF-I closed, transferred to the government of Iraq, or transferred to the U.S. mission 505 bases of all types, from section-sized provincial joint coordination centers to large COBs.[9] Table 9.1 lists the seven types of bases USF-I identified for closure or transfer:[10]

Base Closure Direction, Planning, and Tasks

The USF-I J7 staff orchestrated the complex and time-consuming base transition and transfer process. The USF-I DCG for Support was the overall lead, but several forums were used to focus efforts, synchronize actions, and achieve further guidance

Table 9.1
Types of Bases Closed, Transitioned, or Transferred

BaseType	Size	Description
Contingency operating base	> BCT	Theater command and control and/or logistics hub
Contingency operating site	BCT	Regional command and control and/or logistics hub
Contingency operating location	Battalion	Needs COS or COB for support
Coalition outpost	Company	Outpost, usually in urban terrain
Patrol base	Company	Self-defining, well-fortified position
Joint security station	Company	Joint/combined command-and-control hub with Iraqi Army and Iraqi Police
Provincial joint coordination center	Section	Joint/combined civil operations

---

[9]   USF-I, "CS/TS LOO Update to GEN Lloyd J. Austin, III," September 29, 2010c, p. 24.

[10]   USF-I J7, *Base Closure Smartbook*, February 2011, pp. 46–47. The smartbook notes that the term *forward operating base* (FOB) is a generic term for a COB, COS, or contingency operating location and that FOBs are not identified with any kind of official standard.

as required. Figure 9.1 portrays an overview of the complex base closure planning and preparations required to keep overall OND transition plans on track.[11]

The USF-I J7 staff established the Base Management Operational Planning Team to coordinate closure and transfer with each of the other joint staff sections and with relevant external groups and agencies.[12] The staff followed guidance and standards in the USF-I FRAGOs and internally distributed standard operating procedures.[13] However, the overarching document guiding the closure and transfer process was the U.S.-

Figure 9.1
Base Closure and Transition Synchronization Staffing Efforts

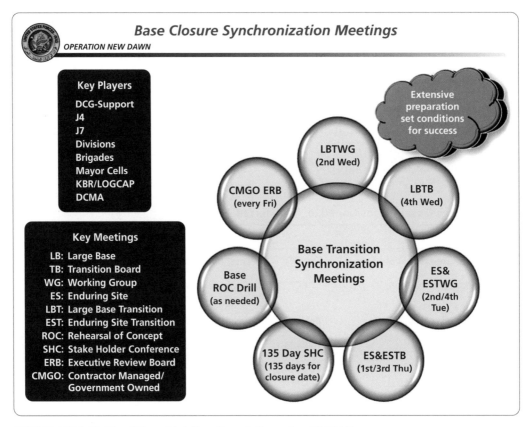

SOURCE: USF-I, J4, "Road Show," briefing, Kuwait, December 18, 2011j.
NOTE: KBR = Kellogg Brown & Root.
RAND RR232-9.1

---

[11] Interview with former USF-I J4 staff officer, North Carolina, February 1, 2013.

[12] USF-I J7, 2011, p. 43. Other groups and agencies included the Army service component at USCENTCOM and the Army and Air Force Exchange Service, which ran the exchanges (retail stores) that existed on most of the mid- to large-size bases.

[13] Specific references are listed on USF-I J7, 2010, p. 48.

Iraq SA, which provided explicit and binding guidance on base and equipment issues ranging from water rights to contracting to airspace control.[14]

USF-I staff published the *Base Closure Smartbook*, which identified the tasks and milestones necessary to close or transfer a base to Iraqi control. Because it was necessary to include both Iraqis and contractors without security clearances in the base closure process, many of the guidelines and standards in the smartbook were subsequently generalized or modified for specific operations. The smartbook, which USF-I J7 contended was a "best practice," described the closure and transfer process as follows:

> The Base Closure Process is a series of actions to prepare, document, and finally close or return bases to the Government of Iraq. The entire process can take between 90 to 365 days, and includes 89 tasks. These tasks are broken down into four parallel processes: Real Estate Management, Environmental Oversight, Property Distribution, and Contracting.[15]

The J7 staff estimated it would take a full calendar year to close or transfer one of the sprawling COBs, just over six months for a COS, and 90 days for any one of the four varieties of relatively small contingency operating locations. Figure 9.2 depicts the parallel closure process for both COSs and COBs. Each of the four overall processes associated with real estate management, environmental oversight, property distribution, and contracting were conducted in parallel to ensure timely closure and transfer and to facilitate coordination across tasks. The number of days required to complete each sequence is depicted in parentheses, with the first number depicting days needed to close a COS, and the second depicting the number of days needed to close a COB. Each one of these sequences reflects one or more tasks from the list of 89 base closure tasks.

Executing the Base Closure Process

While the J7 staff oversaw the planning, direction, and tracking of the closure and transfer process, USF-I directed a number of subordinate organizations to support base closure and enlisted external aid to focus explicitly on executing the closure process.

As military personnel who helped operate bases withdrew from Iraq along with their units, the number of base operations staff at remaining facilities shrank by two-thirds. USF-I thus faced a staff shortage, not only for managing and operating the remaining facilities but also for taking on the increased number of tasks associated with the closure and transfer of the facilities. USF-I instituted a Base Operations Support–Iraq (BOS-I) program to augment base "mayor's cell" staffs with both DoD

---

[14] United States of America and the Republic of Iraq, 2008a.

[15] USF-I J7, 2010, p. 4. The term *return* seems to only be appropriate when coalition forces occupied a former Iraq military facility and is not generalizable to all bases, many of which did not exist prior to the onset of hostilities in 2003. Therefore, this chapter refers to *transfers* instead of *returns*.

Figure 9.2
Contingency Operating Site and Base Closure Process

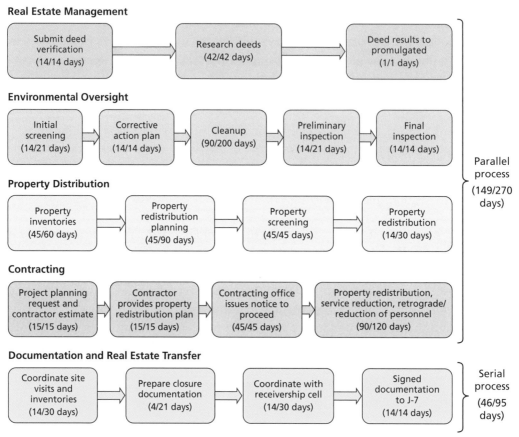

SOURCE: USF-I, 2010c.
RAND RR232-9.2

civilians and contractors to fill positions dealing with such functions as base logistics, public works, and information technology.[16] The BOS-I teams assisted military mayor cells in day-to-day operations and provided critical property management expertise for facilitating inventory, disposition, and transfer of equipment to U.S. government entities and/or the government of Iraq on the larger bases.[17] In the final weeks of the U.S. force presence, BOS-I staff members were instrumental in inventorying equipment and property and preparing to transfer excess materiel to the Iraqi government through the Foreign Excess Personal Property (FEPP) program. As bases in the north closed in

---

[16] Appendix F contains a detailed description of BOS-I (Tab 7, Base Operations Support—Iraq). The appendix contains material JLOC supplied to RAND in December 2011.

[17] JLOC, "BOS-I Program Civilian and Contract Support," December 13, 2011b, pp. 54.

earnest with the decision to "go to zero," many members of the BOS-I staff were relocated to assist with the condensed time line for closure of facilities in southern Iraq.[18]

In addition, at least three types of mobile "teams" traveled around Iraq tackling diverse challenges, including the closure of scrap metal yards, extracting rolling stock (e.g., trucks and tanks) from bases, inspecting buildings for leftover equipment, and assisting Iraqi businesses in obtaining or retaining licenses so that they might continue to conduct business on the bases after transfer to Iraqi control.[19] These teams supported the major subordinate commands that controlled the bases prior to transition, each of which in turn coordinated with the J7 staff, J4 staff, the Joint Contracting Command–Iraq (JCC-I), and a range of other U.S. entities. At various steps in the process, designated U.S. representatives also coordinated with the Iraqi Local Land Registry Office for transfer of land ownership and with the Receivership Secretariat in the prime minister's office, which was the single Iraqi government entity charged with accepting all real and personal property.

Because the process was executed primarily along four parallel tracks, it was complex and difficult to coordinate; proper timing and cross-compartmental coordination of each task became essential to successful transition. Moreover, because U.S. troops and contractors would remain on each base until closure or transfer, it was necessary to retain life-support functions, such as food, fresh water, laundry, and hygiene, for these personnel while USF-I was simultaneously shutting down the institutionalized processes for providing life support. These competing requirements led to a laddered process in which the least essential services were withdrawn first (e.g., base education support), and the most essential services were withdrawn last (e.g., water and hygiene).

In most cases, life-support functions were transferred along with the base. For example, when USF-I transferred COS Garry Owen to Iraqi control on October 15, 2011, they also transferred approximately $4 million worth of equipment and buildings including "living quarters, air conditioners, refueling equipment, several civilian vehicles, water containers, concrete barriers, a dining facility, and a battalion headquarters building."[20] Conditions varied from base to base, but U.S. forces made an effort to move the Iraqis into the new facilities prior to the date of closure. Therefore, even on bases that retained a great deal of life support equipment and housing, the U.S. forces would return to a more expeditionary lifestyle weeks or months prior to transfer. For example, during the complete transfer of Convoy Support Center Scania, the U.S. soldiers on the base gradually shifted into more fieldlike conditions:

---

[18] JLOC, 2011b, pp. 1–2, 4.

[19] USF-I J7, 2010, pp. 37–38. These teams included the Expeditionary Disposal Remediation Team from DLA, the Mobile Redistribution Property Assistance Teams (M-RPATs) under an Army field support brigade, and the Mobile Redistribution Teams (MRTs) under an engineer support command. See the section on reposturing the force for greater detail on the M-RPATs and MRTs.

[20] Richard Vogt, "Warhorse Transfers COS Garry Owen to IA, Closes Historic Chapter," press report, October 27, 2011.

They switched from staying in small, air-conditioned trailers to sleeping in tents with several other soldiers. Instead of hot meals, the soldiers ate Meals, Ready-to-Eat during their entire last month at the base. Their communication media were removed during the final weeks, preventing them from calling home and using the Internet.[21]

However, this did not necessarily mean that the contractors supporting USF-I departed the base or the theater immediately. Units tailored base closure and transition plans to include a sequential drawdown of services, eventually leaving only essential personnel and essential services in place to conduct the final base transition. In many cases, contractors who were willing to stay in Iraq, especially LOGCAP and Army Materiel Command and Life Cycle Management Command contractors, were repositioned to fill gaps presented when other contractors, anticipating an end to their contracts, sought and found new employment elsewhere.[22]

U.S. forces redeployed, transferred, or disposed of a considerable amount of material during the closure and transfer process. However, officials made it clear that the process did not involve the removal of Iraqi state or personal property. As one U.S. Army officer stated, "We're not taking anything that the Iraqis had [owned]. We are only taking stuff [equipment and property] that we put in, we utilized, and when we didn't need it anymore, we took it home."[23]

Environmental Cleanup and Oversight

One of the overarching themes of base closure and transfer was "good stewardship." This concept is clearly conveyed in the smartbook and is most obviously revealed by the inclusion of environmental oversight as one of the four primary LOEs in the closure and transfer process. The stated USF-I policy on environmental oversight was to "mitigate negative environmental impacts on U.S. bases in Iraq."[24] A number of documents established or reinforced this rationale, but USF-I primarily referred to Article 8 of the U.S.-Iraq SA:

> Both Parties shall implement this Agreement in a manner consistent with protecting the natural environment and human health and safety. The United States reaffirms its commitment to respecting applicable Iraqi environmental laws, regu-

---

[21] Emily Walter, "Base Transfer Marks Progress," *Expeditionary Times*, July 28, 2010.

[22] Interview with former USF-I J4 staff officer, North Carolina, February 1, 2013.

[23] Andrew E. Kramer, "Leaving Camp Victory in Iraq, the Very Name a Question Mark," *New York Times*, November 10, 2011.

[24] USF-I J7, 2010, p. 16.

lations, and standards in the course of executing its policies for the purposes of implementing this Agreement.[25]

The J7 staff at Camp Victory included six USF-I environmental managers, and each division retained six contracted environmental response and cleanup teams to address issues identified in the inspection process. The environmental staff was also tasked with identifying and cataloging culturally sensitive sites in accordance with Article 5 of the U.S.-Iraq SA.[26] According to USF-I, the environmental cleanup process identified and assessed over 3,300 environmental sites, such as fuel storage facilities, firing ranges, burn pits, incinerators, and medical waste and scrap metal collection points. Ultimately, USF-I processed over 32 million pounds of hazardous waste. To further improve the environmental standards of the bases, USF-I installed 13 incinerators at a cost of $15 million,[27] and DLA instituted a program to recycle trash and scrap metal at all remaining U.S. bases.[28] Time was allocated in the base transfer process to enable extensive mitigation of any problems that might be found; as Figure 9.2 shows, environmental cleanup is the most time-consuming step in the process, requiring up to 90 days for a COS and 200 days for a COB, or 60–75 percent of the total time allocated for base closure.[29]

Final Status of Closure, Transfer, and Transition

By December 15, 2011, USF-I had closed, transferred, or transitioned all 505 of the bases it had identified in January 2008. OSC-I and DoS (including INL) assumed responsibility for or retained a presence on only 14 bases. While the closure and transfer process took place along specified time lines for each base, the process itself had been ongoing since 2003. Many of the early transitions were rushed, leading to embarrassing setbacks. For example, coalition forces handed the Saddam Palace complex in Tikrit over to the Iraqi government in November 2005, but within days, looters had reportedly picked the complex clean.[30] Few such reports emerged in the wake of base transitions during the last several years of the U.S. military presence, although the process was not without setbacks. As described earlier, poor coordination with Iraqi

---

[25] United States of America and the Republic of Iraq, 2008, p. 16. USCENTCOM Regulation 200-2, Contingency Environmental Guidance, 2012, provided the practical direction and guidelines for environmental oversight.

[26] USF-I J7, 2010, p. 21.

[27] USF-I, 2010c, p. 32.

[28] Paul C. Hurley and John J. Abbatiello, "Responsible Drawdown: Synchronizing the Joint Vision," *Joint Force Quarterly*, No. 59, 4th Quarter 2010, p. 133.

[29] This chart is replicated from USF-I, 2010c, p. 30. Some of the terms within the chart were simplified, and acronyms were removed to help the reader understand the overall process. More specific detail is available in the briefing and in the *Base Closure Smartbook* (USF-I, J7, 2011).

[30] Ellen Nickmeyer, "Tikrit Palace Complex Allegedly Picked Clean," *Seattle Times*, January 13, 2006.

counterparts regarding the handover of a military base in Kirkuk contributed to a tense standoff between Kurdish and ISF in November 2011.[31]

Although the last push toward transition took place in 2011, a conscious and consolidated effort to achieve transition closure and transfer goals had been ongoing for at least two years prior to final handover. By the time Operation Iraqi Freedom transitioned into OND on September 1, 2010, only 92 bases remained under U.S. control.[32] By June 2011, this number had shrunk to about 60, of which 39 housed fewer than 100 personnel.[33] In early November 2011, this number was reduced to 12.[34] Figure 9.3 shows the glide path of U.S. military base closures from December 31, 2010, to the final withdrawal on December 18, 2011.

Figure 9.3
U.S. Military Base Closures (December 2010–December 2011)

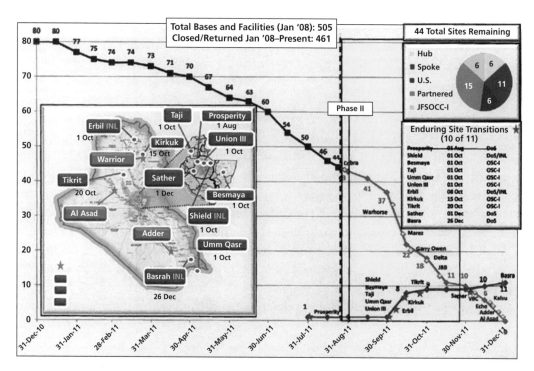

SOURCE: Slide courtesy of USF-I, December 14, 2011.
RAND RR232-9.3

---

[31] Sam Dagher, "Standoff at U.S. Airbase in Iraq," *Wall Street Journal*, November 18, 2011.

[32] Karen Parrish, "Forces in Iraq Pursue 'Methodical, Careful' Exit Plan," press release, USCENTCOM, November 3, 2011.

[33] USF-I, 2011e.

[34] Parrish, 2011.

Equipment Disposal: A Major Logistics Challenge

The withdrawal of troops and equipment from Iraq required a logistical operation of an unprecedented scale. As an Association of the U.S. Army newsletter described it, "If you put all the trucks in a single convoy it would stretch from El Paso, Texas, to D.C. If we stacked every container left in Iraq when the president issued his guidance [on October 21, 2011], the stack would be 51 miles high—nine times the height of Mount Everest."[35]

At the beginning of OND in September 2010, USF-I's roughly 50,000 troops and 60,000 contractors occupied more than 90 bases.[36] These bases contained an extraordinarily large amount of equipment and supplies that would be given to the Iraqi government, provided to the embassy, moved to Afghanistan in support of Operation Enduring Freedom, relocated to a storage facility for prepositioned equipment elsewhere in the region, or shipped out of theater. In the last 15 months of the U.S. military presence, USF-I thus had to account for and either transport 41,000 pieces of rolling stock (vehicles) and 1.8 million pieces of equipment, plus other items that would redeploy along with their units, or transfer the items to DoS or the government of Iraq.[37] The command focus throughout OPORD 10-01 and 11-01 was on units "right sizing" and reducing their footprints on their bases and unit areas, keeping only what was organizationally required or necessary to conduct missions. With so many items moving in so many different directions, the USF-I J4 used JLOC to maintain a logistics common operating picture and to coordinate the drawdown of military equipment and supplies. JLOC normally focuses on sustainment and the status of all commodities of supplies, transportation, and equipment maintenance.[38] Figure 9.4 shows

---

[35] "Equipment Management During Iraq Withdrawal Presents Challenges," *AUSA News*, February 1, 2012.

[36] The actual number of contractors serving in Iraq was never clearly understood because of the nature of some of the contracts. For example, a contract for a particular service, such as laundry, would provide a fee for services. Each company was then responsible for its own employees. While systems were put in place late in the mission to gain accountability for contractors in Iraq, it was understood that the numbers provided to USF-I were, at best, a close estimate. Interview with former USF-I J4 staff officer, North Carolina, January 27, 2013.

[37] Interview with a former USF-I J4 staff officer, North Carolina, January 30, 2011; Branigin, 2010. The majority of the items transferred were nonstandard equipment, and transferring these offered a cost-avoidance benefit to DoD—both the financial and risks to life—by not having to physically transport all of the items. Many of these items were used and had little or no life remaining. Transfers to DoS typically occurred as part of a base transition but also included ISR, force protection equipment, and MRAPs. Transfers to the government of Iraq were managed under two major programs: USETTI and FEPP.

Much of the equipment that had to be disposed of at the end of the U.S. military presence was theater-provided equipment, equipment that was provided to rotational U.S. units as they arrived in theater but that then remained in place when the units redeployed. Examples include large, fixed generators at installations or heavy transportation equipment that could not be routinely deployed and redeployed due to prohibitive intertheater transportation costs.

[38] See the memorandum from JLOC in Appendix F. Prior to ending the disestablishment of USF-I, the J4 provided an unclassified report to RAND to assist in preparing this report. This report is provided in its entirety in

Figure 9.4
Equipment Drawdown During Operation New Dawn

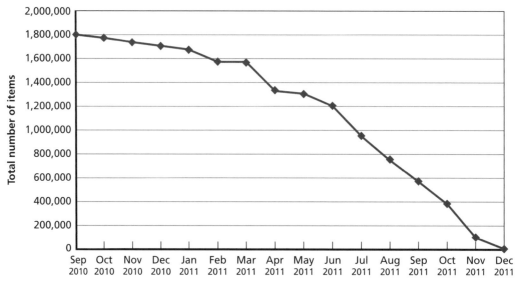

the rate at which equipment was withdrawn from Iraq in the last 15 months of the U.S. military presence.

About two-thirds of the items were shipped out of the country, where military services in the United States, Afghanistan, or elsewhere would continue to use them.[39] However, because of the high cost of shipping materiel from Iraq back to the United States, it was not cost effective to repursue all the equipment; the shipping costs of many items, especially those that had been extensively used, exceeded their residual value. Military units would declare such items excess. USF-I would first determine whether the item could either fill a requirement in theater for another military unit or fulfill a DoS request. If the item was still excess following this initial screening, USF-I would seek disposition instructions from the ARCENT Support Element–Iraq staff, which had been positioned forward and was collocated with the USF-I J4 to facilitate staff interactions between ARCENT and USF-I.[40] USCENTCOM, working with the Army Materiel Command's Responsible Reset Task Force, would then determine whether a piece of equipment was needed by U.S. forces in Afghanistan, other U.S.

Appendix E to this report. In addition to the summary report, USF-I provided in-depth data and assessments regarding commodities management; equipment drawdown; FEPP; USETTI; Asset Visibility; Container Advise and Assist Teams; BOS-I; and the U.S. Mission–Iraq (logistical) Transition. These reports are included as Tabs 1–8 to Appendix F.

[39] Michelle Tan, "Logisticians Shuffle Troops' Equipment," *Army Times*, March 28, 2010.

[40] Telephone interview with an ARCENT Support Element–Iraq staff member, Mali, January 31, 2013.

forces around the world, or stocks of prepositioned equipment.[41] If so, the task force provided disposition instructions telling USF-I where to send the materiel.[42] On occasion, this also included sending items back to the United States, despite shipping costs, to fill requests from state and local governments under the National Association of State Agencies for Surplus Equipment program.[43] If the equipment was deemed not needed by any U.S. entity and was not sensitive, it would be designated for transfer to the Iraqi government through one of several mechanisms that will be discussed in detail later in this chapter.

Many factors influenced USF-I's plans to redeploy personnel and reposture its equipment, but one of the most critical was the delayed decision on a post-2011 enduring troop presence. Although USF-I had begun reposturing equipment two years before the end of its mission, USF-I J4 staff wrote that the "late timing on the decisions regarding the enduring footprint the U.S. government planned to leave in Iraq" was the cause of "major deviation from the projected drawdown." Initial J4 estimates demanded between five-and-one-half to six months to meet retrograde requirements, with the focus completely on leaving. OPORD 11-01 delayed the focus on reposturing until September 1, 2011, which J4 ultimately deemed feasible due to refined and reevaluated staff estimates. However, the delay into October put significant pressure on the very small margins logisticians had built into their planning.[44] Until Washington decided what kind of troop presence (if any) it would leave in Iraq after December 31, USF-I J4 staff explained, "units did not redeploy [to Kuwait or home stations] as bases continued to close, but instead repostured to other locations [within Iraq], thus delaying the equipment retrograde" or turn-in process. A USF-I report stated that the adjustments to OPORD 11-01 made in Change 1

> included replacing 4/1 CAV [Cavalry] Advise and Assist Brigade (AAB) with 4/1 AD AAB instead of [allowing] 4/1 CAV [to conclude] its Combined Security Mechanism (CSM) mission and tour [in Iraq] in August 2011. Employing 4/1 AD AAB allowed USF-I to extend the CSM mission in the USD-N [U.S. Division–North] operational environment (OE). Additionally, the USF-I Main Command Post . . . repositioned . . . from Victory Base Complex in the Baghdad area to Al Asad Air Base . . . in Anbar province vice repositioning to Kuwait prior to Phase II operations. These two major actions, along with the reduction of Phase II operations being condensed to 14 October through 18 December 2011 (originally

[41] Rikeshia Davidson, "AMC Takes Lead in Task Force Aimed at Army Reset Effort," Army.mil, December 21, 2009.

[42] Thomas Richardson, "The Role of U.S. Forces–Iraq with Force Reposturing and the Status of Government-Owned Equipment in Iraq," roundtable with defense bloggers, August 3, 2011.

[43] See the memorandum from JLOC in Appendix F. For more information, see National Association of State Agencies for Surplus Property, "NASASP Mission Statement," website, 2009.

[44] Interview with former USF-I J5 staff officer, Carlisle, Pa., December 29, 2012.

planned to begin on 01 September and to end on 31 December 2011), added significant friction and increased the risk to the completion of all Reposture operations on time. Alternatively, these actions allowed the Commander significant operational flexibility for the other two LOEs.[45]

Thus, while the transition timing was due to residual force decisionmaking, equally important was the commander's intent to make as much progress toward campaign plan objectives as possible to increase the chances for an enduring success in Iraq.[46]

This trend continued until USF-I was informed to "go to zero" in mid-October, forcing the flow of equipment out of Iraq to accelerate.[47] Figure 9.5 shows that, during the last five months of OND, more than 24,000 1st TSC, self-redeployment (SRD),[48] and "door-to-door" (D2D) truckloads were required to move equipment out of Iraq. The SRD vehicles moved in a combat convoy that required provision of ISR throughout the movement, armed escort vehicles, forces providing security along the route of movement, and route clearance by supporting engineers. TSC cargo convoys included MRAP escort vehicles and the enabler support discussed with unit SRD movement. D2D movement was done under Surface Deployment and Distribution Command (SDDC) commercial contracts that used private security contractors.[49] Large convoys would depart military bases every evening under the cover of darkness and quietly make their way out of the country, unseen and unnoticed by the Iraqi populace.

Transfer of Equipment to the Embassy

At the beginning of OND, DoD and DoS established a Washington-based ad hoc senior executive steering group to identify the embassy's logistics and sustainment requirements and determine whether and how DoD could help meet them. The group was cochaired by two senior officials, the Deputy Assistant Secretary of Defense for Program Support and the Deputy Assistant Secretary of State for Logistics Management, which facilitated decisionmaking.[50]

---

[45] See Appendix D for material the USF-I Joint Plans Integration Center supplied RAND in December 2011, p. C3.

[46] Interview with member of USF-I command element, Washington, D.C., December 31, 2011.

[47] JLOC, "J-4 Summary of Equipment Drawdown," December 13, 2011e, p. 9. See Appendix F in this volume.

[48] A combat unit with MRAP vehicles was expected to conduct its own tactical road march out of Iraq. This entailed less risk to force due to the additional combat power in a convoy consisting mainly of MRAPs and because the tactical road march only had to travel one way. TSC convoys were used at times, but this meant the TSC convoy had to travel both into and out of theater; plus, the TSC convoy only had a set number of MRAP vehicles assigned to protect the contracted lift assets executing the transport mission. Interview with a former USF-I J4 staff officer, North Carolina, January 31, 2013.

[49] Interview with USF-I J4 staff officer, Washington, D.C., December 22, 2012.

[50] GAO, 2011, p. 36.

Figure 9.5
Truckloads and Self-Redeploying Equipment Departing Iraq

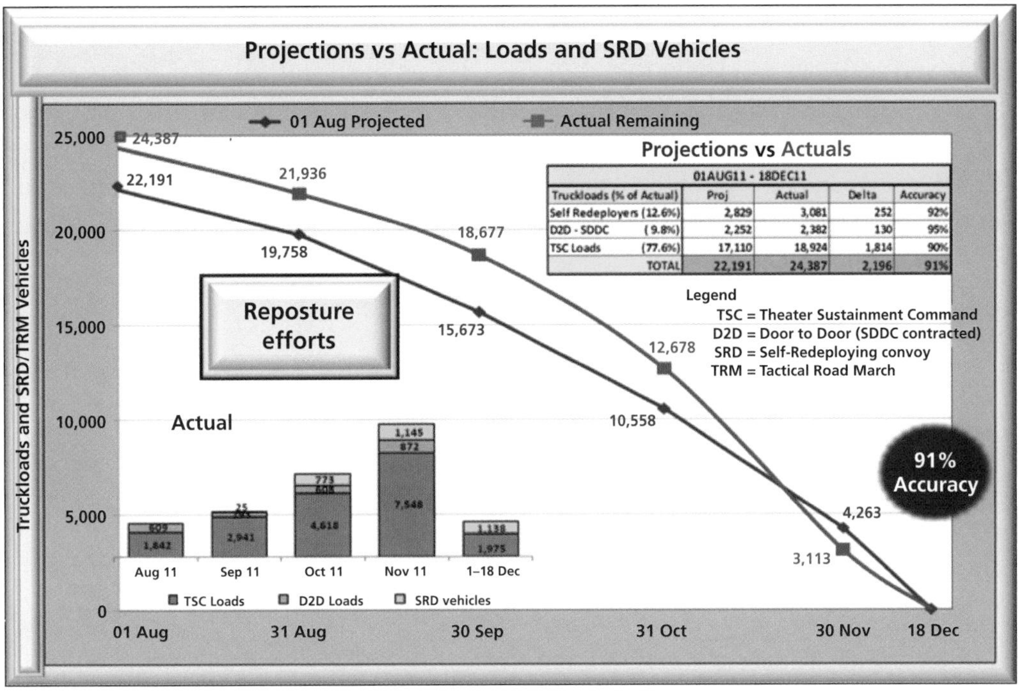

SOURCE: USF-I, J4, 2011j.
RAND RR232-9.5

In Baghdad, USF-I led the interagency Enduring Embassy Support Working Group to identify the embassy's evolving requirements, manage the transfer of equipment, and resolve issues that arose. Equipment was generally divided into "green" equipment (standard military-specific items) and "white" equipment (commercial-off-the-shelf or other nonstandard military items).[51] Between January 1, 2011, and November 25, 2011, USF-I provided DoS with 62,310 items of white equipment (24,172 truckloads) valued at $171,121,835 (before depreciation). Separately, USF-I provided OSC-I with 53,211 items of white equipment (1,176 truckloads) valued at $111,933,345 (before depreciation). Items provided included containerized housing units, generators, concrete barriers and walls, fuel tanks, ambulances, and general-purpose vehicles. Equipment was either left in place at enduring sites that DoS was to

---

[51] "White" equipment could be government owned and operated, contractor owned and operated, or contractor managed and government owned. The latter category can include items acquired by the contractor or items furnished by the government for the contractor's use.

take over or transported, at no cost to DoS, by USF-I.[52] Figure 9.6 provides a graphic representation of the white equipment transferred to embassy and OSC-I.

DoS also requested that DoD provide the embassy with green military equipment to address security, logistics, and life-support requirements that were "of a magnitude and scale of complexity that is unprecedented in the history of the Department of State."[53] Emphasizing the need for specialized military equipment to address Iraq's high-threat environment, Kennedy wrote, "If we do not acquire critical military assets before December, 2011, [DoS] will be forced to use less-effective technology and equipment. . . . As a result, the security of [DoS] personnel in Iraq will be degraded signifi-

Figure 9.6
White Equipment Transfers to the Department of State

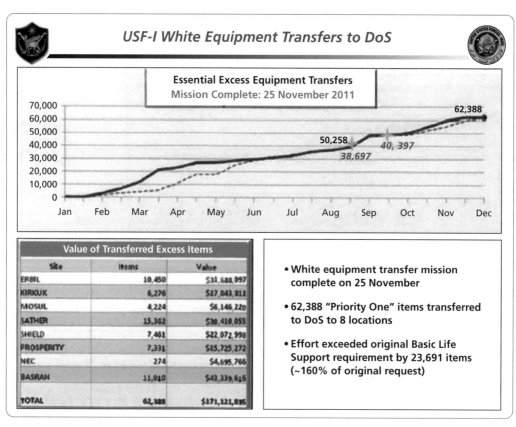

SOURCE: Slide courtesy of USF-I, December 14, 2011.
RAND RR232-9.6

---

[52] Appendix E: USF-I J4-DOSTC, "DOS Transition Cell RAND History Report," December 13, 2011i, p. D-7

[53] GAO, 2011, p. 37.

cantly and we can expect increased casualties."[54] DoS submitted a request for 23,000 items, which—after being assessed by an equipping board comprising representatives from the military services, the Joint Staff, and OSD[55]—was reduced to a total of 3,807 items, which were worth $212 million. DoS reimbursed DoD $10 million for items that DoD considered nonexcess, and DoD loaned DoS (at no cost) several key non-excess items needed for security and personnel protection: 60 Caiman MRAPs; two Giraffe counter rocket, artillery, and mortar radars; and biometric equipment to enable identification of embassy employees and visitors. USF-I considered the remainder, the vast majority of which consisted of medical equipment and supplies, to be excess and it provided to DoS at no cost (except for transportation).[56] See Figure 9.7.

Transfer of Equipment to the Iraqi Government

The primary means of providing equipment to the Iraqi government was the USETTI program, a subset of which was the FEPP program.[57]

U.S. Equipment Transfer to Iraq

Under the USETTI program, USF-I was authorized to transfer non–mission essential military equipment to the ISF to help enhance its capabilities before the U.S. military withdrawal. This program delivered standard military equipment that U.S. forces had used in Iraq, such as HMMWVs, machine guns, communication antennas, generators, and nonstandard equipment, including outdated commercial-off-the-shelf equipment.[58] While the vast majority of items provided to the ISF had been declared excess, nonexcess items could also be transferred to the ISF.[59] As of November 23, 2011, USF-I estimated that, by the end of the mission, it would have provided the ISF with 18,695

---

[54] Kennedy, 2010a (quoted in Richard Lardner, "State Department Wants a Mini-Army in Iraq," *Army Times*, June 14, 2010).

[55] GAO, 2011, p. 37.

[56] JLOC, 2011i, p. 4. See also Frank Kendall, "Frank Kendall, Principal Deputy Under Secretary of Defense for Acquisition, Technology and Logistics, U.S. Department of Defense," in *U.S. Military Leaving Iraq: Is the State Department Ready?* hearing before the Subcommittee on National Security, Homeland Defense, and Foreign Operations, Committee on Oversight and Government Reform, U.S. House of Representatives, 112th Congress, 1st Sess., March 2, 2011a, pp. 38–39.

[57] For details, see Appendix F, Tab 4, "U.S. Equipment Transfers to Iraq (USETTI)."

[58] Interview with former USF-I J4 staff member, North Carolina, January 31, 2013.

[59] The provision to transfer nonexcess items to government of Iraq was authorized by section 1234 of the FY 2010 NDAA (see Public Law 111-84, National Defense Authorization Act for Fiscal Year 2010, October 28, 2009). Authority for the provision to transfer excess defense articles to Iraq, along with many other countries, was provided by section 516 of the 1961 Foreign Assistance Act (see Public Law 87-195, U.S. Foreign Assistance Act of 1961, September 4, 1961).

Figure 9.7
Green Equipment Transfers to the Department of State

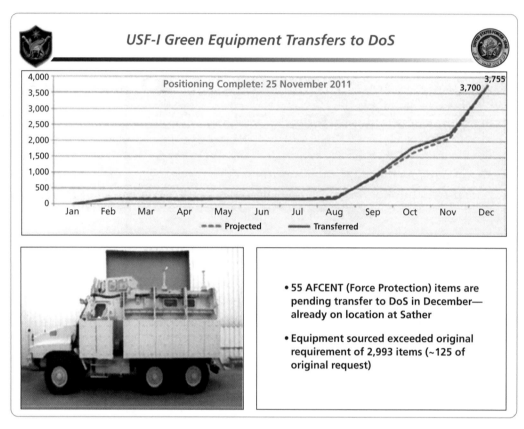

SOURCE: Slide courtesy of USF-I, December 14, 2011.
RAND RR232-9.7

excess items and 1,362 nonexcess items.[60] The total value of equipment provided under USETTI was approximately $640 million.[61]

Figure 9.8 shows the end-to-end USETTI process. From left to right: equipment needs are identified and proper equipment is sourced to meet that need (e.g., the Iraqis need radios, and USF-I identifies radios in its inventory that might be transferred); equipment is screened for viability and suitability; equipment is accepted into the program, and its disposition is changed from U.S. to government of Iraq; equipment is transported to a central facility or a holding area (e.g., the equipment maintenance site at Taji); it is maintained (repaired, cleaned) at this facility; and finally, it is transferred to the Iraqis.

---

[60] JLOC, "USETTI Background Paper," December 13, 2011h, p. 4.

[61] U.S. Department of the Army, "Audit of the U.S. Equipment Transfer to Iraq Program—Phase II," 2011a, p. 13.

Figure 9.8
Issues in the U.S. Equipment Transfer to Iraq Program End-to-End Transfer Process

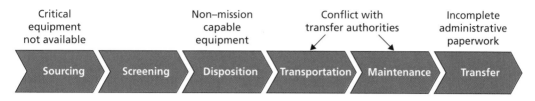

SOURCE: U.S. Department of the Army, "Audit of the U.S. Equipment Transfer to Iraq Program—Phase II," 2011, p.19.

RAND *RR232-9.8*

At the request of USF-I J4, the U.S. Army Audit Agency (USAAA) audited USF-I's execution of USETTI programs (but not USETTI's effectiveness in helping the ISF reach MEC) in 2011. USAAA found that USF-I had only partially put "the proper organizational structures, policies, and processes in place to ensure commands can sufficiently execute the USETTI program in accordance with established transfer authorities."[62] USAAA could not identify a comprehensive joint plan to "integrate the various entities and complex processes involved with transferring equipment under the program."[63] This lack of joint service integration resulted in confusion over roles and responsibilities for implementing the USETTI program and in specific gaps in the process of sourcing, screening, determining disposition, transporting, maintaining, and transferring the equipment destined for the Iraqi government.

Figure 9.8 also depicts the problems USAAA identified in five of the six phases: insufficient equipment was available for sourcing; the original Redistribution Property Assistance Teams (RPATs)[64] accepted non–mission capable equipment (e.g., one audit showed 38 of 1,468 pieces of equipment to be non–mission capable; another showed 20 percent of all equipment to be "grossly damaged"); administrative paperwork was sometimes incomplete or missing during the transportation and transfer phases; and there was some confusion about how to appropriately fund maintenance of equipment prior to transfer to the ISF.[65]

The USAAA report revealed what appeared to be a set of concerning yet unsurprising gaps in coordination and administrative control. USETTI was a unique and complex equipment transfer process undertaken in the midst of ongoing combat operations. The lack of sufficient available equipment might have resulted from a command oversight but, in some cases, might also have resulted from an actual lack of sufficient

---

[62] U.S. Department of the Army, 2011a, p. 18.

[63] U.S. Department of the Army, 2011a, p. 19.

[64] In Iraq, these Army Materiel Command teams relieved units of their excess property and then coordinated the transportation of that excess equipment out of theater.

[65] U.S. Department of the Army, 2011a, pp. 1–19.

excess or transferable equipment of the type required in each request. It is important to note that the delay in beginning the final withdrawal of U.S. forces also meant that they were still using the equipment, so it was not available for transfer to the ISF as early as originally planned.[66] The USAAA report made a number of recommendations for the continued implementation of USETTI through the end of mission, as well as for the implementation of similar equipping efforts being undertaken in Afghanistan.

Foreign Excess Personal Property

The FEPP program was a mechanism, first used in 2007, to provide excess personal property—mostly contractor-managed, government-owned items, such as generators, air conditioners, commercial vehicles, and containerized housing units—to approved Iraqi government entities at the federal, provincial, and local levels. The vast majority went to Iraqi military- and police-related entities, however, including the MOD (60 percent), MOI (15 percent), and ISF (14 percent). From January 2009 to November 2011, USF-I transferred to the Iraqi government almost 4 million items worth $440 million (after depreciation). Transferring property accountability to the Iraqis and leaving the items in theater saved the U.S. government $1.1 billion in transportation costs—more than twice the value of the goods.

Most FEPP items were transferred when the military bases at which they were located were transferred to the Iraqi government. (Items were provided "as is, where is.") In July 2009, the Deputy Under Secretary of Defense for Logistics and Materiel Readiness granted USF-I approval to authorize transfer of up to $5 million of FEPP items along with a base; this limit was later raised to $15 million per base in July 2009 and then to $30 million in October 2009. In early 2011, as the transfer of large facilities became imminent, it was raised to as much as $200 million (depending on the base).[67]

The FEPP program had a tiered approval system. Depending on the value of the item to be transferred (after depreciation), approval could be granted by an O-6 (colonel) commander (up to $10,000), the first general officer in the chain of command (up to $50,000), or the USF-I J4 (up to $1 million). Any item over $1 million had to be approved by the Deputy Under Secretary of Defense for Logistics and Materiel Readiness.[68]

---

[66] Interview with a former USF-I J4 staff officer, North Carolina, January 31, 2013.

[67] JLOC, "Foreign Excess Personal Property (FEPP) Transfers to the Government of Iraq," December 13, 2011c, pp. 1–5. The facilities and their authorized FEPP caps were Joint Base Balad ($200 million); Victory Base Complex ($170 million); Adder ($80 million); Al-Asad ($70 million); Speicher ($65 million); and Marez ($60 million). On November 10, 2011—just five weeks before the final departure of U.S. troops—OSD Logistics and Materiel Readiness authorized an increase to the caps for Al-Asad and Adder to $120 million each.

[68] Alan F. Estevez, Acting Deputy Under Secretary of Defense for Logistics and Materiel Readiness, "Authority to Transfer Foreign Excess Personal Property in Iraq," memorandum for Commanding General, Multi-National Force–Iraq, October 9, 2009.

Transfer of Equipment to Afghanistan

Given ongoing military operations in Afghanistan, some military materiel being shipped out of Iraq was sent to support U.S. troops there. (A great deal of equipment had previously been sent from Iraq to Afghanistan; approximately 40 percent of the equipment that units supporting the 2010 surge in Afghanistan needed came from Iraq.)[69] Although some equipment went by air, most items took one of two routes: (1) north through Turkey and then into Afghanistan from the north using the so-called Northern Distribution Network that passed through Georgia, Azerbaijan, Kazakstan, and Uzbekistan (as well as by boat across the Caspian Sea) or (2) overland to Kuwait and then by sea to Karachi, Pakistan, from where it was shipped by truck into Afghanistan. (See Figure 9.9 for a map.)[70]

Transfer of Equipment Out of Theater

State Agencies

Through a partnership with the nonprofit National Association of State Agencies for Surplus Property, USF-I arranged for excess nontactical equipment to be provided to state and local government agencies at no cost except to cover transportation from Kuwait. By the time the last troops left Iraq on December 18, 2011, USF-I had provided 399 pieces of equipment worth more than $4 million to 11 states, including bulldozers, forklifts, backhoes, generators, musical instruments, and hand tools.[71] This equipment has been used to fight wildfires in Oklahoma, equip a fire department in South Dakota, and provide power to a sewage plant in Alabama, among other things.[72] A potential reason this program was not better utilized was that the associated transportation costs were passed on to the U.S. state and local governments. The idea of reutilizing government purchased items was noble; in reality, however, it was more cost-effective to leave many items behind in theater.

---

[69] Karen Parrish, "Command's Relocation Aids 'Strategic Velocity,'" American Forces Press Service, October 28, 2010. See also Frank Panter, Jr., "Statement of Lieutenant General Frank A. Panter, Deputy Commandant, Installations and Logistics," in *Army and Marine Corps Materiel Reset*, hearing before the Subcommittee on Readiness, Committee on Armed Services, U.S. House of Representatives, 112th Cong., 2nd Sess., March 28, 2012, p. 7.

[70] Stephen Farrell and Elisabeth Bumiller, "No Shortcuts When Military Moves a War," *New York Times*, March 31, 2010.

[71] JLOC, memo to RAND, December 13, 2011f. See also Devon Hylander, "Taxpayers Benefit from Army's Excess Non-Standard Equipment," Army.mil, November 24, 2010, and Lori K. McDonald, "Drawdown Brings Non-Standard Equipment Mission to Sierra," Army.mil, January 18, 2011.

[72] Michael S. Oubre, "Equipment Once Used in Iraq Is Now Helping State and Local Governments," *Army AL&T Magazine*, January 6, 2012. Also Elizabeth Dwoskin and Gopal Ratnam, "For Sale, Cheap: The Things You Need to Invade a Nation," *Businessweek*, December 14, 2011.

Figure 9.9
Iraq-Afghanistan Transit Routes

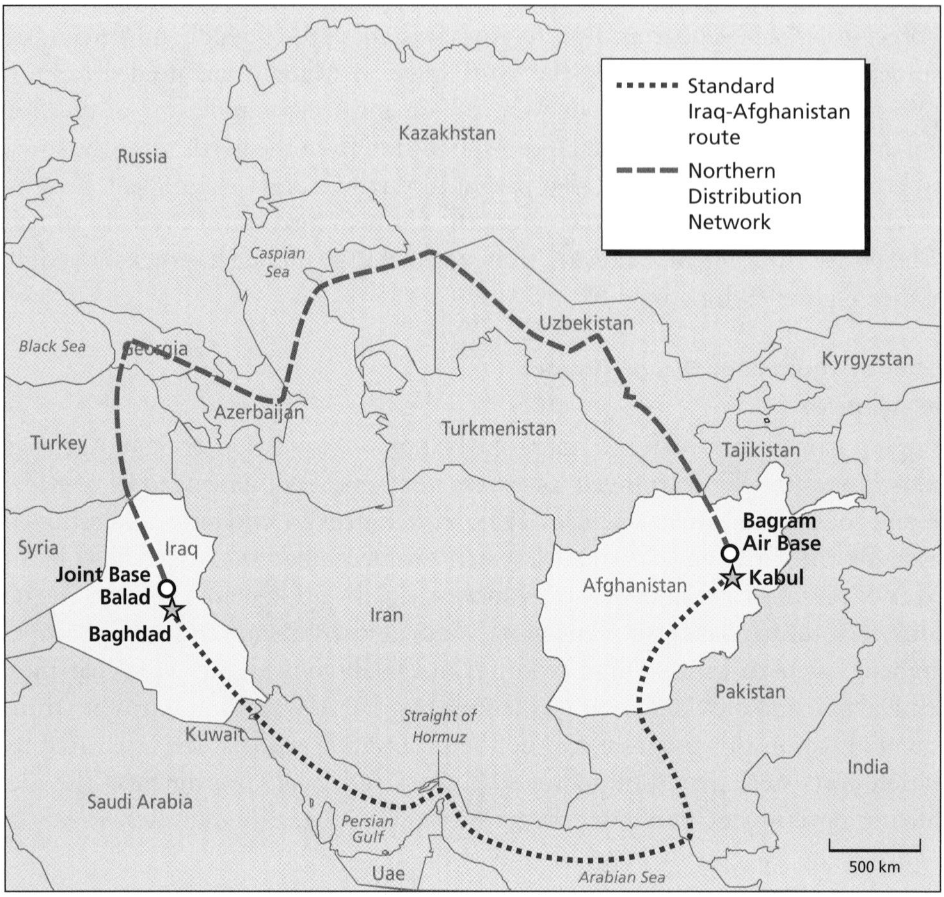

SOURCE: Data from *New York Times*.
RAND *RR232-9.9*

Retrograde to the United States and Other Destinations

Major end items and theater-provided equipment would be brought to Kuwait, where it would pass through a series of maintenance and inspection procedures before it could be shipped onward to the United States.

MRAPs, HMMWVs, and other rolling stock were repaired and upgraded, as needed; washed thoroughly to ensure they were clear of pests or contaminants that could harm U.S. agriculture; and subjected to a U.S. Customs inspection to ensure no contraband was hidden inside. The "wash rack" through which all vehicles had to pass—basically a 70-acre truck wash that could handle 100 vehicles at a time—operated 24 hours a day to keep up with the pace of arriving vehicles.[73]

Some of the materiel withdrawn from Iraq was, after being repaired and upgraded, sent to forward-deployed stocks of equipment prepositioned around the world for the use of units engaged in training exercises or contingency operations. To support the "pivot" toward the Asia-Pacific region directed by the January 2012 Defense Strategic Guidance,[74] the Army began developing plans to send equipment that had been in Iraq to prepositioned stocks in the Pacific.[75]

Operational Maneuver

In addition to transitioning bases and retrograding or transferring equipment, the final phase of OPORD 11-01 included a massive effort to maneuver U.S. military forces out of Iraq. In 2011 alone, USF-I was responsible for redeploying approximately 50,000 troops; coordinating the redeployment of more than 60,000 contract personnel; and processing over 2 million pieces of equipment.[76] Redeploying these troops and equipment from Iraq to the United States (and to other bases around the world) required the combined effort of the entire staff, all of USF-I's subordinate elements, the embassy interagency team, Iraqi government agencies and security forces, contractor support, and a number of other agencies and components external to USF-I, including USCENTCOM and the DLA. Military reserve and National Guard forces also mobilized to support the redeployment.[77]

As Figure 9.10 shows, the operational maneuver of U.S. forces began in late September 2011 and continued at a steady but rapid pace through December 18, 2011, when the last soldier entered Kuwait.

In addition to the operational maneuver of forces, USF-I also had to responsibly redeploy over 60,000 contractors who were deployed to support the force. The overarching guidance to "go to zero" identified a significant gap in how DoD accounted for contractors on the battlefield. As Figure 9.10 shows, USF-I J1's troop drawdown

---

[73] Jason Adolphson, "2nd Bn., 401st ASB Aids Drawdown," *Outpost*, May 2010, p. 4. See also Natalie Cole, "Seeking Sand, Birds, Ammo: Arifjan Wash Rack, Customs Ready Armored Vehicles for Redeployment," Defense Video and Imagery Distribution System, August 12, 2010, and Claire Swedberg, "U.S. Army Deploys 'Soldier-Friendly' System to Track Thousands of Vehicles in Kuwait," *RFID Journal*, November 8, 2011.

[74] DoD, *Sustaining U.S. Global Leadership: Priorities for 21st Century Defense*, January 2012a.

[75] Raymond Mason, "Statement of Lieutenant General Raymond V. Mason, Deputy Chief of Staff, Logistics, G4, U.S. Army," in *Army and Marine Corps Materiel Reset*, hearing before the Subcommittee on Readiness, Committee on Armed Services, U.S. House of Representatives, 112th Cong., 2nd Sess., March 28, 2012, p. 8.

[76] JLOC, 2011e. See also JLOC, "Operation New Dawn Commodity Drawdown," December 13, 2011g.

[77] Andrew Slovensky, "Heavy Metal Takes a Ride to Kuwait," press release, U.S. Forces–Iraq, November 10, 2011.

Figure 9.10
Troop Drawdown

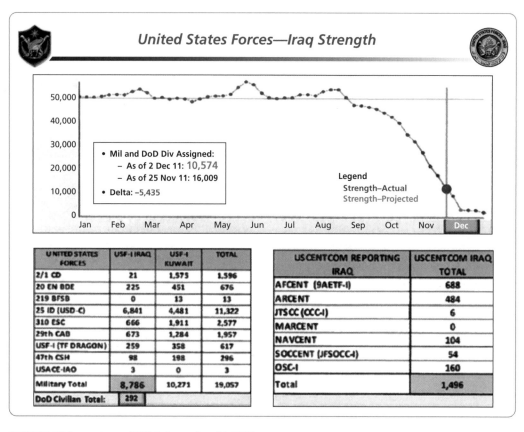

SOURCE: Slide courtesy of USF-I, December 14, 2011.
NOTE: ESC = Expeditionary Sustainment Command.
RAND RR232-9.10

accounted for only the military and government civilian workforce. It quickly became apparent that, although every staff within USF-I used some type of contracting service, no one staff element was responsible for overseeing all the contracts and contractors in Iraq. Coupled with the nature of service contracts, it became even more apparent that the true number of contractors on the ground was an unknown, especially given the complexity of contingency contracting. Contracting operations encompassed what was typically referred to as the "Big 7," with no one contracting agency responsible in its entirety for all: Senior Contracting Officer–Iraq (SCO-I)/JCC-I, DCMA, LOGCAP, Air Force Contract Augmentation Program (AFCAP), U.S. Army Corps of Engineers (USACE), Program Manager/Assistant Secretary of the Army for Acquisition, Logistics and Technology, and DLA. Figure 9.11 portrays the complexity of the contracting environment in Iraq. JCC-I, in coordination with USF-I, established a contracting fusion cell that attempted to account for both the number of contracting personnel

Figure 9.11
Complex Contracting Environment

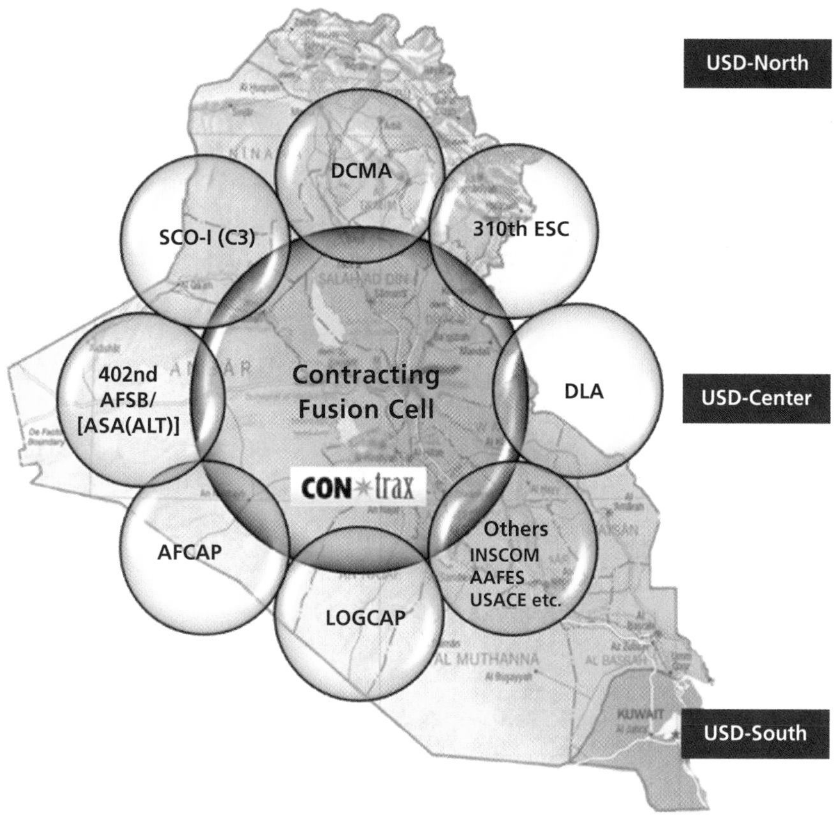

SOURCE: USF-I, J4, 2011j.
NOTES: AFSB = Army Field Support Brigade; INSCOM = U.S. Army Intelligence and
Security Command.
RAND *RR232-9.11*

and their associated equipment, whether contractor-managed, government-owned, or contractor owned and operated. The contracting fusion cell coordinated with SCO-I and developed and published demobilization instructions; worked with USF-I J3 to develop deployment and redeployment operation nodes; maintained close contact with U.S. divisions to ensure contracted services were terminated or transferred according to synchronized time lines; and built a contract database in coordination with the USF-I KM section to assist in transitioning contracting activities to Embassy Baghdad, USCENTCOM, or other government agencies.[78]

Although physically redeploying contractors was contractually incumbent on each contract, USF-I did use military aircraft to transport contractors to designated deployment and redeployment operation nodes located near the Iraqi civilian air hubs.

---

[78] Interview with former USF-I J4 staff officer, North Carolina, February 1, 2013.

Additionally, the Secretary of Defense authorized a change in military airlift procedures in Iraq to allow contractors to fly on a "space required" basis instead of "space available," allowing USF-I more flexibility in coordinating base transfers and drawing down base life support and other contracted services.[79] Figure 9.12 portrays the rapid decline in contractors that also occurred during this period.

Redeployment Planning and Procedures

The redeployment was planned and executed as a combat mission because of the hostile security environment. While maneuvering forces was a complicated operational challenge, it was also a logistical challenge. The USF-I staff had to determine the precise number of convoys, flights, and cargo trucks required to move equipment and personnel from their bases to consolidation and staging areas, onward to ports of embarka-

Figure 9.12
Contractor Drawdown

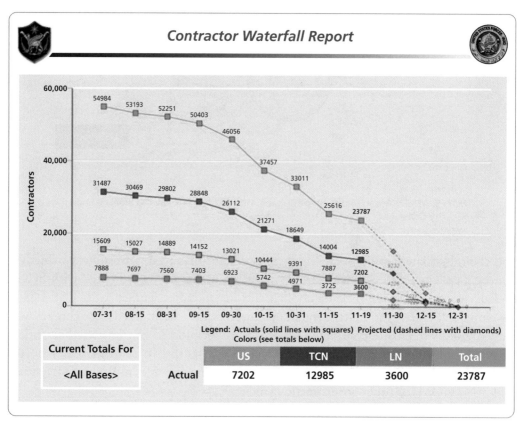

SOURCE: Slide courtesy of USF-I, December 14, 2011.
NOTES: TCN = third-country national; LN = local national.
RAND RR232-9.12

---

[79] Interview with former USF-I J4 staff officer, North Carolina, January 31, 2013.

tion, and then out of theater, all on a fixed schedule, with dwindling resources, and in parallel with the base closure process. Planning also had to take into account the capacities of the out-of-theater receiving stations: How much could they accept, at what pace, and at what cost?

Most troops, sensitive equipment, and vehicles were redeployed through seaports and airports in Kuwait to the south, although many of the 2,382 D2D cargo truckloads for the U.S. division located in northern Iraq and the U.S. division located in the center of Iraq were shipped west to Aqaba, Jordan.[80] Figure 9.13 shows the basis for the transportation planning, including the "sustainment spine" originating in Kuwait that fed the hub-and-spoke logistics system USF-I used. To minimize risk to force by keeping supply lines and redeployment lines short, OPORD 11-01 called for a systematic transfer of bases from north to south and from the edges to the middle of the "spine." However, Change 1 to OPORD 11-01, designed to build flexibility for a future security agreement and 2012 U.S. forces, directed key bases in the north and the west to stay open longer than originally planned, requiring operational and tactical units and assets to keep main supply routes that supported the bases open and clear.[81]

There was a myth at the beginning of 2011 that the entire process was designed around the understanding that the ongoing drawdown would necessitate a reduction in available logistics capacity. However, this was not the case. The reality was that, as time went by, many bases in the north transferred; larger bases partially transferred; and the population at bases was reduced, albeit mainly through reductions in contracted services and requirements because many units remained in theater with Change 1 to OPORD 11-01. These actions had four effects: The supply lines were shorter; sustainment requirements for food, fuel, and other commodities were lower; the demand for logistics equipment, personnel, and vehicles decreased; and more equipment was available, so that the logistics capability actually increased over time. For example, Convoy Support Center Scania—a vital midroute staging and refueling area between Baghdad and Kuwait—closed in mid-2010.[82] However, because of other remaining bases on the sustainment spine and because of the decrease in overall sustainment requirements starting with the so-called responsible drawdown of forces in summer 2010, Scania's closure did not negatively affect 2011 OND operations. USF-I J4's Joint Plans and Integration Cell took specific actions in "killing untrue myths" by breaking them down with facts, socializing the true concept, and then adding more fidelity to the planning factors. By August 17, 2011, the USF-I J4 was on its 17th version of planning factors shared with the U.S. divisions, 1st TSC and ARCENT.[83] Also, tight coordination with service commands—primarily the Army and U.S. Army Reserve

---

[80] GAO, 2011, pp. 4–5.

[81] Interview with former USF-I J4 staff officer, North Carolina, January 31, 2013.

[82] Walter, 2010.

[83] Interview with USF-I staff officer, Washington, D.C., December 22, 2012.

Figure 9.13
Transportation Sustainment Spine in Iraq for Operation
New Dawn

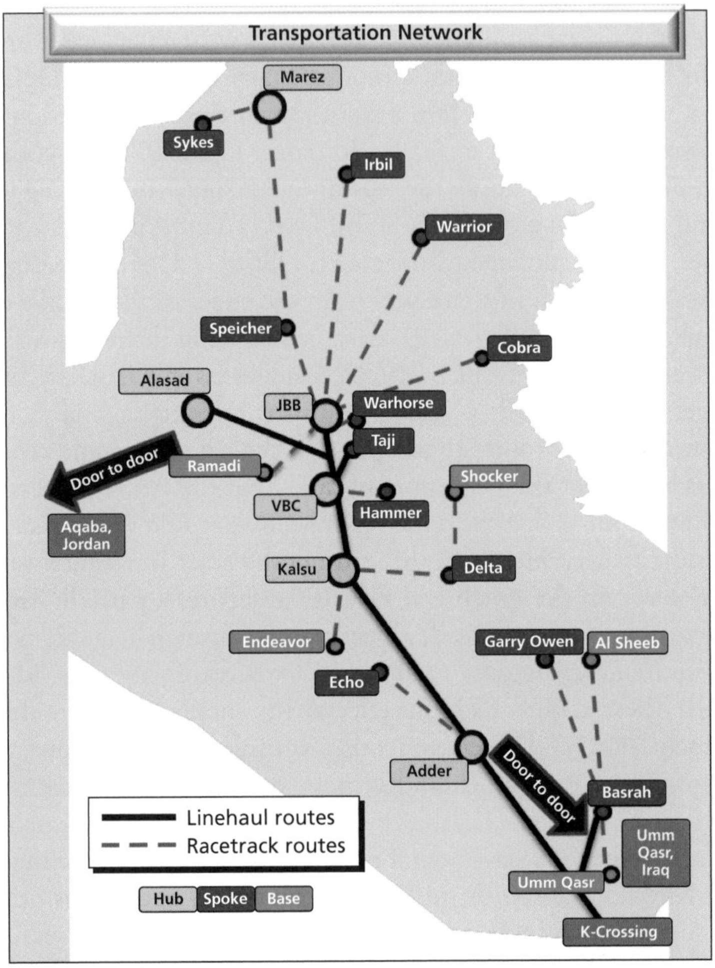

SOURCE: USF-I, J4, 2011j.
RAND RR232-9.13

Command—became necessary to determine which equipment would be used to help the services "reset" from the war (i.e., recompose their tables of equipment) or redeploy to Afghanistan to meet ongoing commitments there.

Over the course of the redeployment process, many of the centralized consolidation and staging functions were repositioned to COS Adder, located in southern Iraq adjacent to the city of Talil. These functions included final equipment consolidation and accountability, customs inspections, cleaning, loading and packaging, refueling, and all other steps necessary to prepare for onward movement.[84] Adder was also the home to an MRT yard, a facility run by the 1729th Field Support Maintenance Company of the Maryland Army National Guard. Equipment identified as excess from across the theater was consolidated at the MRT yard, where the 1729th sorted it, inspected it, and moved it to Kuwait for redistribution or destruction. Some of this equipment (e.g., communications wire, generators, and off-the-shelf equipment) was returned to the United States, while some was sent to Afghanistan to support Operation Enduring Freedom as a means to reduce the costs of the war in Afghanistan.[85]

Redeploying a tank like the M-1 required not only coordinating movement and movement security from its operating base to a port of embarkation but also determining its disposition (who would receive the tank, why, and how); thoroughly inspecting and itemizing its parts; estimating the degree of wear to determine part replacement requirements; removing, itemizing, cleaning, and stowing its standard associated equipment (e.g., radios, machine guns, thermal sights); removing all dirt and dust according to U.S. agricultural regulations; and conducting a customs inspection to ensure the vehicle did not contain any illicit material for reentry into the United States. At the end of the transition process, the heavy equipment transport itself would go through a similar process, probably from a port of embarkation in Kuwait.

The consolidation, accounting, and movement process began at individual bases across Iraq in concert with the base closure process. Larger bases, such as Joint Base Balad, set up RPAT yards, at which units would turn in their theater-provided equipment, which would then be inventoried, retrograded, repaired, and returned to the service for further use.[86] Smaller bases received M-RPATs, which consolidated, cleaned, inspected, and assigned each piece of rolling stock and other heavy equipment for onward movement and redeployment.[87] RPATs and M-RPATs consisting of military and contractors determined which pieces of equipment would be destined for reset and which would be destined for other outcomes, such as refurbishment or decommissioning. This process was supported by Army National Guard quartermaster units, but

---

[84] For example, see Anthony Zane, "Last Stop for Fuel in Iraq: Bulk Fuel Farm Consolidates Fuel as Drawdown Continues in Iraq," press release, U.S. Forces–Iraq, November 7, 2011b. For equipment to be loaded aboard ships or aircraft, additional cleaning, inspection, and administrative processing might have been required at the port of embarkation.

[85] Anthony Zane, "Equipment Leaving Iraq Saves Tax Dollars While Soldiers Return Home, Materials Are Redistributed," press release, U.S. Forces–Iraq, November 5, 2011a.

[86] 1st Theater Sustainment Command, "1st TSC 2010 Initiatives and Highlights," fact sheet, August 2011, p. 9, Not available to the general public. See also Pierre A. Spratt, "402nd AFSB Provides Retrograde Support," Army.mil, June 17, 2011, and David Kline, "Mobile RPAT Mission at COB Basra," *Outpost*, Vol. 1, No. 4, September 2010, p. 5.

[87] 1st BCT, 1st CD, "U.S. Forces Turn in Equipment in Support of Drawdown," press release, U.S. Forces–Iraq, October 30, 2011. In some cases, these groups were simply called MRTs. See, for example, Army National Guard, 224th Sustainment Brigade, "Guard Mobile Redistribution Team Supports Iraq Drawdown," press release, undated.

it also required extensive contractor support and varying amounts of labor from the redeploying units themselves. Self-redeploying units would turn in their MRAPs and other theater-provided equipment items to RPATs located at Camp Arifjan and Camp Virginia, in Kuwait.

Security Considerations for Redeployment

Given that extremists remained capable of attacking U.S. forces with rocket, mortar, IED, and direct-fire attacks, the operational maneuver involved significant risk. Moving troops and equipment from across Iraq required a comprehensive security plan to minimize casualties, reduce the effects on the Iraqi people, and ensure minimum damage to equipment. Balancing the requirements for comprehensive security while simultaneously drawing down the forces that might provide that security was no less challenging than balancing logistics support. To prepare, USF-I created an operational risk management working group to identify mitigation measures. This group met monthly through the end of the transition period.[88] Each unit, including those deploying in support of redeployment, had already established standard operating procedures for movement security that were integrated into the theater security plan.

While theater air support was gradually reduced within Iraq, fixed-wing aircraft continued to provide on-call air support to help secure redeployment and also to help the Iraqis protect their borders. The process of bringing this support to bear was complicated to some extent by complete transfer of control of Iraqi airspace to the Iraq Civil Aviation Authority on October 1, 2011, after which movement of surveillance and reaction aircraft required formal bilateral coordination.[89] However, U.S. military aircraft did continue to fly security and support missions, in coordination with the Iraq Civil Aviation Authority, to facilitate the redeployment during the final months of OND.[90]

Security risks were compounded by the fact that the U.S. security footprint, available surveillance coverage, and quick-reaction forces were diminishing even as the movement of redeploying forces accelerated. To compensate for these challenges, each self-redeployment convoy was planned as a combat mission, and each movement took place under a security bubble of aerial surveillance and IED route clearance. Originating from various bases throughout Iraq, typically in coordination with the final actions of a base transfer, SRD convoy movements to Kuwait were enhanced versions of the standard TSC theater sustainment convoy movements that had taken place daily

---

[88] USF-I, 2011e, p. 7.

[89] Shafran, 2011. See also Chuck Broadway, "U.S. Transfers Airspace to Iraq," U.S. Air Force website, June 6, 2011.

[90] A. M. Lavey, "South Dakota Medevac Team Ready for Action," press release, U.S. Forces–Iraq, October 23, 2011.

throughout the conflict, but SRD movements were typically viewed as distinct combat missions under OPORD 11-1.

"Collapse" of Forces for Redeployment

The concept of operations involving the retrograde of forces was, at its most basic, deliberate collapse of the force on itself, drawing down from north to south, from east to west, and from the outside in toward Route Tampa, the main supply route that goes through Mosul in Northern Iraq; goes through Baghdad in the center of Iraq, COS Kalsu south of Baghdad, and COS Adder in southern Iraq; and exits Iraq at K-Crossing at the Iraq-Kuwait border.[91] Planning for such an operation was extraordinarily complex. In some cases, additional units provided security for units that were conducting the tactical movement. Furthermore, all units and equipment had to keep moving according to a meticulously planned timetable. To ensure the redeployment went according to plan, the USF-I command element relied on the joint plans and operations process, while J3 augmented this through the use of the OSB. This daylong planning meeting, chaired by the USF-I commander, took place monthly during the last year to ensure that all aspects of the transition, base closure, logistics management, and operational maneuver were effectively coordinated. The meeting also provided a forum to track any open decision points affecting the plan. By incorporating a wide range of participants, including the USF-I staff, U.S. divisions, DoS, OSC-I, Contracting Command, ARCENT, and other supporting agencies, the OSB ensured that the battalions conducting movement received priority for resource allocations. The value of the OSB came not from the overly rehearsed briefing to the CG but in the days of staff work prior to the briefing, during which issues were worked out and solutions were generated.[92]

The redeployment also had to be structured so as to prevent security vacuums that could precipitate conflicts. For example, months before the redeployment of U.S. forces from the areas along the Arab-Kurdish faultline, U.S. troops gradually reduced their involvement in the trilateral U.S.–Iraqi Army–Kurdish *peshmerga* CSM to enable Iraqi and Kurdish forces to begin interacting with each other more directly while their American counterparts were still there. American forces transitioned from participating in patrols and checkpoints to engaging in operational overwatch, in which they played more of an observer role, to strategic overwatch, in which they primarily interacted with their Iraqi Army and Kurdish counterparts at regional coordination centers. The senior leadership of USF-I viewed this as an extremely sensitive transition because,

---

[91] Route Tampa was one of the major lines of advance that U.S. forces took in March 2003 during the invasion of Iraq. This major thoroughfare is listed as Highway 1 on civilian maps of Iraq. Figure 9.13 depicts this route graphically.

[92] Interview with former USF-I J5 staff officer, Carlisle, Pa., January 29, 2013.

if not handled properly, it could have led to widespread violence in the north and complete disruption of USF-I's planned redeployment.[93]

As noted earlier, the key to starting the redeployment process of a unit was first relieving it of its missions through a well-thought-out and synchronized transfer of that mission. As is always the case, operations drove the mission, and redeployments were the natural next step to releasing excess forces. The following vignette describes the movement of the 4th Battalion, 9th Cavalry Regiment (4-9 CAV) and the 1st Battalion, 8 Cavalry Regiment (1-8 CAV) of the 2nd BCT, 1st Cavalry Division (2/1 CD).

Movement Security Vignette: Positioning the Strategic Reserve

Of critical concern for General Austin was the ability to maintain an operational strategic reserve should additional combat power be required for force protection during any portion of the operational maneuver.[94] These reserves were located at Camp Adder, Iraq, and Camp Buehring, Kuwait, respectively. The operation reserve consisted of an armored battalion (1-77 Armor Regiment) and would be the first unit called on in response to a security threat that exceeded the capability of an installation or U.S. division to address using their organic reserves.[95] The strategic reserve was drawn from the first brigade to depart Iraq as part of the operational maneuver: the 2nd Brigade of the 1st CD (2/1 CD). While 2/1 CD was the first brigade to exit Iraq, it remained in Kuwait through the end of December to provide overwatch for the movement of the entire force.[96]

The lead battalion of 2/1 CD was 1-8 CAV, which began its redeployment on October 6, 2011. The after-action report describes this process as an extended rearward "passage of lines" operation that began at COS Warhorse in western Diyala province, moved through Joint Base Balad in southern Salah al-Din Province, on to COB Kalsu in Babil Province, then COS Adder in Dhi Qar Province before transitioning into an administrative convoy after passing through the Khabari border point with Kuwait (K-Crossing).[97] Making such a long movement going through areas controlled by both the ISF and other U.S. forces can best be explained as an extended passage of lines. A passage of lines is often considered one of the most complex operations that can be

---

[93] Interview with former USF-I J5 staff officer, Carlisle, Pa., January 29, 2013; Brennan notes, 2009–2011.

[94] The information for this section was drawn from an interview with an officer assigned to 2-1 CD who was responsible for planning the retrograde of his unit, conducting the strategic reserve mission, and developing the after-action report for this operation, December 12, 2011.

[95] Each installation was required to maintain a platoon on alert to respond to immanent threats or attacks against the facility. Likewise, each U.S. division retained a company as its internal quick-reaction force. If the nature of the threat exceeded these capabilities, USF-I had the ability to deploy both the operational and/or strategic reserve.

[96] Interview with USF-I J3 staff officer, Baghdad, December 10, 2011.

[97] 1-8 Cavalry Battalion, 2d Brigade, 1st Cavalry Division, untitled after-action report, November 18, 2011. Movement to K-Crossing took place at a later date.

conducted in combat. What makes one so dangerous is that the moving force has to be concerned not only about making contact with the enemy but also about the actions of friendly forces. Consequently, an enormous amount of coordination must take place in advance of the movement, entailing intensive coordination and cooperation between the unit that is moving and the unit that is being moved through. A poorly coordinated passage of lines can lead to friendly fire incidents or exposure to enemy attack if avenues of approach (or, in this case, likely ambush areas) are left uncovered.[98] The 1-8 CAV report stated that "at no time were we allowed to think of this as an administrative move."[99] The complexity of the mission and the inherent danger to the force can be illustrated by the fact that General Austin personally visited each battalion weeks before their movement to receive a detailed briefing about the operation and to determine whether the battalion commander needed any type of additional support that he did not already have.[100]

The tactical movement 1-8 CAV undertook, called Operation Mustang Cassidy, took place in two phases. In the first phase, 4-9 CAV, another squadron in 2/1 CD, had to move from Balad Ruz, located about 40 miles from the Iran-Iraq border, through 1-8 CAV's area of operations to relocate to Joint Base Balad, where it would remain until the base was closed in late October. For this portion of the operation, 1-8 CAV had to provide a military police escort for the unit going through its sector. The first part of the movement required 1-8 CAV to secure the movement corridors for 4-9 CAV, requiring the employment of all 13 of its maneuver platoons. Planning the operation required weeks of preparation, including conducting clearing and security operations weeks prior to the passage of lines, employing aerial surveillance techniques to identify possible IED sites, and shifting 1-8 CAV's main ground counter-IED effort to focus on the routes that 4-9 CAV would use. Simultaneously, 1-8 CAV increased its efforts to disrupt groups known to be involved in the use of IEDs while ensuring all units involved in the operation were aware of the intentions and capabilities of these extremist groups. Despite these aggressive efforts, an element of 1-8 CAV was attacked by an explosively formed penetrator only two days prior to the movement—the first such attack it had received in 30 days. This attack emphasized the inherent danger in the passage of lines and of the overall redeployment mission.[101]

---

[98] The U.S. Army states that a *rearward passage of lines* "continues [an ongoing] defense or retrograde operation, maintaining enemy contact while allowing for recovery of security or other forward forces." This same section emphasizes that "[c]ounterintelligence analysis provides an assessment of enemy collection against friendly forces, specified by gaps and vulnerabilities, and countermeasures to enemy collection." Field Manual 3-90, *Tactics,* Washington, D.C., Headquarters, Department of the Army, July 4, 2001, pp. 16–31. A *forward passage of lines* takes place when a unit is moving toward the enemy through another friendly unit.

[99] 1-8 Cavalry Battalion, 2011.

[100] Telephone interview with former 1-8 CAV staff officer, December 15, 2012.

[101] 1-8 Cavalry Battalion, 2011.

On the day of the rearward passage of lines, 1-8 CAV utilized all 13 maneuver platoons in the battalion to provide maximum physical presence and to maximize observation of key terrain. A 1-8 CAV section or platoon was positioned at each intersection and other likely IED or ambush points along the movement corridor, and mobile patrols reinforced these static positions. The battalion was supported in the overall security effort by Task Force 2-82 Field Artillery, which provided some overhead surveillance assets and mounted patrols on the two primary routes along the movement corridor. 1-8 CAV conducted what its commander called "deliberate" clearing operations, requiring patrols to dismount and walk through likely IED or ambush areas.

Once 4-9 CAV completed its passage of lines, 1-8 CAV began its own movement. To facilitate this movement with adjacent units, it placed liaison officers at COS Adder (one of the focal points for redeployment movement) to work with the brigade headquarters responsible for the terrain along the 1-8 CAV movement corridor. On the morning 1-8 CAV was scheduled to depart, engineers initiated a route clearance mission moving between two and four hours ahead of the lead element from 1-8 CAV. As the units moved through one sector to another, another unit would provide escort for their movement. And, as they had done for 4-9 CAV, each unit 1-8 CAV passed through surged combat forces to ensure that they were secure during movement. As one staff officer in 1-8 CAV stated,

> It's a matter of pride. We were not going to let 4-9 CAV get attacked while in our sector, and nobody was going to allow us to get attacked while we moved through their sector.[102]

Not only did 1-8 CAV have the support of ISF and U.S. forces along the route, but additional protection was also provided by an armed helicopter escort and a full range of ISR assets that were at the disposal of USF-I. Finally, while not visible, fixed-wing attack aviation was hovering only minutes away should the convoy be attacked.

1-8 CAV was the first of 24 battalions to complete this journey to Kuwait. This tactical movement would be repeated another 23 times by the end of the operation, and the USF-I team met the CG's intent to ensure that the last soldier who departed Iraq had the same level of protection as the first. Figure 9.14 depicts the maneuver battalion waterfall as units exited Iraq.

However, these 24 SRD convoys need to be put into context as a percentage of the reposture total. The SRD convoys consisted of 3,081 vehicles (128 vehicles per convoy on average but less than 8 percent of the overall reposture mission). The 1st TSC managed the movement of 18,924 truckloads in the same period (over 77 percent of the overall reposture mission). The number of vehicles in each TSC convoy was signifi-

---

[102] Telephone interview with former 1-8 CAV staff officer, December 15, 2012.

Figure 9.14
Maneuver Battalion Waterfall

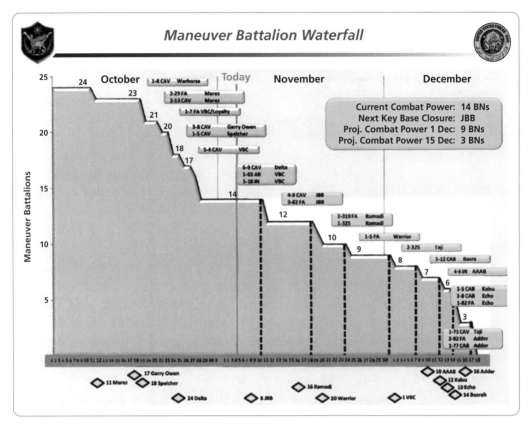

SOURCE: Slide courtesy of USF-I, December 14, 2011.
RAND *RR232-9.14*

cantly smaller than the SRD because the vehicles were mainly of hardened commercial trucking assets supported by MRAP convoy escort teams. An SRD convoy was a lethal force of command and control, MRAPs, and dedicated enabler protection, while the logistics convoys did not have the same dedicated protections or defensive power as the SRD convoys.[103]

How This War Ended

In the months leading up to the last soldier leaving Iraq, USF-I and Embassy Baghdad began to plan for the final transitions. Multiple political and diplomatic events accom-

---

[103]Interview with USF-I J4 staff officer, Washington, D.C., December 22, 2012.

panied the final phase of transition, the final steps in the end-of-mission process for USF-I, and the final handover itself.

War Termination Assessment

Throughout the execution of OPORD 11-01 and 11-01 Change 1, USF-I J5 maintained a running assessment of the transition process. With the decision made to withdraw all forces by the end of 2011, the assessment process was broadened to examine the status of the entire operation. There was not going to be a transition to another force, as there had been between Operation Iraqi Freedom and OND on September 1, 2011. Instead, the United States was ending the war in Iraq, and the warfighting authorities delegated to the USF-I commander would be returned to USCENTCOM consistent with U.S. laws and DoD policy. Moreover, many of the goals and objectives that the operational military headquarters had once pursued would now be pursued through normal state-to-state relations. Thus, the war termination assessment process would seek to identify the goals and objectives assigned to USF-I and evaluate the degree to which each had been achieved.

The assessment started with the identification of assigned goals and objectives during all stages or phases of operations in Iraq. The analysis included guidance found in presidential speeches, the National Security Strategy, the SFA, the Joint Strategic Concepts Plan, the JCP, USCENTCOM's Iraq Transition Plan, OPLAN 1003V, and other documents. Many objectives or goals repeated higher-level guidance, so the list was adjusted to remove redundancy. The assessment process ultimately identified 43 objectives that had been assigned to MNF-I/USF-I. While a fair number of these could justifiably be considered complete, the large majority could best be described as an ongoing effort, and some of these would take years or even a generation to complete. Many of the goals and objectives were not completed because of Iraqi internal divisions and the decisions of Iraqi leaders. Some of the goals for helping to build a better Iraq were undermined by neighboring countries, such as Iran, that continue to arm, train, and equip extremists to conduct violent acts in Iraq. In the end, USF-I identified 18 items that the military needed to track and complete so that it could report to national leaders that USF-I could honorably depart Iraq. The 18 goals were met. While the reasons for not meeting some of the more aspirational goals in Iraq vary, the Iraqis are now fully responsible for their own fate. The challenge for the United States starting in 2012 is how to best help the Iraqis address the shortfalls that existed at the end of the USF-I mission. An analysis conducted by USF-I J5 presented to the senior leadership during a planning meeting conducted in November 2011 captures the challenges confronting Iraq in the future:

> Iraq's issues with political legitimacy (government formation), challenges with stability and self-reliance, delay in building an external defense capability, reliance on energy [oil] revenues, social spending, vulnerability to foreign [Iranian] influence,

reluctance to enforce the law across sect and ethnicity, challenges with regional integration, and the immaturity of governmental processes and systems threaten progress and success in Iraq.[104]

While all the goals and objectives had not been achieved, the mission of USF-I would end when the Secretary of Defense approved the end of operations consistent with the military obligations contained in the U.S.-Iraq bilateral SA. Just as the war started with a political decision, it ended by a political decision. Indeed, President Obama forecast this eventual decision during his Camp Lejeune Speech:

> What we will not do is let the pursuit of the perfect stand in the way of achievable goals. We cannot rid Iraq of all who oppose America or sympathize with our adversaries. We cannot police Iraq's streets until they are completely safe, nor stay until Iraq's union is perfected. We cannot sustain indefinitely a commitment that has put a strain on our military, and will cost the American people nearly a trillion dollars. America's men and women in uniform have fought block by block, province by province, year after year, to give the Iraqis this chance to choose a better future. Now, we must ask the Iraqi people to seize it.[105]

Diplomatic and Ceremonial Activities

Although USF-I held public ceremonies to mark base transfers early in the transition process, doing so created security risks; insurgents, alerted to the fact that the U.S. military presence had been reduced to a minimum, frequently attacked facilities in the days before the handover ceremonies. USF-I thus dispensed with public events, and base officials simply worked with their Iraqi counterparts on the formalities necessary to finalize the transfer of facilities.[106]

In December 2011, however, the U.S. and Iraqi governments marked the end of the combat mission in Iraq with a number of speeches and ceremonies. In early December, Vice President Joseph R. Biden visited Iraq to pay tribute to U.S. and Iraqi troops and to engage with Iraqi leaders on the transition.[107] The most prominent activities took place in mid-December, as transition drew closer and USF-I cased its colors. On December 12, 2011, President Obama and Prime Minister Maliki laid a wreath at the Tomb of the Unknown Soldier at Arlington National Cemetery in Arlington, Virginia. At this event, President Obama referred to Prime Minister Maliki as the leader of a "sovereign, self-reliant and democratic" country. He also stated that the

---

[104]Brennan notes, 2009–2011.

[105]Obama, 2009.

[106]Thom Shanker, Michael S. Schmidt, and Robert F. Worth, "In Baghdad, Panetta Leads Uneasy Moment of Closure," *New York Times*, December 15, 2011.

[107]Mark Landler, "Biden in Iraq to Prepare for Postwar Relations," *New York Times*, November 30, 2011.

United States and Iraq were building a "comprehensive partnership." Prime Minister Maliki echoed this statement:

> [A]nyone who observes the nature of the relationship between the two countries will say that the relationship will not end with the departure of the last American soldier. It only started when we signed in 2008, in addition to the withdrawal treaty, the Strategic Framework Agreement for the relationship between our two countries.[108]

Prime Minister Maliki also stated that Iraq had become "reliant completely on its own security apparatus and internal security" but that it still needed support from the United States regarding intelligence, counterterrorism, training, and materiel. These seemingly contradictory claims make sense when considering the tremendous pressure on Prime Minister Maliki to simultaneously demonstrate Iraqi independence from the United States and reassure Washington that Iraq was committed to a lasting partnership.

President Obama delivered his remarks to U.S. military units at Fort Bragg, North Carolina, prior to the end-of-mission ceremony. He defined the war in Iraq as a success. The President also acknowledged that Iraq had "many challenges ahead," and that it was not a perfect place:

> It's harder to end a war than begin one. Indeed, everything that American troops have done in Iraq—all the fighting and all the dying, the bleeding and the building, and the training and the partnering—all of it has led to this moment of success. Now, Iraq is not a perfect place. It has many challenges ahead. But we're leaving behind a sovereign, stable and self-reliant Iraq, with a representative government that was elected by its people. We're building a new partnership between our nations. And we are ending a war not with a final battle, but with a final march toward home.[109]

Final Handover to the Iraqis

The final, end-of-mission ceremony took place at Baghdad International Airport on December 15, 2011, with Secretary of Defense Leon E. Panetta and Chairman of the Joint Chiefs of Staff GEN Martin E. Dempsey representing the United States. The ceremony included the formal military traditions associated with closing a command (in this case, USF-I). General Austin officially cased the USF-I colors with his senior enlisted advisor, CSM Joseph R. Allen, after which a military color guard formally

---

[108] Barack Obama and Nuri al-Maliki, "Remarks by President Obama and Prime Minister al-Maliki of Iraq in a Joint Press Conference," transcript, Washington, D.C.: The White House, December 12, 2011.

[109] Barack Obama and Michelle Obama, "Remarks by the President and First Lady on the End of the War in Iraq," Fort Bragg, N.C., December 14, 2012.

retired the command and national colors. Despite the ceremony, approximately 4,000 U.S. military personnel remained in Iraq for three more days as they prepared the final convoys and flights out of Iraq.

While Secretary Panetta acknowledged the challenges that lay ahead for Iraq in his remarks at the ceremony, he also expressed a degree of optimism:

> The Iraqi army and police have been rebuilt. Violence levels are down, al-Qaida weakened, rule of law strengthened, educational opportunities expanded and economic growth expanding. And this progress has been sustained even as we have withdrawn nearly 150,000 U.S. combat forces from the country.[110]

Secretary Panetta also drove home the point that the transition marked the beginning of an enduring relationship with Iraq: "This is not the end. This is truly the beginning."[111] Like President Obama, Secretary Panetta refrained from using the term "victory" to describe the end of the conflict. A number of news reports on the ceremony noted that no top-level Iraqi officials had attended despite the fact that places had been reserved for them. Indeed, the failure of so many Iraqi political and military leaders to show up to this final ceremony took USF-I and Embassy Baghdad by surprise. Their failure to participate indicates the extent to which senior Iraqi officials had already moved on to the next stage in Iraq's political development, and their desire to distance themselves from the U.S. military. Before USF-I had cased its colors, Iraqi leaders were already consumed with internal power struggles and maneuvering for political dominance.

President Obama stated during Prime Minister Maliki's visit to Washington that the transition reflected the "normalization of the relationship" between the United States and Iraq.[112] "Normalization" has two sets of implications. From the Iraqi perspective, it signifies that the United States will treat Iraq as an equal, sovereign ally. However, from Washington's vantage point, as Iraq becomes just another "normal" country with which the United States engages, it is likely to recede in importance and receive less attention from military officials and policymakers than it did during wartime.

Indeed, U.S. policymakers had already begun to focus on other issues near the end of 2011. In November 2011, while the redeployment from Iraq was in progress, Secretary of State Clinton explained the coming U.S. "pivot" toward Asia, writing in *Foreign Policy* that "[t]he future of politics will be decided in Asia, not Afghanistan

---

[110] Donna Miles, "Panetta Concludes Iraq Mission Noting Service, Sacrifice," American Forces Press Service, December 15, 2011.

[111] Miles, 2011.

[112] Obama and al-Maliki, 2011.

or Iraq, and the United States will be right at the center of the action."[113] On January 5, 2012, just three weeks after the last American soldier crossed the Iraqi border into Kuwait, the White House released new "Defense Strategic Guidance" stating that the United States will "rebalance toward the Asia-Pacific region." In an introduction to the document, President Obama made explicit that Iraq was no longer at the core of U.S. strategic interests: "[A]s we end today's wars, we will focus on a broader range of challenges and opportunities, including the security and prosperity of the Asia Pacific."[114]

The Last Convoy

The last U.S. combat troops departed Iraq on December 18, 2011, in a 100-vehicle convoy carrying just over 500 soldiers. One of the soldiers to cross the border in this final convoy, PFC Martin Lamb, stated that "[i]t's a good feeling . . . knowing this is going to be the last mission out here. Part of history, you know—we're the last ones out."[115]

On December 18, 2011, with the last soldier across the border in Kuwait, a USCENTCOM FRAGO directed the disestablishment of USF-I and the transfer of all responsibilities for Iraq not under the purview of the Chief of Mission in Iraq back to USCENTCOM. In 2003, OPLAN 1003V had launched the invasion of Iraq and guided operations under MNF-I and USF-I, but this new FRAGO revoked the authorities of 1003V and returned activities in Iraq to the status quo ante.

Summary

During a decade of U.S. military operations in Iraq, 4,475 U.S. service members and 13 DoD civilians had been killed and an additional 32,227 troops had been wounded in action.[116] While a definitive count of the number of U.S. contractors killed and wounded in Iraq cannot be determined because of the different ways in which contracting companies accounted for personnel serving in Iraq, data available through the Office of Workers' Compensation Programs within the U.S. Department of Labor make it clear that at least 1,595 U.S. contractors died while serving in Iraq and another 20,306 received some level of workers compensation for injuries.[117] The last service

---

[113] Hillary Rodham Clinton, "America's Pacific Century," *Foreign Policy*, November 2011d.

[114] DoD, 2012a, introductory letter from President Obama and p. 2.

[115] "Last US Troops Withdraw from Iraq," BBC News, December 18, 2011.

[116] DoD, "List of Casualties in Operation Iraqi Freedom, Operation New Dawn, and Operation Enduring Freedom," as of July 30, 2012b.

[117] U.S. Department of Labor, Office of Workers' Compensation Programs, "Defense Base Act Case Summary by Nation," web page, September 1, 2001–March 31, 2013. See also Neta C. Crawford, "Civilian Death and Injury in Iraq 2003–2011," Costs of War Project, paper, September 2011.

member to be killed in Iraq was U.S. Army Specialist David E. Hickman of Greensboro, North Carolina, who was the only soldier killed during the operational maneuver departing Iraq. Hickman, who was 23 years old at the time of his death, was killed by an IED on November 14, 2011, just one month before the redeployment was complete.[118]

While U.S. military forces have departed Iraq, questions involving whether or not the mission in Iraq can be called a success are likely to be debated for decades to come. What is clear is that the USF-I planned and executed the last phase of OND under conditions of policy uncertainty and operational risk and did so in a manner that responsibly transitioned duties to the government of Iraq, Embassy Baghdad, USCENTCOM, and other U.S. government departments and agencies. However, the war termination assessment conducted during the last two months of the operation identified a large number of goals and objectives that that were left "in progress." This unfulfilled portion of the USF-I mission can be traced back to the optimistic objectives contained in the various versions of the JCP, reflecting Washington's overly optimistic policy goals. Finally, the fact that USF-I planners were conducting this war-ending assessment while simultaneously conducting the final portion of the transition and operational maneuver raises the question of whether the task would have been better accomplished in Washington as part of a broader strategic review relating to Iraq and the region as a whole.

---

[118] J. Freedom du Lac, "In Iraq, the Last to Fall: David Hickman, the 4,474th U.S. Service Member Killed," *Washington Post*, December 17, 2011. The DoD casualty list (DoD, 2012b) counted 4,474 deaths as of its December 17, 2011 update. However, the DoD report states, "personnel injured in OND who die after 31 December 2011 will be included in OND statistics." By February 20, 2013, DoD's list of U.S. military and DoD civilians killed in Iraq totaled 4,488.

Establishment of USF-I

January 1, 2010, Baghdad: USF-I was created by merging three former major commands: Multi-National Force–Iraq, Multi-National Corps–Iraq, and Multi-National Security Transition Command–Iraq.

January 1, 2010, Camp Victory, Iraq: GEN Raymond T. Odierno (USA), Commanding General, MNF-I, and CSM Lawrence K. Wilson, MNF-I, uncase the USF-I colors during the organization's activation ceremony at Al-Faw Palace. Photo courtesy of U.S. Department of Defense.

Operation New Dawn Begins

August 30, 2010, Baghdad: MG William "Burke" Garrett, USF-I chief of staff, welcomes GEN Lloyd J. Austin III back to Iraq. General Austin assumed command of USF-I on September 1, 2010. Photo courtesy of U.S. Department of Defense.

September 1, 2010, Baghdad: General Austin receives the command colors from U.S. Marine Gen. James N. Mattis, commander, U.S. Central Command, at the change-of command ceremony for USF-I. DoD photo by U.S. Navy Petty Officer 1st Class Chaad J. McNeeley.

Operation Proper Exit

Undated: Operation Proper Exit was a warrior care initiative that began in June 2009 that gave wounded warriors a chance to revisit the places they once served in Operation Iraqi Freedom. The initiative gave troops wounded in battle an opportunity to make a "proper exit"—on their own terms—by walking to and from an aircraft and climbing the ramp, rather than being medically evacuated. Photo courtesy of U.S. Department of Defense.

Undated: Between June 2009 and June 2011, Operation Proper Exit returned to Iraq ten times for week-long rotations with different groups of wounded warriors who returned to the places where they had been wounded. Photo courtesy of U.S. Department of Defense.

Iraqi Special Operations Forces

March 24, 2010: Iraqi special operations forces captured four major al-Qaeda in Iraq leaders and killed three others. Those detained included the Minister of Oil for the Islamic State of Iraq, his deputy, and the brother-in-law of Abu Umar al-Baghdadi, Emir of the Islamic State of Iraq. The Detainee Affairs Emir, the Northern Iraq Economic Security Emir, and the Wali of Mosul were killed. Photo courtesy of U.S. Department of Defense.

March 24, 2010: Iraqi special operations forces in action. Training and equipping the Iraqi special operations forces to conduct counterterrorism missions was a priority effort. By the time USF-I departed, the Iraqis were assessed to have a highly capable special operations force. Photo courtesy of U.S. Department of Defense.

Al-Tadreeb Al-Shamil—Iraqis Training Iraqis

January 12, 2011, Nasiriyah: Iraqi soldiers participate in the 25-day Al-Tadreeb Al-Shamil [collective training] course. The training program was conducted by Iraqi cadre at Imam Ali Air Base, near Nasiriyah. Photo courtesy of U.S. Department of Defense.

January 12, 2011, Nasiriyah: This all-encompassing training enabled Iraqi battalion commanders to train on individual and unit-level tasks and assess the unit's strengths and weaknesses. The course culminated with a company or battalion-level exercise, in which leaders coordinated multiple weapon systems and complex movement techniques. These types of exercises focused on conventional tasks and helped build a culture or training in the Iraqi security forces. Photo courtesy of U.S. Department of Defense.

January 12, 2011, Nasiriyah: The course provides experience for Iraqi army commanders in training their units in defense against external military threats and helps develop foundational competency at the company level. By early 2011, the training program was fully under Iraqi control, with the U.S. military providing assistance when requested. Photo courtesy of U.S. Department of Defense.

January 12, 2011, Nasiriyah: Iraqi soldiers conduct a rehearsal for a company live-fire exercise during the final phase of the 25-day Al-Tadreeb Al-Shamil course at the 10th Iraqi Army Training Center. Photo courtesy of U.S. Department of Defense.

Lion's Leap Capstone Combined Arms Training Exercise

Spring 2011: Exercise Lion's Leap was a capstone Iraqi security forces live-fire training event. The Iraqi army, in coordination with the other branches of the Iraqi military, conducted this exercise to demonstrate its increasing capability to defend the nation. The exercise utilized a joint-service approach, integrating the Iraqi army, air force, navy, and special operations forces elements. Photo courtesy of U.S. Department of Defense.

Spring 2011: An Iraqi Mi-171 multimission aircraft conducts a joint mission in support of advancing ground forces. Photo courtesy of U.S. Department of Defense.

Spring 2011: An Iraqi Mi-171 aircraft lands on the objective, allowing Iraqi counterterrorism forces to conduct a training exercise involving capture of a high-value target. Photo courtesy of U.S. Department of Defense.

Spring 2011: An Iraqi M60-AI conducts a combined arms maneuver along with dismounted troops, air force, and special operations forces. Photo courtesy of U.S. Department of Defense.

Training the Iraqi Navy

May 22, 2011, Umm Qasr: When British military forces ended their hands-on mission in Iraq, 81 Royal Navy sailors turned over the task of patrolling the waters of Umm Qasr to the Iraqi navy and marine corps. Photo courtesy of U.S. Department of Defense.

September 26, 2011, Umm Qasr: Iraqi sailors man the decks of one of the Iraqi navy's newly christened Swift-class patrol boats at this port in southern Iraq. The patrol boats are used to secure the vital waterways Iraq uses to transport oil, a significant revenue source for the nation. Photo courtesy of U.S. Department of Defense.

Training the Iraqi Air Force

Undated: Modeled after the U.S. Air Force Academy, the new Iraqi Air Force College at Tikrit Air Base took in its first class of 157 cadets on September 1, 2010. The cadets started a three-year program that lays the foundation for the future of self-sufficient airpower in Iraq with studies in airmanship, English, avionics, and the theory of aviation. The college had been training air force officers since 1972 but had completely shut down by the start of Operation Iraqi Freedom. Photo courtesy of U.S. Department of Defense.

Undated: In late February 2010, Iraqi Ministry of Defense officials drafted a letter of request seeking 18 Block 50/52 global configuration F-16s. The contract was approved for a foreign military sale in September 2011. Photo courtesy of U.S. Department of Defense.

Transitioning Back to the Iraqis

July 14, 2011, Contingency Operating Site Sykes: U.S. Division–North reported completion of the transition of this site in Ninewa Province to the government of Iraq. COS Sykes was the first of the large bases to be returned to the Iraqis. Photo courtesy of U.S. Department of Defense.

July 15, 2010, Camp Cropper Theater Internment Facility: The last major U.S.-run detention facility in Iraq was officially transferred to the government of Iraq. U.S. forces had used the facility, which could hold up to 4,000 prisoners, for the care and custody of detainees since April 2003. Photo courtesy of U.S. Department of Defense.

September 22, 2011, Joint Security Station Deason: 1st Lt. Jon Chandler and Iraqi army Col. Hussein (second from right) go over a property list while conducting a joint inventory as part of transitioning the base back to Iraqi control. Photo courtesy of U.S. Department of Defense.

Redeployment of Equipment and Supplies

September 21, 2011, Contingency Operating Base Warhorse: Soldiers move a power generator as they prepare to transfer the last remaining supplies back home to Ft. Hood, Texas, from this base. Photo courtesy of U.S. Department of Defense.

August 1, 2011, Camp Victory, Iraq: Sgt. Emanuel C. King, a paratrooper, and Eduardo Calvo, a heavy equipment operator, load containers onto flatbed trucks. Photo courtesy of U.S. Department of Defense.

2011: During the last 15 months of the U.S. military presence in Iraq, over 41,000 vehicles had to be accounted for and either transported or redeployed along with their units. Photo courtesy of U.S. Department of Defense.

Iraqi and U.S. Day of Commitment Ceremony

December 1, 2011: The Iraqi government hosted the Day of Commitment ceremony at Al-Faw Palace in Baghdad. The ceremony would be the last of its kind as the U.S. forces continued to draw out of Iraq. USF-I had used the palace as a headquarters until handing it back over to the Iraqi government on December 1, 2011. U.S. Army photo by Staff Sgt. Caleb Barrieau.

December 1, 2011: U.S. Ambassador to Iraq James F. Jeffrey, U.S. Vice President Joe Biden, Iraq President Jalal Talabani, Iraqi Prime Minister Nouri al-Maliki, and U.S. and Iraqi troops salute during the singing of the Iraqi and U.S. national anthems during the Iraqi government's Day of Commitment ceremony in the Al-Faw Palace on Victory Base Complex in Baghdad. U.S. Army photo by Staff Sgt. Caleb Barrieau.

December 1, 2011: Iraqi Prime Minister Maliki talks to Vice President Biden after giving a speech during the Iraqi government's Day of Commitment ceremony. Ambassador Jeffrey (far left) and President Talabani (far right) participated in the ceremony, which commemorated the sacrifices and accomplishments of U.S. and Iraqi service members. U.S. Air Force photo by Master Sgt. Cecilio Ricardo.

Disestablishment of USF-I

December 15, 2011, Baghdad: U.S. Secretary of Defense Leon E. Panetta (center) meets with General Austin (right) and Ambassador Jeffrey (left) before the ceremony marking the end of the U.S. military mission in Iraq. DoD photo by Erin A. Kirk-Cuomo.

December 15, 2011, Baghdad: Secretary Panetta speaks during the ceremony marking the end of the U.S. military mission in Iraq. Panetta told the troops they would leave Iraq knowing their sacrifice helped the Iraqi people. DoD photo by Erin A. Kirk-Cuomo.

December 15, 2011, Baghdad: General Austin and CSM Joseph R. Allen case the command's flag during the ceremony marking the end of the U.S. military mission in Iraq on the former Sather Air Base. DoD photo by Erin A. Kirk-Cuomo.

The Last Military Convoys Leave Iraq

December 18, 2011, Iraq-Kuwait border: Vehicles move out in the early morning hours as the last convoy of U.S. troops leaves Iraq. The 1st Cavalry Division's 3rd Brigade Combat Team, Greywolf, was the final unit to leave Iraq. U.S. Army photo by Staff Sgt. Lynette Hoke.

December 18, 2011, Iraq-Kuwait border: Troops, guests, and media representatives from throughout the world watch as the last convoy of U.S. service members entered Kuwait from Iraq. U.S. Army photo by Staff Sgt. Lynette Hoke.

December 18, 2011, Iraq-Kuwait border: U.S. and Kuwaiti troops unite to close the gate between Kuwait and Iraq after the last military convoys passed through, signalling the end of Operation New Dawn. U.S. Army photo by Cpl. Jordan Johnson.

Mission Complete

December 20, 2011, Joint Base Andrews, Md.: President Obama applauds General Austin, the last commander of USF-I, as he arrives for a ceremony to retire the organization's colors. DoD photo by D. Myles Cullen.

December 20, 2011, Joint Base Andrews, Md.: President Obama; Vice President Biden; Deputy Secretary of Defense Ashton B. Carter; GEN Martin E. Dempsey (USA), Chairman of the Joint Chiefs of Staff; and General Austin salute as the national anthem plays during the ceremony retiring the colors of the U.S. mission in Iraq. DoD photo by D. Myles Cullen.

December 20, 2011, Joint Base Andrews, Md.: General Austin salutes President Obama during the ceremony. U.S. Air Force photo by Senior Airman Perry Aston.

Members of the Command Group for Operation New Dawn

December 2011, Baghdad: Shown are members of USF-I's senior staff (left to right): CSM Joseph Allen, Command Sergeant Major; MG Adolph McQueen (USA), Deputy Commanding General for Detention Operations and Provost Marshall; MG Arthur Bartell (USA), J3; MG Jeffrey Buchanan (USA), J9; Maj Gen Anthony Rock (USAF), Director, Iraqi Training and Advisory Mission (Air Force); BG Jeffrey Snow (USA), J5; GEN Lloyd J. Austin III (USA), Commander; MG Mark Perin (USA), J2; BG Rock Donahue (USA), J7; BG Darsie Rodgers (USA), Commander, Joint Forces Special Operations Component Command–Iraq; RADM Hank Bond (USN), J6; MG William "Burke" Garrett (USA), Chief of Staff.

A New Beginning

April 2011, Camp Victory, Iraq: Al-Faw palace at night. Photo by Rick Brennan.

PART IV

The Aftermath, Conclusions, and Recommendations

CHAPTER TEN

After the Transition

On December 18, 2011, the transition was completed, and U.S. military forces had departed Iraq. As President Obama stated in his speech to soldiers at Fort Bragg just days before the last troops departed Iraq,

> One of the most extraordinary chapters in the history of the American military will come to an end. Iraq's future will be in the hands of its people. America's war in Iraq will be over.[1]

Vice President Biden further asserted that a new era in U.S.-Iraq ties had begun during a speech in Baghdad on December 1, 2011:

> [O]ur relationship, borne on the battlefield and long defined by the imperative of security alone, is now giving way to a new, more normal partnership between sovereign nations seeking to build a future together.[2]

But in many important ways, Iraq in 2012 did not resemble the country that DoD and DoS transition planners expected. A serious internal political crisis erupted immediately. The embassy found that the Iraqis were initially unprepared for and unwilling to facilitate the entry of large numbers of U.S. security and support contractors. Rather than achieving an Iraq that was a supportive partner, Embassy Baghdad found the Iraqi political leadership, and Prime Minister Maliki himself, unwilling to take the most basic measures necessary to enable the effective functioning of the embassy and the newly formed OSC-I.

The political crisis erupted first. As previously noted, the day after the last U.S. troops left the country, Prime Minister Maliki issued an arrest warrant for Vice President Tariq al-Hashimi and several other prominent Sunni leaders. The warrant for Hashimi charged him with directing death squads against internal rivals. Many observers concluded that the prime minister was acting decisively and preemptively

---

[1]   Obama and Obama, 2012.

[2]   Joe Biden, "Remarks by Vice President Biden at Event to Honor U.S. and Iraqi Servicemembers," Aw-Fal Palace, Baghdad, December 1, 2011.

to marginalize his political rivals and consolidate his power and expected this to be the beginning of an extended period of increased authoritarianism. Such an internal political crisis seemed to undermine Iraq's nascent political institutions and threaten the integration of Iraq's Sunnis into the political process.

A range of enduring security challenges keeps Iraq unstable and unsafe. Military planners had concluded in 2011 that, while al-Qaeda in Iraq was a shadow of the force it was in 2006, it would remain capable of conducting mass casualty attacks well into the future, especially if the enablers USF-I had provided were rapidly removed from Iraq.[3] This assessment proved accurate. Soon after the departure of U.S. forces, al-Qaeda in Iraq launched an antigovernment and anti-Shi'a campaign to reestablish itself in areas from which the U.S. military had expelled it. By the third quarter of 2012, violence escalated to levels not seen in more than two years. A total of 4,568 Iraqi civilians were killed by violent attacks in 2012, up from 4,144 during the previous year. While the increase in the number of attacks was small, it was the first increase in civilian deaths since 2009; more important, a number of attacks were executed in multiple locations throughout the country simultaneously, indicating an increased ability to plan and coordinate complex attacks. While it is true that December 2012 recorded the lowest number of violent attacks since 2010, June marked the highest level of violence in the last three years.[4] The evidence from 2012 suggests that Iraq remains a country in a low level of war, and there is little evidence to suggest that the situation will improve anytime in the foreseeable future.

On the positive side, Arab-Kurdish tensions—the one dynamic that many observers feared would explode into violence—did not initially erupt. This is not to suggest that the enduring problems were resolved. Rather, Erbil and Baghdad continue to advance their respective interests without making any efforts to compromise on such key issues as hydrocarbons, revenue sharing, and the status of disputed territories, and both sides have exercised restraint and prevented escalation that would lead to armed conflict. Kurdish forces and the Iraqi Army have come close to conflict a number of times, but Embassy Baghdad and OSC-I were able to successfully mediate each crisis.

Finally, violence conducted by Iran-sponsored militias has dropped off significantly, and preliminary evidence suggests that some of the groups are attempting to integrate themselves into the political process. In fact, a former member of Embassy Baghdad indicated that Prime Minister Maliki was interacting with Qais Khazali, the leader of Asa'ib al-Haq, to enter the political process. Preliminary information suggests that this outreach is working, and some distance has been created between Asa'ib al-Haq and the Iranian IRGC-QF. However, it is still too early to tell whether this will result in the disestablishment of Asa'ib al-Haq as an Iranian proxy.[5] Iran continues

---

[3]   Brennan notes, 2009–2011.

[4]   See "Iraq Deaths from Violence," Iraq Body Count website, January 1, 2013.

[5]   Interview with former member of Embassy Baghdad, Washington, D.C., February 21, 2013.

exert influence throughout Iraq without needing to resort to violent actions through its proxy militias. Further, Tehran has shifted its focus from Iraq as the civil war in Syria threatens its ally, President Bashar al-Assad. This has, however, strained the U.S.-Iraqi relationship because of the willingness of the Maliki government to facilitate Iranian support to the Assad regime despite significant American opposition.

Embassy officials were able to implement diplomatic engagement objectives after the U.S. military withdrawal, but the embassy faced significant operational challenges in multiple areas. The mere size of the U.S. diplomatic and contractor presence became contentious in the initial months of 2012. DoS responded to Iraqi resistance on this issue by promising to reduce the embassy's personnel and contractor footprint by up to 25 percent—a goal that proved difficult to meet. The Iraqi government also resisted the quasi-military security measures used to facilitate the movement of embassy officials around the country. While the embassy adjusted, security risks made it especially difficult for INL's civilian police advisors to travel to remote training sites. As a result, PDP costs skyrocketed; training withered in scope; and the Iraqi government withdrew its support for the program. INL eventually scaled back the program, and by late 2012 it was a shadow of the ambitious plan envisioned in early 2011, with only one-tenth the proposed cadre of advisors.

The embassy returned several of its facilities to Iraqi control, including sites intended for training by PDP advisors and by contractors supporting Iraq's FMS purchases. While these decisions would lead to cost savings over the long run, the U.S. government had already spent hundreds of millions of dollars building, renovating, and upgrading them for U.S. training and assistance efforts that would now either not take place or would take place elsewhere. As the expeditionary embassy and OSC-I retrenched and consolidated, DoS nevertheless insisted that it continue to engage Iraqi government counterparts and advance the U.S. foreign policy agenda successfully.

Internal Political Crisis

Prime Minister Maliki moved to consolidate his power immediately after the U.S. troop withdrawal in December 2011. On December 17, 2011, Maliki called on parliament to vote no confidence in Deputy Prime Minister Mutlaq after Mutlaq had publicly called Maliki a dictator.[6] On December 19—the day after the last U.S. troops left the country—Maliki issued a warrant for the arrest of Vice President Hashimi, accusing him of running death squads that killed police officers and judges and of

---

[6] Arwa Damon and Mohammed Tawfeeq, "Iraq's Leader Becoming a New 'Dictator,' Deputy Warns," CNN, December 31, 2011.

plotting terrorist attacks against Maliki and the legislature.[7] Hashimi fled to the Kurdistan Region, where KRG officials refused to turn him over to Prime Minister Maliki's government. Ultimately, Hashimi found safe haven in Turkey. The Sunni-dominated Iraqiyya bloc threatened to boycott parliament, and Maliki responded by threatening to form a new majority Shi'a government without it.[8] Thus, after eight years of efforts to build a stable representative democracy in Iraq, Prime Minister Maliki took dramatic steps to eliminate his rivals and consolidate his power, starting the very day after the U.S. military presence in Iraq ended.

Within the executive branch, Maliki appointed loyalists to senior levels of the security forces, intelligence services, and other government entities.[9] He also undermined or bypassed the established chain of command in key institutions by, for example, continuing to keep the Defense and Interior portfolios for himself and by issuing orders directly to senior officials whom he installed throughout the military and the bureaucracy. Prime Minister Maliki so weakened Iraq's supposedly independent government institutions—including the Central Bank; the Committee on Public Integrity; and, to a lesser extent, the Independent High Electoral Commission (IHEC)—that they no longer significantly constrain his power. To the contrary, he has, to some extent, used these institutions as tools to advance his agenda. In April 2012, for example, two years after he accused IHEC of fraudulently declaring Iraqiyya the winner of March 2010 parliamentary elections and nine months after he sought a parliamentary no-confidence vote in IHEC,[10] Maliki ordered the arrest of IHEC director Faraj al-Haidari on charges of corruption.[11] Haidari was convicted of the charges in August and given a one-year suspended sentence,[12] but Iraq's highest court reversed the conviction two months later.[13]

In October 2012, Maliki sacked the widely respected and apolitical chief of the Central Bank, Sinan al-Shabibi, and sought his arrest on corruption charges while Shabibi was representing Iraq at an International Monetary Fund meeting in Japan. Maliki was widely believed to be trying to gain control of the bank—despite constitutional provisions saying it is under the Council of Representatives' jurisdiction—to weaken the parliament's powers, gain access to the bank's roughly $65 billion in reserves, and facilitate Iran's ability to acquire dollars through the Iraqi institution.

---

[7]   Ramzi Mardini, "Iraq's First Post-Withdrawal Crisis," Washington, D.C.: Institute for the Study of War, December 19, 2012.

[8]   Tim Arango, "Prime Minister Puts Power-Sharing at Risk in Iraq," *New York Times*, December 21, 2011.

[9]   Joel D. Rayburn, "Rise of the Maliki Regime," *Journal of International Security Affairs*, Spring/Summer 2012.

[10]   "Iraq Ex-Electoral Chief Says Cleared of Graft," Agence France-Presse, October 16, 2012.

[11]   See International Crisis Group, *Iraq's Secular Opposition: The Rise and Decline of Al-Iraqiya*, Middle East Report, No. 127, July 31, 2012, p. 15 (fn 91).

[12]   "Iraq Election Chief Gets Prison Sentence for Graft," Agence France Presse, August 28, 2012.

[13]   "Iraq Ex-Electoral Chief Says . . . ," 2012.

Maliki had also previously tried to use the courts to place the bank and other independent organizations under the Council of Ministers, and thus under his own control.[14]

Given Maliki's success in controlling Iraqi institutions, the remaining brakes on his freedom of action are political rather than institutional. But Maliki has also sought to undercut many Iraqi leaders and political factions who could challenge him. Maliki isolated or strong-armed opponents he deemed irreconcilable (such as Haidari and Shabibi), co-opted those whose loyalty he could win over, and promoted those who will work with him. The prime minister has simultaneously weakened and divided the Sunni opposition, for example, by buying off individual Sunni leaders with the prospect of jobs, money, and other benefits while marginalizing or eliminating key rivals and challengers, such as Hashimi.[15]

Within the Shi'a community, Maliki's primary political rival is Shi'a cleric Muqtada al-Sadr, whose following among Iraq's urban poor gives him widespread support and the ability to muster large antigovernment street protests. However, according to a public opinion poll conducted for the National Democratic Institute, Maliki's popularity rose among disaffected Shi'a in early 2012—primarily at Sadr's expense.[16] Given the decline of Iraq's other leading Shi'a political party, ISCI, few alternatives to Maliki's Islamic Da'wa Party, which dominates his governing State of Law coalition, remain in the Shi'a political spectrum.

Enduring Security Challenges

Iran and the Conflict in Syria

The departure of American troops from Iraq seemed to fulfill one of Iran's primary objectives: to ensure U.S. forces left the country without maintaining permanent or semipermanent military bases. Iran continues to maintain extensive influence in Iraq in the wake of the U.S. withdrawal. The leadership in Tehran has successfully embedded itself in the Iraqi political process, developing influence among Iraqi Shi'a political figures (both in and out of government), clergy, and foundations. Tehran provides Maliki with extensive support, and Maliki behaves as if he believes he needs Tehran's

---

[14] Sam Dagher and Ali Nabhan, "Iraq Dismisses Central Bank Chief Amid Investigation," *Wall Street Journal*, October 16, 2012. See also Adam Schreck, "Iraq's Ousted Bank Chief Professes His Innocence," Associated Press, November 9, 2012, and Ali Latif, "Iraq's Central Bank Governor Is Removed Under Cloud," *Azzaman* (Iraq), trans. Sahar Ghoussoub and Naria Tanoukhi, October 15, 2012.

[15] In September 2012, Hashimi was sentenced to death in absentia, and the following month the Baghdad government stopped paying his $60,000 monthly salary. See "Iraqi VP Tariq al-Hashemi Sentenced to Death," BBC News, September 9, 2012. See also "Iraq Stops Paying Salary of Convicted Fugitive VP," Associated Press, October 10, 2012.

[16] Greenberg Quinlan Rosner Research, "A Major Shift in the Political Landscape: Report on the April 2012 National Survey," June 2012, p. 6. See also Greenberg Quinlan Rosner Research, "A Major Shift in the Political Landscape: Graphs for the Report on the April 2012 National Survey," May 2012.

continued backing for his own political survival. However, Iran also engages other Shi'a actors—such as Muqtada al-Sadr and ISCI—to ensure that Maliki does not irreversibly dominate either the Shi'a political spectrum or the government as a whole. Iran benefits from a weak Iraq that relies on, and fears, the more powerful and established Shi'a neighbor.

Nevertheless, the Maliki government is far from becoming an Iranian puppet, as many commentators have warned, and it is unlikely to become one. Iraq is certainly the weaker of the two states, and it can only do so much to fend off pressure from its much stronger neighbor. Iraq poses no military threat to Iran, and it will not be able to defend itself against Iranian aggression for many years. However, Maliki has demonstrated his willingness to assert Iraq's independence, protect its political and economic interests, and secure its territory. Far from being subservient or ideological vis-à-vis Tehran, Baghdad's approach seems to be businesslike and pragmatic.

But by far the most important component of the U.S.-Iran-Iraq relationship is the key role that increasing Iraqi oil exports have on enforcing the oil sanctions imposed by the 2012 NDAA, which essentially requires the U.S. President to certify that alternative oil supplies are available to continue pressing for cuts in world consumption of Iranian oil. According to Ambassador Jeffrey, Iraq's oil production level is one of the key components of the U.S. strategy in this regard. Thus, whether deliberately or not, Iraq is a major U.S. ally in the most successful pressure strategy employed against Iran since the Tanker War.[17]

That said, Maliki is likely to work with Iran when doing so either advances his domestic political position or entails few risks. Since the end of OND, for example, Iraq has refused to address U.S. concerns about Iran. Indeed, in September 2012, as civil war raged in neighboring Syria, U.S. officials denounced the Iraqi government for permitting Iran's Revolutionary Guards to ship arms and munitions to the beleaguered Syrian military through Iraqi airspace and overland through Iraqi territory.[18] Maliki denied Iraq was permitting Iran to supply the Assad regime and refused to interfere with Iranian flights, although his government did deny overflight permission for a North Korean aircraft en route to Syria.[19] While U.S. forces were in Iraq, they were largely responsible for defending Iraqi airspace. However, with the departure of USF-I, Iraq assumed full control, which led to the Iranians being able to use Iraqi airspace without interference.[20] Even if the Maliki government was willing to stand up to Iran

---

[17] Ambassador James Jeffrey, email to Charles Ries of RAND, January 14, 2013.

[18] Gordon and Trainor, 2012, pp. 677–678.

[19] Suadad al-Salhy, "Iraq Blocks Syria-Bound North Korean Plane, Suspects Weapons Cargo," Reuters, September 21, 2012. See also Louis Charbonneau, "Exclusive: Western Report—Iran Ships Arms, Personnel to Syria via Iraq," Reuters, September 19, 2012.

[20] Given the current state of the Iraqi air defense and radar system, the country does not yet have the ability to monitor all its airspace. Consequently, it would have difficulty being able to identify violations should they occur

on the matter, it lacks the military capacity to prevent Iran from using its airspace (a dynamic that transitioning U.S. forces anticipated would pose a serious risk). Iraq has no air force or air defense capacity with which to compel Iran to respect its sovereignty, and, as previously noted, Iraqi officials had stated that the country would not be able to defend its airspace until at least 2020.[21]

As Iran has focused increasingly on propping up the Assad regime in Damascus, the IRGC-QF continues operations to maintain influence in Iraq, Lebanon, Gaza, and Afghanistan.[22] According to an embassy official, it appears that Tehran has scaled down its direct military support for some of its proxies in Iraq to provide more expansive support to support the Syrian government. However, one Iraqi extremist group, Kata'ib Hezbollah, continues to maintain the closest of relationships with the IRGC-QF and is suspected to be conducting operations in Syria at Tehran's behest.[23] This suggests that USF-I's internal assessment that Kata'ib Hezbollah is a direct action arm of the IRGC-QF, working directly for Major General Qassem Soleimani,[24] was indeed accurate. The potential collapse of the Syrian regime would greatly undermine Iran's ability to project power in the region. If and when this comes to pass, Iraq will become even more important to Tehran, not only as a buffer between themselves and potential Sunni adversaries, but also because it will be through Iraq that Tehran will seek to exert its influence throughout the Levant and broader Middle East. Iraq's far greater strategic importance would drive the Iranian government to increase the scope and scale of its activities and influence throughout the country, although not necessarily through the use of violence.

Even as the Maliki government facilitated Iranian assistance to the Assad regime, other elements of Iraqi society took competing positions on the conflict in Syria. Iraqi Sunnis supported the Syrian opposition, smuggling fighters and weapons across the border,[25] raising the question of whether a future Sunni-dominated regime in Damascus might someday decide to return the favor and help Iraqi Sunnis overthrow the government that is oppressing them. For its part, the KRG provided training and support to Syria's Kurds,[26] in part to increase their influence as the dominant faction of the pan-Kurdish movement but also perhaps at the urging of Turkey, which hoped that Erbil might moderate the influence of anti-Turkish elements in the Syrian Kurd-

---

in uncovered areas. Interview with former USF-I J5 staff officer, Carlisle, Pa., January 6, 2013.

[21] Patrick Markey, "Iraq Says Signs Contract for 18 F-16 Fighter Jets," Reuters, October 18, 2012.

[22] Interview with nongovernment expert on Iraq and Iran, Arlington, Va., October 18, 2012.

[23] Interview with an officer assigned to OSC-I between October 2011 and June 2012, Arlington, Va., October 1, 2012.

[24] Brennan notes, 2009–2011.

[25] "State Dept.: Al-Qaeda in Iraq Fighting in Syria," CBS/Associated Press, July 31, 2012.

[26] Jane Arraf, "Iraqi Kurds Train Their Syrian Brethren," al-Jazeera, July 23, 2012.

ish community.[27] This mix of ethnic and religious tension, with loyalties that do not run first to a central authority in Baghdad, does not build confidence in a strong, stable, self-reliant Iraq that has a strong, enduring partnership with the United States. Regional conflict and violent extremism, both of which would affect U.S. national interests, have fertile ground.

Violent Extremist Organizations

An independent analysis conducted by Iraq expert Michael Knights concluded that Iraqi-on-Iraqi violence increased 18 percent in the roughly six months after the departure of USF-I.[28] The situation continues to worsen. SIGIR reported in October 2012 that July through September 2012 was the most violent period in Iraq for more than two years, with 850 Iraqi civilians killed and more than 1,600 wounded.[29] Most of this increased violence appears to have been instigated by Sunni extremists. Throughout 2012, violence was consistently higher in the Sunni west of Iraq, the Kurdish north, and Sunni/Kurdish North-Central Iraq, where Sunni extremists are most active, than in the predominantly Shi'a southern areas.[30] In July 2012, al-Qaeda in Iraq launched its "Breaking the Walls" campaign to retake areas from which the U.S. military had driven it.[31] By the fall, al-Qaeda in Iraq attacks had increased from 75 to 140 per month. Similarly, the number of al-Qaeda in Iraq fighters more than doubled between October 2011 and October 2012, from roughly 1,000 to 2,500.[32]

With the departure of U.S. forces, the Sunni extremist group JRTN had to rearticulate and refocus its *raison d'etre*—it had "succeeded" in forcing the United States to leave Iraq, so it needed a new calling. It appears its focus shifted toward the Iraqi government, launching attacks that demonstrated the government's inability to provide security and thus undermining its legitimacy. During the first eight months of 2012, JRTN was second only to al-Qaeda in Iraq in the number of violent attacks launched.[33] Despite its change in focus, JRTN is the only Sunni extremist group in Iraq to have grown in size, strength, and public support since the U.S. surge in 2007.[34]

---

[27] Renad Mansour, "Iraqi Kurdistan & the Syrian-Kurd Pursuit of Autonomy," Al-Jazeera Center for Studies, September 24, 2012.

[28] Gordon and Trainor, 2012, p. 690.

[29] SIGIR, *Quarterly Report to the United States Congress*, October 30, 2012g.

[30] Olive Group, "Weekly Security Update: Regional Activity (Iraq) 22 Jan 2012–4 Nov 2012," *Iraq Business News*, November 7, 2012.

[31] Lara Jakes, "Wave of Bombings Kills 26 Across Iraq," Associated Press, October 1, 2012. Also "Iraq Attacks Kill 110 in Deadliest Day in 2 Years," Associated Press, July 23, 2012.

[32] Qassim Abdul-Zahra, "Al-Qaeda Making Comeback in Iraq, Officials Say," *USA Today*, October 9, 2012.

[33] Interview with OSC-I staff officer, Arlington, Va., October 1, 2012.

[34] Knights, 2011.

The SOI has not turned to violence since the U.S. redeployment—at least not in a coordinated, organized fashion—but the Maliki government's continued marginalization of Sunnis could inspire this group to remobilize. As U.S. troops prepared to depart the country, the SOI—forsaken by the United States and mostly spurned by the Iraqi government—remained, in the words of *New York Times* reporter Andrew E. Kramer, "a loose end left by the United States."[35] Indeed, after U.S. military forces left the country, the fate of SOI members became just one component of broader U.S. concerns regarding the Maliki government's marginalization of Sunnis and its failure to integrate the SOI into the security apparatus, as it had promised to do. Failure of the Iraqi government to follow through with its commitment will further exacerbate Sunni perceptions that they are being excluded from playing a part in the future of Iraq, leave tens of thousands of armed Sunni unemployed, and potentially facilitate the resumption of a Sunni insurgency.

There is little evidence that Shi'a militias have responded in kind to Sunni-led political violence since the departure of USF-I.[36] To some extent, this is because Iraqi Shi'a viewed Maliki's increasingly authoritarian anti-Sunni measures as contributing to their security, which may be one of the factors that led Maliki's popularity to rise during 2012. However, should the government prove unable to keep Iraq's Shi'a population safe, Shi'a militias may step in and engage in their own campaign of violence, which would undermine the government's legitimacy in its primary constituency.

Arab-Kurd Tensions

Baghdad and Erbil made little progress in 2012 to resolve outstanding questions related to federalism, including the Article 140 process, the status of disputed territories, and control over hydrocarbon development and revenues. Continued failure to settle these disputes certainly makes the possibility of armed or military conflict more likely; however, both sides have found ways to benefit from the status quo, making it entirely possible that these issues may never be fully resolved.

In the meantime, the potential for violence between the ISF and Kurdish *peshmerga*—whether deliberate or inadvertent—remains real, particularly since the trilateral ISF-Kurdish-U.S. CSM ended, and what remains has devolved into somewhat of a formality. Since the U.S. withdrawal, the Iraqi Army and KRG *peshmerga* units have nearly come to blows over rights to former U.S. military bases and the authority to patrol the Syrian border (normally a federal responsibility) in disputed territories

---

[35] Andrew E. Kramer, "U.S. Leaving Iraqi Comrades-in-Arms in Limbo," *New York Times*, December 13, 2011.

[36] Interview with an officer assigned to OSC-I between October 2011 and June 2012, Arlington, Va., October 1, 2012.

under the de facto control of the KRG.[37] In both cases, senior officials in Baghdad, Erbil, and Washington had to intervene to resolve the disagreements.

Article 140 and Disputed Territories

Negotiations over the implementation of Article 140 of the Iraqi Constitution remain at a stalemate. Bureaucratic impediments, legal obstacles, and uneven implementation are hampering the resolution of property disputes and resettlement rights in areas from which Saddam Hussein evicted Arabs and Kurds.[38] No steps have been taken to prepare for a census. The Article 140 Committee the Iraqi parliament established in 2006 became paralyzed by debates over the committee's structure and membership.[39] Although Kurdish leaders publicly refuse to make any concessions regarding Article 140, they are working to reinforce the de facto control they have already gained over some disputed territories through domination of local political institutions, the deployment of Kurdish security forces, and the issuance of oil exploration contracts with increasingly significant multinational oil companies, whose willingness to enter into such agreements suggests a growing sense that the KRG will maintain control of these areas. As negotiations continue, these developments will increasingly strengthen the KRG's negotiating position. However, even if the status of the disputed territories is never fully resolved, the KRG will have achieved many of its territorial, demographic, and economic objectives.

Budget Disputes

Erbil and Baghdad also continue to negotiate over the extent of the central government's financial support to the KRG. Baghdad continues to deduct "sovereign expenses" for such things as border security from the 17 percent of the federal budget to which the KRG is entitled, an arrangement that the KRG refuses to accept. Similarly, Erbil is trying to get the federal government to pay costs associated with 100,000 active *peshmerga* fighters and 90,000 *peshmerga* veterans. However, Baghdad continues to argue that the maintenance of a regional militia is the region's own responsibility and has agreed to cover only the costs of 30,000 members of the Kurdish Regional Forces who have an affiliation with the Iraqi Army.

Hydrocarbons

As Figure 10.1 shows, Iraq's oil and gas production reached near highs of 3 million barrels per day (bpd) in 2012, and Iraq overtook Iran as the second-highest oil pro-

---

[37] Dagher, 2011, and Patrick Markey, "Analysis: Syrian Border Standoff a New Front in Iraq-Kurdish Rift," *Reuters*, August 8, 2012.

[38] Nawzad Mahmoud, "Property Claims Law Fails Thousands of Kurdish Families," *Rudaw*, August 27, 2012. See also Hevidar Ahmed, "Arabs Claim Kurds Aren't Implementing Article 140," *Rudaw*, December 14, 2011.

[39] Adnan Hussein, "Six Years Later, Questions Linger About Effectiveness of Article 140 Committee," *Rudaw*, September 13, 2012.

Figure 10.1
Iraqi Oil Production, 1970–2012

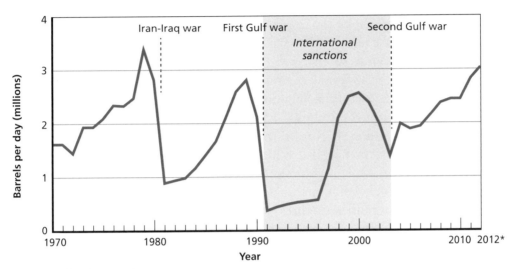

*Based on first five months.
SOURCE: International Energy Agency data.
RAND *RR232-10.1*

ducer in the Organization of Petroleum Exporting Countries.[40] The Kurdistan Region alone generated approximately one-tenth of the total. The International Energy Agency (IEA) conservatively estimated in late 2012 that Iraq is likely to double its production to 6 million bpd by 2020 and reach 8.3 million bpd by 2035, by which point Iraq will have generated $5 trillion in oil revenues.[41]

Despite lengthy and protracted negotiations, the country still lacks a hydrocarbon law. Negotiations remain stalled over the core questions of how much control the Kurds and Baghdad would respectively have over exploration, production, and revenues. While the central government's control over resources in the south and center of the country are not in doubt, it has become conventional wisdom that Baghdad and Erbil must create a durable legal framework that protects multinational companies' investments if the KRG is to maximize its potential production. Indeed, IEA wrote that an agreement on hydrocarbons could enable the KRG's production (currently about 250,000 bpd) to triple by 2020 and almost quintuple by 2035.[42]

In practice, although negotiations do continue, both sides seem content to proceed without a hydrocarbon law. Indeed, both sides have had rapid increases in production

---

[40] Javier Blas, "Iraq's Oil Output Overtakes Iran's," *Washington Post*, August 10, 2012. See also Guy Chazan and Javier Blas, "IEA Predicts Boom for Iraq Oil Industry," *Financial Times*, October 9, 2012.

[41] IEA, *Iraq Energy Outlook*, October 9, 2012, pp. 49, 56–57, 59–60, 107.

[42] IEA, 2012, pp. 49, 60.

and revenues despite the lack of a legal framework. The Kurds have managed to sign 50 independent oil and gas contracts—including with such "majors" as ExxonMobil, Chevron, Total, and Gazprom—indicating that multinational companies are willing to invest in the region despite the risks.[43] The potential returns on investment in the north are so great that ExxonMobil decided to pursue agreements with the Kurdistan Region even though doing so effectively took the company out of the running for future contracts in areas under the central government's control. Further throwing its lot in with the KRG, ExxonMobil announced in November 2012 that it would sell off its 60 percent stake in the West Qurna–1 field in southern Iraq.[44]

The greatest limitation on Kurdish oil production is the absence of an independent infrastructure to export its oil. While the KRG began trucking small amounts of oil to Turkey in late 2012, it also began negotiating the construction of a pipeline directly to Turkey, which would permit large-scale exports without having to rely on the central government's infrastructure.

The KRG's successful development of its hydrocarbon sector will strengthen Erbil's finances, contribute to the region's economic development, and further enhance the KRG's relations with neighboring Turkey. It also strengthens Erbil's negotiating position vis-à-vis Baghdad on a wide range of issues, including the terms of potential hydrocarbon laws and control of disputed territories. That said, as the KRG earns more oil revenues and as it manages an increasing number of oil projects in disputed territories under its control, the likelihood of a comprehensive hydrocarbon agreement decreases further.

Security

Arab-Kurd tensions did not explode into violence in 2012, as some had predicted. That said, Sunni extremist groups engaged in an organized campaign of violence in the Kurdish-controlled disputed territories. Attacks on police stations, police officers, and other low-level targets in these areas were common. In the third quarter of calendar year 2012, more than 100 senior government officials were targeted for assassination nationwide; outside Baghdad, the largest number of such attacks occurred in Kirkuk province.[45]

As of early 2012, the nature and extent of U.S. involvement in a continued CSM-like mechanism was still a subject of interagency debate.[46] The embassy had yet to determine which officials would participate in the senior-level "political/security meet-

---

[43] Hevidar Ahmed, "Oil Companies Are Security Guarantee for Kurdistan, Experts Say," *Rudaw*, August 30, 2012.

[44] Hassan Hafidh, "Iraq Says Exxon Seeks Bids on Oil-Field Stake," *Wall Street Journal*, November 8, 2012.

[45] SIGIR, 2012g, p. 73.

[46] Interview with Deputy Assistant Secretary of State for Near Eastern Affairs Barbara Leaf, Washington, D.C., January 13, 2012.

ings" envisioned in the transition plan,[47] and it had not yet decided whether—despite limited capabilities—it would seek to play a more hands-on role in CSM processes.[48]

The embassy has only limited influence to continue U.S. involvement in a joint security mechanism. It can contribute to policy-level coordination and transparency through senior-level working groups and other forums, as outlined in the transition plan. Although the embassy may be able to maintain this level of engagement within existing resources, such strategic and infrequent engagement would provide little insight into the dynamic between Iraqi Army and Kurdish forces. Moreover, such an arrangement would not provide U.S. officials with early warning of tensions on the ground or an institutionalized way to defuse a crisis at the operational level before it escalated.

Embassy officials have acknowledged that continuing the CSM is the best way to ensure that ethnic violence does not erupt in disputed areas, that Iraqi and Kurdish security forces do not inadvertently engage in hostilities, and that the internal boundaries disputed between Erbil and Baghdad do not change through the use of force, but the United States is not in a position to support or direct CSM activities, as previously envisioned.[49]

Capacity-Building

Security Cooperation

U.S. officials never expected the Iraqi Army to be fully capable of performing all missions by December 2011, although most USF-I officials and other knowledgeable commentators felt the ISF had achieved the capabilities necessary to maintain internal security and address threats from violent extremist organizations.[50] OSC-I was to manage future training and capacity-building efforts for the ISF and coordinate in-country and rotational trainers to support ISF modernization and training goals in an effort to produce a capability for the Iraqi military to defend its borders from external aggression.

However, OSC-I was ill-prepared on January 1, 2012. As discussed in Chapter Eight, the belief that U.S. troops would lead most training forced OSC-I to fundamentally reassess how it could accomplish its assigned mission, given constrained resources. First, OSC-I lacked the personnel, funding, and other resources necessary to conduct its expanded mission. Exacerbating this resource deficit, DoD initially failed to provide OSC-I with the funding and other authorities needed for the mission. For

---

[47] Interview with U.S. embassy officials, Baghdad, June 28, 2011.

[48] Interview with USF-I J35 Enduring Operations, Baghdad, June 25, 2011.

[49] Interview with U.S. embassy official, Baghdad, June 28, 2011.

[50] This section relies heavily on interviews with USF-I officials, June 2011; interview with Stuart Bowen, SIGIR, November 2011; and Gompert, Kelly, and Watkins, 2010.

example, while USF-I continued to obligate ISFF money until the end of mission on December 19, 2011, the Secretary of Defense did not designate OSC-I as the executor of ISFF until February 16, 2012, leaving no one authorized to obligate ISFF money for two months.[51] Similarly, DoD failed to ensure that Congress added OSC-I's operating budget to the FY 2012 Continuing Appropriations bills, forcing a last-minute scramble to ensure the office received necessary funding.[52]

Nevertheless, the U.S. military—through OSC-I—continues to build the capacity of the ISF through training programs that include the following ongoing and future initiatives:

- The U.S. Coast Guard is providing maritime security advisory teams to help the Iraqi Navy develop riverine and coastal border security patrol forces based in Basra.[53]
- The Iraqi Navy is developing a partnership with U.S. Naval Forces in the region to facilitate bilateral or regional exercises that develop, among other things, operational skills to work collaboratively on mutual interests.[54]
- The Iraqi Air Force has already sent the first group of 27 pilots to the United States for F-16 pilot training, in mid-2012.[55] Moreover, the United States is funding Iraqi Air Force aircraft maintenance and English language training at a facility in Jordan through 2012 and 2013.[56]

However, the withdrawal of U.S. troops ended sustained training programs. The departure of USF-I also deprived the ISF of irreplaceable enablers, particularly in such areas as logistics management, intelligence support, ISR, maintenance support, and airspace control. The ISF will need several more years before it can undertake these missions effectively on its own. Furthermore, OSC-I and the three training programs served as the core of an ambitious expansion of military cooperation quickly hammered together after the October Iraqi decision not to grant immunity to U.S. military personnel, resulting in the departure of U.S. forces two months later. In addition to the traditional OSC-I functions listed earlier, this cooperation was to include continued U.S. Naval Forces Central monitoring and protection of Iraq's oil export terminals and economic lifeline in the Gulf, various types of counterterrorist cooperation and

---

[51] SIGIR, 2012b, p. 3.

[52] Josh Rogin, "Congress Fails to Reauthorize the Pentagon's Mission in Iraq," *Foreign Policy*, October 1, 2012. See also Kevin Baron, "Pentagon Extends Iraq Mission Funding," *Foreign Policy*, October 2, 2012.

[53] SIGIR, 2011f, pp. 64–65.

[54] Interview with former OSC-I staff officer, Washington, D.C., July 20, 2013.

[55] "Iraqi Pilots Begin Flight Training in U.S.," *Defense World*, September 3, 2012. See also DoD, "Contracts," Press release no. 659-12, August 7, 2012c.

[56] "Military Spending: 11,175,067," *Time*, August 14, 2012.

intelligence sharing, USCENTCOM exercises, and U.S. facilitation of closer military-to-military relationships between Iraq and its neighbors. This plan was laid out to an appreciative Prime Minister Maliki in December 2011 during his visit to Washington. The expansive plan relied not only on OSC-I but also on DoD, USCENTCOM, U.S. Special Operations Command, the Intelligence Community, and DoS DS programs and resources to provide a broad range of assistance. In addition, the aforementioned plan to have U.S. military personnel continue monitoring the CSMs remained an important operational element of this holistic approach. According to Ambassador Jeffrey, much of this plan has been implemented, albeit with significant modifications.[57]

Police Training—The Gutting of the Police Development Program

Despite ambitious initial plans for a multiyear nationwide civilian police training program, DoS's PDP proved to be impractical once U.S. troops departed and without congressional funding support. PDP planners overestimated their ability to operate in a high-threat environment. The program was predicated on advisors' ability to travel frequently to remote training locations throughout the country. After the military withdrawal, when transportation proved both dangerous and exorbitantly expensive, the program proved impossible to implement on a large scale. Moreover, despite being located across the street from Iraq's national police headquarters in Baghdad, INL trainers were "unable to cross [the] street without heavy security and have largely ceased any outside movement."[58] An embassy security official told SIGIR that police trainers had much greater freedom of movement during the U.S. military presence, when USF-I could provide security, transportation, and other support.[59]

As described in Chapter Seven, the original PDP plan, estimated to cost $1 billion annually, envisioned 350 police advisors who would visit 28 training sites throughout Iraq by both ground and dedicated air transportation assets. Over the course of 2010 and 2011, the estimated costs of security, life support, and transportation rose even as it became clear that DoS would not receive all the funds it requested. INL repeatedly scaled the program back to a fraction of its intended size—from 350 advisors, to 200, then to 190. As it did so, the potential benefits of the program dwindled, while its overhead and security costs grew.

When the department submitted its FY 2013 budget request to Congress in February 2012, it was still aiming to implement the 190-advisor plan in full despite the fact that, according to an INL official, INL never developed a timetable for progressing from phase one to phase two.[60] DoS asked for $850 million to fund a program

---

[57] Email comment from Ambassador James Jeffrey to Charles Ries, January 11, 2013.

[58] Karen DeYoung, "State Department Seeks Smaller Embassy Presence in Baghdad," *Washington Post*, February 8, 2012.

[59] SIGIR, 2012d, p. 13.

[60] Interview with DoS INL official, Washington, D.C., January 20, 2012.

consisting of "approximately 190 advisors, based in three hub cities (Baghdad, Basra, and Erbil), who would travel to approximately 30 Government of Iraq critical 'spoke' sites."[61] The request made no mention of the fact that the PDP was currently being implemented in phases.

However, DoS stuck to this plan only briefly after the transition. Despite this ambitious proposal, some observers of the PDP were describing the program as being "on life-support"[62] and likely to be "a short-lived program."[63] By March 2012, INL had reduced the PDP footprint to a mere 50 advisors,[64] and rumors that the program would be scrapped led the embassy to issue a statement asserting that "the U.S. Embassy in Baghdad and the Department of State have no plans to shut down the PDP in Iraq that began in October 2011."[65] By July 2012, INL further reduced the number of advisors to 36 and the program budget to $111 million—roughly one-eleventh of the program's original size. One-half the advisors would continue to serve in Baghdad, and the other half would be based in the permissive city of Erbil, capital of the Kurdistan Region, despite the fact that the capabilities of the Kurdish police forces already far exceed those of their federal counterparts.[66]

The PDP's costs became increasingly unsustainable. In DoS's FY 2013 budget request, support costs increased to 94 percent of the total program budget, with the cost per advisor doubling to $4.2 million.[67] SIGIR Stuart Bowen pointed out to a congressional subcommittee that the PDP's budget was about equal to 15 percent of the entire MOI budget, with which it paid salaries and support costs for a force of 650,000 people.[68] Thus, the services of the PDP's small handful of advisors and mentors would cost about the same as 97,000 Iraqi police officers.

PDP spent a great deal of money on facilities, some of which were quickly abandoned. INL began construction at BPAX (formerly called Joint Security Station Shield), before securing Iraqi permission to use the facility, which was only belatedly granted, on December 11, 2011. Even then, Iraq granted permission only for a single year. Given the myriad difficulties executing the program, the PDP was an easy target

---

[61] Secretary of State, 2011b, p. 175.

[62] DeYoung, 2012.

[63] Interview with senior DoS official, Washington, D.C., February 17, 2012. See also SIGIR, 2011e, pp. 9–10.

[64] Arango, 2012b. See also SIGIR, 2012d, p. 17.

[65] Embassy of the United States, Baghdad, "U.S. Embassy: 'Police Development Program Is a Vital Part of the U.S.-Iraqi Relationship,'" press release, May 13, 2012b.

[66] Interview with Assistant Secretary of State for International Narcotics and Law Enforcement William Brownfield, Washington, D.C., August 23, 2012. One month earlier, SIGIR had reported that the PDP would be reduced to slightly lower levels—36 advisors, split evenly between Baghdad and Erbil. See SIGIR, 2012d, executive summary.

[67] SIGIR, 2012d, executive summary and pp. 19, 22.

[68] Bowen, 2011b.

for cuts when DoS decided to reduce the overall footprint of the embassy in early 2012. In March 2012, DoS decided to return BPAX to Iraqi control despite the fact that INL had already spent $108 million on construction of training facilities and housing for two-thirds of all PDP personnel. Given that BPAX was to serve as the primary PDP training facility, the early handover of the facility severely undermined the viability of the program.[69] By July, DoS had decided to close the PDP hub in Basra, after spending $98 million on its police training facilities, because the MOI decided it no longer wanted to conduct training at that location. SIGIR reported in July 2012 that the closure of PDP facilities at BPAX and Basra "brings the total amount of de facto waste in the PDP—that is, funds not meaningfully used for the purpose of their appropriation—to about $206 million."[70]

Despite DoS's efforts, the Iraqi government saw little value in the PDP. As noted in Chapter Seven, Senior Deputy Interior Minister Asadi complained in late 2011 that the PDP's enormous overhead and security costs meant that only a small fraction of the program's funding was spent on training Iraqi police officers. After only six months of training, the MOI issued a sweeping criticism of the program. In a formal written assessment of the PDP's second quarter (January 1, 2012, to March 30, 2012), the MOI reported that "the PDP program is not well organized and lacks leadership, police advisor teams work separately and their efforts are not coordinated, U.S. government PDP funds are not cost beneficial, and police advisory services are 'subpar.'" In May 2012, Asadi himself called the program "useless."[71] A SIGIR audit of the PDP in July 2012 stated that the program "powerfully underscores" a key lesson learned from years of reconstruction experience in Iraq: "that host country buy-in to proposed programs is essential to the long-term success of relief and reconstruction activities."[72]

Given these planning and management challenges—not to mention senior Iraqis' comments that the program was a waste of money—members of Congress who were critical of reconstruction efforts in Iraq found the PDP to be an easy target. Rep. Gary Ackerman, the senior Democrat on the House Foreign Affairs Committee's Subcommittee on the Middle East and South Asia, responded to Bowen's testimony on INL's plans for the PDP by stating, "No metrics, no milestones, no money."[73] In May 2012,

---

[69] SIGIR, 2012d, p. 13. See also Arango, 2012b; Walter Pincus, "Troops Have Withdrawn from Iraq, but U.S. Money Hasn't," *Washington Post*, June 27, 2012a; and Stuart W. Bowen, Jr., "Statement of Stuart W. Bowen, Jr.," in *Assessment of the Transition from a Military- to a Civilian-Led Mission in Iraq*, Subcommittee on National Security, Homeland Defense, and Foreign Operations, Committee on Oversight and Government Reform, U.S. House of Representatives, 112th Cong., 2nd Sess., June 28, 2012.

[70] SIGIR, 2012d, executive summary.

[71] SIGIR, 2012d, pp. 10, 12.

[72] SIGIR, 2012d, executive summary.

[73] Rep. Gary Ackerman, statement, in *Preserving Progress in Iraq, Part III: Iraq's Police Development Program*, hearing before the Subcommittee on the Middle East and South Asia, Committee on Foreign Affairs, U.S. House of Representatives, 112th Cong., 1st Sess., November 30, 2011, p. 36.

the Senate Appropriations Committee, noting "the largely unsuccessful implementation of the police development program in Iraq," cut all funding for the PDP from its version of the FY 2013 DoS and Foreign Operations appropriations bill.[74] Although a senior INL official described the PDP's high support costs as "the cost of doing business in Iraq,"[75] many members of Congress clearly did not think the program—particularly given its many flaws—was worth the price.

The remarkable implosion of the PDP demonstrates that host-nation buy-in is absolutely essential for reconstruction programs to succeed. The United States should have been willing to walk away from police training when it became clear that the Iraqi government was not an interested and engaged partner, but Washington believed in the value of police training far more than the Iraqis did.

The Expeditionary Embassy in Practice

Although the embassy remained able to conduct its primary mission of engaging senior Iraqi government officials on U.S. foreign policy priorities, the U.S. Mission in Iraq faced a number of operational challenges after the U.S. military withdrawal. In contrast to U.S. officials' assumptions, the security situation failed to improve over time. Indeed, as noted earlier, violence increased over the course of 2012. The Iraqi government, eager to assert its sovereignty, objected to the size of the U.S. embassy presence and to the quasi-military nature of its security operations. In response, DoS pledged to reduce the size of the embassy staff. A combination of security threats, lack of Iraqi support, and budget pressures limited the embassy's ability to manage training and assistance programs in the wake of the U.S. military's departure, with the PDP—the United States' largest remaining civilian reconstruction initiative in Iraq—being the most notable casualty. Finally, although the transition of administrative functions, such as life support and contract oversight, went smoothly, DoS was unable to assume full responsibility for these tasks as quickly as it had hoped. In late June 2012, DoD's Iraq Transition Coordinator informed a congressional subcommittee that "U.S. Army LOGCAP IV support will continue through Calendar Year 2013"—15 months longer than originally planned—"to provide the Department of State sufficient time to implement its phased approach to build its own capability."[76]

---

[74] U.S. Senate, Committee on Appropriations, "Department of State, Foreign Operations, and Related Programs Appropriations Bill, 2013: Report to Accompany S. 3241," Report 112-172, 112th Cong., 2nd Sess., May 24, 2012, p. 19.

[75] Interview with DoS INL officials, Washington, D.C., January 20, 2012.

[76] Verga, 2012.

Embassy Security

Perhaps more significantly, however, transition planners made optimistic assumptions about the security environment in Iraq and about the embassy's ability to operate in that environment without direct U.S. military assistance. Although the mission and USF-I had engaged in extensive planning to ensure the security of embassy facilities and personnel, U.S. officials were optimistic in assuming (1) that the security situation in Iraq would likely get better and (2) that, despite years of ups and downs in the bilateral U.S.-Iraqi relationship, Iraqi political leaders would welcome a large-scale, highly visible, proactive American civilian presence after the departure of U.S. troops. Neither of these assumptions proved accurate. As one unnamed senior administration official told the *Washington Post* in early February 2012, "I don't want to say we miscalculated, but we initially built a plan based on two things that have not played out as we had hoped. One was the politics, and the other was security."[77] In fact, this comment could be made regarding the entire Iraqi operation from 2003 onward.

Almost immediately after the transition, the Iraqi government made clear to the embassy that it would no longer tolerate aggressive and highly visible U.S. security measures. Iraqi officials insisted the embassy's use of surveillance drones was an affront to their country's sovereignty, with acting Interior Minister Adnan al-Asadi stating in January 2012: "Our sky is our sky, not the U.S.A.'s sky."[78] Similarly, a senior DoS official claimed that the 60 MRAP vehicles USF-I had provided the embassy were so politically unpalatable to the Iraqi government as to be "useless."[79] As of March 2012, the Iraqi government had refused to issue registration permits for these military vehicles. Consequently, despite paying for maintenance, driver training, and other related expenses, the MRAPs were being used as physical barriers at gates at the consulate in Basra.[80] By June 2012, Embassy Baghdad dropped plans to use both the MRAPs and "little bird" helicopters because of Iraqi objections.[81]

Similarly, the failure to fully comprehend the challenges associated with how the mission might operate in a hostile political environment created severe problems when the sheer size of the embassy became a political albatross. Mere weeks after the military's departure, DoS was forced to consider drastic cuts to embassy staffing levels and the potential closure of embassy facilities, raising questions back in Washington about whether the transition process needlessly spent billions of dollars on construction, security measures, and outside contracts. The Obama administration and mem-

---

[77] DeYoung, 2012.

[78] Schmitt and Schmidt, 2012.

[79] RAND discussion with senior DoS official assigned to the U.S. Mission in Iraq, Washington, D.C., May 7, 2013.

[80] Interview with senior DoS official, Washington, D.C., February 17, 2012.

[81] Kennedy, 2012. See also Walter Pincus, "Iraq Transition Raises Thorny and Expensive Questions," *Washington Post*, July 2, 2012b.

bers of Congress, already sensitive to wasteful spending, high debt levels, and grow-ing budget deficits, will continue to question whether the embassy can advance U.S. interests in Iraq sufficiently to justify the enormous expenditures. Clearly, part of this calculus will depend on the extent to which Iraq demonstrates is willingness to be the type of partner envisioned when the SFA was signed in 2008. At this point, the record is mixed, and it is unclear whether the government of Iraq shares the desire to be a reli-able partner in the region and the war on terrorism. Nowhere is that more clear than in Baghdad's relationship with Tehran and in the government of Iraq's unwillingness and/or inability to resist Iran's use of Iraqi airspace to export weapons and munitions to Syria, Lebanon, and other countries in the region.

Embassy Size

Although the U.S. mission's outsized support and security contingent was intended to ensure that core diplomatic staff would be able to do their jobs more effectively, the embassy's sheer size became a hindrance to its effectiveness in the months after the transition. Politically, a senior DoS official claimed in early 2012, "Iraqis have no stom-ach" for the large numbers of contractors employed by the embassy, especially those conducting security functions.[82] According to the GAO, the Iraqi government initially refused to issue visas for any U.S. security contractors during the first four months of 2012.[83] A senior DoS official commented that the slow pace represented a conscious Iraqi decision to exercise its sovereignty by regulating the size and nature of the U.S. presence.[84]

The combination of increased tensions between the U.S. and Iraqi governments and the desire of the Iraqi government to assert its sovereignty in its dealings with the United States caused the size of embassy Baghdad to become a political and operational liability. The problem was exacerbated by the fact that the embassy expanded in the first months after the transition beyond what it planned for. According to SIGIR, as of April 2012, the embassy employed 16,973 contractors. With the 1,369 U.S. govern-ment employees included, the embassy's total presence was 18,342 people, more than 2,000 (or 12.5 percent) above the planned ceiling of 16,000 total staff.[85] Table 10.1 shows planned and actual embassy staff levels over time.

---

[82] Interview with senior DoS official, Washington, D.C., February 17, 2012.

[83] Courts, 2012.

[84] Interview with Deputy Assistant Secretary for Near Eastern Affairs Barbara Leaf, Washington, D.C., January 13, 2012.

[85] Bowen, 2012. SIGIR notes that DoS reported far smaller contractor numbers than these, which come from DoD's Synchronized Predeployment and Operational Tracker (SPOT) database. It appears likely, however, that the DoS numbers are incorrect. After U.S. Ambassador to Iraq nominee Robert Beecroft testified in September 2012 that the embassy had reduced its staff levels by approximately 2,000 (to 13,000–14,000), DoS acknowl-edged that the reporting Ambassador Beecroft had relied on had undercounted certain staff categories. See SIGIR, 2012g, p. 5.

Table 10.1
Evolution in Embassy Staffing

Plan or Actual	Date	U.S. Government Staff	Contractors	Total	Ratio
Plan	March 2011	~2,833	~14,166	17,000[a]	1:5[b]
Plan	October 2011	~2,666	~13,334	16,000[c]	1:5
Actual	January 2012	1,490	12,300	13,790[d]	1:9
Actual	April 2012[e]	1,369	16,9733	18,342	1:12.4
Plan for end of FY 2013	May 2012[f]	Not specified	Not specified	11,500	N/A
Actual	July 2012[g]	1,235	13,772	15,007	1:11
Actual	October 2012[h]	1,075	14,960	16,035	1:14

[a] Secretary of State Hillary Clinton, reply to question posed by Sen. Lindsey Graham (Clinton, 2011a).

[b] Secretary of State Hillary Clinton, reply to question posed by Rep. Norm Dicks (Clinton, 2011b).

[c] Interview with Kathleen Austin-Ferguson, Executive Assistant to the Under Secretary of State for Management, Washington, D.C., October 25, 2011.

[d] SIGIR, 2012c, p. 33. These numbers, which DoS provided to SIGIR, are likely low; the following month, the *New York Times* reported that embassy personnel levels had reached "nearly 16,000 people"; see Arango, 2012.

[e] SIGIR, 2012c, pp. 5, 33. DoS had reported only 11,386 contractors to SIGIR, for a total embassy staff of 12,755 (a ratio of 1:9.3). However, SIGIR suggests that this significantly undercounted the number of contractors, stating that DoD's SPOT database showed 16,973 contractors alone supporting embassy operations as of April 2012. See also Bowen, 2012. DoS's subsequent acknowledgement that its later reporting undercounted certain categories of contractors suggests that SIGIR was likely correct in its assessment of the April data. After U.S. Ambassador to Iraq nominee Robert Beecroft testified in September 2012 that the embassy had reduced its staff levels by 2,000 (to a total of between 13,000 and 14,000), DoS informed SIGIR that Beecroft's testimony was incorrect. See also testimony of Robert Stephen Beecroft, hearing of the Committee on Foreign Relations, U.S. Senate, September 19, 2012.

[f] Courts, 2012.

[g] SIGIR, 2012d, p. 29.

[h] SIGIR, 2012f, p. 5.

DoS had to adjust the size of the embassy in response to both Iraqi sensibilities and U.S. congressional opposition to the enormous costs in comparison to other embassies. Less than two months after the transition from a military-led to a DoS-led presence in Iraq, DoS spokesperson Victoria Nuland stated, "what we have here is an embassy structure that was built for a different time and that relied a lot on expensive contracting for a whole range of reasons."[86] Other senior DoS officials announced that they would "right-size" the embassy's staff and establish "a more normalized embassy presence," primarily by hiring more local staff and procuring more goods and services

---

[86] Victoria Nuland, State Department Daily Press Briefing, Washington, D.C., February 7, 2012.

locally so as to reduce the embassy's reliance on third-country national contractors.[87] DoS officials also considered reducing core mission program staff by 25 percent across all agencies, which would reduce the number of movement security and life-support contractors needed.[88] According to the GAO, by May 2012, "State had a plan to reduce the presence to 11,500 personnel at 11 sites by the end of fiscal year 2013."[89]

By July 2012, the embassy reportedly had shed one-sixth of the total staff it had employed in April. The 15,007 employees remaining consisted of 1,235 U.S. government employees, representing a 50-percent reduction from the mission's original staffing plan, and 13,772 contractors.[90] In October 2012, it reported higher numbers to SIGIR: 16,035 total, including 1,075 U.S. government employees and 14,960 contractors.[91] At these numbers, the ratio of U.S. government employees to contractors was 1:14, less than one-half of the 1:5 ratio DoS's original plan had envisioned.

The mission also planned to eliminate three large facilities (BPAX by the end of 2012, OSC-I headquarters at Union III by mid-2013, and the Prosperity support annex by the end of 2013) by returning them to the Iraqi government and consolidating activities at the U.S. embassy main compound or other facilities around the country. While these steps will reduce the embassy's operational budget over the long term by cutting the large support and security teams needed to run and protect them,[92] they highlight the financial cost of failing to take Iraqi political constraints into consideration during the planning process. As noted earlier, the United States had already spent hundreds of millions of dollars on construction at BPAX, Union III, and Prosperity before deciding to return them to the Iraqi government. Furthermore, DoS estimated that the movement of staff from these sites to the main embassy compound would require $115 million in facility upgrades and expansions a mere three years after the embassy was completed at a cost of $700 million.[93]

Closure of Kirkuk Consulate Demonstrates Transition's Myriad Challenges

DoS closed its consulate in Kirkuk in July 2012, a mere seven months after the U.S. military's departure. The establishment of a diplomatic facility in Kirkuk had been deemed critically important to monitoring ethnic tensions there, which are widely seen

---

[87] Kennedy, 2012; SIGIR, *Quarterly Report and Semiannual Report to the U.S. Congress*, July 30, 2012e, p. 30; and Thomas Nides, "Rightsizing U.S. Mission Iraq," briefing, Washington, D.C.: U.S. Department of State, February 8, 2012. Also Nuland, 2012. Also Arango, 2012a. Also DeYoung, 2012.

[88] Interview with DoS Iraq Transition Coordinator Ambassador Patricia Haslach, Washington, D.C., February 17, 2012.

[89] Courts, 2012.

[90] SIGIR, 2012e, p. 29.

[91] SIGIR, 2012g, p. 5.

[92] Kennedy, 2012, p. 4.

[93] Pincus, 2012b.

as one of the most likely flashpoints for political violence in Iraq.[94] By all accounts, the post had been successful in doing so. On the eve of the closure, one senior DoS official stated that the consulate "had incredible access" to leaders of all of Kirkuk's communities (Kurdish, Arab, Turkmen, and Christian). "All of them," the official added, "looked at the U.S. as a sort of honest broker. Without a permanent presence there, I'm not sure we can continue in that role."[95]

By October, DoS reported that it was able to monitor events in Kirkuk effectively from Erbil, noting that the relatively calm security environment in the north enabled the consulate's five diplomatic staff to visit Kirkuk two or three times per week and that operating from the seat of the KRG enabled U.S. officials to appreciate the views of the two primary Kurdish parties regarding Kirkuk. That said, the logistical complications of traveling from Erbil made it harder for the consulate to organize meetings on short notice or to arrange visits beyond the Kirkuk Government Building. Thus, while the embassy is able to manage its diplomatic relationships in Kirkuk from afar, the arrangement has some shortcomings.[96]

The closure of the consulate in Kirkuk was driven by several diverse factors that capture the difficulty of operating a diplomatic mission in Iraq and of the U.S. transition planning effort as a whole:

- *The Iraqi government did not buy into U.S. efforts.* Although the United States decided to conduct training for Iraqi pilots and for aircraft maintenance and support personnel in Kirkuk, the Iraqi government evidently was not committed to the U.S. plan. The government's decision to move these training functions to Tikrit left no reason for OSC-I to operate a facility in Kirkuk, which then made it impossible for DoS to maintain that facility. Consulate staff relocated to Erbil in July 2012 in preparation for OSC-I's plan to return the Kirkuk site to Iraq in September 2012.[97]
- *Budget pressures limit diplomatic outreach opportunities.* Funding for the construction of a permanent diplomatic post in Kirkuk, which would have cost several hundred million dollars, remained "dubious, at best," according to a senior DoS official.[98] Closing the facility and consolidating DoS operations at Erbil eliminated the need to secure funds from Congress for future construction and for security and support operations.

---

[94] For more information on ethnic tensions in Kirkuk, see Hanauer and Miller, 2012, and Hanauer, Martini, and al-Shahery, 2011.

[95] Written correspondence with senior DoS official, July 2012.

[96] Email from senior DoS official, October 9, 2012.

[97] SIGIR, 2012e, p. 31. Also written correspondence with senior DoS official, July 2012.

[98] Written correspondence with senior DoS official, July 2012.

- *A large U.S. footprint is politically toxic in Iraq.* It is difficult to reduce the overall U.S. footprint in Iraq by 25 to 30 percent without cutting core diplomatic staff and thus significantly affecting programmatic functions. The most effective way to cut nondiplomatic staff is to close an entire facility because this eliminates the need for the support and security staff that a self-sustaining facility requires. OSC-I could have relocated its trainers to Tikrit to accommodate the Iraqi government's decision on training locations, but then DoS would have had to assume the entire cost of maintaining the Kirkuk site's security and support functions. DoS's inability to do so was the primary reason Kirkuk was made a DoD-led site in the first place.

- *Security threats undermine the viability of diplomatic outposts.* According to SIGIR, "U.S. facilities in Kirkuk had been subject to regular indirect fire attacks since they opened."[99] It would only be a matter of time before such attacks killed or wounded OSC-I or consulate staff, which would lead officials in Washington to demand expensive security upgrades and/or question the continued viability of the facility.

- *Unanticipated changes can stretch the capacity of other U.S. mission sites.* Kirkuk consulate staff moved to Erbil, which, according to SIGIR, then had to "prepare additional containerized housing units that [would] serve as living quarters and office space for those personnel relocated from Kirkuk." Similarly, the planned consolidation of personnel from U.S. mission facilities at FOB Union III, BPAX, and Prosperity onto the main embassy compound would require $115 million worth of additions and upgrades to the embassy's housing, office space, and support infrastructure (power plant, water and sewer systems, telecommunications networks, etc.)[100]

- *Everything about the U.S. presence is affected by—and seen as a commentary on— Iraqi domestic politics.* Monitoring ethnic tensions on a key fault line was relatively easy from Kirkuk itself, where U.S. officials could build relationships with local leaders without drawing attention to their efforts. In the absence of a local U.S. presence, every trip a U.S. official makes to Kirkuk will be seen either as a barometer of local political dynamics or U.S. interest in the city. Given Kirkuk's disputed status, the future efforts of U.S. diplomats to engage Kirkukis will be fraught with complex political challenges. Although Erbil is closer to Kirkuk than Baghdad, and is thus a logical choice from a logistical point of view to use as a base from which to cover Kirkuk issues, the decision to include Kirkuk in the territorial purview of the U.S. consulate located in the seat of the KRG suggests that the United States views the city as being more under the dominion of the KRG than of the central government in Baghdad.

---

[99] SIGIR, 2012e, p. 31.

[100] Pincus, 2012a.

Summary

Iraq in 2012 is not what U.S. officials and planners had envisioned. As had been the case since planning for Operation Iraqi Freedom began in earnest during 2002, over-optimistic assumptions drove the planning process, leading to goals and objectives that either were unachievable or were unachievable given the amount of time, manpower, and funding available. The expeditionary embassy was essentially a unit "left in contact" without the level of support it needed to operate in the Iraqi security environment. Likewise, OSC-I, initially envisioned as one component of a larger U.S. residual force, was given an impossible mission. Although the optimistic strategic and operational goals and objectives of the 2010 JCP officially terminated with the departure of USF-I, national policy and overarching strategic goals for Iraq remained unchanged. Thus, in 2012, Embassy Baghdad and OSC-I were challenged to pursue the same goals and objectives that had proven illusive for a USF-I force of 50,000. While it is true that the story of the U.S. experience in Iraq has rested on a series of transitions, policymakers failed to anticipate the transformational character of the final transition and the political fallout the departure of USF-I would have in Iraq. It is now over ten years from the start if the Iraq war, and nearly two years since the departure of U.S. forces from Iraq, and we are still unable to answer the question raised by then MG David Petraeus in 2003, "how will this end?"

CHAPTER ELEVEN

Conclusions and Recommendations

Conclusions

At its peak in 2011, the transition that is the focus of this book involved virtually every military and civilian American stationed in Iraq along with hundreds in Washington, in Kuwait, at USCENTCOM headquarters, and elsewhere. The result—the withdrawal of U.S. forces after an eight-year presence during which both Americans and Iraqis worked diligently to place the country on a path toward stability—marked a pivotal point in U.S.-Iraqi relations. While the future of the relationship is uncertain, all U.S. officials who contributed to the successful execution of the transition and the safe departure of U.S. forces should be justly proud of their contributions to U.S. national interests.

Iraq began its posttransition future with a raucous political system, an uneven security force structure, and (fortunately) a growing economy. Iraq's situation reflects the myriad contributions Americans and other coalition allies have made since 2003. The Iraqi people and their elected representatives must address a wide range of challenges in the years ahead, a task made more difficult by the internal struggle for power among competing groups and factions; high levels of corruption; acquiescence to Tehran on some issues affecting regional security; and an unwillingness on the part of the Maliki government and the Iraqi political system to address critical issues associated with reconciliation, reintegration, and repatriation of the Sunni minority.

Transitions in Iraq over more than eight years of U.S. involvement involved functional and institutional changes for both the military and diplomatic components of the U.S. Embassy in Iraq. During the first six years, the military assumed responsibilities for many activities required to conduct COIN and nation-building operations. During the last two years, and most notably during the final 12 months, USF-I effectively transitioned responsibility for activities to Embassy Baghdad, USCENTCOM, and other U.S. government agencies. In almost no case, however, did the transition of responsibilities to the ISF or other elements of the government of Iraq go as planned. In large part, this was because most U.S. transition planning relied on assumptions and performance measures that proved overly optimistic. Transitions were also affected by the highly dynamic internal situation in Iraq that was marked by a dangerous security environment, requiring changes to key transition objectives and extension in the time lines necessary for its completion.

It is tempting, therefore, to see the transition overseen by OND as just one point in a continuum of transitions. But the last transition was a much more fundamental reorientation of the U.S. relationship with Iraq than any previous transition because it resulted in the end of the presence of U.S. military forces in Iraq. This final transition transferred all remaining U.S. military activities to the government of Iraq, the U.S. embassy in Baghdad (and OSC-I), and USCENTCOM or terminated them. In some cases, activities were terminated because the United States could not continue them without military forces on the ground. In most instances, however, the transition meant that the activity transformed into something smaller in scope and scale or that the U.S. role definitively ended. In some cases, the rapid pace of the final transition created a void, such as between Arab and Kurd forces, that was impossible to fill with embassy officials alone, changing Iraqi politics in the aftermath (as described in Chapter Ten).

Tackling the Challenges of the Transition

The signing of the SFA and the associated SA can be viewed as the final terms of settlement of the U.S. war in Iraq. The SA established the terms of reference guiding the conduct of the U.S. military in Iraq, including ending the combat mission and moving out of the cities in 2009 and the retrograde of U.S. forces in 2011. However, what both policymakers in Washington and planners in Iraq largely missed was that the years following the signing of the SFA/SA should have been viewed as a period of political and diplomatic change that would result in a new U.S.-Iraq relationship, requiring a fundamental reassessment of U.S. policy and strategic goals in Iraq.

The transition was a carefully planned, deliberate handover of responsibilities for activities the U.S. military had previously conducted. The overarching strategic and policy goals that the 2007 to 2010 JCPs established remained constant. The optimistic strategic goals of the 2007 JCP were based assumptions relating to time and resources. However, the 2010 JCP left these optimistic goals and objectives in place, including the time lines for accomplishment, even though the force was dwindling, resources were more constrained, and time was running out. It was not until October 2011, during a war termination assessment, that military planners fully recognized that the goals and objectives of the JCPs would not be viable once USF-I was no longer on the scene.

Just as the redeployment of USF-I had unforeseen political ramifications within Iraq, the end of U.S. military presence in Iraq would also critically affect what the United States would be able to achieve in Iraq, especially in the short term. What gradually developed between 2008 and 2011 was a widening gap between established policy and strategic goals and the means and resources available to achieve them. A fundamental reassessment of the U.S. strategy for Iraq given the overall situation in Iraq and in the region might have identified the mismatch between resources and objectives. No such review took place, leaving Embassy Baghdad and the newly formed OSC-I with the challenge of trying to accomplish ambitious strategic and policy goals with insufficient resources to accomplish the desired ends. If viewed as a process, the transitions that accompanied the end of OND were part of a phase change that should have required a new strategic vision with achievable goals.

But to the extent the administration focused on the Iraq transition, its emphasis was on the mission, size, and immunities for a potential residual force, not on what would likely occur in the aftermath of the U.S. military withdrawal. This focus on the end of the decade-long military mission in Iraq, rather than the beginning of a new "strategic relationship," almost certainly distracted from the administration's intended message that the United States sought to remain engaged in Iraq over the long term and undermined the administration's efforts to secure resources for DoS's post-2011 role.

The Importance of Process

Military and interagency structures of OND facilitated a level of U.S. civil-military cooperation unmatched in other recent complex contingency operations. The structures should be considered for use in future operations that demand decisionmaking in a dynamic environment characterized by uncertainty.

While the organizational structure, processes, and procedures that Embassy Baghdad and USF-I developed addressed the specific transition-related challenges in Iraq, several key principles emerged that transcend this operation. First, the transition planning process was developed as a single effort that the embassy and USF-I managed jointly. Second, the "transition plan" was both a plan and a process. From the outset, planners secured input from stakeholders at all levels of the command and the embassy to ensure that all military and civilian officials involved with the transition shared the same goals, objectives, time lines, and overall vision. Third, the initial plan evolved and changed during the execution phase through a layered process through which leaders at all levels received frequent assessments of the transition as it related to their areas of responsibility, made decisions consistent with their authority, kept higher levels of command informed of transition progress, and highlighted critical issues that required higher-level involvement.

The transition management process in the field relied on joint embassy–USF-I working groups; the ECG, chaired jointly by the Deputy Chief of Mission and USF-I Chief of Staff; and the Core Group, chaired jointly by the ambassador and USF-I CG (see Chapters Six through Nine). This architecture provided a mechanism for key civilian and military decisionmakers to manage transition activities while simultaneously executing the operational retrograde of U.S. forces from Iraq. Although such an extensive committee-based coordination mechanism worked well to synchronize multiple military stakeholders, it was a burden on the much smaller Embassy Baghdad staff.

Securing effective Iraqi participation in planning and executing the transition proved to be one of the greatest challenges to U.S. authorities working the relevant issues in Baghdad. Other than for base transfers, no effective forums were created for senior U.S. embassy, USF-I, and Iraqi authorities to plan for the transition of USF-I activities to the Iraqi government. According to Ambassador Jeffrey, the United States typically decided what the Iraqis needed and merely informed them of American decisions.[1] Thus, the United States never gained official Iraqi buy-in for the plans and programs that would be carried after the departure of USF-I. U.S. officials in Iraq mitigated this shortcoming largely by seeking to coordinate with Iraqi authorities informally. However, for future transition contingencies, more-formal mechanisms that include host-country participation should be pursued relentlessly, keeping in mind security concerns as the host nation learns about U.S. and coalition plans.

Security Challenges and ISF Capabilities

The transfer of effective capabilities for security was fundamental to the success of the transition. The U.S. experiment in Iraq would likely have been condemned as a failure if, following the departure of U.S. forces, insurgents had toppled Iraqi political institutions or even had the insurgency returned to its 2006–2007 levels. Therefore, USF-I greatly emphasized appraising transitional security challenges; accelerating training and equipping efforts; and mitigating the threat that al-Qaeda in Iraq, Iraqi Sunni extremists, and Iranian-backed Shi'a extremists posed to give ISF the wherewithal to succeed on their own (see Chapter Six).

Longer-term capacity-building and security assistance programs—most notably, the procurement of weapon systems, aircraft, materiel, and associated technical assistance through the FMS program—were transitioned to OSC-I. These efforts would be important elements in realizing the goals of the USCENTCOM country plan finalized late in the drawdown process and would allow the United States to continue to assist the ISF to build its capacity for both internal security and external defense.

By the time U.S. troops withdrew, the ISF—despite the serious continuing shortfalls (see Chapter Six)—had demonstrated the capability to handle most of the internal security threats that violent extremist organizations pose. This is especially true for Sunni violent extremist groups; the government of Iraq has shown a willingness and ability to target them. However, such Sunni groups retain the capacity and will to strike at government and Shi'a civilian targets, posing a long-term threat to Iraqi stability.

An equal security challenge for the Iraqi government is garnering the political will to take on Iranian-sponsored Shi'a extremist groups, such as Kata'ib Hezbollah and Asa'ib al-Haq, both of which receive weapons, equipment, training, and funding

---

[1] U.S. Ambassador to Iraq James Jeffrey, RAND Corporation Roundtable Discussion, Arlington, Va., March 14, 2012.

from Iran. Unfortunately, doing so would apparently both threaten the political coalition that keeps the Maliki government in power and antagonize Tehran, which uses the groups to alter Iraqi political dynamics by dialing the level of violence up or down. Although the groups appear to have been less active since the departure of U.S. troops, they will no doubt remain able to engage in sectarian attacks as long as they continue to receive support from Tehran.

The American strategy for BII was largely political and economic and had thus always been the primary responsibility of Embassy Baghdad. However, USF-I efforts to neutralize Iranian-supported violent extremists, conducted as part of its "Countering Malign Iranian Influence" plan, were primarily kinetic and could therefore not be transitioned to the U.S. embassy. Instead, the mission of thwarting Iranian-backed extremists within Iraq was transferred to the government of Iraq, which has not embraced the task thus far.

The Expeditionary Embassy

U.S. transition plans envisaged an expeditionary embassy of unprecedented scope and scale to maintain U.S. influence in Iraq in the years following the final transition in an effort to help the Iraqis maintain their security, political, and economic gains (see Chapter Eight).

Transition planners made overly optimistic assumptions about the sustainable scale of the posttransition embassy, the functions it could undertake, and the environment in which it would operate. Although the mission engaged in extensive planning to ensure the security of embassy facilities and personnel, U.S. officials assumed that the security situation in Iraq would only get better, not worse. Similarly, despite years of ups and downs in the bilateral U.S.-Iraqi relationship, transition planners optimistically presumed that Iraqi political leaders would continue to welcome a large-scale, highly visible, proactive American civilian presence after the departure of U.S. troops.

Neither assumption turned out as planned (see Chapter Ten). As a result, as the security situation deteriorated after the departure of U.S. forces at the end of 2011, it was difficult for embassy officials to travel, even with extensive security precautions. Despite being based across the street from Iraq's national police headquarters, DoS's reduced cadre of police trainers was unable to travel even that far without significant security and, as a result, was largely prevented from any movement outside its facility, which was soon transferred to the Iraqis.

However, the single biggest obstacle to standing up an effective diplomatic mission in Baghdad may have been the lack of political support on the banks of both the Potomac and the Tigris. In Baghdad, the Iraqi government quickly made clear that it did not want a large-scale, highly visible official American presence in the country. Senior Iraqi officials objected to the embassy's proactive security initiatives and pressured the United States to reduce the embassy's 17,000-person footprint. As a result, mere weeks after the military's departure, DoS was forced to plan drastic cuts to embassy staffing levels and consider whether to close some embassy facilities (see Chapter Ten). This raised concerns in Washington as to whether the transition process needlessly spent billions of dollars on construction, security measures, and outside contracting for a presence that was not sustainable.

Moreover, after eight years of outsized U.S. military influence, Iraqi officials eagerly asserted Iraq's sovereignty in 2012 in ways that complicated U.S. goals. For example, while under the umbrella of a large-scale military presence, the embassy was able to conduct personnel and logistics activities under very liberal SOFA procedures that did not require independent agreements with or specific approval from the government of Iraq. Once the military departed, however, the embassy had to make a dramatic shift to conduct operations under normal diplomatic protocols governed by the Vienna Convention legal regime. Complicating this shift was the fact that Embassy Baghdad was 10 times larger than even a "normal" huge American embassy and included extraordinary operations, such as air, security, and convoy logistics. The sheer magnitude of the operation was a surprise to the Iraqis and led to confusion and lack of Iraqi preparedness in instituting new procedures for requesting visas, obtaining contractor work permits, clearing imports, and dealing with other routine matters from which the embassy was previously exempt. A more concerted effort to engage Iraqi officials in transition planning might have generated a greater and more-rapid measure of host-nation support (or at least alerted the United States to the potential for future hurdles), although the contemporaneous political gridlock at senior levels of the Iraqi government meant there were no clear, empowered interlocutors with whom the embassy could collaborate. The focus of both sides on the political debate over a residual force made Iraqi planning all but impossible. Eventually, literally facing possible meltdown of the U.S. mission in late December 2011 to February 2012, the Iraqis managed to put in place, with much embassy support and assistance, workable, if cumbersome, procedures.

In Washington, the executive branch never secured congressional support for an expeditionary embassy. Many members of Congress doubted that DoS would, without the benefit of military assistance, be able to provide adequate security to conduct its operations, and balked at the high costs of the security measures DoS considered necessary. Furthermore, the political and economic environment was not hospitable to an undertaking as ambitious and costly as the one DoS proposed to lead in postwar Iraq. In an era defined by low growth, high debt levels, and growing deficits, Congress was reluctant to allocate the billions in current and future funding needed to establish and operate such an expeditionary embassy. The unavailability of funds to build and operate high-cost embassy outposts in Mosul and Diyala forced revision of the embassy's original plans for diplomatic outreach in the Iraqi provinces. Congress was also unwilling to fund the construction of permanent facilities in Erbil, Kirkuk, and Basra. The frequent use of CRs, instead of full-year appropriations, in the period also hindered

DoS's ability to plan and budget for the U.S. mission's long-term construction, security, and life-support requirements.

However, in retrospect, the skepticism of appropriators about plans for a large civilian footprint seems prescient. Appropriators were more in tune with the concerns of influential Iraqis than officials directly involved in the military-to-civilian planning and doubted the feasibility of an extraordinarily robust U.S. civilian presence in Iraq from 2012 forward.

Notwithstanding these challenges, the embassy was generally prepared to assume the lead U.S. role in Iraq in December 2011. However, the long-term success of the expeditionary embassy created by this transition process is not guaranteed as long as Iraq remains dangerous and politically unstable. Unless the embassy's operating environment improves, DoS may have to revisit a number of central tenets of the transition effort that consumed the past several years.

Assessing the Transition

The transition of the U.S. relationship with Iraq represented a historic turning point in modern Iraqi history and in U.S. policy in the Middle East. Among the accomplishments during OND were the reposturing of tens of thousands of U.S. troops and contractors under difficult operational and security conditions; the closure or transferal of more than 90 bases and outposts; the transfer or movement of 41,000 vehicles; and the 1.8 million pieces of transportable equipment safely moved over long distances to rear-area logistical hubs in Kuwait, Jordan, and Turkey.

But other factors complicated transition planning and may have contributed to suboptimal outcomes and affected the readiness of Embassy Baghdad and OSC-I to seamlessly pursue U.S. goals and objectives. As the United States looks ahead to analogous transitions in other countries, particularly the still unfolding changes in mission for the International Security Assistance Force in Afghanistan, several such factors may come into play again and thus merit special attention.

The Uncertainty of the End State

The transition was dogged from the outset by the ambiguity of Iraqi and U.S. interest in a follow-on U.S. troop presence (see Chapters Seven through Nine). When transition planning began in earnest in 2009 and 2010, neither the U.S. administration nor the Iraqi government had a clear position on the desirability of a follow-on presence. Moreover, neither side could agree on the missions that such a force would undertake, which meant that the number of troops it would require varied considerably. It was also unclear whether Iraq would be prepared to enter into a second security agreement to provide any remaining U.S. troops with protections that would be a precondition for an enduring military presence.

U.S. transition planning was complicated by the necessity for a two-track effort to plan for both possible outcomes. Given the President's clear statement at Camp

Lejeune that "I intend to remove all U.S. troops from Iraq by the end of 2011,"[2] USF-I and embassy planners prepared for a complete withdrawal of U.S. forces. At the same time, however, contingency planning efforts considered a range of possible follow-on options and projected what equipment, facilities, and contracted services might be required in country to support them.

Iraqi officials similarly tried to straddle the fence. Driven by overwhelming popular opposition to an enduring U.S. military presence, Iraqi leaders from all parts of the political spectrum issued public statements opposing the continued deployment of U.S. forces—even though many of the same leaders privately confided to U.S. officials that they believed an enduring presence would contribute positively to internal security and hoped the two governments would find a way to extend the mission.[3] While Iraqi political groups continued to debate the issue vigorously in summer 2011, U.S. officials were anxious to resolve the debate, as evidenced by Secretary of Defense Panetta's admission that "I'd like things to move a lot faster here, frankly, in terms of the decisionmaking process. Do they want us to stay, don't they want us to stay?" followed by his famous, apparently heartfelt outburst, "Dammit, make a decision."[4]

Ultimately, in August 2011, all major political parties except Muqtada al-Sadr's Sadrist Trend agreed in principle to support an appropriately scaled "training" mission but refused to grant the immunities and protections on which U.S. officials insisted.[5] Unable to reach an agreement on immunities that both sides could accept, President Obama announced definitively on October 21, 2011, that "the rest of the troops would come home by the end of the year."[6]

The uncertainty about whether there would—or would not—be a follow-on presence affected transition planning in important ways. For example, planning guidance initially established June as when U.S. forces would have to initiate the gradual withdrawal of U.S. forces and the transition of responsibilities to Embassy Baghdad and OSC-I. This timing would have enabled U.S. military and civilian officials to work through any unanticipated challenges while USF-I remained in Iraq. This earlier timing would also have focused the entire USF-I and DoS effort toward the transition instead of continuing to balance competing efforts with limited staffs and resources, specifically time. To allow time for the U.S. and Iraqi governments to negotiate a follow-on agreement, the decision point to execute the final phase of OPORD 11-0111-01, dealt with in two major changes to the order, continued to shift "to the right," with October

---

[2] Obama, 2009.

[3] Marisa Cochrane Sullivan, "Obama's Iraq Abdication," *Wall Street Journal*, July 28, 2011. Also Abdul-Zahra and Santana, 2011.

[4] Bumiller, 2011.

[5] Arango and Schmidt, 2011.

[6] Barack Obama, "Remarks by the President on Ending the War in Iraq," Washington, D.C.: The White House, Office of the Press Secretary, October 21, 2011.

15—the date by which "the laws of physics" required the redeployment to begin to be completed by December 31—becoming the final deadline for a decision. When time ran out and the President decided to proceed with the redeployment, what had initially been envisioned as a gradual withdrawal of forces became a steep waterfall. When USF-I, OSC-I, and the embassy identified unanticipated transition-related challenges in November and December, there was little USF-I could do to assist in resolving them because of the immense requirement to reposture the force and exit Iraq in a responsible manner within two months. Perhaps more important, given the requirements for USF-I to conduct the operational maneuver out of Iraq, USCENTCOM should have assumed responsibility for providing assistance and support to OSC-I during the last months of the operation.

- The waterfall of U.S. force departures was designed to keep as many forces in country as long as possible both to continue with the advise and assist mission and to preserve options for the President.[7] USF-I identified forces, locations, and equipment that might be involved in a follow-on mission and released them for final disposition only in the last stages of the transition. As General Austin put it:

  Quite frankly, we're not pushing the Iraqis to ask us for help. All we're saying is if they are going to ask us for help, [they should know] that sooner is better for us because it will not cause us to disassemble things that we might have to spend money to reassemble at a later date.[8]

- Confronted with the uncertainty, DoS decided to contract for specialized medical capabilities to provide "golden hour" treatment for civilians potentially injured in attacks (including multiple facilities and rotary-wing aircraft). It would not have done so if there had been agreement on a prolonged military presence in country because the military would almost certainly have provided similar capabilities to cover all U.S. personnel.
- After an extended internal debate about whether DoS would contract directly for life-support services, the administration decided to continue to provide such services under DoD's global LOGCAP contract vehicle for at least one fiscal year, which may have been the only practical short-term alternative.[9] However, this decision also kept open the option of rapidly bolstering the U.S. mission to sup-

---

[7]   Ambassador Jeffrey explained that USF-I also wanted to keep significant forces in the north committed to the CSM along the line between Arabs and Kurds as long as possible to mediate any possible disputes. Interview with Ambassador James Jeffrey, Arlington, Va., March 12, 2012.

[8]   Jim Garramone, "Austin Gives Insight into Drawdown, Possible Aid to Iraq," American Forces Press Service, July 11, 2011.

[9]   Interview with DoS Transition Coordinator Ambassador Patricia Haslach, July 15, 2011.

port any remaining troops—an arrangement that would have reversed the eight-year pact under which the military provided life support to the embassy.

- Similarly, DoD agreed to buttress DoS's contract oversight capabilities with the assignment to Baghdad of 50 DoD civilian auditors from the Defense Contracting Agency (see Chapter Eight). This helped DoS oversee the LOGCAP services contract and other contracting vehicles, but the decision also would have eased support for an enduring presence had one needed to be supported after January 1, 2012.

Planning for OSC-I was somewhat delinked from other transition planning processes, and until midyear 2011, it did not receive much senior-level attention outside Baghdad (see Chapter Five). A security cooperation plan—crucial for guiding normal defense relationships with partners around the world—was not finalized until late 2011. Observers have suggested that this lacuna was due to the expectation of many in USF-I, USCENTCOM, and DoD that some sort of enduring security presence would ultimately be agreed with Iraq, sharply changing the mission of OSC-I itself to include many training and exercise responsibilities that are normally outside the scope of an office of security cooperation.[10]

And finally, a USCENTCOM exercise program that included Iraq was not developed until late 2011—well outside the normal window for developing, scheduling, and funding this portion of the security and cooperation program.[11]

The "Political Transition"

The careful process of "binning" activities to be transitioned, the 2010 and 2011 JCPs, and the execution of activities under OPORDs 10-01.4, 11-01, and 11.01 Change 1 ensured comprehensive oversight of the activities being transitioned, ensured adjustment to changing circumstances and opportunities, and fostered close civilian-military cooperation throughout.

It is perhaps ironic, therefore, that a consequential factor related to the success of the transition was an erstwhile LOO that was not transferred at all: the political aspect of the U.S.-Iraq relationship.[12] The transition plan did not include elements of the political LOO because the 2010 JCP made clear that these were already the responsibility of the embassy, leaving no political tasks to transition. The embassy had the lead for political engagement with the government of Iraq. However, the decision to delink these activities from the transition process had unintended consequences: Transition plans failed to identify and assign measures to support this critical component of the

---

[10] Interview with former DoD official, January 17, 2012.

[11] Interview with USF-I J5 staff officer, Baghdad, July 1, 2011.

[12] As described in the study, OPORD 10-01 actually dropped references to the political relationship as a LOO, on the basis that USF-I had little direct effect on it. Instead, the 2010 OPORD approved three LOOs: (1) strategic partnership, (2) operations, and (3) civil support/theater sustainment.

overall mission, which likely contributed to the U.S. failure to anticipate how the rapid withdrawal of U.S. forces would affect the already fragile Iraqi political arena.

That is not to say that the embassy was unaware of the fact that the withdrawal of forces would likely have unpredictable political consequences within Iraq. For example, in January 2011, Ambassador Jeffrey delivered a presentation at the Washington Transition Conference hosted by USCENTCOM in which he discussed what he considered the "five M's of transition": money (budget and authorities), missions (what the embassy would have to do with a focus on USF-I's binning process), months (time available before December 2011), management (the tools available to do this along with the overall magnitude of the operation), and Maliki (shorthand for the actions the government of Iraq would have to take to transition from an embassy operation under what was essentially a SOFA to an embassy operating under the Vienna Conventions).[13] However, planning for this political transition was not a focus either in Baghdad or in Washington.

The primary U.S. political objectives for Iraq in 2010 and 2011 were interrelated: to help ensure the success of government formation following the 2010 national elections, with a broad-based, stable government; to ensure Iraq's security and territorial integrity; and to preserve and enhance a strategic U.S. relationship with Iraq. As Ambassador Jeffrey put it to the Senate Foreign Relations Committee in February 2011:

> We have today a historic opportunity and a critical window to help Iraq emerge as a strategic partner and a force for stability and moderation in a troubled region. We cannot afford to let the gains we have sacrificed so much for slip away. The President has clearly articulated our vision for partnership with Iraq. We seek there a country that is sovereign, stable and self-reliant, with a government that is just, representative and accountable, that denies support and safe haven to terrorists, is able to assume its rightful place in the community of nations, and contributes to the peace and security of the region.[14]

It had been evident to planners that the transition, especially the planned complete departure of USF-I, would affect the Iraqi political process, influencing various groups in divergent ways. For the United States, the challenge was, as Ambassador Jeffrey stated at his confirmation hearing, to "reinforce in words and deeds that the withdrawal of U.S. combat forces in no way signals a lessening of our commitment to Iraq."[15] Vice President Biden's November trip to Iraq for a meeting of the U.S.-

---

[13] Email correspondence between Ambassador James Jeffrey and Charles Ries, January 11, 2013.

[14] Jeffrey, 2011a, p. 7. (Ambassador Jeffrey was referring to President Obama's February 2009 Camp Lejeune speech on Iraq.)

[15] James F. Jeffrey, "Statement by Ambassador James F. Jeffrey: Senate Foreign Relations Committee," July 20, 2010.

Iraq Higher Coordinating Committee (as provided for under the SFA) and meetings with Iraqi political leaders were parts of this reassurance effort,[16] a process that culminated with President Obama's invitation to Prime Minister Maliki to visit Washington December 12–13, 2011.[17]

A prosecuting judge and MOI personnel precipitated one of Iraq's most significant political crises to date by seeking the arrest of Vice President Hashimi and other Sunni leaders, which in turn led Sunni leaders to boycott Iraqi political institutions (see Chapter Ten). Prime Minister Maliki disclaimed responsibility for the arrest warrant; however, he was most likely aware of it and its significance. An important question about this crisis is whether it was precipitated or aggravated by the final withdrawal of U.S. forces and, if so, whether the transition could have been managed in such a way as to attenuate such political repercussions. As is typical in such political analysis, one can never know for sure what the counterfactual would have been. Nevertheless, the dramatic Iraqi political events that followed the transition were sobering reminders of the limitations of transition preparations.

In Iraq, even when U.S. forces were at peak levels, crippling political crises had emerged. Hashimi pulled Iraq Islamic Party ministers out of the Council of Ministers in 2007 over slights from the Dawa-led government; the De-Ba'athification Commission's disqualification of Mutlaq and other prominent Sunnis from running in the 2010 national elections caused an acute crisis in December 2009–January 2010; and, most notably, it took a rancorous nine months after the March 2010 election to form a broad-based Iraqi government (and even then, no ministers were chosen for the MOD and MOI, hindering U.S. forces' ability to coordinate the transition).

In each of these political disputes, the United States sought to deploy its influence and leverage to resolve the crisis and promote reconciliation among the parties. In the context of the history of Iraq since the 2003 invasion, the 2012 political challenges are not totally out of character for the still-developing, but regrettably brass-knuckles, political culture of the country. In this respect, it may be argued that, even though it constrained U.S. levers of influence, USF-I's departure from Iraq was not singularly responsible for setting back Iraq's political development. However, the actual consequences of the complete withdrawal of U.S. military forces and the failure to come to an agreement on a smaller residual force will remain difficult to assess, even in the long run.

---

[16] Amy Dudley, "Vice President Joe Biden: 'In America, and in Iraq, the Tide of War is Receding,'" blog, Washington, D.C.: The White House, December 2, 2011. See also United States of America and the Republic of Iraq Higher Coordinating Committee, "Joint Statement by the United States of America and the Republic of Iraq Higher Coordinating Committee," Washington, D.C.: The White House, Office of the Press Secretary, November 30, 2011.

[17] "Obama and Maliki Back Iraq Post-War Future," BBC News, December 12, 2011.

Iraqis and the Scale of the U.S. Civilian-Led Presence

A final aspect complicating the transition in Iraq was the evolution in attitudes toward the U.S. presence. The U.S. "occupation" of Iraq and its symbols (e.g., up-armored Humvees or MRAPs, frequent helicopter transits, aerostats, U.S. personnel in body armor and helmets, and T-walls) had always been distressing to the Iraqis, who are strong nationalists across the political spectrum.[18] Memories of the 2007 Nisour Square incident, in which a U.S. contractor protective detail killed Iraqi bystanders, hardened Iraqi opposition to the United States' aggressive and highly visible security posture.

While the formal occupation of Iraq legally ended in 2004 with the establishment of the interim Iraqi government, U.S. and British forces continued to operate in Iraq with the legal authorities of an occupation forces in accordance with UNSCRs through the end of 2008. For Americans, the rights and obligations of the U.S. military as an occupying force ended *de jure* on the implementation of the SA on January 1, 2009, and to a more tangible extent on July 1, 2009, when U.S. forces moved out of Iraqi cities in accordance with the SA's provisions. (Thereafter, U.S. forces were permitted to reenter urban areas only with prior notice and Iraqi security force escorts.) Even so, official Americans and many classes of contractors (as well as equipment and supplies) routinely entered the country into December 2011 without inspection by Iraqi authorities. Helicopters and aerostats remained highly visible. To many Iraqis, therefore, Iraq did not fully regain its sovereignty until the last U.S. forces left the country December 18, 2011.

As the December 2011 transition approached, senior American and Iraqi policymakers focused on discussing the scope, privileges, and immunities for a possible follow-on U.S. military training mission, as described earlier. But also on the to-do list was the need to secure land-use agreements for the U.S. government to be able to construct and utilize facilities on ten of the 11 enduring locations for a civilian-led presence.[19] (Land title for the embassy compound in Baghdad was secured before its construction began years earlier.) Yet Iraqi policymakers—distracted by internal political crises, without permanent Ministers of Defense and Interior, and under no deadline pressure themselves—proved serially unwilling to authorize the land use the U.S. diplomatic mission needed. Such a bifurcation of incentives could also apply in future stabilization mission transitions.

In 2011, the Ministry of Foreign Affairs authorized the establishment of U.S. consulates in Erbil and Basra via diplomatic notes, although in the Iraqi system, the Ministry of Foreign Affairs itself does not control land use. The MOI similarly authorized the use of land adjacent to its headquarters in the Rusafa district of Baghdad (formerly FOB Shield) for PDP use. But otherwise, the United States entered 2012 without land use agreements for sites the embassy needed for transportation, warehousing,

---

[18] Interview with Ambassador James Jeffrey, Arlington, Va., March 12, 2012.

[19] Interview with Ambassador James Jeffrey, Baghdad, June 29, 2011.

contractor support, FMS training and support, refugee processing, and OSC-I offices. Almost all these facilities were originally built or improved with U.S. military or civilian construction funding.

The lack of land use agreements, however, was but a symptom of a broader political problem affecting the follow-on U.S. civilian presence: a widespread Iraqi allergy to the size of the envisaged U.S. civilian footprint. Ambassador Jeffrey bluntly described this public sentiment in March 2012: "Iraqis hate us for having occupied the country for eight years, and they don't want to see us around anymore."[20]

Beginning in November 2011, Iraqi officials received large numbers of applications for visas and work permits for embassy contractors. More than 100 contractors were even detained at Baghdad International Airport in early January 2012 when they attempted to enter with purportedly insufficient documentation.[21] Iraqi media replayed for domestic audiences U.S. press coverage of the embassy's 16,000-strong staff and planned aggressive security measures—including the use of aerial surveillance drones and thousands of armed security contractors[22]—which fostered Iraqi perceptions that the United States intended to continue acting as an occupying country whose officials would do as they pleased.[23] Facing such Iraqi opposition to the scale of the U.S. civilian presence—not to mention congressional skepticism about its high cost—DoS announced five weeks into 2012 that it would significantly reduce (or "right size") the number of staff and third-country national contractors working at the embassy (see Chapter Ten).[24]

Implications for Future Transitions: Key Insights and Recommendations

The following insights and recommendations are presented as strategic- and policy-level lessons learned that should be considered by military planners and policymakers when crafting strategies for transitions and posttransition relationships. These insights and recommendations relate to relations with host countries, priorities for security assistance, and more technical-level recommendations on civil-military coordination and cooperation, planning horizons, contracting, and KM. To be successful, all of

---

[20] Interview with Ambassador James Jeffrey, Arlington, Va., March 12, 2012.

[21] Michael S. Schmidt and Eric Schmitt, "Flexing Muscle, Baghdad Detains U.S. Contractors," *New York Times*, January 15, 2012.

[22] DoS had reportedly requested weapon permits for 3,000 third-country nationals. Interview with senior U.S. government official, Washington, D.C., November 28, 2011.

[23] Schmitt and Schmidt, 2012.

[24] Arango, 2012a. See also Nides, 2012.

these elements of successful transitional planning require long lead times and high-level commitment.

The Iraq experience illustrated that transition from a U.S. presence dominated by a major military command to one managed by a U.S. embassy is not just a transition in scale but also in kind. A fundamental transformation of the mission took place in Iraq. While a programmatic approach to what can and should be transitioned from military to civilian organizations (to include an office of security cooperation, within the embassy) is necessary, planning should start by identifying U.S. strategic goals for the era after the transition and only then considering how a civilian-led embassy can be set up to accomplish these goals. An approach that transfers functions "as is" from the military to the embassy may not be as effective as an approach that plans a fundamentally new mission from scratch and only looks at functional transfers once this new planning foundation is established.

Recommendation 1

Policymakers should initiate a multiagency planning process under the direction of the White House national security staff well in advance of the anticipated transition to (1) define enduring U.S. interests in the country, (2) establish realistic goals and objectives that an embassy operating under the requirements and limitations of the Vienna Convention can achieve, (3) assess follow-on military presence and resources required to achieve desired objectives, and (4) identify authorities that the embassy and its Office of Security Cooperation will require to operate within the country.[25]

An embassy-led presence is fundamentally different from a military-led mission and must be designed to be consistent with global U.S. foreign and security policy interests and with the requirements and limitations of the Vienna Convention operating framework. The Iraq experience illustrated that a transition from a U.S. presence dominated by a major military command to one managed by a U.S. embassy is not just a matter of scale but also of kind. A fundamental transformation of the mission took place in Iraq. While a programmatic approach to what can and should be transitioned from military to civilian organizations (including an Office of Security Cooperation, within the embassy) is necessary, planning should start by identifying U.S. strategic goals for the era after the transition and only then considering how a civilian-led embassy can be set up to accomplish these goals. An approach that transfers functions "as is" from the military to the embassy may not be as effective as an approach that plans a fundamentally new mission from scratch and only looks at functional transfers once this new planning foundation is established.

---

[25] After reviewing the draft manuscript, Ambassador Jeffrey prepared a personal assessment for us, which we have enclosed as Appendix A at his request. Email correspondence between Ambassador James Jeffrey and Charles Ries, January 11, 2013.

Recommendation 2

Policymakers should secure support from relevant congressional committees on the nature and likely cost of an enduring civilian-led mission well in advance of the departure of military forces, in the context of U.S. foreign and security policy goals and in conjunction with normal budget planning cycles.

Transitions in force posture do not always imply transitions from war to peace. After the U.S. military departure, Embassy Baghdad was expected to begin performing functions that no other U.S. diplomatic post in the world must undertake. The embassy and OSC-I sites needed to operate in an insecure environment with limited force-protection capabilities and restricted movement options. Of more than 180 bilateral U.S. embassies in the world, Embassy Baghdad is the only one to have a "sense and warn" radar system, aerial surveillance drones, or a fleet of MRAPs for quick-reaction rescue of personnel *in extremis.*

In July 2010, the independent, congressionally chartered Commission on Wartime Contracting identified 14 "lost functionalities" to be expected with the departure of U.S. forces. USF-I and Embassy Baghdad identified an additional seven critical functions that the military performed that the U.S. embassy would need to assume. While such requirements and related funding authorizations were small from a DoD perspective, they were enormous from a DoS perspective because of the department's more modest resources.

Normally, U.S. executive branch agencies begin developing their budgets two years ahead of time. Furthermore, DoS and USAID face far greater political obstacles than DoD does in getting large appropriations or supplemental appropriations to cover contingencies. Moreover, if planning efforts do not anticipate all costs, DoS and USAID do not have anywhere near the flexibility that DoD has to reprogram funds within existing budgets to meet needs. Planning should, therefore, include options driven by different potential funding levels, and budget proposals should incorporate foreseeable requirements.

As responsibility shifts from DoD to DoS, it is also important to make sure the U.S. embassy has all the legal authorities it needs to operate after the drawdown. This did not occur during the Iraq transition and caused a number of problems that came to the surface immediately after the departure of USF-I. These included the initial inability of contractors to enter Iraq to support embassy operations, the requirement to gain Iraqi government approval for the movement of food and other goods into Iraq, the inability of the chief of OSC-I to obligate funds, and the requirement to license embassy vehicles to operate in Iraq—to name just a few. These challenges regarding authorities highlight the transformational nature of the transition from a DoD to a DoS mission.

In the Iraq transition, congressional committees consistently rejected the Obama administration's requests for increased funding to support expanded embassy operations in Iraq. In retrospect, congressional committees appear to have correctly anticipated U.S. domestic—and Iraqi public opinion—constraints on U.S. posttransition presence better than the administration (and Baghdad-based) planners did.

Recommendation 3
Policymakers and military transition planners should initiate work early with the host nation to identify posttransition requirements and to reach firm agreements with the host nation to ensure the smooth transition and success of posttransition U.S. presence. The parameters of the scope and functions of the U.S. presence should be identified early, and, when possible, agreements should be crafted to support U.S. and host-nation needs, possibly even accommodating future variations in the footprint to build flexibility into plans and programs. Such dialogues should be buttressed by outreach to other political interest groups and should be integrated with public diplomacy efforts.

In the future, posttransition circumstances and programs will depend on the security situation, U.S. objectives, and agreements with the host nation regarding the residual U.S. footprint after the military force draws down. This will require extensive engagement with senior host-nation officials and other political interests and extensive public-diplomacy efforts to ensure wider understanding of U.S. goals and objectives. For a variety of political and practical reasons, such consultations do not appear to have been systematically undertaken in the case of Iraq.

Recommendation 4
Military and civilian planners both in theater and in Washington should make a fundamental reassessment of campaign goals and objectives well before the departure of forces, recognizing that previously established campaign goals likely will not have been achieved by the end point of the transition process. Therefore, planning should rigorously prioritize efforts in advance to set the critical conditions for the success of the organizations that will assume some of the military force's responsibilities rather than aim to achieve all the goals and objectives established during the campaign planning process. In particular, with respect to the crucial task of training security forces, *minimum essential capability* for host country forces is the "good enough" functionality required to fulfill basic responsibilities, not equivalence to U.S. forces' capabilities.

In Iraq, the mantra was that OND's JCP was conditions based and time constrained. In reality, the JCP was conditions based and resource constrained, with time being the limiting factor. In building the ISF, the successive U.S. military transition commands operated on the basis of defining, then seeking to help the Iraqis achieve, a "minimum essential capability" for each discrete function (air, naval, special forces, combined arms, etc.) it was assumed that Iraq would need to maintain its security and sovereignty. While USF-I stopped using the term *minimum essential capability* in late 2010, when it assessed that the goals were unachievable in the time remaining, the focus remained on providing a minimal capability deemed as a necessary foundation for future development of an external defense capability. Once the security agreement made it clear that U.S. training would come to a close (or at least change significantly) at the end of 2011, the United States changed its aspirations for the ISF. Instead of striving to develop requirements-based competencies, U.S. officials worked to develop the minimum capabilities that would permit U.S. forces to depart, defined in practice as whatever was possible by the end of the time-constrained U.S. military presence. In the end, the capabilities identified were driven more by the reality of time available rather than the achievement of the goals established in the JCP (and approved by both DoS and DoD). However, at no time were the JCP goals and objectives modified to meet the time and resource constraints. Moreover, neither policymakers nor planners conducted the type of campaign plan reassessment that was necessary to establish achievable goals and objectives during and after the period of transition.

Recommendation 5

Military planners should make institution-building a priority effort to ensure that the progress made through training, advising, and assisting will be sustained after the transition. In planning for sustainable host-country posttransition security, the human resource functions of recruitment, training, and professionalization are more important than providing equipment and modernization. Institutional capacity must ensure that the equipment provided can be successfully used and maintained after the departure of U.S. forces.

Of greater concern than achieving tactical and operational skills competencies, however, was whether the ISF would continue on the path to professionalization. The U.S. military training and advisory mission focused significant effort on individual and small unit skills necessary to conduct tactical operations. However, much less effort was placed on creating the type of institutional capacity that would ensure the continuation of this training by the ISF after the departure of U.S. forces. While there were some success stories, at the time of the transition all ISF elements had serious institutional deficiencies in their training capabilities and thus in their abilities to sustain the process of recruiting, training, and fielding professional military and police forces.

Recommendation 6

Prior to fielding equipment packages for a host nation, military planners should critically assess the long-term capacity of the partner nation to independently sustain the equipment and systems after the departure of U.S. military, contractors, and funding. Planning for sustainable host-country posttransition security, the life-cycle management of the equipment, and the capacity and capabilities of the host country are just as important as the intended purpose of that equipment.

The gap in the ISF components' ability to sustain their equipment and systems with a mature logistics system was well known. While USF-I DCG (A&T), ISF, and other agencies worked to close this gap prior to USF-I's transition, the complexity of the equipment, numerous variants, and logistics management programs and processes made the task that much greater. The departure of U.S. military advisors, contractors, and funding exposed a lack of Iraqi capacity to independently sustain much of the modern equipment and systems the United States had provided. Nowhere is this more evident than in logistics and maintenance, where both institutional failings and cultural norms have worked together to impede progress.

Recommendation 7

Pretransition planning should be launched several years ahead of the transition deadline, led jointly by a general officer and a senior civilian, staffed with capable planners who are not involved in current operations, and granted all necessary authorities. Moreover, effective transition planning must proceed on the basis of seamless top-level collaboration between the senior military and senior civilian in country working together in partnership.

Civil-military cooperation in Iraq throughout the last transition was exceptionally good, and this is clearly one of the major reasons it went as smoothly as it did. The cooperation was due in large part to the commitment of the ambassador and the USF-I commanding general. They made it clear that they would take all key decisions together and demanded comparable cooperation from their subordinates. The cooperation was also a result of the increasing convergence of the core missions of USF-I and Embassy Baghdad. In particular, after U.S. forces left the cities in summer 2009, USF-I's combat mission was largely limited to counterterrorism and force protection. As a result, the primary mission of USF-I from that point forward was to set the conditions for the Iraqi government and U.S. embassy to succeed after USF-I departed. Such unity of effort is critical to a successful transition.

The Iraq security assistance transition effort began as a small cell in MNSTC-I well in advance of the actual transition, with a planning team that was not involved in current operations and could therefore focus on the long-range planning. However, the planning team did not have access to senior-level officials, the authority to task MNF-I (later USF-I) or USCENTCOM staffs for support and information, or a direct civilian counterpart at the embassy. These shortcomings caused challenges and delays in the cell's ability to plan for posttransition security assistance and security cooperation at the level of detail required.

Recommendation 8

A single office to manage all contracts and contractors should be established in theater early in the operation. The USCENTCOM Contracts Fusion Cell established for Iraq is a model that could usefully inform other U.S. efforts to develop

and maintain a common operating picture for the state of contracts, as well as to coordinate with and among contract owners. In addition, a mechanism must be developed to ensure individual accountability of all contractors in country to help facilitate their departure along with the military forces they support.

Since the U.S. military relied heavily on contractors, it spent a great deal of time planning for the demobilization and redeployment of tens of thousands of civilians, as well as uniformed military. Planners need to account for contract provisions for ending services and getting contractors and their equipment out of the host nation in ways that support and do not hinder the military drawdown. However, at no point in the eight-year operation did the military ever have an accurate accounting of the individual contractors who were in country. This was largely due to the fact that contractors are paid for services provided, not on the basis of the number of individuals it takes to provide the services.

Thus, the military headquarters from the initial outset of the contingency operation must oversee, manage, and prepare to terminate or hand over contracts managed by several organizations (e.g., Corps of Engineers, Logistics Civil Augmentation Program, USCENTCOM Contracting Command) both during and after the transition. Doing this well requires in-depth knowledge of major U.S. government contracts in the host nation, a designated staff lead, and a commitment to transfer that knowledge to the succeeding embassy-led team.

Recommendation 9

Transition planners should engage host-nation officials in planning for use of third-country contractors following departure of U.S. forces because immigration restrictions and political constraints may limit an embassy's ability to use contractors for specific support functions.

The remaining civilian presence in Iraq is also dependent on contractors for security and base support and, if agreed, to provide training and other services to the Iraqis. New contracts had to be in place to support the embassy well before the transition. Embassy Baghdad's reliance on private-sector support also required awareness of host-nation political sensitivities to large numbers of foreign contractors. Contractors cannot be a staffing solution unless the host nation agrees.

Recommendation 10

Future transition efforts should undertake a systematic knowledge management survey and ensure that all databases (military and contracted civilian), key leader engagement logs, assistance project files, and other vital information remain accessible to the follow-on civilian mission.

As responsibility for many functions is handed over from U.S. forces to civilian officials, there is a danger that critical information could be lost. Furthermore, it is important for staff planners and operators to get the right information when shortfalls cannot be made up with manpower and money. As a result, a robust knowledge management effort is very important during and after the drawdown period. The knowledge management effort that Embassy Baghdad and USF-I developed provided a means of identifying and ensuring access to a wide variety of data without collecting it all in one location.

Recommendation 11

Policymakers and commanders in future transitions should resist the temptation to delay final decisions on ending operations to such an extent that rapidly retrograding forces create a power vacuum like the one that may have occurred in Iraq. A more-gradual "waterfall" of troops, contractors, and equipment not only would have been more logistically manageable but might also have contributed to greater political stability in Iraq.

The delay in making the final decision regarding residual forces had multiple causes, the most important being the inability to reach a U.S.-Iraq consensus on the mission of, size, and protections afforded to any U.S. Title 10 military personnel that might remain beyond 2011. Moving the decision point forward from June to October 2011 not only created a monumental logistical challenge associated with the sharp retrograde of military personnel, contractors, and equipment and the accelerated handover of military bases to the Iraqis but also likely exacerbated a power vacuum in Iraq that Prime Minister Maliki and others immediately exploited to gain power over political adversaries. Although the exact motivations are not known, the government of Iraq initiated preemptive measures that had not been attempted previously (e.g., arrest warrants for Vice President Tariq al-Hashimi, Deputy Prime Minister Saleh al-Mutlaq, and others) in the immediate aftermath of the retrograde of U.S. forces. A more-gradual departure of U.S. forces might have reduced both the opportunity and incentives to make such sudden and destabilizing moves.

Recommendation 12

Policymakers, commanders, and planners should use the lessons derived from the final two years of USF-I and its transition efforts to inform critical decisions and time lines required to end large-scale military operations successfully in the future.

Making a decision to go to war is profound. Wars often change combatant countries' internal political and social dynamics and affect regional and international security. How a war is fought will contribute to the postwar security environment. Finally, history shows us that the most important part of a war is how it ends, for that will set the stage for what is to follow. Despite the importance of understanding how wars end, this topic has received far less attention from historians, social scientists, and military strategists than other phases of war. The preponderance of literature about war focuses on how and why wars begin and, once initiated, how battles and campaigns are fought.

This study on how the United States ended the war in Iraq is a first attempt to bridge the gap in strategic and policy thinking regarding how wars end. DoD, the broader national security community, and academia should use these lessons learned to conduct policy relevant research and analysis, including the development of joint doctrine that focuses on the strategic and operational aspects of how wars end.

Summary

It took roughly two years to wrap up a long-term, countrywide military presence in Iraq that, at its peak, involved over 160,000 American troops; a comparably sized army of supporting contractors; and 505 bases and outposts. Political, operational, bureaucratic, and fiscal challenges arose from both the U.S. and Iraqi sides, but Embassy Baghdad was prepared to undertake its primary diplomatic missions when U.S. forces departed. More than a year later, and despite hardships and political upheaval, the embassy continues to manage a multifaceted bilateral relationship that advances the shared political, economic, and strategic interests of both nations—something that was not possible a mere ten years earlier.

It has often been said that all conflicts are *sui generis*. Each conflict has its own set of dynamics that are unlikely to be replicated elsewhere. Each transition therefore must be planned for given the unique opportunities and constraints associated with the particular conflict at hand. However, while the transition process will vary, the key lessons learned from Iraq should be used to inform policymakers and military planners as they devise future transition plans for operations given the particulars of the specific conflicts; U.S. interests; and a broader assessment of the ends, ways, and means necessary and/or available to advance the U.S. interests. The USF-I transition process was uniquely developed for Iraq, but the policy and strategic lessons learned provide important data points that can inform how to end future conflicts.

Comments of Ambassador James Jeffrey

After reviewing the draft manuscript, Ambassador Jeffrey requested that the following personal assessment be added to the report. (Email correspondence between Ambassador James Jeffrey and Charles Ries, January 11, 2013.)

My personal view from three years "in-country" including witnessing the CPA-Embassy and USF-I embassy transitions (and the MACV [Military Assistance Command, Vietnam–to]–U.S. embassy transition in 1973), is that, while Recommendation 1's finding that a transition from military to embassy is one not of degree but of kind, i.e., "fundamentally different," is on target, the problem for COIN is even deeper. In a phase IV Stability Operation, an embassy assuming that violence is still present or potential will not only have to assume extraordinary paramilitary security measures (one of the major components of Embassy Baghdad's "expeditionary" status) but also the huge "nation-building" political and economic goals that are attendant to the U.S. military's conducting COIN operations as specified in the Army/Marine Field Manual 3-24. However, its not just that an embassy is not capable of taking on these functions (i.e., almost impossible to have the resources, the "country-wide presence" or the security to protect such a presence that the military had); its that the move from a military to civilian led operation reveals the fundamental flaws with broad political-economic nation building as part of COIN operations including during Phase IV. It is apparent to me that, even with the military in country, and scores of billions of dollars in aid programs, real sustainable political and economic macro reform is almost impossible to achieve. But as long as the military is there,

1. The huge "inputs" mask the lack of "outputs."
2. The doctrine "requiring" such an effort is so pervasive as to blind observers, including me, to the lack of long-term, sustainable change for the better.
3. Certain short-term or specific "retail" successes achieved—in Iraq, the security forces notably, along with government capacity to at least operate, and the oil sector, albeit much done by the Iraqis themselves—give hope that greater success is "around the corner."

4. The American public and political class's demand for a "home run" transformational or transcendental "success" to balance the inherent doubts about engaging in a COIN "war of choice" discourages realistic expectations and "proof of results."

5. The "strategic weight" and necessity of the American military presence in the eyes of the local government and elites render them "temporarily complacent" about massive U.S. social tinkering with their state and society.

But once the U.S. military is gone, and a civilian presence without any of the listed attempts to "maintain the nation building momentum," it quickly becomes apparent that the "emperor has no clothes." To put it another way, it's not just that an embassy cannot do the same nation-building goals as the U.S. military does, it's that the fundamentally different nature of the embassy (and of the U.S. commitment, vis-à-vis hundreds of thousands of boots on the ground) leads to the removal of the blinders that impede us in seeing that, even with a huge ground uniformed presence real sustainable nation-building is usually (largely) a chimera. The fundamental difference is not in capability (while the military with 150,000 troops and generous funding can do lots of temporary or little things and persuade lots of locals "temporarily" it is no more capable of transforming a state, building a new or different nation, than is an "expeditionary" embassy), but rather in camouflage. As long as the military is busy throughout the country we can kid ourselves that we are making a long-term transformational difference in the country. But while the military is necessary to provide force protection to protect the force that is out amongst the population providing "force protection for civilian work," is it really having fundamental long-term effects? When the military departs, so does the façade of transformational progress.

Joint Campaign Plan—Base Document

USF-I supplied the material in this appendix to RAND to make it widely available.

AMERICAN EMBASSY – BAGHDAD

MULTI-NATIONAL FORCE – IRAQ

2010 JOINT CAMPAIGN PLAN

The entirity of Base Document through page 26 of
the JCP is **UNCLASSIFIED**

AMERICAN EMBASSY- BAGHDAD
BAGHDAD, IRAQ
APOAE 09870

23 November 2009

It has been almost a year since the release of the 2009 Joint Campaign Plan (JCP) and six months since we adjusted its terms of reference to incorporate President Obama's policy guidance. Our vision remains an enduring strategic partnership between the United States and a sovereign, stable, self-reliant Iraq; an Iraq that is committed to just, representative, and accountable governance, and contributes to regional peace and security. Working with our Iraqi partners, our whole of government efforts have allowed the Iraqi people to move toward a promising future. As our mission and U.S. presence in Iraq continue to evolve- underpinned by the sacrifices and tireless efforts of civilian and military members of the U.S. Government (USG)- we must recognize the importance of how we transition this mission in the advancement of our interagency objectives.

The success of our efforts to date and the resulting rapid progress in Iraq's security and stability has led to this most recent JCP revision. The previous plan guided us well through several major changes. The implementation of the Security Agreement (SA) and Strategic Framework Agreement (SFA) fundamentally altered our relationship with the Government of Iraq (Got), moving us closer to the kind of cooperative strategic relationship that we share with other regional and global security partners. In particular, the SA set the stage for the Got's resumption of security responsibilities in cities, towns, and localities throughout Iraq. The 2009 JCP also guided our support to the Gol as it carried out successful provincial and regional elections, reinforcing emerging democratic political norms that are a model for the region. In addition to these major milestones, the JCP oversaw daily USG efforts, incrementally improving Iraqi governance and essential services in villages and neighborhoods across the country.

This latest JCP accounts for the improved conditions across Iraq and the growing capacity of its government. This is an evolutionary rewrite. We have added assumptions with the JCP's focus shifting to the concept and realities of "transition", particularly in Stage 2 (January 1, 2010 through December 31, 2011) of the campaign. Stage 2 goals are attainable and measurable; Stage 3 (January 1, 2012 and beyond) goals remain aspirational. This JCP is aligned with the key principles and cooperative areas outlined in the SA and SFA, is in general agreement with the Embassy's Mission Strategic Plan for Fiscal Year 2011, and will be an invaluable tool for developing future Mission Strategic Plans and United States Forces-Iraq's OPLAN 10-01.

Iraq and our mission supporting it are entering a time of accelerated transition. We will increase the proportion of our security operations devoted to building stability, a reflection of improving conditions across the country brought about through years of focused efforts by American and Iraqi forces. The near future will also see further SFA implementation and bilateral cooperation through the transfer of enduring functions to the American Embassy-Baghdad. These major advances in our campaign have driven the changes documented in this updated JCP. The USG will continue to leverage the full range of our national power to execute this updated plan, jointly orchestrated by U.S. civilian and military authorities. Our updated plan is based on three stages of execution:

Stage 1 – Transition to the New Strategic Environment (January 1, 2009 – December 31, 2009). Stage 1 culminates in the run-up to the national elections. This stage reflects the new operating environment consistent with the authorities contained within the Security Agreement. U.S. combat forces have withdrawn from cities, villages, and localities. A second national election will be held in 2010 as the GoI strengthens its political and military capabilities to deter threats against its sovereignty, political independence, territorial integrity, and constitutional federal democratic system. This stage is witnessing a transition from a USG lead to a partner and, ultimately, to an over-watch posture performing advise, train, assist and enable functions in support of the Iraqi Security Forces.

Stage 2 – Transition to a Stronger Bilateral Relationship (January 1, 2010 – December 31, 2011). This stage envisions Iraq's evolutionary development as a sovereign, stable nation and the planned drawdown of U.S. military forces, ending the combat mission by August 31, 2010, with the exception of targeted counter-terrorism operations. This stage will be characterized by U.S. forces performing training, enabling, and advising functions while helping to enable the Iraqis and assist where necessary.

Stage 3 – Iraq, Strategic Partner (January 1, 2012 and beyond). This stage is characterized by Iraq's evolving and maturing relationship with the United States into a long-term strategic partnership. The defining characteristic of this stage will be Iraq's normalization of relations with the international community.

The road ahead, mapped out in this JCP, contains obstacles to our joint success. The most dangerous are the drivers of instability. The struggle for power and resources, long a part of Iraqi political life, poses the most immediate and dangerous threat to Iraqi progress. Insufficient GoI capacity, a product of decades of neglect and corruption under the previous regime, threatens to limit the speed of Iraq's progress. Both violent extremist groups and malign external interference endanger the newfound and hard-won progress and freedoms won by the Iraqi people. While these drivers of instability are real, they are also manageable. This plan is the key to overcoming these and other obstacles; it provides direction and focus to our interagency efforts. The force moving us in that direction is unchanged, and it is irresistible: the continuing dedication of every civilian and military member of the USG engaged in building a brighter future for the Iraqi people.

Ambassador Christopher R. Hill
Chief of Mission
American Embassy – Baghdad

General Raymond T. Odierno
Commanding General
Multi-National Force-Iraq

CONTENTS

Annex A   Political LOO
Annex B   Economic and Energy LOO
Annex C   Rule of Law LOO
Annex D   Security LOO
Annex E   Strategic Risk
Annex F   Transition
Annex G   Strategic Messaging
Annex H   Intelligence
Annex I   Strategic Engagement
Annex J   Resources and Sustainment
Annex K   Campaign Management and Assessment
Annex L   Balancing Iranian Influence
Annex M   Countering Violent Extremist Organizations
Annex N   Cultural, Education, and Scientific Cooperation
Annex O   Glossary of Terms, Abbreviations, and Taxonomy
Annex P   Amendments to the JCP

EXECUTIVE SUMMARY

This version of the Joint Campaign Plan (JCP) moves our campaign forward, reflecting a new era of transition brought to fruition through the tireless efforts of civilian and military members of the U.S. Government (USG), working in concert with our Iraqi and Coalition partners. Our whole of government effort, guided by previous JCPs, has allowed us to help the Iraqi people pursue their future as a sovereign, stable and self-reliant nation.

President Obama's Camp Lejeune speech and the forthcoming National Security Strategy provide direction implemented in this JCP. This plan is guided by the bilateral Security Agreement (SA) and Strategic Framework Agreement (SFA) and has one overarching goal: produce a long-term and enduring strategic partnership between the United States and a sovereign, stable, and self-reliant Iraq that contributes to the peace and security of the region. This goal is further broken down into goals within three stages. The first stage encompasses the past through December 31, 2009. The second stage, which is the primary focus of this plan, begins on January 1, 2010, around the time of the national elections, and extends through December 31, 2011, the completion date of the Responsible Drawdown in accordance with the SA. The third stage is our aspirational future for Iraq.

Two subordinate documents will flow from the JCP: the State Department's Mission Strategic Plan (MSP) 2012 and United States Forces-Iraq's OPORD 10-01. A comprehensive, interagency strategy, originating with President Obama's speech and the National Security Strategy, reinforced by the SA and SFA, will result in a more focused MSP 2012 and OPORD 10-01.

The JCP is sufficiently detailed, yet flexible enough to be adjusted in the midst of a rapidly evolving strategic environment. It is America's transition document for Iraq. It provides:

- A discussion of strategic context, and the place of the current JCP with respect to the recent history of the U.S. relationship with Iraq;
- A description of the ends, ways, and means of this campaign;
- Three envisioned stages with enumerated goals for each:
  - "Transition to the New Strategic Environment" (through December 31, 2009)
  - "Transition to a Stronger Bilateral Relationship" (through December 31, 2011)
  - "Iraq, a Strategic Partner";
- Four Lines of Operation (LOOs): Political, Economic and Energy, Rule of Law, and Security;
- Specific, integrated JCP goals for each campaign stage, which are further detailed by well-defined conditions, objectives, tasks, and measures of effectiveness/metrics developed along the LOOs; and
- Clearly articulated strategic priorities and risks in achieving JCP goals and objectives.

STRATEGIC GUIDANCE

PREAMBLE

The Strategic Context: Why Iraq Matters

Iraq occupies a central place in the Arab and Muslim worlds. It hosts Shi'a Islam's holiest sites and has a multi-sectarian and multi-ethnic population. Endowed with substantial energy reserves, Iraq could play an increasingly influential role in the global economy. Geographically, Iraq occupies a critical strategic position in a challenging region, bordering the important countries of Jordan, Kuwait, Saudi Arabia, Turkey, and Syria as well as representing the frontier between the Arab and Persian worlds. For the first time in thirty years, Iraq has the promise to play a positive and stabilizing role in the region.

Recent security gains have increased stability in Iraq, but consolidating these gains and ensuring long-term stability require continued strengthening of Iraqi institutions to respond better to the needs of Iraqi citizens. We must continue to help the Government of Iraq (GoI) close the gap between the expectations of its citizens and its own capacity to satisfy their needs. Iraq's leaders must also work across sectarian lines and address major issues, securing a consensus for a common vision for the future of Iraq.

Under the Security Agreement (SA), Iraq has assumed responsibility for its internal security. As we responsibly drawdown U.S. military forces, our security partnership will grow to emphasize training and cooperation, maturing in the long-term into a relationship analogous to that enjoyed with other regional friends and allies. The Strategic Framework Agreement (SFA) serves as the foundation for this long-term partnership. Cooperation under the SFA spans many areas for forging a strong bilateral relationship: political and diplomatic, economic and energy, culture and education, health and environment, information technology and communications, law enforcement and judicial, and defense and security. GoI and U.S. Government (USG) officials may also agree to use SFA fora to share information on U.S. efforts to transition enduring missions and to solicit GoI officials' feedback on those efforts. Using cooperative efforts under both the SA and SFA, we envision an Iraq with capable defense forces, embracing confidence and security building measures, participating in cooperative security programs with its neighbors, and playing a positive and multifaceted role in regional diplomacy.

History

This Joint Campaign Plan (JCP), like its predecessors, focuses the efforts of the United States on strengthening the GoI and narrowing the divide between its capacity and the needs of its citizens. Together with the GoI, we have made immense progress in providing a secure environment for the GoI to mature into a modern state that can bring durable stability and prosperity to the Iraqi people. Through this JCP, and the continuing efforts of the civilian and military personnel dedicated to this mission, the USG) is firmly committed to enabling the GoI to close the gap with a government that is sovereign, stable, self-reliant, and contributing to regional peace and security.

Iraq has made great progress – often at great cost – since 2003 in achieving domestic security, gaining political stability, rejoining the international community, establishing the rule of law, rebuilding essential services for its people, and restoring economic infrastructure. The USG, in a whole of government effort, has worked in partnership with the GoI and the Iraqi people in their quest to build the Iraqi nation.

Achieving domestic security is the most visible, and arguably the essential, precondition for enduring progress in building Iraqi civil society. Security in Iraq has improved dramatically since the "surge" of 2007, but these gains are fragile and reversible. Nonetheless, there has been a dramatic decline in violent incidents, falling from nearly 2,000 per week to a tenth of that number today. This quantitative measure is matched by the qualitative improvement in life for Iraqi citizens, who have resumed a degree of normalcy and are more secure in their streets and in their homes. The expansion of the Iraqi Security Forces (ISF) has been pivotal to this change. The Iraqi Army has grown from only 27,000 in 2004 to over 240,000 today, and the police have grown from 83,000 to over 400,000 in the same period. These units, trained by coalition forces, are primed to secure their nation. Half of the Iraqi Army's units are fully prepared to conduct internal security operations on their own with only enabler support from U.S. forces. The ISF demonstrated this growing competence when it secured Basra during Operation "Charge of the Knights" in 2008, with a continued trend of independent operations growing daily. This emergent capacity led to the successful transition of security responsibility for Iraq's cities, villages, and localities from coalition to Iraqi forces on June 30, 2009. Despite attempts to reignite sectarian violence, insurgent attacks have failed to smother this rebirth of civil life in the ISF-protected streets of Iraq's cities.

The improved security environment has provided the space for Iraq's government to mature toward a model of representative and open government for the region. Iraq has made major strides forward toward political stability through two national and three provincial elections. The first open, free, and fair election in the living memory of Iraqis took place in December 2005, only two months after the adoption of an Iraqi constitution. This great success set the stage for the continuing development of a dynamic political life. This system showed further maturation during the January 2009 provincial elections in which issues-based voting displaced sectarian allegiances. Today, nearly 300 political parties are maneuvering within this representative system in preparation for the upcoming national elections. Iraq's political development is continuing beyond the ballot box, as the Council of Representatives has begun to hold the ministries accountable for their performance, bringing to light corrupt practices and broadly demonstrating the checks and balances essential for a representative democracy. Iraq's political freedom is perhaps clearest when held in contrast to therecent turmoil in Iran. Unlike Tehran's suppression of dissent, the issues facing Iraq, and all the political parties' platforms for the nation's future – including those at odds with the current government – are freely debated by the Iraqi public and reported in the Iraqi press.

9

Establishing the Rule of Law is inseparable from the development of Iraq's domestic political life. Shaking off the corrupt practices of the previous regime and bringing Iraq's justice system into the 21st century in the face of an ongoing insurgency have been particularly difficult, but the GoI and the USG have remained strongly committed to strengthening the Rule of Law. The USG has completed 64 courthouse and correctional facility projects, including several Rule of Law complexes, which provide a secure environment for trying Iraq's most dangerous criminals. The USG has also been instrumental in transporting judges around the country in a traveling judge program, reducing the justice system's case backlog. Additionally, the USG routinely assists Iraqi authorities in obtaining witnesses to aid in the smooth progress of their cases. With an eye toward modernization, the USG has built several forensic labs and worked closely with police and judicial officials to integrate the most updated methods into Iraq's justice system. These are all part of a continuing effort that is building an Iraqi legal structure with Iraqi standards, protecting the people's freedoms and civil liberties, and earning their trust and respect.

With the same determination with which the GoI is building its internal political capacity, it is rejoining the international community. Iraq has named nearly sixty ambassadors and senior representatives to missions abroad. It is working to reduce its foreign debt and to join the World Trade Organization. Regional and international actors are responding to Iraq's efforts, with over a dozen regional states and nearly fifty other international actors represented in Baghdad. While Baghdad's ties to the rest of the world retain room for improvement, it is actively reaching out to its neighbors, aided by the good offices of the United States.

In addition to building enduring representative governmental institutions, the GoI is moving steadily forward in providing for the basic needs of its people. The GoI, with the close and continuous aid of the USG, has made major progress in rebuilding and improving essential services across the broad spectrum of the Iraqi people's needs. Electricity was unreliable in 2003, with Baghdad receiving as little as three hours a day. Today, after the completion of 42 substations and 11 overhead line projects, the majority of provinces receive over 12 hours of electricity a day, with some close to continuous generation. The rehabilitation of almost a dozen sewage plants and the completion of over 800 water projects have dramatically improved the quality of life for over 5.2 million people, bringing clean water to many of them for the first time. Iraqi healthcare has similarly advanced. From a country with endemic measles, respiratory infections, tuberculosis, cholera, and malaria, which affected 3 of every 10 children under five years of age, the USG and GoI have made critical developments to the country's health system. The USG has completed 44 hospital rehabilitation projects and 133 new healthcare centers, bolstering Iraq's system to the point where it can annually treat 6.6 million patients in hospitals and a further 4.6 million in Primary Healthcare Centers. A focused immunization program has reduced measles, the leading cause of vaccine- preventable childhood fatalities, by 90 percent, and there have been no communicable disease outbreaks since 2004. A similar effort has improved education, the essential building block of an enduring democracy, across the country. The USG has repaired or constructed over 1,100 schools, giving hundreds of thousands of students quality facilities in which to learn. The USG and GoI have also nurtured partnerships between five U.S. and ten Iraqi universities, allowing 1,500 Iraqi faculty and students to participate in workshops, training, conferences, and courses in Iraq, the Middle East, Europe, and the United States.

The long-term future of the Iraqi people will depend on a firm economic foundation built on a solid infrastructure. The GoI and the international business community are repairing and expanding Iraq's long-neglected and war-torn infrastructure. Iraq's oil exports, which stood at 1.48 million barrels per day (MBD) in June 2004, today average above 2 MBD, reaching 2.1 MBD during some months. Iraq has also held its first oil auction, which saw the largest field in the country successfully contracted out, showing Iraq's commitment to long-term reintegration with the world energy industry. Furthermore, the USG and GoI have worked closely to rebuild the transportation infrastructure, with five major airport-related and seven seaport-related projects. The USG has also completed 266 road and expressway and 112 railroad renovations. Together, these advances are building a solid foundation for the future expansion of a diversified Iraqi economy, tied into the global economic system.

Iraq has traveled a long way in six years. The price has been high, in lives and resources, paid by Iraqis, Americans, and our coalition partners. Iraq's development has been remarkable, but remains a work in progress.

Today

Our mission in Iraq is transitioning to a strong bilateral relationship. We are now operating in an environment of increased stability and growing Iraqi capacity that is the result of many years of focused effort, as detailed above. In this period of maturing Iraqi sovereignty, our cooperation is within the framework established by the SA, which most recently saw the ISF proudly assume security responsibility for Iraq's cities, villages, and localities. In the larger picture, the SA is moving the USG from leading Iraq's security, to partnering, and increasingly to overwatch. These changes indicate the success of our efforts thus far. We are transitioning to a stronger bilateral relationship, as Iraq develops – with continued USG mentorship – into an increasingly sovereign and stable nation. Our responsible drawdown of forces is underway, putting us on track to end our combat mission by August 31, 2010.

Challenges

Our work is by no means complete. Despite the very real advances of recent months, building a stable and sustainable Iraq across the many dimensions that compose a modern state remains difficult and requires the focused effort of all USG entities. Several specific challenges continue to hinder development in critical areas.

The most critical challenge to the success of our efforts to see a democratic, tolerant, prosperous, and secure Iraq is unresolved ethno-sectarian tensions. These tensions manifest themselves in potential clashes over the distribution of political power, wealth, essential services, and social goods. Today's most visible example is the relation of Arabs and Kurds, seen in the problem of Disputed Internal Boundaries (DIBs), which in 2009 has been a major flashpoint. It must be contained in the likely election and subsequent government formation period of the first half of 2010 to enable the creation of a stable government that includes all key elements of the political process. Moreover, achieving national unity will require greater integration between the Kurdish and central government, including the merging of Peshmerga forces into the ISF. Exploiting this and other seams, a weakened but desperate Al Qaeda in Iraq (AQI) strives to return Sunni and Shi'a groups to the cycle of ethno-sectarian violence that tore Baghdad and the country apart in 2006-2007. AQI's efforts, coupled with the danger of irresponsible behavior by political elites, could lead to a post-election perception of marginalization among the Sunni population (a challenge to the creation of a national basis for politics represented most clearly by the integration of Sons of Iraq into state institutions). Should such a situation arise, it could lead to reversals in the reconciliation trends and the accompanying fall in violence.

At a more fundamental level, the institutions of the GoI and many provincial governments are immature at best, corrupt in the worst cases. Their management and planning, technical operations, finance and budgeting, and promulgation of policy guidance all have failed to take hold with depth and certainty. It is not enough for us to continue to try to present government ministries of provincial governments with models for success; we must identify those key institutions where it is best to focus our time and resources. Iraq's government, like all young democracies, is struggling to define enduring democratic norms consistent with its culture, history, and potential role in the region. The role of sectarianism in government is particularly daunting – ministers who cannot be fired because of their political role, as opposed to ministers responsible to the government for performance, limit much of the effectiveness of our own engagement. Similarly, as noted above, the reconciliation process centered on the Sons of Iraq, and also increasingly encompassing Shi'a groups like the Asa'ib Ahl Haqq (AAH), while a powerful force for the future of democracy, is incomplete. Full reconciliation, and full responsibility of public institutions to the needs of a diverse public, will require continued major effort and development – a challenge as American assets diminish in coming years.

12

Endemic corruption, a huge weakness in the Rule of Law, plagues many aspects of Iraqi life, from daily transactions in the marketplace to major trade and industrial concerns. International investors remain skeptical as long as laws and practices fail to protect investments. This is more than merely a question of leaders who line their pockets. A closed and suspicious mindset, based on the precepts of "oil socialism" under Saddam Hussein's years of rule, and years of uncertainty, leads many in Iraq to resist the transparency necessary for progress. Lack of a dependable and transparent legal framework not only hampers international investment, but undermines public trust in existing institutions. Normal guarantors of the Rule of Law – a local police capacity and functioning judicial systems, trusted by the population – do not yet fully exist. Notably, local police are not improving at the same pace as the rest of the Iraqi Security Forces. Similarly, judges, judicial staff, and defense attorneys still remain in short supply to deliver a fully functional justice system as shown by high case backlogs. These remain priorities for us in the time period of this JCP.

Iraq's economy is not diversified and thus remains fragile, dependent on oil revenues and increased exports. Oil production and export depend on more foreign investment which, as we have noted, is hampered by close-minded attitudes and corruption among Iraqi leaders. Iraq requires a better investment climate, not just to serve the needs of the energy industry, but to allow it to leverage its advantages of population and location and lessen its vulnerability to the swings of commodity markets. In addition, water supplies, a problem for the entire region, require infrastructure development, improved relations with key neighbors, and with luck, the end of a prolonged drought. Without water, Iraq's agricultural prospects will remain bleak. Banking and finance, which could mobilize Iraqi talent and energy in the private sector, are dominated by statist thinking, and private capital still finds it difficult to function in this environment.

Finally, the long-term effects of Iraq's international isolation – not just in recent years but as a direct result of Saddam Hussein's decades long policies – remain a key challenge. This will be the focus of our educational and scientific exchanges: to rebuild the elite structures that can lead the country into a modern society after years of trauma. In addition, our political efforts to engage Iraq as part of our broader regional strategies – in security and in energy – as it reengages with its neighbors and the broader international community in a constructive manner must be a top priority if we are to see Iraq as a positive actor in the region and beyond. This will take serious efforts to breakdown the legacy of years of mistrust between Iraq and its neighbors.

These are but a few of the serious challenges we face: political fragility, economic recalcitrance, and legal immaturity. They force us to set our tasks in realistic and sober terms. We cannot merely look for "more of the same." Our approach must be creative and patient. Achieving Rule of Law must take into account the need for institutional changes that only long-term projects can deliver. Finally, our economic efforts must be based on blunt honesty with our Iraqi partners, assessing for them and with them the need for change.

13

Tomorrow

Today's ally will become tomorrow's strategic partner under the Strategic Framework Agreement. 2010 will be a pivotal year. National elections will test and reinforce the political process that the GoI and the Iraqi people have built with our help. The drawdown of U.S. forces will see the reduction of our current 11 Brigade Combat Team structure down to approximately 50,000 troops in six Advisory and Assistance Brigades, tailored to support our whole of government capacity-building effort across Iraq.

More broadly, 2010 will see the transition of our enduring missions, guided by the SFA and jointly orchestrated by AMEMB-B and USF-I. Security coordination with Iraq will normalize, falling under the purview of an Office of Security Cooperation. The USG's priority will be to help the ISF maintain its internal security and develop a foundational self-defense capability, while seeking support to aid Iraq's regional reintegration.

We will also seek to rebuild Iraq's strategic depth by strengthening Iraq's relationships with the United States and its regional neighbors. This will depend on Iraq's diplomatic and economic reintegration with the international community and global economy, processes that will build on the foundations established over the past six years.

The path from today's transition to tomorrow's Iraq, a cornerstone of regional stability, runs through this Joint Campaign Plan. It will call on the synchronized efforts of the civilian and military elements of the U.S. government. It will demand the continued courage and commitment of our Iraqi partners, and of each civilian, Soldier, Sailor, Airman, Marine, and Coast Guardsman to building a brighter future for the Iraqi people.

Given this vision of future U.S.-Iraqi relations, the JCP details specific ends, ways, and means, better enabling the USG to assist Iraq in becoming a strategic U.S. partner, meeting the needs of the Iraqi people, and contributing positively to the international community.

ENDS (STEADY-STATE)

Our efforts will produce a long-term and enduring strategic partnership between the United States and a sovereign, stable, and self-reliant Iraq that contributes to the peace and security of the region.

WAYS

Understanding that the Strategic Framework Agreement serves as a roadmap for the long-term and enduring strategic partnership, the Campaign Strategy integrates the efforts of the AMEMB-B, USF-I, and other U.S. and international partners across four lines of operations (LOOs): Political, Economic and Energy, Rule of Law, and Security. Working with our Iraqi partners and using available resources through the end of 2011, the USG will increasingly transition the proportion of effort to stability operations. This transition will vary geographically across Iraq, depending on local conditions. The USG will also continue implementation of the SFA by transitioning enduring functions to AMEMB-B and other U.S. and international entities. Additionally, we will assist the GoI to:

- Develop the capacity and legitimacy of its governing institutions;

- Develop professional security forces capable of defending the Iraqi people and neutralizing terrorist threats;

- Neutralize hostile forces operating in Iraq, including AQI, Shi'a and Sunni extremist groups, and former regime elements;

- Transition the proportion of effort to stability operations, although it will vary geographically across Iraq depending on local conditions;

- Balance malign external influence; and

- Address the drivers of Iraqi instability and the prioritized risks associated with them and apply efficient use of limited resources through increased commitment to representative government, effective institutions, political accommodation, and just governance.

MEANS

The United States will comprehensively and deliberately apply the full range of the elements of national power to support Iraq's continued development, while seeking synergy with the application of GoI and international resources. Developments in the strategic environment during the responsible drawdown period will require the USG to continuously reevaluate and reprioritize its efforts and application of increasingly limited resources.

INTERAGENCY STRATEGIC PLANNING

Since 2003, the USG has been engaged in a monumental undertaking in Iraq. Our efforts will prove to be transformational and historic if we focus on retaining gains to date and continue progress towards the goals articulated by the President.

CAMPAIGN STAGES

Transition between stages is both condition- and time-based. Specifically, if all conditions are met we will transition to the next stage; otherwise we will transition on the dates as articulated by stage below.

Stage 1 – <u>Transition to the New Strategic Environment</u> (January 1, 2009 – December 31, 2009). Stage 1 culminates in the run-up to the national elections. This stage reflects the new operating environment consistent with the authorities contained within the Security Agreement. U.S. combat forces have withdrawn from cities, villages, and localities. A second national election will be held in 2010 as the GoI strengthens its political and military capabilities to deter threats against its sovereignty, political independence, territorial integrity and constitutional federal democratic system. This stage is witnessing a transition from a USG lead to a partnership and, ultimately, to an overwatch posture wherein U.S. forces performs train, advise, enable, and assist functions in support of the ISF.

Stage 2 – <u>Transition to a Stronger Bilateral Relationship</u> (January 1, 2010 – December 31, 2011). This stage envisions Iraq's evolutionary development as a sovereign, stable nation and the planned drawdown of U.S. military forces; ending the combat mission by August 31, 2010 with the exception of targeted counter-terrorism operations. This stage will be characterized by U.S. forces performing training, enabling and advising functions while helping to enable the Iraqis and assist, where necessary.

Stage 3 – <u>Iraq, Strategic Partner</u> (January 1, 2012 and beyond). This stage is characterized by Iraq's evolving and maturing relationship with the United States into a long-term strategic partnership. The overriding characteristic of this stage will be Iraq's normalization of relations within the international community.

JCP GOALS BY STAGE

Stage 1: January 1, 2009 – December 31, 2009

GOAL 1: Develop a legitimate and participatory governance that:
GOAL 1A: Demonstrates broad political participation and accommodation
GOAL 1B: Reflects operational government institutions
GOAL 1C: Provides essential services
GOAL 1D: Improves security of judicial personnel
GOAL 2: Develop an Iraq that is capable of internal and external defense that:
GOAL 2A: Builds capacity to protect the population nationwide
GOAL 2B: Builds capacity to protect critical infrastructure nationwide
GOAL 2C: Partners with USF to sustain pressure on networks in order to neutralize AQI, terrorists, violent extremists, and irreconcilables
GOAL 2D: Develops an increasingly competent, capable, and professional ISF

and security-related ministries that are increasingly guided by the Rule of Law
GOAL 3: Establish an economy that has begun to diversify that:
GOAL 3A: Improves flow of oil
GOAL 3B: Improves agriculture
GOAL 4: Develop a sovereign, self-reliant Iraq that contributes to peace and stability in the region that:
GOAL 4A: Counters malign external influence
GOAL 4B: Advances external political accommodations
GOAL 4C: Develops a strategic partnership with the United States

Stage 2: January 1, 2010 – December 31, 2011

GOAL 1: Advance toward an ethical, competent, and participatory governance that:
GOAL 1A: Demonstrates continued broad political participation and accommodation
GOAL 1B: Reflects increasingly effective government institutions
GOAL 1C: Provides reliable access to essential services
GOAL 1D: Establishes an effective and impartial legal system
GOAL 2: Achieve an Iraq, capable of internal defense and developing a foundational defense against external threats, that:
GOAL 2A: Continues to improve GoI capacity to protect the population
GOAL 2B: Continues to improve capacity to protect critical infrastructure nationwide
GOAL 2C: In addition to partnering with USF, builds its own capacity to sustain pressure on networks in order to neutralize AQI, terrorists, violent extremists, and irreconcilables
GOAL 2D: Maintains internal security with a developing foundational capability and capacity to deter and defend against external threats through an increasingly professional ISF and security-related ministries
GOAL 3: Build a diversifying economy that has begun to:
GOAL 3A: Significantly increase petroleum and gas production
GOAL 3B: Leverage hydrocarbon profits to stimulate other sectors of the

national economy
GOAL 3C: Build an agricultural sector that is trending towards regional competitiveness and sustainability

GOAL 4: Achieve a sovereign, self-reliant Iraq that contributes to peace and stability in the region and beyond that:
GOAL 4A: Degrades the negative impacts of regional malign actors
GOAL 4B: Pursues Iraqi interests constructively in the region and the international community
GOAL 4C: Engages in a long-term strategic partnership with the United States

Stage 3: January 1, 2012 and beyond

GOAL 1: Achieve and sustain an ethical, competent, and participatory governance that:
GOAL 1A: Practices participatory governance and political accommodation among all political factions, which participate peacefully in the democratic process
GOAL 1B: Maintains public confidence in government institutions
GOAL 1C: Provides reliable access to essential services for all Iraqis
GOAL 1D: Operates an equitable/effective legal system that ensures the state is subject to/complies with the law

GOAL 2: Achieve an Iraq that is capable of internal and external defense that:
GOAL 2A: Protects the population and infrastructure through mature law and order processes
GOAL 2B: Protects critical infrastructure nationwide
GOAL 2C: Deals with low levels of violence and extremism, competent and capable of maintaining internal security
GOAL 2D: Maintains internal security; demonstrates self reliance and an ability to deter and defend external threats with a professional ISF and security-related ministries

GOAL 3: Advance and sustain a diversifying economy that:
GOAL 3A: Realizes potential as one of the top petroleum and gas exporters in the world (sustained by hydrocarbon revenue)
GOAL 3B: Achieves a competitive, sustainable, and globally integrated

agricultural sector
GOAL 4: Advance and sustain a sovereign, self-reliant Iraq that contributes to peace and stability in the region that:
GOAL 4A: Is able to neutralize negative influences of malign regional actors
GOAL 4B: Contributes positively in the region and the international community
GOAL 4C: Demonstrates that it is a state with strategic depth; at peace with its neighbors and a U.S. ally in the long-term fight against international terrorist groups

EXECUTION

SITUATION

General

The JCP supports the President's policy objectives for Iraq. The strategic environment in Iraq continues to be dynamic. Several factors will affect future execution of our mission and set the conditions for an enduring strategic partnership between the United States and Iraq. These include the responsible drawdown of American forces from Iraq, continued implementation of the Strategic Framework and Security Agreements, successful national elections with a corresponding peaceful transfer of power, and ongoing security challenges with violent extremist organizations.

Stability is improving but is not yet enduring. As we face 2010 and beyond, the USG will use a whole of government approach to build Iraq's governing organizations as legitimate, representative, and effective institutions serving all Iraqi citizens. The GoI has demonstrated increased good governance, although Iraq remains a nascent state. National unity and sustainable stability depend on an undeveloped national vision and a fragile economy buffeted by global economic market forces.

Although the priority of JCP efforts will be focused on near-term security, stability, and civil well-being, we must continue to foster broader GoI capacity in foreign relations and international partnership. Presenting an increasingly stable post-war Iraq and promoting trade will greatly benefit the USG, GoI, the region, and the wider international community. Consequently, we will continue to help Iraq lift the sanctions imposed by United Nations Security Council Resolution (UNSCR) 661 (economic sanctions against Iraq) and to ameliorate supporting resolutions, such as UNSCR 687 (missile and weapons of mass destruction), which perpetuate the former regime's negative image. As Iraq has rejoined the international community, it has acceded to major treaties pertaining to the non-proliferation of nuclear, chemical, and biological weapons. Ensuring Iraq can comply with treaty stipulations will assist in both making the world safer and lifting United Nations restrictions on Iraq.

Transition

In 2010 and 2011, the U.S. Government is fundamentally changing its interaction with the GoI, moving toward an enduring relationship between sovereign nations. The GoI continues to improve its capability to provide security, essential services, effective governance, and a functioning legal system for its citizens. As it does so, American military and civil authorities will shape and encourage these developments, and in doing so, reshape our own interagency efforts. In line with this steady progress, and in keeping with our bilateral agreements, the U.S. footprint will change dramatically in the next two years.

The transition annex outlines some of these key changes. At the center of the transition effort, the military has begun the process of canvassing, categorizing, and defining a "handover" process of functions it now performs. This "handover" exercise will prioritize whether these functions can and should be turned over to American civilians, Iraqi entities, multilateral or private institutions – or terminated. The Embassy will take on many of the tasks which, up to now, have been performed by the military, such as establishing an Office of Security Cooperation that will address many of the issues currently tackled by Multi-National Security Transition Command-Iraq and a Defense Attaché's Office that will take on representational and military intelligence functions now covered in large part by Multi-National Force-Iraq. Outside of Baghdad, the Provincial Reconstruction Teams (PRTs) will not only reduce in number but transform in function, differentiating among themselves based on U.S. policy priorities in different parts of Iraq. As the military draws down, the Embassy's primacy in political, economic, and rule of law lines of operation will become manifest, as planning for key leader engagement and influence is transformed by systematic joint Embassy-USF-I prioritization; the military's Embassy presence, through J-9, will diminish and focus on key areas of common messaging and provincial outreach. As offices like the Iraq Transition Assistance Office close their doors and as program funding falls off sharply, project management authorities will focus more on oversight of existing projects and less on initiation of new ones, reinforcing the overall message that a sovereign Iraq must now take control of its own development, from infrastructure to institutional reform. The relative importance of civilian assistance agencies, such as United States Agency for International Development (USAID), will increase as Defense Department programs such as Commander's Emergency Response Program wind down.

Many of the elements of this transition depend on future national guidance. Without pre-supposing this guidance, it is important that military and civilians embed transitional changes throughout the tasks of the JCP in 2010 and 2011, so that each line of operation defines its tasks not only by measuring progress in substantive areas but recognizing the impact of anticipated structural changes (funding, personnel, physical plant). Further, as the JCP guides this process internally among the American participants, strategic messaging, public affairs, and key leader engagement must reflect these developments in communication with Iraqi authorities and the Iraqi people: that a sovereign, independent, secure, and prosperous Iraq has taken control of its institutions, and that a flexible, responsive, and supportive American presence has adjusted to do all it can to support efforts to ensure our common success.

Responsible Drawdown

A sovereign, stable, and self-reliant Iraq is essential to the U.S. strategic vision for Iraq. The only enduring security solution is the creation of viable representative governmental institutions, ones which achieve internal and external political accommodation. It is therefore vital that the United States maintain our political and diplomatic support for Iraq while this young democracy asserts its sovereignty. This Joint Campaign Plan requires the focused efforts of both USF-I and AMEMB-B for its successful execution. The drawdown progressively transitions functions and responsibilities from the military to civilian agencies. Maintaining continuity of operations over this period is vital. The intent of the JCP is to seamlessly guide this transition while continuing progress toward long-term U.S. goals.

Center of Gravity

The center of gravity remains the independence, competence, and capacity of Iraq's government institutions and the accountability of those institutions to the Iraqi people. To sustain confidence in the democratic process, and to promote a stable and prosperous future, the Iraqi government must demonstrate its ability to meet the needs of the Iraqi people in a competent, non-sectarian manner.

Risk and Drivers of Instability

The greatest risk to the accomplishment of campaign goals is the failure of the Iraqi state. "Failure of the State" refers to the GoI's inability to develop sustainable capacity, capability, and constitutional governance. "Drivers of Instability" and the potential for critical miscalculations can greatly impact the progress and direction of the campaign, and potentially derail the plan. Thus, an understanding of the root causes of instability, their impact on the operational environment, how they can lead to "Failure of the State", and possible mitigations remain essential. The "Drivers of Instability" and their associated indicators, "Critical Miscalculations", and possible mitigations are discussed in depth in Annex E.

ASSUMPTIONS

1. The bilateral Security Agreement between the United States and Iraq will remain in effect through December 31, 2011.

2. External military aggression by conventional forces will not significantly threaten Iraqi sovereignty while U.S. Forces are operating in or near Iraq.

3. Sufficient resources, especially funding and personnel (both military and civilian), will be available to pursue the goals and objectives set out in this JCP.

4. The Department of State will continue to staff PRTs at appropriate locations throughout Iraq in order to achieve desired effects through at least December 31, 2011.

5. USAID will maintain programs at national, provincial, and local levels until at least December 31, 2011.

6. U.S. combat operations will end by August 31, 2010.

7. The January 2010 Iraqi national election will be credible and legitimate in the

eyes of the international community.

8. The GoI will not develop an internal plan for post-election transition to new government.

9. The ISF will maintain or improve on current security levels.

10. Sufficient GoI resources, especially funding and skilled personnel, will be available to pursue the goals and objectives set out in this JCP.

11. The new Iraqi government will be a strategic partner.

12. The Strategic Framework Agreement between the United States and Iraq will remain in effect.

MISSION

In order to develop a strategic partnership with a sovereign, stable and self- reliant Iraq that is a contributor to peace and stability in the region and beyond, the USG employs integrated political, economic and energy, rule of law, and security means to support Iraq in:

- Developing increased commitment to a just, representative and accountable government through effective institutions, and political accommodation;

- Countering regional negative influence, building relations with neighbors and other gulf countries, and contributing constructively to the international community;

- Providing essential services and employment opportunities for all Iraqis, and developing a self-reliant and diversified economy;

- Developing and sustaining an effective and impartial legal system coupled with independent institutions that ensure the state is subject to, and complies with, the law; and

- Protecting the Iraqi population, securing critical infrastructure, and developing a professional security sector capable of providing for Iraq's security.

CONCEPT OF OPERATIONS

Purpose

Building on the success to date in protecting the Iraqi population, the United States will continue to assist the Republic of Iraq in developing governing institutions that are just, representative, and accountable, are responsive and effective, are based on the rule of law, and serve the Iraqi people. Our efforts will produce a long-term and enduring strategic partnership between the United States and Iraq that provides for Iraqi internal stability and contributes to regional security – a matter of vital national interest to both nations.

Method

The USG will continue to encourage the international community to help the GoI increase its capacity by building capable and accountable national and provincial institutions that bridge ethnic and sectarian divides and balance negative external influence. The campaign will transition towards stabilization and enabling civil authorities. This will take place throughout Iraq as local conditions permit. The United States will work with the international community to develop of Iraqi capabilities, lift international sanctions, and promote a legitimate and accountable GoI. Our strategy will support Iraq across four integrated LOOs: Political, Economic and Energy, Rule of Law, and Security. Transparent, just, representative and accountable government is the key to protecting and nurturing the campaign center of gravity; building it is our task. Nonetheless, U.S. and Iraqi forces must work together to improve security conditions, allowing continued advances along all LOOs.

Coordinating Instructions

The U.S. strategy through 2011 will be to support Iraq across four integrated LOOs: Political, Economic/Energy, Rule of Law, and Security. The LOOs all support the development of the Iraqi government's capacity, promoting good governance, fostering international partnership and enabling increased security. The USG must coordinate and integrate its efforts to achieve campaign objectives. The USG will support improving national, provincial, and local governmental capacity, executing budget and capital projects, and developing infrastructure. The GoI's maturing sovereignty following the expiration of UNCSR 1790, and later redeployment of U.S. troops from Iraqi cities, villages and localities fundamentally changed the operational conditions. These dynamic circumstances will continue as an all-U.S. forces mission (USF-I) succeeds the Multi-National Force-Iraq coalition, the responsible drawdown continues, full-spectrum operations decisively shift toward stability operations, and civilian agencies assume enduring functions.

Integration of the Lines of Operation

All LOOs (Political, Economic and Energy, Rule of Law, and Security) and supporting activities must work in concert to achieve U.S. goals. The campaign's success is contingent on deliberate and coordinated pursuit of LOO activities. For example, changes in the security environment can powerfully influence the success of PRT capacity-building efforts. LOO owners must be aware of the operational environment, and agile enough to take advantage of emergent opportunities. LOO owners must regularly evaluate the impact of their LOO on the other LOOs, and the effect of the other LOOs on their own.

Prioritizing Resources

The new operational environment is characterized by changing authorities and shifting resources. All agencies must eliminate duplication. We must identify which activities we can discontinue, which can be passed to the GoI, and which can be transitioned to other governmental or non-governmental organizations. Only then can we decide what the USG must do and what resources are required. Successive iterations of the JCP have sought to refine and prioritize the activities undertaken in the various LOOs, and this JCP provides further focus. The goals, conditions, and objectives in this version of the JCP reflect the priorities defined by the LOO owners and focus efforts where results will be delivered. This plan, and other planning documents including but not limited to the Mission Strategic Plan and Operational Orders, must be mutually supporting and reinforcing.

RESOURCES AND SUSTAINMENT

USF-I, in coordination with the AMEMB-B, must logistically support U.S. and Iraqi forces throughout the Iraqi Joint Operating Area. Both forces, with the ISF in the lead, will be able to support the JCP's enduring goal in maintaining national and regional stability, building a long-term partnership. USF-I must remain constantly aware of the fluid operating environment, shifting resources as needed to support all LOOs. As USF-I conducts sustainment operations, it must maintain close links to other organizations that oversee seaports, airfields, basing facilities, and lines of communication with neighboring countries. This infrastructure will help build ISF capacity, and serve the Iraqi people by laying the foundation for a stable and regionally-integrated economy.

COMMAND AND CONTROL

- The U.S. Chief of Mission and the Commanding General, USF-I jointly lead the United States Government efforts in the execution of this plan. The AMEMB-B is the supported USG agency for governmental activities, with Department of Defense, specifically USF-I, as the supporting agency.

- The Chief of Mission is the personal representative of the President of the United States in Iraq and is responsible for execution of all U.S. non-military activities in Iraq that have not been specifically assigned to USF-I.

- The Commanding General, USF-I commands all military forces within Iraq and is the senior U.S. military representative.
  - USF-I is under the Operational Control (OPCON) of Commander (CDR), U.S. Central Command.

25

o USF-I provides support to the GoI, AMEMB-B, the UN Assistance Mission in Iraq, NATO Training Mission – Iraq, and other international organizations.

o For U.S. Department of Defense purposes, USF-I is the supported command for building the ISF within Iraq. All other component commanders and civilian agencies are supporting organizations.

- Command Posts
  o The Chief of Mission operates out of the U.S Embassy.

  o The CG, USF-I, commands and controls operations from Headquarters (HQ), USF-I located at Victory Base Complex. As a consequence of the need for close liaison, consultation, and coordination with the GoI and USF-I will maintain a forward HQ collocated with the Ambassador and an office in the AMEMB-B.

JCP Annex F—Transition

USF-I supplied the material in this appendix to RAND to make it widely available.

ANNEX F

TRANSITION

Reference: See Base Document dated Nov 2009.

1. **SITUATION.**

A. Vision: This Annex addresses the effort to create a framework that enables United States Forces-Iraq (USF-I) and American Embassy – Baghdad (AMEMB-B) to methodically transfer enduring relationships and crucial programs and projects to a non- USF-I entity. The broad transition process at work within the U.S. Government (USG) agencies in Iraq will mark the shift from an effort grounded primarily in the U.S.-Iraqi bilateral Security Agreement (SA) to one based increasingly on the Strategic Framework Agreement (SFA). Consequently, throughout Stage 2 of the Joint Campaign Plan (JCP) – which is 2010-2011 – all aspects of the military and civilian effort in Iraq will focus on transition.

In each of the lines of operation (LOOs), and as detailed in each of the annexes of the plan, the impact of transition from military to civilian lead – and ultimately to greater exercise of Iraqi sovereignty – has a role that permeates all other tasks. This annex is somewhat different from the other annexes, as it is more descriptive of a process that is common to all and illustrates key areas that will change in 2010 and 2011; after this introduction, it will go into some detail on the exercise of defining tasks that must move from the military to civilian, Iraqi, or other (third-country, multilateral, or non- governmental organization (NGO)) leads.

As USF-I takes shape in the first months of 2010, and then addresses the responsible drawdown in earnest from March to August, AMEMB-B will make adjustments accordingly. The Multi-National Security Transition Command-Iraq (MNSTC-I) will become the Iraqi Train and Assist Mission (ITAM) and the Iraqi Security Assistance Mission (ISAM) within the broader USF-I structure. In corresponding fashion, the Embassy's Political-Military Affairs Section will continue to monitor the work on training and assistance while standing up an Office of Security Cooperation (OSC) that will be able to assume these tasks beginning in summer 2011. The OSC's personnel and mandate will be defined by its tasks rather than created, a priori, by a policy decision; but in general terms, the Embassy and USF-I expect that by summer 2011, a staff of 100-300 OSC personnel (mainly military personnel assigned to Iraq under Chief of Mission authority) will take over a subset of the training and assistance functions currently overseen by MNSTC-I with the help of the Political-Military Affairs Section. The Embassy's Office of Transition Assistance will oversee some budgetary and programmatic elements of this change.

Similarly, in the summer of 2010, a Defense Attaché Office (DAO) will begin to take on the liaison and military intelligence function now performed by a number of

F-1

elementsof USF-I, to include tasks from J-2 and J-3 through some of the protocol tasks of J-9. With the maturation of this DAO, and the creation of an OSC (both during the "school year" of 2010-2011), AMEMB-B expects that the Political-Military Affairs Section can downsize significantly, and set as its target integration as a much smaller sub-unit into the Political Affairs Section, perhaps as early as fall 2012.

Police training is one of the most fully-developed examples of transition. Currently performed by MNSTC-I, this set of programs concentrates on teaching the fundamentals of police work to a broad spectrum of candidates from Iraqi law enforcement institutions. More than one thousand U.S. military and contractors have trained Iraqi law enforcement officers in a wide variety of skills, making the Iraqi police an effective force for order. Working together, U.S. military and civilian authorities have decided that future police training should build upon, rather than merely continue, what has been achieved up to now. This is not only because the Embassy, under the Department of State International Narcotics and Law Enforcement Affairs (INL) Bureau's lead, has different skills than the current U.S. military-led trainers, but also, because all have realized that teaching executive skills to police leaders is the next step in ensuring a successful program. The focus of the State Department INL police capacity-building program, which will be up and running in summer 2011, will be teaching the executive oversight of police institutions necessary to sustain proper supervision and governance of police; the U.S. military's program will decrease and ultimately finish its task by the end of 2011. This illustrates two key points: future work will be downsized because U.S. military resources (necessary to the previous task, now largely achieved, of teaching fundamental skills) will no longer be available; but also, because the more focused, smaller project of the Department of State under INL lead will reflect the needs of a more sophisticated and developed police institution – in place thanks to MNSTC-I's successful efforts. This institutional transition also will parallel the shift from primarily Iraq Security Forces Funds (ISFF)-based funding for Iraqi police training activities to a funding stream built primarily on International Narcotics Control and Law Enforcement (INCLE) funds.

The J-9 Directorate will be the USF-I commander's direct link for civil-military partnership with AMEMB-B, providing support in the political and economic areas, civil capacity, and closely coordinating strategic messaging. Once a new Iraqi government is formed in 2010 after the anticipated January elections, J-9 will phase out its support of the Embassy's monitoring of the Council of Representatives and Ministries. In addition, J-9 economic development functions, including support of business development, agricultural reform, and associated water resources management, will greatly reduce in scope during 2010 and phase out completely by August 31, 2011. The same will happen to J-9's Public Finance Management Action Group (PFMAG), which will end its support of Embassy financial and budgetary assistance to Iraqi ministries in 2011. In Stage 2, in the context of responsible drawdown, J-9 will also ensure unity of effort among tactical military units and the Provincial Reconstruction Teams (PRTs). Many ofJ-9's current efforts to assist the government of Iraq at the national ministry level will reduce, giving way to Embassy-only execution.

The PRTs, for their part, also will go through a transition in 2010-2011. The current 23 PRTs and embedded PRTs will reduce to 16 PRTs by the summer of 2010: one regional PRT based in Irbil serving Kurdistan, and one in each remaining province. Depending on policy decisions in Washington that we expect to be completed by the end of Stage 1, we will know how many of these 16 PRTs will remain by the summer of 2011 and begin to plan on the gradual phasing out of those that will not remain. These could be as many as 7 posts and as few as 2; the Office of Provincial Affairs (OPA) at AMEMB-B is projecting that the current "one size fits all" profile for the 16 PRTs will evolve into specialized profiles (in function and thus in staffing) by the end of the 2010-2011 period. That is to say, some "presence posts" may end up being primarily commercial platforms, and others, as political eyes-and-ears locations, depending on need. OPA itself with downsize, depending on the need for its support as directed by Washington decision-makers.

Additionally, the Iraq Transition Assistance Office (ITAO), the successor to the Iraq Reconstruction Management Office (IRMO) created to oversee and implement almost $18 billion in reconstruction work throughout Iraq, will conclude its operations in May 2010. The remaining ITAO programs and projects will be transferred into OPA and will be concluded by mid-2011. Future civilian assistance will focus on building the capacity of the Government of Iraq (GoI) to budget for and manage reconstruction activities, improved democracy and governance, and encouraging economic growth. Some small reconstruction assistance at the provincial level may continue with the Quick Response Fund (QRF) administered by OPA through Fiscal Year 2011.

Another important aspect of AMEMB-B transition will be the Mission's institution of International Cooperative Administrative Support Services (ICASS) procedures starting virtually in 2011 and formally coming into practice in 2013. This process could affect non-State Department U.S. civilian representation at post and could result in a growth of State Department officers in such areas as Economic Affairs and INL sections. Additionally, normalized consular relations will almost certainly result in increased State Department consular staffing to support increased non-immigrant visa issuance, immigrant visa work, and American Citizen Services expansion. Thus, as the number of contractors working in Iraq declines, the number of Foreign Service employees may increase or at least reapportion themselves to new tasks.

Programmatic transition and development assistance transformation and conclusion will be a baseline activity throughout Stage 2 as large numbers of Commander's Emergency Response Program (CERP) and other Defense Department-financed projects will be concluded or transferred to the GoI or international community with a residual package being assumed by the U.S. interagency community.

F-3

Therefore, at the end of Stage 1, a key task will be defining in a systematic way how both USF-I and AMEMB-B will sort their way through tasks currently performed by the military so that the future direction of these activities can be determined authoritatively. This disposition process (flow chart in Appendix 3 – Transition Process) will canvass the military systematically, inside Baghdad (where many projects are focused on ministries or other central institutions) and outside Baghdad (where, aided by Brigade Combat Teams and overseen by PRTs, projects generally focus on provincial or district authorities and institutions) to assess the strategic value of projects, provide a means by which to prioritize their importance, and then decide whether U.S. military, U.S. civilian, Iraqi governmental or non-governmental, or third-country/multinational organizations should take on such a project; in many cases, this disposition exercise may recommend termination, if the GoI decides the project is not a priority and does not commit to its sustainment.

The disposition process will be divided into several phases: (1) define; (2) measure; (3) analyze; (4) implement; and (5) assess.

The "define" and "measure" phases concluded by late October 2009. Following the establishment of appropriate working groups staffed by civilian and military subject matter experts, the "analysis phase" will begin. After review by the Interagency Planning Group, the Executive Core Group composed of senior USF-I and Embassy leaders will determine disposition. This recurring cycle will continue until disposition of all cases is complete.

Once validated by senior leaders, case "implementation" will start, beginning with a general outreach program to senior GoI and international community leaders and then – possibly in part through Strategic Framework Agreement structures – individual case presentations to potential recipients will start. We foresee this process commencing no later than March 2010. The goal is to complete the "implementation" phase by October 2011 with a significant percentage of the cases complete by June 2011.

As cases are implemented, a subsequent assessment process will be undertaken to confirm that case disposition is tracking as originally intended and to develop lessons learned to provide feedback to the continuing adjudication and validation.

More generally, we should point out that much of the transition addressed here is "internal" transition: the structural and programmatic adjustments to a smaller U.S. military presence, the restructuring of AMEMB-B, and the adjustment of people and programs to the new needs of a sovereign Iraq. Nevertheless, there is an element of transition that is "external" that needs greater work, and this will be the task of the substantive offices of the Embassy, all transition-based offices, the new J-9, and indeed the American leadership throughout the common Mission. This external transition involves the refinement of Key Leader Engagement (KLE) policies between the leadership of USF-I and the Embassy, so that the transfer of authorities from American to Iraqi lead is well articulated and communicated to the Iraqis. This is not simply a question of talking points but of a reassessment of who meets with whom and how often and an even deeper coordination of public message between the AMEMB-B's Public Affairs Section and the J-9. As noted above, for example, when we have decided, though the disposition process, American priorities for projects aimed at Iraq's democracy, prosperity, and security, the next task will be to communicate these priorities – and to assist the Iraqis to make clear their own intentions – with a common outreach project that will describe for elites and common Iraqis alike how we will continue to work as partners throughout this period and into Stage 3 (beyond the end of 2011).

This massive undertaking of transitioning U.S. military efforts in Iraq is inextricably linked to the responsible drawdown of troops that will be completed by the end of 2011. However, it must be viewed from a longer perspective. The creation of positive effects through the transition is essential to the condition of Iraq in 2012 and beyond. It is incumbent upon those involved in this effort to always strive to consider a vision of a stable and secure Iraq that is a viable and peaceful sovereign nation. While each will undoubtedly have their own interpretation of this vision, only through collective energy and drive will a positive outcome result. This outcome cannot be precisely predicted or calculated, but it can be accomplished. Each separate transition of an activity needs to be treated as an integral part of Iraq's future. These activities should not be managed as simple business transactions but rather as complex interactions between entities that will entail thorough and focused execution.

Transition is not simply a passing of responsibility for current activities from USF-I to AMEMB-B. Rather, it is a "whole of government" approach towards setting conditions for the success of the GoI and the Iraqi people. Therefore, this approach will necessitate a great amount of work but the reward will be commensurate. It will require sharp awareness of the overarching vision and scope while we develop manageable deliverables. We can visualize this as the passing of batons in a relay race. Passing even one baton is difficult and takes precise teamwork. Nevertheless, multiple races can be run by multiple teams with most batons passed successfully, and even those dropped can still be picked up and taken across the finish line. Each USF-I activity is a baton to be transferred, transformed, completed or terminated.

B. Environment:

   (1) The influences of the many cultures present in this environment will play a preeminent role in the conduct of this endeavor. Obviously, there are the general aspects of the Iraqi and other national cultures that need to be considered. Beyond the national cultures, we must consider and analyze the effects of the various subcultures of tribes, organizations, institutions, etc. Careful thought should be given to the approach towards each transition as to how will the recipient or recipients respond to the event. There may be second or third order effects that if not included in analysis could cause a specific transition to be more difficult than necessary if not

impossible or in the worst case lead to negative results. Additionally, the differences between cultures and sub-cultures need to be recognized in order to find the most effective combination to resolve a particular problem set. These differences do not only exist eternally to the USG but internally as well. As the whole of government approach is developed, it is necessary that the traits of organizations and individuals be optimized in order to maximize strengths and minimize weaknesses for each unique situation. In a sense, the USG must develop the best team possible to run each of the various relay races which will be run over different lengths and courses.

(2) A major factor that can only be conjectured upon now is the nature and composition of the Iraqi government post the 2010 election. Current relationships will certainly change as the new government is seated. The formulation of transition teams based upon current relationships with GoI representatives will be problematic at that point. Moreover, other entities may not have finalized their policy towards Iraq during this period. This does not mean that transition of activities during this period will not occur or need to, just that the bulk will not begin until approximately June or July of 2010. At that time it will be vitally important that careful attention is paid to developing relationships with the GoI that are conducive to transition. While the political philosophy of this new government remains to be seen, it is most likely safe to assume that many of its members will be new to their jobs, if not government service in general. It will be necessary to proceed methodically and judiciously in order to avoid placing undue stress upon individuals and organizations during this initial developmental stage for the GoI.

(3) During this same time period, USF-I will be drawing down the amount of military forces in Iraq. This will naturally lead to changes in the environment as well. These changes will occur at the micro as well as the macro levels. USF-I will be dedicating much of its energy and assets to current operations plus the significant logistical effort required to drawdown forces. Moreover; individuals and specified units will continue to depart on schedule as their tours are concluded. Great care will need to be taken to ensure that knowledge is not lost simply due to an inevitable departure. Military members and units must not take an approach of simply marking time until their departure and therefore allow activity to just fade away. Transition must be treated as a mission to be accomplished with the requisite attention to planning, executing, and assessing that a mission entails. Conversely, it will be incumbent upon other USG representatives to recognize the need for urgency demonstrated by the military and incorporate this need into their analysis and approach to transition. The approaching departure of troops can be leveraged in a positive manner during negotiations with possible recipients as long as all are aware and prepared.

C. Potential Participants:

(1) GoI

(2) AMEMB-B – including U.S. Agency for International Development

(3) U.S. Central Command (CENTCOM)

(4) Inter-governmental Organizations (IGOs) – United Nations / World Bank / International Monetary Fund / North Atlantic Treaty Organization / Gulf Cooperation Council

(5) Other nations' governments

(6) Non-governmental Organizations (NGOs)

(7) For-profit entities

2. **APPROACH.** The objective of this initiative is to transition all JCP tasks and USF-I programs, projects and relationships (collectively known as "activities") no later than December 31, 2011. Basically, this will involve the assessment of the existing JCP tasks and transferring USF-I enduring activities to a non-USF-I entity. The success of this effort will be measured in the ability to transition the tasks and activities without jeopardizing the conditions and goals specified in the JCP. This endeavor will require a collaborative approach that is inclusive and leverages the unique attributes of all entities involved. Such a methodology necessitates interagency participation and solutions to aspects of this complex undertaking. Moreover, it is imperative that international governmental and non-governmental organizations be engaged whenever it is practical and facilitates an uninterrupted continuation of functions and programs. Proactive engagement by and with the GoI will be absolutely vital to a successful transition. Consideration of Iraqi expectations will continue to be an integral part of the transition program and determination of enduring projects. This review will require both subjective and objective analysis in order to make the most effective decisions that continue to enhance the goals of the Joint Campaign plan. Resource limitations will be considered and even more importantly, the expertise of experienced diplomatic and military leaders must be leveraged.

A. **Transition Scope.**

    (1) **In-Scope**

        (a) A thorough analysis and disposition of all JCP tasks. This will entail an extensive assessment that will help identify enduring JCP goals. Understanding the enduring goals that are crucial to the success of the JCP will assist in properly adjudicating each USF-I activity.

        (b) A thorough analysis and disposition of all tasks, programs, projects and relationships that are currently owned, performed or managed by a USF-I entity. It also includes activities where USF-I is the supporting organization to a lead entity (e.g. AMEMB-B).

i. Tasks – Those that are considered enduring and necessary under the JCP and require disposition to a non-USF-I entity prior to the redeployment of the USF-I owner. Examples: Economic Development Strategy; OPA Operations and Planning Support.

ii. Programs – A planned and coordinated group of activities designed to achieve an operational and/or strategic objective that is typically long-term or enduring. Examples: Commanders Emergency Response Program projects; Freedom of Information Act Program.

iii. Projects – Typically construction in nature (e.g., U.S. Army Corps of Engineers Gulf Region Division projects), ventures that have been awarded but require continuing manpower support due to redeployment of the initial lead unit.

iv. Relationships (government or civil) – Key relationships with Iraqi officials that set the conditions for achieving JCP goals. Example: CJ9 relationship with the National Investment Commission.

(2) **Out-of-scope**

(a) All activities that are force internal (e.g., providing religious support to the force).

(b) All programs or projects that will be completed before September 1, 2010. (c)

All activities outside the Iraq Joint Operations Area (IJOA).

3. **METHODOLOGY.**

A. **Framework.** The framework for transition is structured around three dimensions:

(1) Organization – Identifies all the key teams and working groups involved in the transition effort.

(2) Methodology – Outlines the sequence of phases that will govern the effort and identifies the deliverables required for each phase.

(3) Process & Tools – Delineates all of the critical processes that will be necessary to methodically adjudicate each USF-I activity identified for transition. The processes will include the various types of committee reviews as well as the method for selecting a course of action. Tools such as decision trees, weighted-scoring model and risk matrices will also be used to supplement specific steps of the disposition process.

B. **Organization.**

(1) **Planning Groups**

    (a) Membership – Personnel from AMEMB-B Political-Military Affairs Section and USF-I.

    (b) Function – Ownership for developing the planning, methodology, process and tools associated with the transition effort.

(2) **Working Groups**

    (a) Membership – USF-I subject matter experts (SME) who own a task, program or project designated for transition. AMEMB-B SMEs who will assist in the transition or assume future ownership of the activity.

    (b) Function – Manage the details involved in transitioning a task, program, project and relationship.

    (c) In accordance with Annex K, the Joint Campaign Working Group is the coordinating body that meets weekly to discuss key transition issues and synchronize activities between USF-I and AMEMB-B membership and functional organizations. This body shall list all tasks to be transitioned.

C. **Methodology.**

    (1) **Define Phase**

        (a) Purpose – Provide the scope and charter for the transition initiative as well as define the various terms that will be invoked throughout the process.

        (b) Deliverables

            i.   Term Definitions – Similar to taxonomy, this list will help to minimize misinterpretation of documents and intent. Appendix 1 refers.

            ii. Organizational Structure – A document with a comprehensive list of groups, teams and cells along with their membership and functions within the scope of transition.

            iii. Transition Planning Timeline – A timeline depicting major milestones and events.

            iv. Scope (In/Out) – A critical piece of the Define Phase, the scope of Transition will drive and shape the type of deliverables in the succeeding phases. A properly scoped effort will enable leadership to effectively focus on the key issues.

            v. Problem Statement – A succinct statement that is measurable and time bound. Addresses the key issue of what the transition effort will tackle.

vi.   Goal Statement – A brief phrase that captures the transition's objective.

vii.   Key Stakeholder Analysis – A brief study to improve understanding of what is important to the key stakeholders involved in the Transition effort.

(2) **Measure Phase**

(a) Purpose

i.   To gather data regarding the effectiveness of a particular task in achieving the desired JCP conditions for the associated LOO.

ii.   To quantify the volume of tasks, programs, projects and relationships that will be adjudicated in the transition process

(b) Activities

i.   Canvass all USF-I agencies for a comprehensive list of tasks, programs, projects and relationships.

ii.   Develop a metrics package to accurately capture the volume of activities that must undergo transition.

(c) Deliverables

i.   Determine number of tasks and their USF-I owners

ii.   Determine number of programs and their USF-I owners

iii. Determine number of projects and their USF- I owners

iv.   Determine number of relationships and their USF-I owners

v.   JCP LOO Linkages (LOO, Conditions, Objectives & Tasks)

vi.   Security Framework Agreement (SFA) linkages

vii. Transition Dashboard (see Appendix 2)

(3) **Analyze Phase**

(a) The purpose of this phase to is to identify which current Multi-National Force-Iraq (MNF-I) actions will need to continue in order to achieve the 2010 JCP objectives. As the Transition planning effort serves to bridge the gap between current activity and future actions, the analysis will identify the relevancy and effectiveness of each activity's contribution to achieving those objectives. The resulting series of recommendations regarding the disposition of each activity will inform the 2010 JCP and the OPORD 10-01.

(b) Planning Team Activities

    i.   The Transition planning team will assess the effectiveness of each JCP activity applying measures of performance and effectiveness along with the initial feedback gathered from the information canvassing effort.

    ii. The approach to conducting the analysis will involve multiple inter-agency working groups operating simultaneously to determine whether to transfer, transform, complete or terminate each activity. The composition of each working group will be unique to the particular area of emphasis or line of operation to which it is assigned to analyze. Respective SMEs from various military and embassy staff sections and organizations will collaboratively assess each 2009 JCP task and associated programs, projects, and relationships for disposition.

    iii. Transition working group sessions – Multiple forums for inter-agency working groups to perform the necessary assessment of the JCP tasks and activities.

(c) Deliverables

    i.   2009 JCP task disposition recommendation – After evaluating the measures of performance and comprehending the level of effectiveness, determine if the task should continue to exist as is, undergo modification, or terminate.

    ii. Associated activity disposition recommendation – A decision on whether to transfer or terminate an existing MNF-I/USF-I program, project, and relationship associated with a particular JCP task.

    iii. Transition planning timeline that includes identification of designated recipients of specific activities, who will continue to perform them.

(4) **Implement Phase**

(a) Purpose – Execute the steps required to transfer or terminate an activity to a non-USF-I entity. Once the decision has been made, the owning USF-I entity must undertake the preparatory steps in executing the decision.

(b) Activities – Interfacing with future owners of USF-I activities and effecting the transfer of responsibilities. Part of the analysis of the future recipient of the USF-I activity involves understanding their capability, capacity and limitations. Once this assessment has been performed the following key events should be factored into the transition planning:

    i.   Preliminary Key Leader Engagement – Once the receiving non-USF-I organization has been identified, the initial plan for interacting with the leader of that entity must be developed by the lead US government agency. This, in effect, requires a "salesmanship" approach whereby the benefits of owning the transitioned activity are emphasized. This interaction (or series of interactions) must be carefully orchestrated to enable the US government agency to "close the deal".

ii. Resourcing – This event will require determining the level of manpower, materiel and funding necessary to enable the future recipient to effectively perform the activity. Consideration must be given to modifying a unit's (or a portion of the unit) redeployment date in order to properly effect the transfer. Additional funding entails submission of the request well in advance to ensure timely allocation (e.g., including the request in the AMEMB-B's Mission Strategic Plan).

iii. Training – Prior to affecting the handover, significant training may be required by the end user. This is particularly true in cases where equipment (military or civilian) is involved as part of the transfer. USF-I units must take into consideration the number of trainers (including interpreters), total training time and other logistic factors.

iv. Phasing – Availability or resource constraints on the part of the recipient may limit the amount of interface required to properly transition the activity. Therefore, the transition may have to conduct in a phased approach whereby key milestones are met prior to beginning the next phase.

v. Data Transfer – A program designated for transfer may involve volumes of data needed by the end user to properly perform the activity. Part of the planning must involved the tasks required to transfer the data, whether in the form of hard copy documents or "soft copy" files.

(c) Deliverable – Transition of each activity. The transition handover focuses on establishing requirements and standards to properly execute an approved transition, including activities required to enter, transit, and exit a transition from a USF-I entity to a non-USF-I entity. This is similar to a conventional Change-of-Command event, which prescribes certain formal and legal activities involved with transitioning overall responsibility from one entity to another, to include administrative and operational attributes.

(5) **Assess Phase**

(a) Purpose – Determine the level of effectiveness of the transition effort from a macro as well as micro-level. Understand if the desired JCP or non-JCP condition is being achieved.

(b) Activities

  i.   Initial assessments is performed by USF-I agencies with regard to the efficacy of the transition effort.

  ii. Following USF-I's departure and beyond 2011, the assessments are performed as part of the standard AMEMB-B's host-country reports.

(c) Deliverable – A general assessment of the effectiveness of each activity that was transferred to its new owners and its impact on the JCP conditions.

  i.   Transition reporting consists of requirements to keep local and senior U.S. civilian and military leaders adequately, routinely, and deliberately informed of the "transition" progress. Local leadership consists of the USF-I Commanding General and the AMEMB-B Ambassador, while the external leadership consists of CENTCOM, Joint Chiefs of Staff, Congress, National Security Council, and the President of the United States.

  ii. Reporting ensures that all levels of leadership are kept abreast of the dynamic and evolving "state of transition" as the U.S. military enters an historic drawdown period. It informs them of the methodical and deliberate process USF-I is undertaking to responsibly terminate or transition all activities.

D. **Tools.**

  (1) Canvassing Spreadsheet – Used to query all USF-I agencies for JCP-related tasks and activities.

  (2) Transition Dashboard – A metrics package to track the progress and status of the transition effort.

4. **ADMINISTRATION AND SUSTAINMENT.** See Base Document dated Nov 2009.

5. **COMMAND AND CONTROL.** See Base Document dated Nov 2009.

Appendix 1 – Transition Definitions
Appendix 2 – Transition Dashboard
Appendix 3 – Transition Process

APPENDIX 1 to ANNEX F

Transition Definitions

Reference: See Base Document dated Nov 2009.

1. Tasks – a set of activities designed to accomplish a specified action or achieve an objective. **Does not** involve construction, large scale funding or basic military tasks.

2. Programs – A planned and coordinated group of activities designed to achieve an operational or strategic objective that is typically long-term or enduring. Examples: Commander's Emergency Response Program projects; Freedom of Information Act Program.

3. Projects – A large or major undertaking, especially one involving funding, personnel, equipment or materiel. A project may have been launched as part of a larger program. U.S. Army Corps of Engineers Gulf Region Division infrastructure projects are a prime example.

4. Relationships – Key relationships with Iraqi governmental or civil officials that set the conditions for achieving Joint Campaign Plan goals.

5. Activities – A term that represents the aggregate of tasks, programs, projects, and relationships.

6. Transfer – A shift of ownership and responsibility for an <u>entire</u> activity to a non- USF-I entity (e.g., GoI, AMEMB-B, UNAMI).

7. Transform – A change in scope was required in order to transition an activity.

8. Complete – Conclusion of the activity in accordance with its stated objectives.

9. Terminate – End or cancelation of the activity short of its stated objectives.

10. Activity Prioritization – Relative importance of activity with respect to achieving the overall US strategic goals of a sovereign and stable Iraq.

   a. Essential – Must have in order to achieve success.

   b. Important – Should have in order to enable success.

   c. Desirable – Would contribute to success.

   d. No future requirement – No further contribution towards success.

APPENDIX 2 to ANNEX F

Transition Dashboard

Reference: See Base Document dated Nov 2009.

1. **Introduction.** The purpose of the Transition Dashboard is to report on the overall progress of the transition effort. All of the metrics depicted in this annex was generated using sample data.

2. **Metrics.**

   A. Transition Progress – Entire Period. This graph measures the total volume of functions, programs, projects and relationships that have been transferred or terminated ("transitioned") over the entire reporting period. It is a graphical representation of Transition's Goal Statement.

   B. Transition Progress – Monthly. This chart measures the total volume of functions, programs, projects and relationships that have been transferred or terminated ("transitioned") over a selected reporting period by month and segmented by category.

   C. Transition Progress by Category. This is a snapshot in time of the transition progress. This chart depicts what has been adjudicated (transferred or terminated) and how many more are awaiting adjudication.

   D. Total Activities by USF-I Agency. This chart is a snap shot in time and captures the total amount of activities by USF-I agency. This is the first summary data set that can be derived from the canvass FRAGO.

   E. Summary of Adjudicated Activities. This chart is a snap shot in time and summarizes the total number of activities that have been adjudicated by entity/agency.

APPENDIX 3 to ANNEX F

Transition Process

Reference: See Base Document dated Nov 2009.

1. **Background.** The process map below highlights the major steps involved in adjudicating JCP Tasks and USF-I Activities.

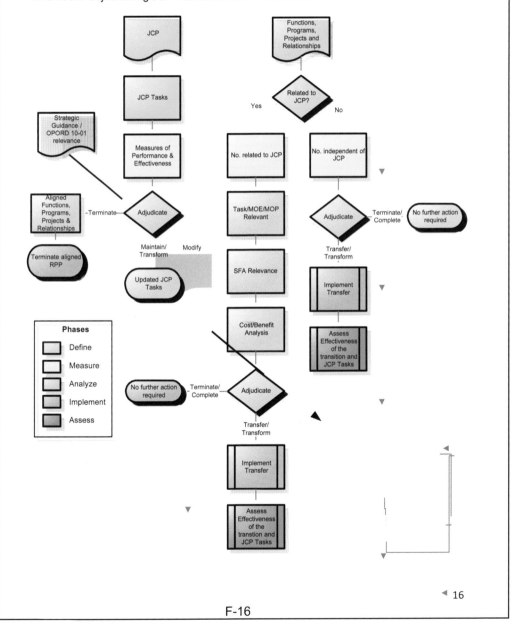

USF-I J4 Joint Plans Integration Center Input

USF-I supplied the material in this appendix to RAND to make it widely available.

HEADQUARTERS
UNITED STATES FORCES – IRAQ
CAMP BUEHRING, KUWAIT
APO AE 09330

USF-I-J4-JPIC   13 December 2011

MEMORANDUM FOR RECORD

SUBJECT: USF-I J4 Joint Plans Integration Center (JPIC) Input for the In-Progress
Unclassified RAND Study on "How War Ends"

Background:

The transition from Operation Iraqi Freedom (OIF) to Operation New Dawn (OND) on 01
September 2010 marked the end of combat operations in Iraq. Long-range planning efforts for
Operations Order (OPORD) 11-01 occurred shortly thereafter from October – December 2010.
United States Forces Iraq (USF-I) published OPORD 11-01 on 06 January 2011 and it went into
effect 06 February 2011. The stated mission was as follows: US Forces-Iraq conducts Stability
Operations, supports the US Mission in IRAQ, and transitions enduring activities to set
conditions for an enduring strategic partnership that contributes to regional stability beyond
2011.

Essential tenets of the plan were established to ensure that the reposture and
retrograde operations occurred within the timeline established by the Security Agreement
while simultaneously providing the Commander with operational flexibility. These tenets, some
key assumptions, and some key dates were essential to the plan and included:

- Utilize a logistics spine built along the center of the Iraqi Joint Operational Area
  (IJOA) based on a distribution network consisting of six hubs and twelve spokes.

- Drawdown the bases from smaller bases to spokes and then spokes to hubs.

- Drawdown from north to south and also from east to west from the outside of the
  IJOA back toward the logistics spine.

- Maximize self-redeployments and tactical road marches for Explosively Formed
  Projectile (EFP) proof Mine Resistant Ambush Protected (MRAP) vehicles to facilitate
  base security through FOB transition and to minimize surface lift requirements.

- Maximize door-to-door (D2D) contracted support with Surface Deployment and
  Distribution Command (SDDC) to lessen theater convoys through the southern
  ground line of communication (SGLOC) to Kuwait.

C-1

- Khabari Crossing (K-Crossing) and the SGLOC will be the main port of embarkation for surface retrograde.

- NLT 31 July 2011, award the Logistics Civil Augmentation Program (LOGCAP) contract for LOGCAP IV base life support (BLS) services in support of United States Mission – Iraq (USM-I) in order to facilitate the transition from USF-I LOGCAP III base life support (BLS) services prior to the expiration of the Security Agreement and the LOGCAP III contract.

- Execute Phase II of OPORD 11-01 from 01 September through 31 December 2011 with prioritization to retrograde operations and operational maneuver vice strengthen the Iraqis and conduct transitions.

- All retrograde movements are operational moves with coverage and coordination.

- NLT 30 September 2011, retrograde the Stored Theater Provided Equipment Iraq (STPE-I) heavy combat contingency set.

- NLT October 2011, transition Combined Security Mechanism (CSM) check points and coordination centers from trilateral operations into bilateral activities between the Kurds and Iraqis.

- NLT 01 November 2011, all USF-I forces will be south of the Samarra/COB KAUFMAN line

- NLT 02 December 2011, all USF-I forces will be cleared from and south of Baghdad.

Also essential to OPORD 11-01 was the operational framework with three major Lines of Effort (LOE) and thirteen supporting objectives that the Commander monitored during weekly Commander's Update Briefs (CUBs) and monthly assessments reviews. The USF-I J4 was the lead staff agent for two of the objective areas: (#6) USM-I enabled to conduct missions; and, (#12) Equipment retrograded or transitioned. While the J3 has staff responsibility for objective (# 10) Forces postured and supported to conduct operations, the J4 was heavily involved with all aspects of achieving this objective.

The USF-I Knowledge Management (KM) section developed the OPORD 11-01 Rehearsal of Concept (ROC) Drill database as a way of capturing critical data points for both operations and logistics. The equipment data entry requirements were developed to help the USF-I J4 staff determine surface truck lift requirements because the initial assumption was that theater common user land transport (CULT) truck lift capacity would be the limiting factor (LIMFAC).

As USF-I implemented OPORD 11-01, there was no single section or agency within USF-I involved with contractor oversight and contract drawdown responsibility. The seven major agencies involved in contracting actions in the IJOA included: LOGCAP, AFCAP, DCMA, DLA, SCO-I, PM ASALT, and others to include AAFES, INSCOM, USACE, etc. There were over 2,000 contracts with contracting officers and contracting officer representatives (CORs) situated in various locations around the globe.

Annex V of OPORD 11-01 detailed numerous tasks requiring termination or transition in order to conclude USF-I operations. The J4 had one task; it involved transitioning the key leader engagement (KLE) relationship with the Government of Iraq (GOI) Minister of Transportation to the US Embassy (USEMB) / USM-I. The USF-I J4 efforts with the GOI Minister were focused on the GOI developing a multi-modal transportation and communications system.

Finally, the retrograde of equipment is not a driving force in and of itself. It is completely dependent on units no longer requiring that equipment for mission purposes. Thus, the key to retrograde operations is an operational one – units must be released from their mission so that the equipment becomes non-mission essential (NME). Once this occurs, the unit can clear its area on a base and retrograde its organizational equipment back to home station, turn-in theater provided equipment (TPE) to mobile or fixed redistribution property assistance teams (RPATs), dispose of it through Defense Logistics Agency Disposition Services (DLA-DS), or transfer it to another element, such as USM-I or the GOI.

Discussion

USF-I published OPORD 11-01 Change 1 in May 2011 and made several significant adjustments to the original 11-01 plan. These adjustments included replacing 4/1 CAV Advise and Assist Brigade (AAB) with 4/1 AD AAB instead of 4/1 CAV concluding its Combined Security Mechanism (CSM) mission and tour in August 2011. Employing 4/1 AD AAB allowed USF-I to extend the CSM mission in the USD-N operational environment (OE). Additionally, the USF-I Main Command Post (CP) repositioned within the IJOA from Victory Base Complex (VBC) in the Baghdad area to Al Asad Air Base (AAAB) in Anbar province vice repositioning to Kuwait prior to Phase II operations. These two major actions, along with the reduction of Phase II operations being condensed to 14 October through 18 December 2011 (originally planned to begin 01 September and to end 31 December 2011), added significant friction and increased the risk to the completion of all Reposture operations on time. Alternatively, these actions allowed the Commander significant operational flexibility for the other two LOEs.

The execution period for OPORD 11-01 and OPORD 11-01 Change 1 experienced many significant events and actions; some added friction and risk to mission, whereas others added capability and capacity. The key issues that added friction included:

- Extend the CSM mission and delay transition of FOB WARRIOR to OSC-I until 01 December 2011. Risk to force included unsecure GLOCs in the USD-N OE. USF-I mitigated this risk by extending the reach of enablers and by resecuring GLOCs in order to maintain sustainment and retrograde convoys.

- 4/1 AD's deployment to the IJOA to replace 4/1 CAV instead of 4/1 CAV reaching end of mission, resulted in USF-I maintaining the same force structure in the IJOA, five AABs and an AAB HQ, through October 2011. Risk to mission included a smaller window of time to redeploy forces, thus creating a steeper waterfall for personnel redeployments than originally anticipated.

- The STPE-I equipment did not retrograde as planned in September 2011. Instead, USF-I required additional theater lift and effort in August and September to transition the legacy STPE-I combat vehicle models and modernize them with M1A2 tanks and M2A3 Bradley Fighting Vehicles (BFVs). The modernized STPE-I fleet then retrograded to Kuwait in October and November 2011. This action increased risk to force by essentially doubling the amount of theater convoys in order to transport both the legacy and modernized STPE-I fleets. Of note is that USF-I did not require the STPE-I contingency fleet form mission support in 2011.

- The LOGCAP IV contract was originally anticipated to be awarded 31 July 2011 and was awarded nearly on time on 01 August 2011. However, one of the losing vendors protested the award results on 11 August and all work associated with transitioning services on enduring bases ceased until the protest concluded 28 September - a 48 day delay. The risk to mission manifested itself as a condensed timeline to transition services on existing LOGCAP III bases as well as bases that receive support from non-LOGCAP vendors. Additional risk to mission developed when USF-I changed the end of mission date from 31 December to 18 December 2011.

- There were several base closure and transition date changes for hubs and spokes.

  o The base closure date for Al Asad Air Base (AAAB) changed from 15 November 2011 and after several changes ultimately became 10 December 2011. The risk to force again included unsecure GLOCs. USF-I mitigated this risk by extending the reach of enablers and resecuring GLOCs in order to maintain sustainment and retrograde convoys. Additionally, this also meant that four major hubs would be open in December as opposed to the original plan for three.

  o Alternatively, the transition date for Joint Base Balad (JBB) moved from its original closure date of 30 November 2011 and ultimately became 08 November 2011. This enabled theater convoys to better deconflict competing retrograde surface moves for JBB and Victory Base Complex (VBC) – slated to close on 02 December 2011 – for the two largest USF-I bases.

USD	Base Name	Est Return Date	Anticipated End State
	ENDEAVOR	15-Sep-11	30th IA BDE
	SHOCKER	16-Sep-11	Point of Entry Transition Team
	AL SHEEB	17-Sep-11	10th DBE
	MINDEN	17-Sep-11	Point of Entry Transition Team
Center	AL RASHEED	30-Sep-11	Rusafa Area Cmd and 11th IA
North	WARHORSE	30-Sep-11	5th IA, 18/5 IA, 21/5 IA
North	CRUZ-MORRIS	30-Sep-11	Diyala Ops Center, CSA Combined Coord Ctr
North	HABUR GATE	30-Sep-11	MoI
Center	FALCON	30-Sep-11	2FP TAC
Center	HAMMER	1-Oct-11	OSC-I, 9th IA
Center	TAJI	1-Oct-11	OSC-I, NTM-I, 9th IA
ITAM-P	SHIELD	1-Oct-11	INL, NTM-I, 6th IA
RAOC	UNION III	1-Oct-11	OSC-I, NTM-I, 6th IA
Center	FALLUJAH	1-Oct-11	1st IA
North	COBRA	8-Oct-11	2/1 IA, 4/1 IA, 8/3 DBE
North	SYKES	8-Oct-11	10, 11, 12/3 IA, 6/2 DBE
North	ERBIL	8-Oct-11	DoS, INL, KRG
North	DELTA	12-Oct-11	8th IA BDE HQ
	GARRY OWEN	14-Oct-11	TBD
North	MAREZ	15-Oct-11	DoS, 7/2 IA, 26/7 IA, 3rd FP Div
North	WARRIOR	15-Oct-11	DoS, 10th IA, Kirkuk IP
	UMM QASR	15-Oct-11	OSC-I, Iq Navy
North	BAYJI O.R.	18-Oct-11	Iraqi Oil Protection Police Force, MoI
North	SPEICHER	20-Oct-11	IqAF Academy, MoD, OSC-I (T)
North	KAUFMAN	22-Oct-11	Al Askiri FP BDE
Center	AL ASAD	15-Nov-11	7th IA
Center	RAMADI	16-Nov-11	DBE, 1/1st IA
North	JBB	30-Nov-11	OSC-I, IqAF, MoD
Center	VBC	1-Dec-11	DoS, NTM-I, ISOF, 6th IA, TIFRC
	ADDER	14-Dec-11	IqAF, MoD
	KALSU	15-Dec-11	8th IA
	ECHO	16-Dec-11	5th DBE
	BASRAH	16-Dec-11	DoS, INL, Provincial Gov, IA

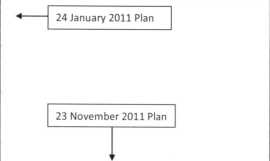

24 January 2011 Plan

23 November 2011 Plan

MSC	Base Name	Est. Transition Date	Anticipated End State
USD-C	AL MUTHANA	30-Nov-11	54/6th IA
USD-C	VBC	2-Dec-11	DoS, PM
USD-C	AL ASAD	10-Dec-11	7th IA
USD-C	KALSU	12-Dec-11	8th IA
USD-C	ECHO	17-Dec-11	8th IA
USD-C	BASRAH	14-Dec-11	DoS, Provincial Gov, 14th IA
USD-C	ADDER	18-Dec-11	IqAF, 10th IA

- Several AABs turned in viable EFP-protected MRAPs to the RPAT yards instead of conducting self-redeployment by tactical road march (TRM) and turning them in at Camp Virginia, Kuwait as planned. This action added risk to force by requiring unprogrammed theater convoys to travel the GLOC both north and south to pick up the loads and retrograde them to Kuwait vice the AAB unit executing TRM traveling only one way, south. Additionally, an MRAP has better protection than a truck and can defend itself better. Finally, there was risk to mission by the added lag time of turning in the EFP MRAPs and then awaiting the Transportation Movement Request (TMR) to be filled and the assets retrograded to Kuwait.

- A significant amount of USF-I and United States Division (USD) effort was spent developing viable plans for military forces to conduct possible follow-on missions Post-2011. Although USF-I planners expected a final decision on this contingency NLT 31 July 2011, providing the Commander additional time and space for a decision as late as possible created considerable friction between the operations and logistics communities within USF-I. The friction manifested itself with units retained in theater for missions. These same units kept all of their equipment in anticipation of future operations rather than "thinning the lines" in anticipation of Reposture. Further, multiple changes to base closure and transition dates required more flexibility from the supporting assets and required units to cancel their scheduled D2D bookings because of uncertainty in meeting contractual timelines. Ultimately, the USF-I Commander announced on 14 October 2011 to execute the program of record plan OPORD 11-01 Change 1 FRAGO 1058, which directed USF-I to complete Reposture and "go to zero" on or about 15 December 2011.

- Tied to the decision to "go to zero" was another relocation of the USF-I Main, this time moving it from AAAB in the IJOA to Camp Buehring, Kuwait as a USF-I Tactical Command Post (TAC).

- The USF-I end of mission (EOM) ceremony slated for 15 December 2011 at SATHER Air Base also added significant risk to force and mission by requiring USF-I forces to support the ceremony and maintain elements in Baghdad beyond the originally planned date of 01 December 2011.

As stated earlier, there were many key events and actions that added capability and capacity to the logistics mission that directly supported the retrograde operation. These included:

- USF-I, in coordination with CENTCOM, established the Contracting Fusion Cell (CFC) in March 2011 and developed the Knowledge Management (KM) CONtrax database in May 2011 to track contracts, contractors, and their associated equipment was functional by May 2011. Though there was not a USF-I J-staff element that assumed responsibility for all contracts, utilizing the CENTCOM Contracting Command (C3) element in Iraq as the Office of Primary Responsibility (OPR) provided subject matter expertise that was essential to the retrograde, demobilization and transition of contracts and the associated contract workers.

- The Program Manager (PM) community also worked with KM to develop the Materiel Enterprise Transition Common Operating Picture (METCOP) database to track PM contractors and their associated equipment.

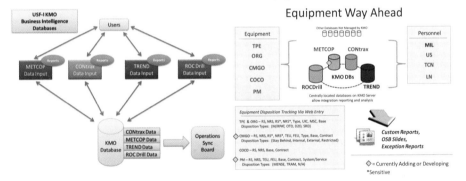

- Deployment and Redeployment Operations (D/ROPs) cells established nodes in three different areas to support the contract personnel drawdown plans. Each USD developed a D/ROPs site vicinity of a major commercial airport in order to facilitate the contractors redeploying out of the IJOA. The CFC maintained visibility and worked solutions to mitigate any friction points that arose.

- Contractors are typically only authorized "space available" travel on military aircraft (MILAIR); however, exceptions to policy were requested from the Secretary of Defense (SECDEF) levels to allow contractors to fly "space required" status in the IJOA IOT enable contractors to fly to the D/ROPs nodes IAW their demobilization timelines. This initiative enabled USF-I to program and allocate MILAIR assets to support the retrograde of all personnel - military, DOD civilians, and contractors.

- Army Central Command (ARCENT), as the supporting command to USF-I for retrograde operations, expanded its theater convoy contract, Heavy Lift 6.5 (HL6.5). The new contract, Heavy Lift 7 (HL7), made additional flat bed and Heavy Equipment Truck (HET) equivalent commercial trucking assets available in sufficient quantities to relieve most of the surface transportation friction, even with the shortened timelines for the retrograde operations.

- ARCENT also used military trucking assets to expedite clearing the USD-N OE by operating Task Force Hickory out of Joint Base Balad (JBB). TF Hickory executed a "ring route" system and transloaded cargo from the more northern bases and systematically retrograded the equipment out of theater using the commercial HL6.5 and HL7 assets on throughput missions.

- USF-I developed several battle rhythm events that specifically supported the Reposture effort and operational maneuver during Phase II operations. The J4 Transportation Section hosted or co-chaired several of these meetings to synchronize transportation operations and provide the J4 with situational awareness of transportation movements.   The events included the Main Supply Route/Alternate Supply Route Working Group (MSR/ASR WG), the Sustainment Synchronization Working Group (SSWG), the Equipment Drawdown Synch Board (EDSB), and the Movement Synchronization Working Group (MSWG). These meetings provided action officers and decision makers a common operating picture (COP) for near-term synchronization and execution of near-term Reposture operations with velocity.

Event	STANDARD TIME	METHOD	M	T	W	TH	F	Sa	Su
MSWG	1030-1045	Adobe Breeze	x	x	x	x	x	x	1400-1415
SSWG	1045-1130	Adobe Breeze	x	x	x	x	x	x	1415-1500
EDSB	1930-2015	SVTC	x	x	x	x	x	x	x
MSR/ASR WG	N/A	Adobe Breeze	absorbed into the SSWG						

Lessons Learned

There were several lessons learned during the planning and execution of OPORD 11-01.

BATTLE RHYTHM: The daily logistics battle rhythm events that the J-Staff held during OPORD Phase II operations were vital in keeping the retrograde operations coordinated and synchronized. The SSWG/EDSB review should remain at the action officer level, with the exception of the two MCB commanders (53rd MCB in the IJOA and 330th MCB in Kuwait). Their attendance was not a hindrance but instead enhanced the discussion and more importantly, the execution of movements. The EDSB became a focus point and forum for senior leaders (USF-I DCG-S and J4) to gain detailed information from the ground level concerning retrograde operations. This information was used to brief the CG on the retrograde status. The recommend action is to maintain this battle rhythm event while conducting similar operations.

KNOWLEDGE MANAGEMENT: First, a comprehensive concept for a knowledge management automated system should include data fields that support planning and contingency planning efforts and it must also include data fields to capture and track actual execution glidepaths. Critical staff participants and enablers include J1, J33, J35, J4, J5, J7, MSC G3/G4/G5, PM ASALT, Contracting Command, LOGCAP, DCMA and key Department of State (DoS) representatives if the Reposture mission also includes transitioning tasks to the DoS.

Secondly, USF-I J4 and our Lean Six Sigma (LS6) representative did numerous comparisons of the 11-01 ROC Drill database equipment information to the Consolidated Property Book Listing (CPL) – essentially a combination of Organizational and Theater Provided Equipment (TPE) information available on Sharepoint to all users. The 11-01 ROC Drill database was based on generic information, Rolling Stock and Non-Rolling Stock which was further distinguished as Sensitive or Non-Sensitive. In hindsight, a better way would have been to integrate the CPL information with the 11-01 ROC Drill database so that specific LIN numbers and equipment would correlate with the Reposture plans. This type of detailed information would have made planning easier to track for units, and definitely would have made tracking of the execution of the Reposture effort easier for all.

Finally, the 11-01 ROC Drill database must include more than just the equipment retrograde. Other categories to capture include commodity retrograde (Class II/IIIP/IX, CL V), AAFES retrograde, Program Manager (PM) item retrograde, and any Contractor Managed Government Owned (CMGO) equipment retrograde. Two areas that USF-I did not capture initially in the planning mission analysis were convoy escort team (CET) requirements for LOGCAP sub-contractors who would need to retrograde their Contractor Owned Contractor Operated (COCO) equipment as well as the equipment repositioning to other bases. Repositioning efforts within the Iraq Joint Operational Area (IJOA) included numerous enabler units and assets as well as LOGCAP CMGO and materials required to fully enable enduring bases for post-OND operations under Department of State (DoS). Although USF-I had the capacity to support all of these movements, identifying the requirements earlier would have allowed for better use of resources and prioritizations. In the end, the intent of KM tools is to help you "see yourself" better in order to make informed decisions. The 11-01 ROC Drill database and ROC drills themselves accomplished this goal to a certain degree, both could have been better tools, given the right emphasis and effort earlier in the planning process.

REPORTING: Commanders take responsibility to report the execution of their plan against their projections. During USF-I's Operation Iraqi Freedom (OIF) Responsible Drawdown of Forces (RDoF), the twenty-nine (29) major subordinate commands (MSCs) briefed their personnel drawdown plans and efforts weekly from July-September 2010 which enabled USF-I leadership to immediately see when commanders were off-glidepath and to make adjustments accordingly. OPORD 11-01 and the USF-I J33 did not require commanders to report their progress in relation to established metrics. Instead, the USF-I staff captured reports periodically and subjectively through other venues such as the J35 Future Operations (FUOPs) led Operational Synchronization Board (OSB), the J7 led Large Base Transition Boards and the USF-I Deputy Commanding General – Operations (DCG-O) Command Post of the Future (CPOF) updates. Reporting requirements for units and bases, whether they were 180-days out from an action or within the final 30-days of completion of a task were similar, however, the reporting fidelity should become more refined as the termination or transition date approaches.

Commanders take responsibility for the fidelity of their Reposture plans and the data that both staff and external supporting agencies use to develop Reposture support plans and analysis. One of the key lessons learned was that units did not refine or validate their Reposture plans until the unit executing Reposture physically arrived in the theater of operations; units with steady-state missions who would redeploy and not have to execute Reposture, did not put forth the same amount of effort to make definitive Reposture plans to set up the follow-on unit for success. Regardless, the point to make here is that there must be top-level leadership involvement in the details of the Reposture plans to ensure that they are solid and comprehensive.

DECISIONS, DECISION POINTS (DPs) and ASSUMPTIONS: Several decisions and decision points proved critical to the OPORD 11-01 operations and ultimately to the risk mitigation to both force and mission for USF-I. There were also several key assumptions tied to specific dates and times that also did not turn out to be facts. Staffs at all levels must be able to forecast the near-term and the long-term 2nd and 3rd order effects of decisions, decision points that get passed without an associated decision and critical assumptions. The ultimate lesson to learn regarding retrograde operations is to ensure that the original OPORD plan and mission have a robust, sound method to execute with metrics to measure performance along the way.

- USF-I did not develop metrics to transition bases from USF-I to USM-I in detail until mid to late September 2011. Most Office of Security Cooperation Iraq (OSC-I) base transitions, as well as OSC-I's fully operational capable (FOC) date, were 01 October 2011. September 2011 was too late to establish performance metrics and execute responsible transition of bases to OSC-I and DoS.

- The mid-August 2011 decision to modernize the STPE-I fleet proved to be an excessive effort, given the timing of the decision and the follow-on decision on 14 October 2011 to execute the complete retrograde and Reposture of all forces. Though 2nd and 3rd order effects and delays to ongoing intra-theater surface transportation lift requirements were identified, the modernization of STPE-I was given priority for movement in September 2011. The modernized fleet retrograded out in late October/early November 2011 when the GLOCs, Convoy Support Centers (CSCs) and USF-I bases in the USD-N OE were closing, transitioning, or supporting unit equipment turn-in and redeployment operations.

ENEMY, POLITICS and TIME: One final observation for military forces and the countries that employ them is that certain elements are not controllable by the military forces. The enemy always gets a vote and his actions will shape the operation over time. The military force ultimately reports to the Commander-in-Chief (the President of the United States (POTUS)), and is only one of the four elements of national power (DIME – diplomatic, information, military, and economic) that leadership employ or terminate. Finally, time is finite and once national leaders determine a definite date to conclude a war/operation, the planning goals should relate to both the early completion of the retrograde operation as well as the completion of as many transition tasks as possible.

Conclusion

There are several overarching take-aways from the planning and refinement actions in support of OPORD 11-01. First and foremost is that there is a significant amount of effort in reposturing a military force and its corresponding contracted support out of an operational environment that is not completely secure. Logistically, the challenges in terminating tasks and closing bases proved to be just as significant and complex as the tasks and bases that transitioned to USM-I. Second, time is a fleeting and finite resource that can not be reclaimed. Every day that decisions or Reposture efforts are delayed may provide the Commander operational flexibility, however, these same delays alternatively take away all flexibility during the drawdown. Third, fixing responsibility and developing clear, concise metrics that are measured and reported often by the unit that is responsible for the Reposture task is essential. Finally, Reposture and the retrograde of forces is an operation that must be planned by the J5 staff, executed meticulously by the J3 staff and operators, and lastly supported by the rest of the staff to include the J1, J4 and J7. Often times, the Reposture effort was pushed toward the supporting staff as opposed to the J3/operators leading and synchronizing to ensure mission success.

USF-I J4 Department of State Transition Cell RAND History Report

USF-I supplied the material in this appendix to RAND to make it widely available.

HEADQUARTERS
UNITED STATES FORCES – IRAQ
CAMP BUEHRING, KUWAIT
APO AE 09330

USF-I-J4-DOSTC   13 December 2011

MEMORANDUM FOR RECORD

SUBJECT: USF-I J4 Department of State Transition Cell RAND History Report

Summary:

DoD USF-I J4 facilitated the transfer of green and white equipment to The Department of State (DoS) using various levels of effort and coordination. USFI-J4 used internal staff assigned to the Department of State Transition Cell (DoSTC) as a Liaison. Additionally, LOGCAP, DCMA, and OSD Equipping Board, J4 Asset Visibility participated in Contractor Managed Government Owned (CMGO) equipment transfer.

The DoSTC was housed at Victory Base and later relocated to the IZ, Baghdad at FOB Union III. Logistically, close proximity to DoS allowed direct interaction and planning with Department of State. There, the DoSTC and J7 participated in State Department meetings weekly.

Enduring Embassy Support Working Group and Board

USF-I J4 and USF-I J7 hosted the "Enduring Embassy Support Working Group and Board." The Board's purpose was to serve as the principal synchronization forum for USF-I and USM-I to provide transition updates, identify and resolve issues, and discuss future actions (Dos/INL, USM-I, and USF-I). Each session contained working actions and expected outcomes. Attendees included General Officer (GO) and Senior Executive Service (SES) representation from USF-I, DCG A&T, USF-I J4, USF-I J7, Embassy Baghdad Ambassador, DoS Counselor to the Ambassador, Regional Security Officer (RSO) Baghdad, Embassy Overseas Building Office (OBO), Central Contracting Command (CCC), LOGCAP, DCMA, and USF-I J8. Each enduring site Lead (Military and Civilian) participated via VTC. DoD and DoS Action Officers across various staff sections sat in attendance at each session.

This avenue was used to track and report updates on each enduring base status along with progress on various topics such as fuel farm, waste water, ROWPU's, Postal, CHU delivery, population, transition, and friction points. The forum allowed detailed Q&A

discussions along with decisions which were made on the spot. This vehicle proved to be a forum in which all attendees could voice issues, concerns, and receive a solution.

Department of State relied upon the DoSTC as a liaison, to source equipment to meet their requirements at enduring sites for fuel, BLS, housing, waste, etc. The Cell's focus was transferring "Priority One" equipment to support construction for Base Life Support (BLS) at enduring sites, and delivery of a Concept of Support which provided DoS a roadmap of procedures and processes for sustainment.

End State

Equipment sourcing began on 1 January 2011 and successfully concluded on 25 November 2011. In total, 24,172 truckloads of 62,310 "Priority One" equipment items valued at $171,121,835 (before depreciation) were successfully transferred to Department of State. CMGO equipment transferred to DOS included Containerized Housing Units (CHU), generators, fuel tanks, laundry facilities, ablution units, T-walls, Jersey Barriers, Ambulances, ROWPU's, and Non-Tactical Vehicles (NTV). See Table below for site-by-site layout of white equipment transfers. All white equipment transfers were completed at no charge to Department of State.

OSC-I also required CMGO equipment. In total, 1,176 truckloads of 53,211 "Priority One" equipment items valued at $111,933,345 (before depreciation) were successfully transferred to OSC-I enduring locations. CMGO equipment transferred to included Containerized Housing Units (CHU), generators, fuel tanks, laundry facilities, ablution units, T-walls, Jersey Barriers, Ambulances, ROWPU's, and Non-Tactical Vehicles (NTV). See Table below for site-by-site layout of white equipment transfers. All transfers were synchronized with base-closure.

Table A: DoS Enduring Site CMGO Equipment Transfers

Enduring Site	Erbil	Mosul	Sather	Shield	Prosperity	NEC	Basrah	Kirkuk	Total Equipment Value per Site	Total Truckloads
Site Value	$31,688,097	$6,146,220	$30,410,055	$22,072,998	$15,725,272	$4,695,766	$43,339,616	$17,043,811	$171,121,835	
Total Transferred Items	10,450	4,224	15,362	7,461	7,331	274	11,010	6,276	62,388	24,172

*Totals reflected are prior to depreciation

Table B: OSC-I Enduring Site CMGO Equipment Transfers

Enduring Site	Tikrit	Al Asad	Besmaya	Taji	Umm Qasr	Union III	Total Equipment Value per Site	Total Truckloads
Site Value	$20,138,459	$642,600	$22,216,648	$42,352,127	$4,986,411	$21,597,100	$111,933,345	
Total Transferred Items	6,864	960	13,599	19,895	1,375	10,518	53,211	1,176

*Totals reflected are prior to depreciation

Contract Managed Government Owned (CMGO):

The majority of USFI-J4 CMGO transfers occurred in conjunction with USF-I J7 Base Transition Plan. DoS developed requirements and USF-I J4 provided DoS with Contract Managed Government Owned (CMGO) "white" excess list to enable transfer. CMGO equipment was located, blocked, braced, and relocated to final destination to meet DoS requirements which supported base life support in theater at no cost to Department of State. Enduring sites retained equipment that resided at existing locations where entire bases were transitioned over to Department of State. As non-enduring bases closed, equipment was de-scoped and relocated to support enduring base and their tenants. Additional equipment was nested and made a part of maintenance for LOGCAP's sustainment support package at enduring sites post 2011.

1. Quarterly History Reports. The J4 produced quarterly history reports to create a record of the status of all transfers made to the Department of State and OSC-I.

2. Relevant reporting charts, graphs slide presentations or charts that depict the transition of equipment and supplies since the beginning of Operation New Dawn used for the transitioning to USM-I (OSC-I and DOS).

a. Green Equipment Requirements and Transfer Process.

In support to the enduring mission in Iraq, the Department of State requested a total 3,807 items of standard or "green" property from the Department of Defense. The slides below provide an overview of the six types of equipment included in this request. As depicted in the upper left quadrant of the slide, force protection equipment, such as radars and biometric systems accounts for 78% of the $212 million dollar total value of the Department of State requirements. In terms of the number of line items, however, force protection equipment makes up roughly 5% of the total. As depicted in the lower right quadrant the greatest variety of equipment by far is in the medical category, which makes up 71% of the total line items requested.

As equipment is sourced, it is associated with one of three transfer mechanisms: excess defense articles, non-excess sales from stocks, and non-excess loans. DoS will only pay for sales from stocks, and their total cost will be approximately $10 million dollars, or less than 5% of the total value of all items transferred. The loaned items are two Giraffe radars, 60 Caiman Plus MRAPs, and all the biometrics equipment.

All equipment sourced by USF-I will be transferred to the Department of State only after it is no longer needed to support our mission. The transfer process depends on the type of sourcing. For information technology and communications equipment, the type of sourcing is non-excess sale from stocks. Once the State Department transfers the funds through a Military Interdepartmental Purchase Request or MIPR, the Communications-Electronics Command or CECOM will ship the equipment from its stateside warehouses. None of the USF-I IT or communications equipment will be transferred to DoS.

The MRAPs, Giraffes, and biometrics equipment are non-excess loans to DoS. Once DoD and DoS agree on the language of the loan, USF-I will laterally transfer the equipment to a property book account already created for this purpose.

The remaining items are excess to DoD. Once USF-I no longer needs them, we are authorized to transfer them to DoS. We will document all transfers on DA Forms 3161 and DoS will pick them up on their property accountability system, providing a clear audit trail for each transaction.

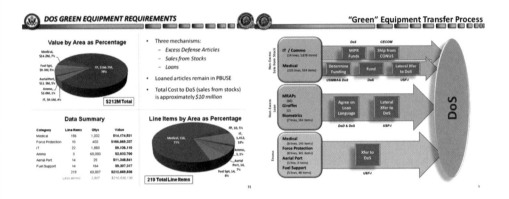

b. Green Equipment Transfers Status Reporting to USF-I Commander and J4 from July to November.

In total over 3,000 pieces of equipment will be transferred or loaned of which 90% of the transactions are complete.

USF-I have transferred Medical Equipment, Information Technology, Force Protection and Communications equipment. Medical equipment transferred in November with only two bases pending property transition to the contracting health solutions (CHS) medical contractor, Tikrit and Besmaya. In total, medical equipment covers over 70% of the standard equipment expected to transfer in support of the Department of State. Green Equipment transfers completed sites include Kirkuk, Sather, Prosperity, Erbil, Shield, Union III and Taji.

Green equipment requirements have also increased over time. The initial requirement of 2,993 has increased to 3,289. The majority of the remaining line items pending transfer will do so between November and December. In September, the OSD Equipping board approved the transfer of additional items such as Force Protection equipment. The Board has a pending request of several medical and fuel support components; we expect final approval of those items this month. Standard equipment loaned to USM-I will be maintained by the Army Materiel Command's enduring "FIRST" maintenance contract.

Green Equipment Transfers:

Medical:

- Transferred approximately 1, 300 pieces of green medical equipment valued at over $16M items.

- Transferred Medical Communications and Informatics (TeleMedicine / TeleRadiology/CT Scanners) after routing the request through the OSD Equipping Board for approval

- Synchronized the transition of medical care and equipment between USF-I and Department of State for successful establishment of Diplomatic Hospitals and Clinics per enduring site

- "White" Ambulances were transferred to Department of State for use to support the medical contract with government furnished equipment (GFE)

Class III: Fuels

- Transferred "white" Fuel Tanks SCAT/Smart Tanks to support enduring base requirements of JP8, DF2, and MOGAS requirements

- Department of State requested LOGCAP retain 35 - 350GPM Pumps for sustainment to include items connected on existing fuel farms

- Coordinate with OBO and J7 to identify requirements/consumption

- Fuel Meetings were held with DoS Fuel Representatives, OBO, DLA, LOGCAP, KBR, Site Leads/DOL on Fuel Storage and Equipment Requirements
    - Fuel and Power Summit
    - Weekly Fuel Meetings at UIII
    - Embassy Baghdad, DLA, and USF-I J4 oversees fuel delivery, vendors, routes, DoS Logistic Management Control Center

- DoS supported embedded representation to learn the process IAW the Concept of Support

Force Protection:

- HQDA EXORD 156-11 authorized ARCENT Loans of Caiman's to Department of State
- ARCENT loaned 60 Caiman and 2 Giraffes
- Transferred Backscatter (ZBV) Vans
- Loaned Sense and Warn Systems
- Binoculars
- Night Vision Goggles (NVG)

Information Technology:

- OSD approved transfer of biometrics, radios, and COMSEC devices
- DoS received PRC-152 (Sale from Stock)
- Loaned Radios Netted Iridium's
- Telephone Switch – CS 1000

Ammo: DoD transferred over 60K pieces of Ammo to DoS as approved by OSD

Aerial Port:

- OSD Equipping Board approved transfer of 31 Aerial Port Equipment items from the Airforce
- Tow Bar / Tow Vehicle
- Maintenance Stands
- Loaned Tactical Meteorological Observing System (TMOS)

Base Life Support (BLS):

- Air Force approved transfer of 485 Base life support items to DoS in support of emergency projects at Sather Airfield
  - Shelter
  - Reefer Condensers/Air Conditioners
  - Dome Shelter
  - Refrigerator Unit

c. White Equipment Transfers from the Essential Excess Equipment (EEE)

(See Reference Material CHU and TWALL Report)

CHU and T-Wall Movement Plan completed in August.

By the beginning of November USF-I transferred over 52 thousand items, valued at over $156 million dollars surpassing the original requirement of 38,697 items set in January 2011. DoS requirements increased from 40,000 in July to over 52,000 pieces in month of October. The transfers included base life support items such as Containerized Housing Units (CHU), Ablution units, Generators, laundry units, ROWPUs, T-walls, Jersey Barriers, and storage tanks to name a few. All transfers were synchronized with base-closure and transition actions at no-cost to the Department of State. CHU and T-Wall Movement Plan completed in August 2011.

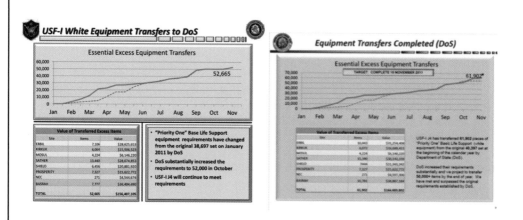

d. **Status of Basic Life Support at OSC-I and DOS Enduring Sites August/November after transfer dates**

e. Service Timelines and BLS Contract Transition by Enduring Locations

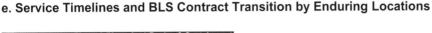

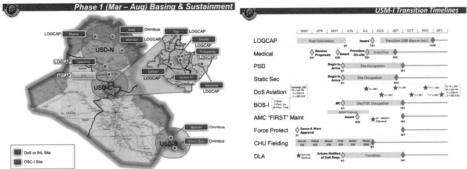

3. Tactics, Techniques And Procedures (TTP) Transition Meetings. In order to effectively manage the transition support activities between USF-I and Embassy Baghdad six separate groups were established as follows:

- **DoS Management Staff Meeting:** This is a weekly meeting chaired by the Embassy's Deputy to Management (Transition). The meeting was designed to synchronize requirements and deconflict transition efforts between USF-I and Embassy Baghdad. Attendees include: J4 Director, J7 Director, DoS Regional Management Officer DoD Excess, C3 procurement analyst, and OSD (AT&L).

- **Embassy Support and Enduring Base Transition Working Group:** This bi-weekly meeting was co-chaired by the J4 Director of of Department of State Transition Cell and the J7Enduring Base Director. The purpose of this meeting was to resolve issues and identify future actions required to transition enduring bases to OSC-I and Embassy Baghdad. Attendees include: J4 DoSTC Director, J7 Enduring Base Director, USEMB-B Staff, INL, OSC-I(TT), USF-I DCG-S, USF-I Staff, USDs, UNAMI, NTMI , CENTCOM, ARCENT, ACCE/AEW, TSC/AFSB, VBC Garrison, IZ RAOC, DCG A&T, C3, DCMA.

- **Ad-Hoc Executive Steering Group:** This interagency group met bi-weekly by teleconference between Washington and Baghdad. The meetings were co-chaired by OSD and DoS. This meeting served as the mechanism to receive strategic and policy guidance and updates and to resolve issues and identify future actions regarding bases that will have an enduring OSC-I or DoS presence after the end of the USF-I mission. Those in attendance in Baghdad included appropriate representatives from the following organizations: USEMB-B, OSC-I (TT), J1, J35, J4 Director, J7 Director, INL, USF-I DCG-S, USF-I Staff, USDs, UNAMI, NTMI , CENTCOM, ARCENT, ACCE/AEW, TSC/AFSB, DCG A&T, C3.

- **Embassy Support and Enduring Base Transition Board:** This biweekly meeting was co-chaired by the J4 Director, J7 Director, Director of Management Embassy-Baghdad, and the director of OSC-I-TT. This meeting provided senior level USF-I and USM-I oversight of the base transition process in an effort to synchronize and deconflict all basing issues. The Board established collective priorities and served as the mechanism to obtain executive guidance and decisions from the Commander and Ambassador. Attendees included: J4 Director, J7 Director, USEMB-B, INL, OSC-I(TT), USF-I DCG-S, USF-I Staff, USDs, UNAMI, NTMI , CENTCOM, ARCENT, ACCE/AEW, TSC/AFSB, VBC Garrison, IZ RAOC, DCG A&T, C3, DCMA.

- **OSD Joint Equipping Board:** This weekly meeting conducted by teleconference was chaired by OSD (AT&L) and was the approval board for standard green transfers to the Department of State. Attendees included: J4 Director, J7 Director, DoS Regional Management Officer DoD Excess, C3 procurement analyst, OSD (AT&L).

- **OSC-I Transition Working Group Teleconference:** This was a daily meeting chaired by the OSC-I JOC and was designed to track actions regarding the transition from USF-I to USM-I with a focus on the OSC-I mission. Attendees included: J4 Director, DCG A&T, OSC-I (TT), OSC-I Staff.

4. VIP Briefings to dignitaries from Department of Defense and the Department of State

- **03 MAY11:** USM-I Transition Briefing to Dep Asst SecDef Gary Motsek and Dep Asst SecState William Moser an overview and update of the transition from USF-I to USM-I

- **24 MAY 11**: USM-I Transition Briefing to provide the Principal Deputy Under Secretary of Defense (AT&L)Hon. Frank Kendall an overview of the transition from USF-I to USM-I

- **19 SEP 11**: USM-I Transition Briefing to Principal Deputy Under SECDEF (AT&L) Hon. Frank Kendall

Joint Logistics Operations Center Input

USF-I supplied the material in this appendix to RAND to make it widely available.

HEADQUARTERS
UNITED STATES FORCES – IRAQ
CAMP BUEHRING, KUWAIT
APO AE 09330

USF-I-J4-JLOC   13 December 2011

MEMORANDUM FOR RECORD

SUBJECT: Joint Logistics Operations Center (JLOC) Unclassified Input to RAND

Introduction

In the last year of U.S. military presence in Iraq, Operation New Dawn (OND), an unprecedented operation, took place with multiple and simultaneous lines of effort: security and stability; train, advise, and strengthen the Iraqi Security and Police forces; inter-departmental transitions (DoD to DoS); disposition and retrograde of over 2.42 million pieces of equipment; closure and transfer of 80 bases back to the Iraqis; and the redeployment of over 110,000 personnel (military/contractors) out of Iraq. Touching all these lines of efforts were the logisticians from USF-I J4 staff to the 310th Expeditionary Sustainment Command (ESC) down to the supply sergeant on the ground with each unit.

While the ESC and unit level logisticians were charged with executing tasks, the U.S. Forces-Iraq (USF-I) J4 was responsible for maintaining total sustainment situational awareness of the Iraqi Joint Operation Area (IJOA). This entailed ensuring compliance with policies, procedures, and plans directing the continual sustainment support to forces; the transition of sustainment support to the Department of State; and the movement and reposture efforts during Operation New Dawn (OND). Daily multi-commodity pushes consisting of such things as food, fuel, and parts and the subsequent backhaul of retrograde equipment equated to hundreds of thousands of pounds of "stuff" being moved daily in Iraq.

A historical look at "repceposturing the force with honor and success" requires an in depth review of the logistics and transportation planning, tactics, techniques and procedures (TTPs), friction points, and risk mitigation efforts required to execute the responsible drawdown of forces and equipment from Iraq. The J4 assessment includes the following related subjects:

Discussion

Commodities Management. At its peak, sustainment operations supported eighty (80) bases within the Iraqi Joint Operations Area (IJOA) requiring commodities ordering, distribution and oversight for approximately 50,000 military service members and 60,000 contractors (U.S., local national (LNs) and third country nationals (TCNs)). Sustainment operations required over 46,000 pieces of rolling stock and more than double that amount of power generation and base life support. See Tab 1 for details.

E-1

Equipment Drawdown. Over 1.8 million pieces of equipment required dispositioning as forces withdrew or repositioned. Equipment Fielding and unit rotations brought an additional 550K pieces of equipment into the IJOA during OND which resulted in total disposition being just over 2.42 million pieces of equipment. The three major categories of equipment were Theater Provided Equipment (TPE) with 928K; Organizational equipment (ORG) with 823K; and Contractor Managed Government Owned (CMGO) with 669K. For TPE roughly 30% went to KU for re-utilization in accordance with CENTCOM priorities; 35% went to CONUS for re-set, re-build, and re-issue; and 35% went to DLA Disposition Services as obsolete, unserviceable, or requiring de-mil. CMGO and select TPE items was transferred to the Government of Iraq (GoI), the U.S. Embassy enduring mission in Iraq, and other entities such as the National Association of State Agencies for Surplus Property (NASASP) and LOGCAP operations in Afghanistan. See Tab 2 for details.

Foreign Excess Personal Property (FEPP). This process, under which the U.S. transferred over four million pieces of non-standard equipment, e.g., furniture, containerized housing units (CHU), civilian model generators, and non-tactical vehicles (SUVs, buses, construction equipment), became the primary method of transferring this type of excess property to the Government of Iraq (GOI). Equipment transfers using the FEPP authority was redistributed in accordance with the CENTCOM commander's priorities and was generally tied to base closures. This process included a Cost Benefit Analysis (CBA) to evaluate the cost savings garnered by not transporting everything from base closures back to CONUS. In total 4.2 million items were left on the 80 bases transferred back to the GoI. Fair Market Value (FMV) of these items came to $586.5 million. The calculated transportation costs for these 4.2 million items came to $1.74 billion. The end result was a $1.15 billion saving to the USG and taxpayers. See Tab 3 for details.

U.S. Equipment Transfers to Iraq (USETTI). This program, under which the U.S. transferred over 37,000 pieces of equipment to the GoI, represented three percent of all equipment transferred under a comprehensive Foreign Military Sales (FMS) equipping strategy to enable Iraqi Security Forces (ISF) to achieve Minimum Essential Capabilities (MEC). Three authorities governed these transfers: USC Section 1234 (Non-Excess Defense Articles), USC Section 516 (Excess Defense Articles), and USC Section 710 (FEPP). See Tab 4 for details.

Asset Visibility. As a critical enabler to the responsible drawdown of forces, the Asset Visibility (AV) team tracked 100% of Theater Provided Equipment (TPE) and Organizational (ORG) home station equipment and cross-leveled equipment to fill emerging requirements in the IJOA. The primary objective of the AV team was to provide the Commander accurate, real-time visibility of equipment in the IJOA as it was repostured and cross leveled to satisfy unit shortages or fulfill operational needs statements (ONS). See Tab 5 for details.

Container Advise and Assist Teams (CAAT). The IJOA had approximately 48,000 containers listed in the system of record for tracking as of March 2011. Commanders at all levels were required to maintain accurate accounting of containers and their serviceability status within the system of record. The CAAT provided the expertise necessary to ensure proper inventory management and dispositioning of containers in support of the reposture efforts and movement of equipment within the IJOA. At its peak, there were thirty-four (34) members of the CAAT. See Tab 6 for details.

Base Operations Support-Iraq (BOS-I). The BOS-I Program evolved as the result of a necessity to provide Division Commanders with increased capabilities to operate and close bases as military resources declined during stability operations. BOS-I sourced and provided a wide range of subject matter experts to augment the Mayor's Cells at all bases throughout the IJOA, particularly as the military forces withdrew from bases. At its peak, there were 254 DoD civilians and contractors deployed across the IJOA, many of whom repositioned to larger bases as the drawdown accelerated and larger bases were in the window for closure or transition. See Tab 7 for details.

The US Mission-Iraq (USM-I) Transition. The USM-I ("the Mission") was and is still currently charged with furthering US interests within Iraq upon the withdrawal of U.S. forces. USM-I consists of the Department of State (DoS) mission, and the Office of Security Cooperation -Iraq (OSC-I). Collectively, the Mission is made up of the Embassy compound and eleven satellite sites geographically dispersed across Iraq. Two separate cultures with very different management methods proved challenging and resilient during the period of transition which set the conditions for success in Iraq, post-2011. Executing these reposture efforts with increased velocity as the drawdown accelerated in the last two months of Operation New Dawn (OND) proved challenging, and their assessment provides insights into future reposture planning and execution by the United States. See Tab 8 for details.

National Association of State Agencies for Surplus Property (NASASP): This program was a collaborative effort between numerous States, DoD agencies, and USF-I with the purpose to retrograde excess non-tactical equipment from Iraq back to the U.S. The program supported the needs of state and local governments. At the conclusion of operations in Iraq on 18 December 2011, 399 pieces of equipment ranging from band instruments to bull dozers, representing over $4 million in taxpayer dollars, were provided back to states like Alabama and Wisconsin. NASASP continues to harvest surplus equipment from Iraq from Defense Logistics Agency locations in Kuwait and back in the U.S.

Conclusion

Phase II began on 01 SEP 11 with decisive operations being retrograde and redeployment of USF-I personnel and equipment, while simultaneously conducting operational maneuver and transitioning tasks to enduring partners. The USF-I J4 was responsible for oversight and compliance of the movement and reposture efforts during Operation New Day. The J4 Joint Logistics Operations Center (JLOC) was the primary J4 staff element used to successfully manage the drawdown of commodities, services, equipment and conducting equipment transfers through the series of programs and management tools previously discussed and reviewed in this chapter. One of the key and very successful venues to calibrate the aforementioned programs was the weekly JLOC update (see chart 1).

JLOC Update: Chart 1

Joint Logistics Operations Center (JLOC) Update

- Purpose: to provide a logistical operations picture to the USF-I J4 and Staff. Synchronize efforts and anticipate requirements. Provide a LOG COP

- Frequency: weekly on Mondays

- Participants: JLOC commodities, 310th ESC, DoS-TC, BOS-I, CAAT, FEPP, USETTI, C3, and LNOs

- Venue: VTC

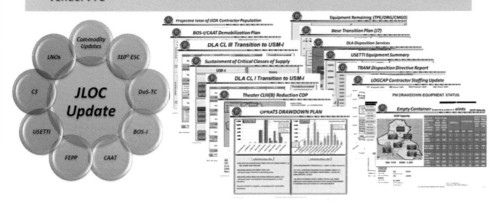

The JLOC Update was the venue where staff representatives from commodity management; Foreign Excess Personal Property (FEPP) transfers, United States Equipment Transfer to Iraq (USETTI); Container Management, Asset Visibility, Base Operations Support – Integration (BOS-I), the Department of State Transition Team, and liaisons met to brief their program status, adjust actions based on developing trends, and synchronize efforts with all interested parties.   It was intended to be used as a form of a logistics common operating picture and could adjust actions accordingly to meet changing mission requirements and regulate programs to stay on the desired course towards end-state and end-of-mission goals.

Once all commodities, services and equipment programs were on glide-path to attain successful end-of-mission metrics, the JLOC added an additional emphasis area for war termination. The J4 JLOC focused on the final close out of essential tasks and the transitioning of any residual tasks to ARCENT. The actual transitioning of tasks and the tracking of tasks transferred were managed by a series of weekly in-progress-review (IPR) updates to the J4.   All sections within the JLOC listed daily battle rhythm of tasks that would cease by end-of-mission or would transfer. The tasks that would transfer were identified, given a corresponding transfer timeline, and a point of contact or agency that the task would transfer to (see chart 2).

Transition Summary: Chart 2

Asset Visibility Transition Plan

Tasks	Status & Agency	Transfer Timeline	POC/REMARKS
Asset Vis TPE	DoS/OSC-I	1 NOV 11	
Asset Vis TPE (ETP for property book items/ONS/ESD sourcing/remission units)	ARCENT	1-10 DEC 11	
Remission Units	ARCENT/OSC-I/USFOR-A	1-10 DEC 11	
Asset Vis (Org PB support)	1st TSC/ARCENT	1-10 DEC 11	
Asset Vis (CMGO green assets process)	DoS/OSC-I	1-30 NOV	
TPE-Planner/TRAM	DoS/OSC-I	1 NOV 11	
TPE-Planner/TRAM	1st TSC/ARCENT	1-10 DEC 11	
DOS Transfer Reconciliation	402D DOS LNO	1 DEC 11	
DCG A&T Close Out	OSC-I	1 NOV 11	
FLIPLS closed or transferred	1st TSC/ARCENT	15 NOV – 1 DEC 11	
Units C2 under USF-I TPE H/R cleared	ARCENT	15 NOV – 1 DEC 11	
1-1 (1-7) CAV AV/TRAM Transition	1ST TSC/ARCENT	1-5 DEC 11	
Records management	ARCENT	15 NOV – 1 DEC 11	
Lessons Learned captured	S: 1 DEC 11	15 NOV – 1 DEC 11	
Contractor Demob	GDIT	10-15 DEC 11	

Transfer Not Started	Progressing & on track	Transfer Complete	Off Track

1

The J4's role in supporting both Iraq's transition from war to a strategic partnership and USF-I's ending of military operations was monumental. Successes were directly attributable to JLOC's oversight of the programs discussed in this chapter. Through the programs, USF-I retrograded more than 2.15 million pieces of equipment, including 39,000 wheeled vehicles that were part of the re-posturing and transition of forces. FEPP transactions processed from September 2010 through November 2011 transferred over 4.19 million pieces of equipment with a fair market value of approximately $580.5M from USF-I to the Government of Iraq. USETT-I transfers consisted of 35,160 pieces of equipment including vehicles, and communications equipment for Iraqi Security Forces (ISF) to assist them in establishing a level of operational capability necessary for the security and defense of Iraq. Support to Operation Enduring Freedom included the disposition of more than 68,400 pieces of equipment: most notably MRAPS. More than 1,110 pieces of equipment were transferred under the National Association of State Agencies for Surplus Property (NASASP) to support the needs of United States state and local governments. The equipment disposal program sold more than 153.4 million pounds of unserviceable scrap metal which benefited the U.S and Iraq governments. And finally, the meticulous and carefully calibrated de-scoping of commodities and services resulted in minimal waste and precision balance between services and support to operations. When operational requirements ended, so did services. The collective management of these programs were operated and executed within the spirit of maintaining the command's emphasis on responsible stewardship and holistically supported reposturing the force with "Honor and Success".

Subject: Operation NEW DAWN Commodity Drawdown (TAB 1)

Background: Operation NEW DAWN began on 1 September 2011 operating from USF-I OPORD 10-01.4 supporting over 80 bases within the IJOA. From a commodities perspective, this translated into supporting approximately 49,000 Service Members and 60,000 contractors (US, LN, TCN), over 46,000 pieces of rolling stock and more than double that amount of power generation and base life support with food, water, fuel, ammunition, and repair parts. This support was provided by 71 feeding establishments (MKT Supported Sites and Contracted Dining Facilities), 16 bulk fuel sites, four Ammunition Supply Points (ASP), four Mortuary Affairs Collection Points (MACP), six Supply Support Activities (SSA), and one Class IV yard.

USF-I OPORD 11-01 was released 6 January 2011 directing operations through end of mission. Phase II of USF-I OPORD 11-01 focused on the final drawdown of forces while maintaining the flexibility to respond to any manner of contingency.

Class I

Phase II began in September, 2011 with 37 feeding establishments at 30 bases supporting a headcount of approximately 49,000 Service Members and 51,000 contractors. 10 feeding establishments were transitioned during this month with one DFAC in Irbil transferring to USM-I control. During October, the headcount reduced to approximately 37,000 Service Members with approximately the same number of contractors. Feeding establishments reduced to 20 including the largest transition to USM-I of 10 DFACs thereby completing all USM-I transitions. November began with approximately 26,000 Service Members and 32,000 contractors. Eight feeding establishments were on schedule to transition by the end of November and the remaining three during December.

No specific challenges were presented during the transition/closure of USF-I feeding establishments. Strategic level suppliers maintained excellent sustainment levels without incident. As the population decreased, ordered quantities also decreased with most feeding establishments reducing to expeditionary ration cycles (A-M-A to M-M-M) at approximately 20 days from transition. As a note, Joint Base Balad (JBB) increased its MRE consumption by 54% during the first two weeks of implementing the expeditionary feeding cycle requiring an order of approximately 13,000 cases of MREs. Based on JBB's consumption data, remaining bases ordered appropriate quantities of MREs based on their anticipated population prior to changing their ration cycle to avoid any shortfalls.

E-1-1

Class III

Phase II began with nine bulk fuel sites with 19 million gallons of storage capacity and a stock objective of 14 million gallons across all three grades of fuel (JP8, Diesel Fuel (DF), and Motor Gasoline (MG)). Contingency Operating Site (COS) Marez bulk fuel farm transitioned to rolling stock storage on 15 September and Contingency Operating Base (COB) Speicher on 19 September relying on daily pushes from both Joint Base Balad (JBB) and the Northern Ground Line of Communication (NGLOC) to sustain operations. Due to a DLA bulk fuel contract issue, the NGLOC suppliers were not able to sustain the COS Marez, COB Speicher, and JBB demands. Southern Ground Line of Communication (SGLOC) requirements were increased to support JBB allowing Marez and Speicher to receive the bulk of the NGLOC supply via direct delivery. Habur Gate, the NGLOC port of entry to Iraq from Turkey, officially transitioned on 30 September eliminating the US forces presence that maintained movement control operations. The lack of a movement control team at Habur Gate meant USF-I fuel trucks would no longer be collected into convoys to meet the contracted escorts for onward movement. This event spurred the requirement for an expediter to remain at Habur Gate to ensure USF-I fuel convoys were linked up with the contracted escorts. On 1 October, COS Taji transitioned to USM-I along with the remaining fuel stocks for continued support of US Forces located at COS Taji. COS Marez continued to conduct tanker to tanker transfer operations until its transfer to the Government of Iraq (GoI) on 11 October. COB Speicher followed suit on 18 October, transitioning to USM-I control.

Due to approximately 900,000 gallons of bunker storage capability, JBB was able to reduce their bag farm footprint on 1 October relying on bunker storage and eventually tanker to tanker transfer support until base closure on 8 NOV. During this period, JBB relied on fuel support from Kuwait due to USM-I intended exclusive use of the NGLOC for fuel support.

After the closure of the JBB bulk fuel site, USF-I forecasted the requirement of additional direct fuel delivery from the two sources supplying the IJOA, the Kuwait SGLOC vendor and a Baghdad based JP8 contractor, rather than relying on the hub and spoke method. This requirement was phased in with Al Asad Air Base and Victory Base Complex (VBC) being a test for the Baghdad based JP8 contract supplemented by pushes from the southern hub, COB Adder, supplying DF and MG. The Baghdad based JP8 contract first supported the direct deliveries to Al Asad Air Base with product only. Transportation to Al Asad Air Base was provided by an independent internal contracted vendor. After three weeks, the Baghdad based JP8 supplier proved able to transport via its own assets where it continued to support until time of this report. VBC JP8 was transported via LOGCAP transportation until 22 November from the Baghdad based vendor as its location was adjacent to the bulk fuel site. DF and MG continued to be supported by COB Adder direct pushes.

E-1-2

Early in October, USF-I identified that the two week order to delivery timeline for DLA-Energy fuel to be processed and transported to bulk sites within the IJOA was too long of a lead. Consumption was decreasing so much that by the time the ordered amount arrived in Kuwait, it was exceeding the actual requirement and causing Kuwait stocks to exceed its holding capability. To prevent this occurrence, USF-I initiated placing orders directly to the Theater Sustainment Command (TSC) to fill requirements which allowed the TSC to adjust its orders directly through DLA. Essentially, TSC was directly supporting the IJOA rather than only being a conduit to DLA.

With support coming directly from TSC and the Baghdad JP8 vendor, USF-I was able to more accurately control stocks at each of its four remaining bulk fuel sites.

Figure 1 below depicts the drawdown of the on-hand balance from the beginning of Operation NEW DAWN (OND) through November 2011. Figures 2-4 depicts the theater on hand balance with corresponding theater issues for Phase II.

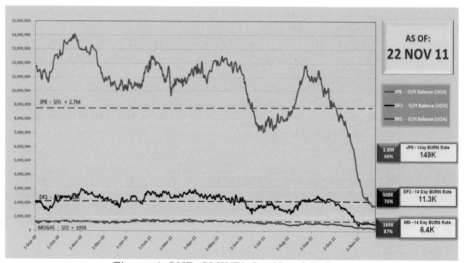

Figure 1 OND CLIII(B) On-Hand Balance

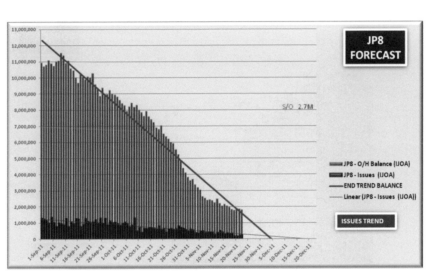

Figure 2 Phase II JP8 On Hand Balance vs. Issues

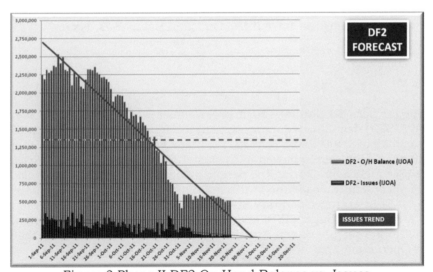

Figure 3 Phase II DF2 On Hand Balance vs. Issues

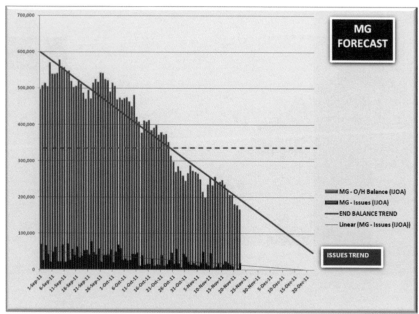

Figure 4 Phase II MG On Hand Balance vs. Issues

Class V

During the period 1 January 2011 to November 30, 2011, United States Forces – Iraq (USF-I) demilitarized, cross-leveled to Afghanistan, or retrograded approximately 3,400 Short Tons (ST) of Class V munitions to Kuwait, of which, 2,000 short tons were removed in the last 120 days leading up to November 30, 2011 (See Figure 5). During the period 1 August 2011 to 30 November 2011, USF-I successfully closed the four remaining Ammunition Supply Activities (ASA) at Contingency Operating Station (COS) Marez, Victory Base Complex (VBC), Joint Base Balad (JBB) and COS Adder on schedule with minimal impact to the six Advise and Assist Brigades (AAB) and enablers in the Iraq Joint Operations Area (IJOA).

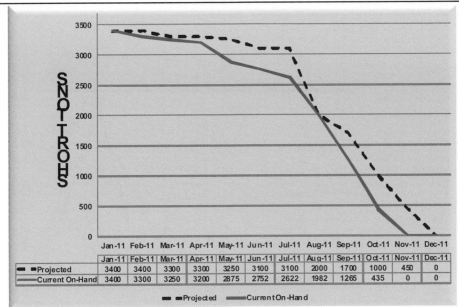

	Jan-11	Feb-11	Mar-11	Apr-11	May-11	Jun-11	Jul-11	Aug-11	Sep-11	Oct-11	Nov-11	Dec-11
■ ■Projected	3400	3400	3300	3300	3250	3100	3100	2000	1700	1000	450	0
Current On-Hand	3400	3300	3250	3200	2875	2752	2622	1982	1265	435	0	0

Figure 5 OND CL V Operations Summary

The United States Army Corp of Engineer (USACE) sponsored Coalition Munitions Disposal (CMD) Program, operating out of the Iraq Besmaya Combat Training Center (BCTC), comprehensively disposed of 663 ST of Condition Code H (CCH) (***See Note 1 below)ammunition and Explosive Remnants of War (ERW) for the period 1 January 2011 until End of Mission (EOM) 30 October 2011 (See Figure 6). Approximately 60 percent (373 ST) of these munitions were destroyed within the final 90 days of operation, averaging 124 ST per month from 1 August to 30 October 2011.

The CMD Program started in December 2008 when USF-I, formerly Multi-National Corps-Iraq (MNC-I), employed USACE Huntsville (the supporting contracting organization) to destroy retrograded Condition Code H (CCH) ammunition, and properly inspected and prepared Explosive Remnants of War (ERW), in order to provide cost effective disposition and improve safety in the operational environment.   The CMD Program, operating at an estimated 150 ST per month capacity, destroyed over 3,460 ST, averaging 102 ST per month, of CCH & ERW over a 34 month period.

Proven successes in Iraq prompted United States Force Afghanistan (USFOR-A) to establish a similar Munitions Disposal program in Afghanistan to support the planned 2014 Force Reposture.

***Note 1: CCH ammunition is defined as any unserviceable ammunition that is uneconomical to repair.

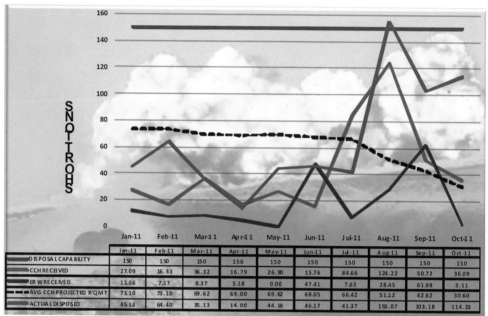

	Jan-11	Feb-11	Mar-11	Apr-11	May-11	Jun-11	Jul-11	Aug-11	Sep-11	Oct-11
DISPOSAL CAPABILITY	150	150	150	150	150	150	150	150	150	150
CCH RECEIVED	27.09	16.33	36.32	16.79	26.90	15.76	84.66	124.22	50.72	36.09
ER WRECEIVED	12.06	7.17	8.37	5.18	0.00	47.41	7.63	28.45	61.98	3.11
AVG CCH PROJECTED R'QMT	73.10	73.10	69.62	69.00	69.62	68.05	66.42	51.12	42.62	30.60
ACTUAL DISPOSED	45.13	64.40	35.13	14.00	44.18	46.17	41.37	155.07	103.18	114.23

Figure 6 OND Coalition Munitions Disposal (CMD) Operations

Ammunition Supply Activity (ASA) closures commenced 1 July 2011 in accordance with USF-I OPORD 11-01 (see above chart).

Marez Ammunition Transfer Holding Point (ATHP) – 31 July 2011
Liberty ATHP – 31 October 2011
JBB Ammunition Supply Point (ASP) – 31 October 2011
Adder ASP – 30 November 2011

310 Expeditionary Sustainment Command (ESC) conducted Synchronization Working Groups (SWG) 90 days prior to ASA closure start dates. The intent to synchronize efforts between the United States Division (USD) G4s, Brigade Support Battalion (BSB) Support Operations Officer (SPO), Support Brigade (SB) SPOs, ESC SPO Munitions Branch (MB), USF-I J4 and ARCENT G4/1st TSC. The SWG focused on ASA closure and execution; Support Courses of Action (COA) for units remaining in the USD Area of Responsibility (AOR) post ASA closures; supported unit tasks and responsibilities (ALL units); and Four Corners Operations (Kuwait) for Self-Redeploy units.   These working groups

greatly improved communications between all impacted parties and significantly minimized the chaos associated with significant changes to the support infrastructure.

In an effort to consolidate stocks as much as possible prior to ASA closures, USF-I directed units to validate mission requirements and inventory stocks for serviceability and to identify stocks either on property books or excess to mission requirements. Although USF-I did not achieve the goal of 100 percent compliance the directive did result in increased accountability and at least 200 STs or better consolidated in the ASAs for retrograde or demilitarization. The majority of these munitions were Found on Installation (FOI) as the units, either through omission or neglect, did not fully bring ALL munitions to record upon receipt from preceding unit or did not turn in unauthorized munitions, per the HQDA mandated Total Ammunition Management Information System (TAMIS), upon assumption of their AO. Although the policy and guidance come from Corps and Division levels, it is incumbent upon the subordinate commanders to enforce these policies through a variety of ways to include Techniques, Tactics and Procedures (TTP) implemented by the BSB SPOs that include informal Technical Assistance Visits (TAV) to each of their supported units.

The ESC established a firm stop-service date for routine customer transactions approximately 30 days prior to closure to eliminate competing circumstances, such as non-emergent unit turns / issues beyond the stop service dates that diverted manning and generate accountability challenges. Units were encouraged to conduct thorough mission analyses and submit final operational requirements (to include training) to the SB at least 60 days prior to the customer stop service date to ensure stock availability and issue to the unit prior to this date. The focus was on the units with Counter IED and border security missions. Simultaneously, non self-redeploying units were encouraged to identify and turn in those munitions not required for EOM. In most cases, units required nothing more than force protection and a small contingent package for the unforeseen emergency. This reduction, as mentioned in the previous paragraph, contributed to a significant consolidation of stocks for retrograde. The 310 ESC conducted manning studies (ASP focused) and concluded that relief-in-place operations between Ordnance Companies must take place no later than 60 days prior to the ASA start closure date in order to gain positive control of ALL stocks and take ownership of execution planning. They also concluded that an ASP need to maintain a FULL Heavy Lift Platoon to close out an ASP. The organic BSB SPO ATHP staff (if fully staffed) proved adequate for the mission of closing the ATHPs. Despite staffing shortfalls and other competing circumstances, (i.e., non-emergent bulk unit turn in AFTER established customer stop service dates, transportation delays, SAAS COMS, etc) the ASAs closed within directed timelines, without serious injury and loss of life and equipment.

E-1-8

Class V Contingency Operations

To sustain emergent contingency operations in light of ASA closures, USF-I J4, in coordination with USF-I J3 and USD G3/G4s developed Combat Configured Loads (CCLs) to be built and staged at a specified threshold date. The intent is to have a ready sustainment package (built on a 463L Pallet) in Kuwait that could be flown STRAT-AIR to a unit within 24-72 hours of notification. Packages are configured to the primary weapons systems that MAY be employed in the event of any established scenarios.

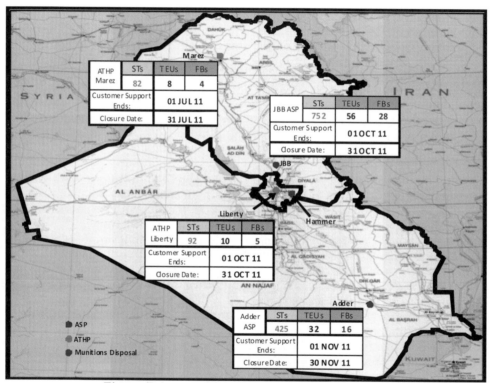

Figure 7 Ammunition Supply Activity Closures

E-1-9

SUBJECT: Equipment Drawdown (Tab 2)

As Operation New Dawn (OND) began, the USF-I J4 felt it was necessary to develop a tracking system that would portray the responsible reposture of equipment and track the weekly progress. From September 2010 until late February 2011, USF-I used the following two charts as a means to satisfy the tracking requirement and briefed these charts on a weekly basis to the J4 and the DCG-S.

Chart one depicts the green rolling stock, white rolling stock and non-rolling stock equipment remaining in the IJOA. The collected data pertaining to all of the Theater Provided Equipment (TPE) and Organizational (ORG) equipment is pulled from the Property Book Unit Supply Enhanced (PBUSE) System. All of the data pertaining to Contractor Managed/Government Owned (CMGO) equipment was pulled from Kellogg Brown & Root's (KBR) logistical database, referred to as Strategic Tactical Enterprise Asset Management (STEAM). For all other contracted agencies, the data was consolidated and submitted twice a month by Defense Contract Management Agency (DCMA) or CENTCOM Contracting Command (C3). As we transitioned from OIF to OND, the total pieces of equipment across the IJOA was just over 1.8 million items made up of over 40,000 pieces of rolling stock and approximately 1.77 million pieces of non-rolling stock. The total pieces in the IJOA, reflected on the chart above as of 19 Feb 11, stood at 1.48 million items consisting of over 37,000 pieces of rolling stock and approximately 1.45 million pieces of non-rolling stock. At that particular time, USF-I reduced on-hand quantities by 3,064 pieces of rolling stock, and another 316,174 pieces of non-rolling stock.

Chart 1

UNCLASSIFIED

Equipment Summary

		Bal. on 1 Sep	Bal. on 19 Feb	Progress to Date	Change from Last Week
Green RS	ORG	4,519	4,117	402	30
	CMGO	996	1,132	(136)	0
	TPE	13,542	11,465	2,077	20
White RS	ORG	1,027	1,178	(151)	13
	CMGO	12,936	12,051	885	0
	TPE	7,162	7,169	(7)	20
NRS	ORG	544,496	586,574	(42,078)	(7,077)
	CMGO	504,267	299,795	204,472	0
	TPE	219,035	278,019	(58,984)	2,022
	CIF/IPE	507,917	295,153	212,764	(3,127)
Total		1,815,897	1,488,554	319,244	(8,099)

- No major reductions expected until Spring of 2011
- RIP/TOAs and XSAPI fielding will cause small adjustments

Objective
1.8M
pieces

As of: 19 FEB 11

POC: MAJ Brian Steele @ DSN: 318-485-3060

E-2-1

Chart two depicts the historical trends of the on-hand balances of TPE, Organizational, and CMGO property within the IJOA. As of 19 Feb 11, most of the increases in the TPE balances were attributable to the XSAPI fielding that took place, which was at times offset by unit retrogrades. The organizational property trend line showed peaks that coincided with multiple unit RIP/TOAs. CIF/IPE warehouse stocks showed intermittent decreases in order to meet quarterly reduction goals published by USF-I. Finally, the CMGO trend line showed a number of drastic decreases tied to base closures and the transfer of equipment to the Government of Iraq via FEPP.

Chart 2

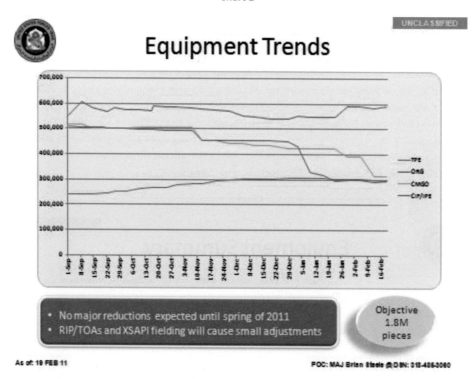

USF-I OPORD 11-01 was released 6 Jan 11 and covered USF-I operations from January 2011 through End of Mission (EOM). USF-I directed a series of Rehearsal of Concept (ROC) drills IOT ensure synchronized execution between USEMB-B, USF-I, Major Subordinate Commands (MSC) and separates. The OPORD 11-01 ROC Drill database was established by USF-I Knowledge Management (KM) section as a common way to capture, share and display information for both operations and logistics. With regards to equipment, USF-I J4 developed the equipment data points IOT determine the amount of surface lift required to retrograde out the Reposture equipment. The raw data in the 11-01 ROC Drill database enabled the J-Staff to extrapolate out the unit Reposture plans and brief the equipment Reposture requirements. By utilizing the KM data input by units within the IJOA for the March 2011 Sustainment ROC Drill, USF-I J4 was able to produce a new tracking mechanism. The new chart depicted three categories; Projected Reduction (blue line), Actual Amount Remaining (red line), and Monthly Retrograde (green boxes). The red line and green boxes were updated by conducting weekly data pulls for the information, and require no further explanation. The blue line, however, requires some clarification.

Its initial development required two approaches. The first approach concentrated on determining the drawdown of ORG and TPE, which was formed by simply using the on hand numbers (reported in PBUSE) at the end of February 2011 and subtracting the quantities of equipment the reporting units identified for retrograde during each month (March – December) of OND.

The second approach focused on the drawdown of CMGO. This approach required a little more effort and attention, but utilized the same train of thought. Historical FEPP data provided USF-I J4 with a planning factor utilized throughout OND for disposition of CMGO. The historical data illustrated that roughly 87% of equipment on CMGO property books fell into the category of items eligible for transfer to the GoI via FEPP, transfer to the GoI to satisfy USETT-I requirements, or transfer to USM-I for use during an enduring presence. The remaining 13% of equipment, mainly restricted or sensitive items, would require retrograde out of the IJOA. USF-I J4 applied these percentages to equipment located at each base in order to determine the amount of equipment that would require transportation assets during reposture efforts. Next, we analyzed the closure/transition date of each base, which was provided by the USF-I J7. Starting with the on hand quantities for end of month February, USF-I J4 subtracted the quantities of equipment (each month) tied to base closure/transition. Upon completion, the two approaches were combined thus creating the blue projection line.

 As a caveat, the projected glide path was never intended to be a measurement of our transportation capabilities, as misconstrued by many, but merely a visual tool for us to monitor the units' progress and to ensure they were adhering to the plans developed from OPORD 11-01. Any deviation throughout OND was easily captured, thus enablUSF-I J4 the ability to engage the units for resolution.

Chart three depicts the first iteration, of many, of the new tracking chart for the reposture of equipment during OND; dated 4 March 2011. All information prior to March was backdated and portrayed by utilizing historical records. As we transitioned from OIF to OND, the total pieces of equipment across the IJOA was just over 1.8 million items made up of over 40,000 pieces of rolling stock and approximately 1.77 million pieces of non-rolling stock. The total pieces in the IJOA as 4 March stood at 1.52 million items made up of over 36,000 pieces of rolling stock and approximately 1.49 million pieces of non-rolling stock. At this time, USF-I had retrograded approximately 637.5K pieces of total equipment out of the IJOA since Sep 10. We experienced a decrease in equipment quantities in all categories that week except ORG NRS. Recent RIP/TOAs attributed to the increase in ORG NRS quantities. CIF quantities decreased by more than 54K items since the previous week due to warehouse right sizing.

**It appeared as if USF-I was behind glide path at that point in time, but the projected line reflected end of month quantities and the on hand quantities were from 4 March.

Chart 3

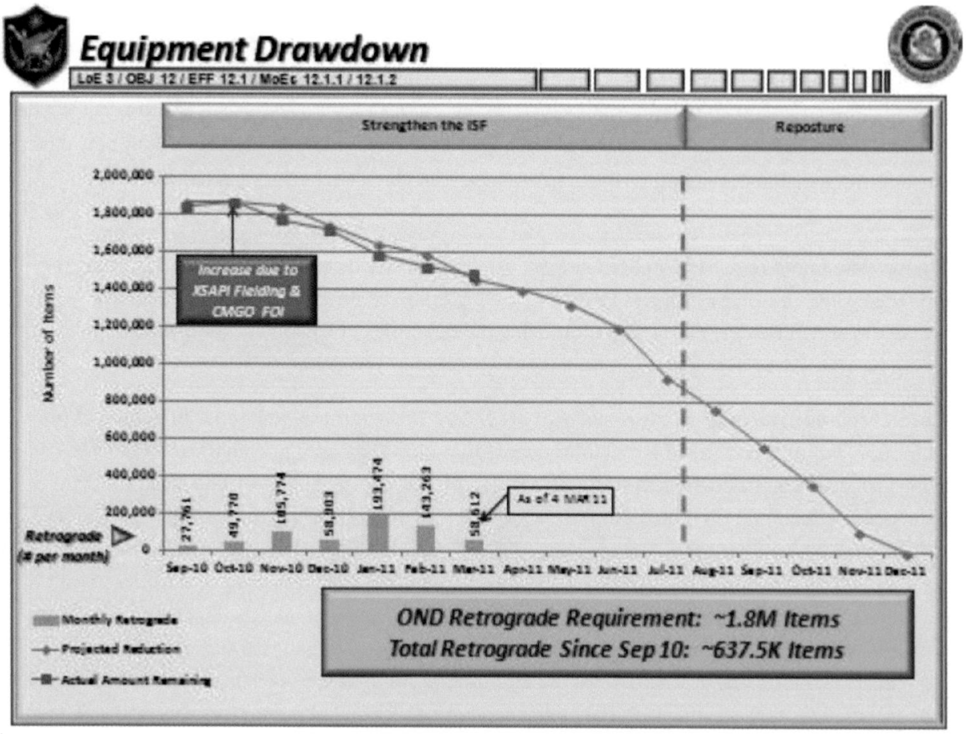

Equipment Drawdown

OND Retrograde Requirement: ~1.8M Items
Total Retrograde Since Sep 10: ~637.5K Items

From March to Late July, USF-I remained on glide path and only deviated from our drawdown plan a handful of times. These slight deviations usually resulted as a direct correlation to unit RIP/TOAs and new equipment fieldings, which eventually corrected themselves as property books were reconciled and updated accordingly. Also during this period, we refined our reporting products numerous times to include more detailed tracking charts as a means to provide more accurate data for the USF-I Commanding General and other interested parties (CENTCOM, DA, etc..). The charts segregated the data into four main equipment categories (Total, TPE, ORG, and CMGO) and then broke out even further to show three sub-categories of equipment (Total, Rolling Stock (RS), and Non Rolling Stock (NRS)) for each main category. In essence, twelve charts (four slides) were compiled:

1. Chart four: Combined (TPE, ORG, CMGO) Total, Combined RS, Combined NRS

2. Chart five: TPE Total, TPE RS, TPE NRS

3. Chart six: ORG Total, ORG RS, ORG NRS

4. Chart seven: CMGO Total, CMGO RS, CMGO NRS

Chart four depicts the progress made by USF-I with regards to all three categories of equipment from January through July. The total pieces of Theater Provided Equipment, CIF/IPE, Organizational, and CMGO equipment across the IJOA was approximately 1.86 million items made up of approximately 40,000 pieces of rolling stock and approximately 1.82 million pieces of non-rolling stock. As of 29 July 11, USF-I had repostured approximately 1.27 million pieces of equipment since September 2010. There were approximately 986.6 thousand items remaining to reposture by 31 December.

With regard to rolling stock, USF-I had repostured approximately 14 thousand pieces of rolling stock items since September, which left approximately 30.8 thousand pieces to reposture by 31 December.

For non-rolling stock, USF-I repostured approximately 1.26 million pieces of non-rolling stock items since September, which left approximately 955.8 thousand pieces of non-rolling stock items to reposture by 31 December.

By the end of July, USF-I was ahead of glide path and did not experience nor anticipate any issues at that time.

Chart 4

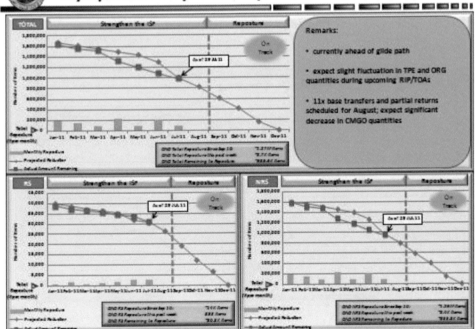

Chart five depicts the progress made by USF-I with regards to TPE reposture from January through July. As we transitioned from OIF to OND, the total pieces of Theater Provided Equipment and CIF/IPE equipment across the IJOA was approximately 747.7 thousand items made up of just over 20,700 pieces of rolling stock and approximately 727 thousand pieces of non-rolling stock. As of 29 July 11, USF-I repostured approximately 640.3 thousand pieces of theater provided equipment and CIF/IPE since September 2010. There were approximately 280.5 thousand items remaining to reposture by 31 December.

With regard to rolling stock, USF-I had repostured approximately 7.4 thousand pieces of TPE rolling stock items since September, which left approximately 15.2 thousand pieces to reposture by 31 December.

For non-rolling stock, USF-I had repostured approximately 632.9 thousand pieces of TPE and CIF/IPE non-rolling stock items since September, which left approximately 265.4 thousand pieces of non-rolling stock items to reposture by 31 December.

By the end of July, USF-I was ahead of glide path and did not experience nor anticipate any issues at that time.

Chart 5

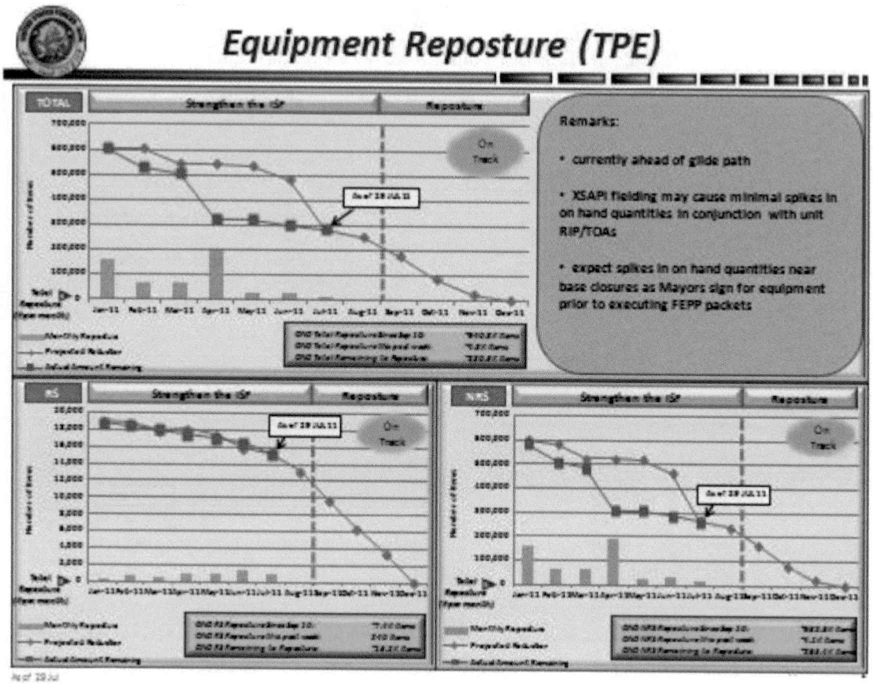

Equipment Reposture (TPE)

E-2-6

Chart six depicts the progress made by USF-I with regards to ORG reposture from January through July. As we transitioned from OIF to OND, the total pieces of Organizational equipment across the IJOA was approximately 550 thousand items made up of just over 5,500 pieces of rolling stock and approximately 544.5 thousand pieces of non-rolling stock. As of 29 July 11, we had repostured approximately 257.9 thousand pieces of organizational property since September 2010. There were approximately 383.3 thousand items remaining to reposture by 31 December.

With regard to rolling stock, USF-I had repostured approximately 3.7 thousand pieces of ORG rolling stock items since September, which left approximately 3.4 thousand pieces to reposture by 31 December.

For non-rolling stock, USF-I had repostured approximately 254.2 thousand pieces of ORG non-rolling stock items since September, which left approximately 379.9 thousand pieces of non-rolling stock items to reposture by 31 December.

By the end of July, USF-I was ahead of glide path and did not experience nor anticipate any issues at that time.

Chart 6

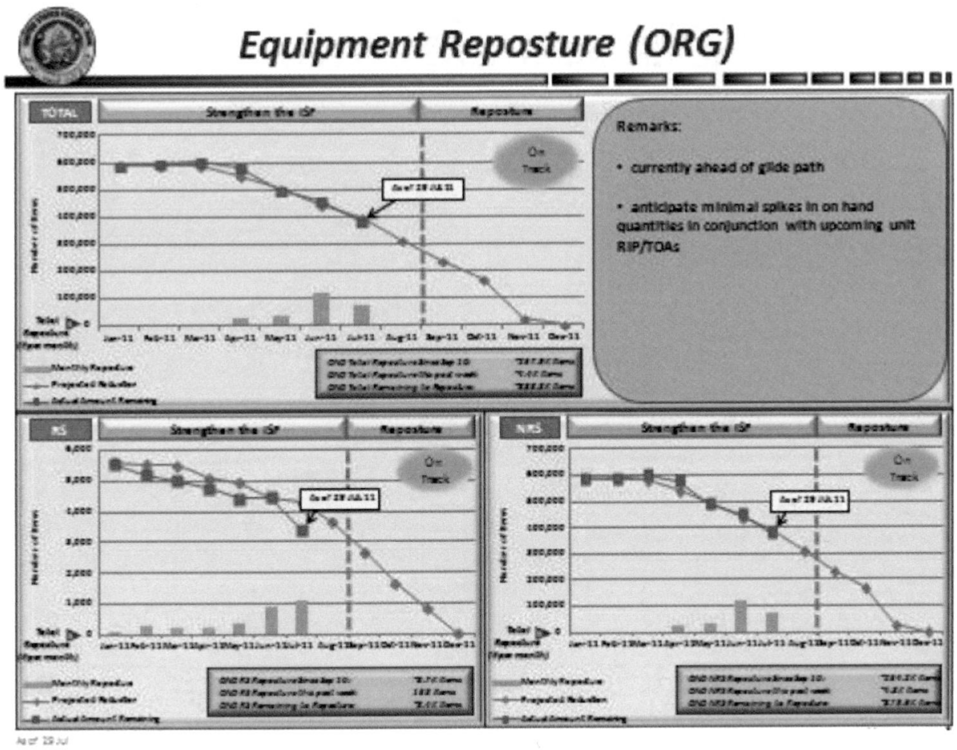

Equipment Reposture (ORG)

Chart seven depicts the progress made by USF-I with regards to CMGO reposture from January through July. As we transitioned from OIF to OND, the total pieces of CMGO equipment across the IJOA was just over 518K items made up of just under 14,000 pieces of rolling stock and approximately 504K pieces of non-rolling stock. As of 29 July 11, we had repostured approximately 371.3 thousand pieces of CMGO property since September 2010. There were approximately 322.8 thousand items remaining to repposture by 31 December.

With regard to rolling stock, USF-I had repostured approximately 2.9 thousand pieces of CMGO rolling stock items since September, which left approximately 12.3 thousand pieces to repposture by 31 December.

For non-rolling stock, USF-I had repostured approximately 368.5 thousand pieces of CMGO non-rolling stock items since September 2010, which left approximately 310.5 thousand pieces of non-rolling stock items to repposture by 31 December.

USF-I appeared to be off track, but the deviation was due to corrupted property book data reported from LOGCAP. The error was corrected the following week and USF-I remained on track.

Chart 7

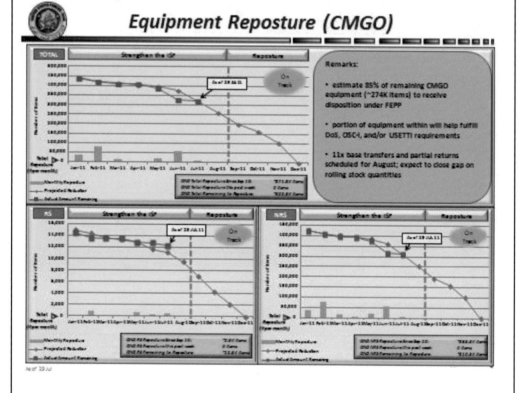

USF-I first encountered significant deviations from the glide path as mid August approached. Two major factors contributed to these deviations, one of which was beyond the control of USF-I. The first factor was the decision to deploy 4/1 AD and a couple of other smaller support units. The issue, as a result of that decision, was the fact that those units, and associated equipment, were not factored into the OPORD 11-01 ROC Drill KM Database. 4/1 CD developed their drawdown plan with the assumption they would not be replaced, and all of the data input into the database reflected their plan of reducing down to zero. This decision alone resulted in approximately 60K TPE and ORG property book items remaining in Iraq that would have otherwise repostured out of the IJOA. Additionally, a significant amount of CMGO property remained which was used in support of services provided to the units by contracting agencies.

The second factor that caused major deviation from the projected drawdown was late timing on the decisions regarding the enduring footprint the US Government planned to leave in Iraq in order to assist the GoI with additional training and advisement following OND. It was during this time that USF-I J4 noticed that the on hand quantities of equipment began to plateau. In order to maintain operational flexibility, units did not redeploy as bases continued to close, but instead repostured to other locations, thus preventing the decrease in equipment quantities. This trend continued until the final decision on our end-state was made in mid October, at which time we were informed to go to "ZERO".

Charts eight (TPE/ORG/CMGO), nine (TPE), ten (ORG), and eleven (CMGO) illustrate the significant deviations USF-I experienced between August and mid October in direct correlation with the contributing factors mentioned above.

The TPE chart (Chart Nine) illustrates that USF-I continued to reposture equipment during this time, yet the reposture efforts did not maintain the same velocity as was projected. The lack of velocity was, in part, directly related to the above conversation. Another factor contributing to the lack of decrease in TPE quantities stemmed from the Mayor Cells signing for CMGO equipment, from the contracting agencies, in order to place the items on FEPP lists. However, the reason USF-I continued to experience a minimal decrease in equipment quantities attributed to the identification of non-mission essential and/or excess equipment by the units on ground.

Clearly, as portrayed in Chart Ten below, the Org property suffered the most during this time as base closures did not equate to off ramping units. The quantities of ORG property remained the same for a little over two months (end of July until mid October) as units repostured within the IJOA. Further analysis into the ORG equipment showed USF-I J4 that a large percentage of remaining items were tied directly to boots on ground; as Soldiers depart, so will the equipment.

The CMGO chart (Chart Eleven) proves the validity of USF-I J4's plan on the reposture of CMGO equipment in conjunction with base closures. As bases closed during this time period, the equipment quantities decreased in unison.

Chart eight depicts the deviation experienced by USF-I with regards to all three categories of equipment reposture from August through mid October. As of 14 October 11, USF-I had repostured approximately 1.68 million pieces of equipment since September 2010. There were approximately 743.2 thousand items remaining to reposture by 31 December.

With regard to rolling stock, USF-I had repostured approximately 24.8 thousand pieces of rolling stock items since September, which left approximately 22.2 thousand pieces to reposture by 31 December.

For non-rolling stock, USF-I had repostured approximately 1.65 million pieces of non-rolling stock items since September, which left approximately 721 thousand pieces of non-rolling stock items to reposture by 31 December.

Total:	RS:	NRS:
Projected – 580,198	Projected – 20,240	Projected – 559,958
Actual – 743,247	Actual – 22,233	Actual – 721,014
Delta – (163,049)	Delta – (1,993)	Delta – (161,056)

Chart 8

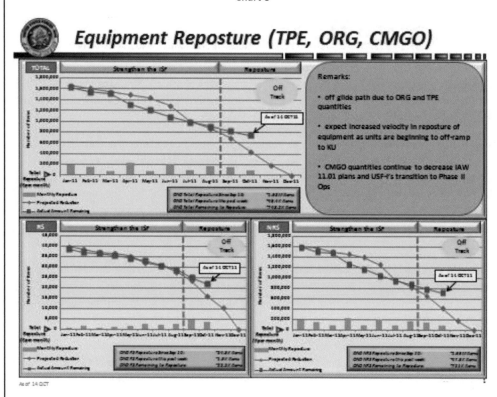

E-2-10

Chart nine depicts the deviation experienced by USF-I with regards to TPE reposture from August through mid October. As of 14 October 11, USF-I had repostured approximately 749.7 thousand pieces of theater provided equipment and CIF/IPE since September 2010. There were approximately 178.7 thousand items remaining to reposture by 31 December.

With regard to rolling stock, USF-I had repostured approximately 13.2 thousand pieces of TPE rolling stock items since September, which left approximately 10.7 thousand pieces to reposture by 31 December.

For non-rolling stock, USF-I had repostured approximately 736.5 thousand pieces of TPE and CIF/IPE non-rolling stock items since September, which left approximately 168.1 thousand pieces of non-rolling stock items to reposture by 31 December.

Total:	RS:	NRS:
Projected – 137,098	Projected – 8,250	Projected – 128,848
Actual – 178,730	Actual – 10,656	Actual – 168,074
Delta – (41,632)	Delta – (2,406)	Delta – (39,226)

Chart 9

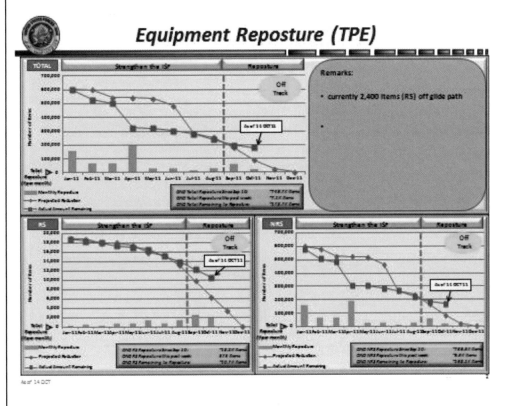

E-2-11

Chart ten depicts the deviation experienced by USF-I with regards to ORG reposture from August through mid October. As of 14 October 11, USF-I had repostured approximately 435.6 thousand pieces of organizational property since September 2010. There were approximately 362.6 thousand items remaining to repasture by 31 December.

With regard to rolling stock, USF-I had repostured approximately 5.5 thousand pieces of ORG rolling stock items since September, which left approximately 2.5 thousand pieces to repasture by 31 December.

For non-rolling stock, USF-I had repostured approximately 430.1 thousand pieces of ORG non-rolling stock items since September, which left approximately 360.1 thousand pieces of non-rolling stock items to repasture by 31 December.

Total:	RS:	NRS:
Projected – 205,865	Projected – 2,221	Projected – 203,644
Actual – 362,597	Actual – 2,465	Actual – 360,132
Delta – (156,732)	Delta – (244)	Delta – (156,488)

Chart 10

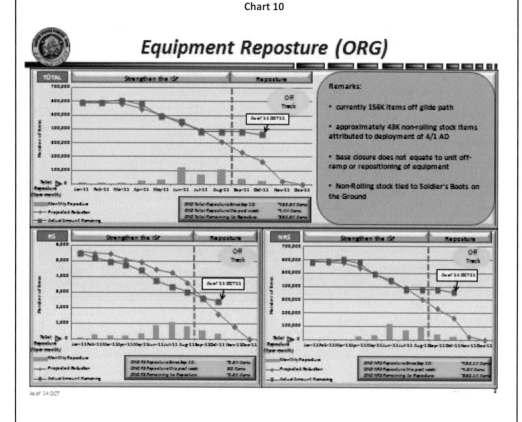

Chart eleven depicts the deviation experienced by USF-I with regards to $CMGO$ reposture from August through mid October. As of 14 October 11, USF-I had repostured approximately 492.2 thousand pieces of CMGO property since September 2010. There were approximately 201.9 thousand items remaining to reposture by 31 December.

With regard to rolling stock, USF-I had repostured approximately 6.1 thousand pieces of CMGO rolling stock items since September, which left approximately 9.1 thousand pieces to reposture by 31 December.

For non-rolling stock, USF-I had repostured approximately 486.1 thousand pieces of CMGO non-rolling stock items since September 2010, which left approximately 192.8 thousand pieces of non-rolling stock items to reposture by 31 December.

Total:	RS:	NRS:
Projected – 237,235	Projected – 9,769	Projected – 227,466
Actual – 201,920	Actual – 9,112	Actual – 192,808
Delta – 35,315	Delta – 657	Delta – 34,658

**It appeared like USF-I was off track, but the blue line depicted end of month quantities, not mid month. The numbers above explain the true story as of 14 October 2011.

Chart 11

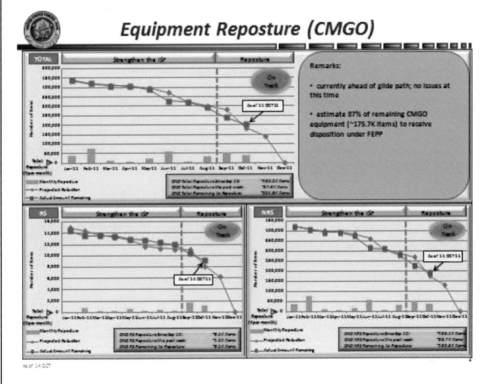

USF-I J4 re-evaluated reporting techniques following the mid October decision and made a number of changes to the charts. Key changes, which will be discussed in further detail below, included; removing the historical progress from January through May, adding a stacked bar (purple) on top of the repostured bar (green) to depict items in Kuwait, removing the blue projected line, and adding pie charts to depict key buckets of equipment.

USF-I removed the months January through May in order to highlight the progress made leading up to, and during the waterfall period of Phase II. June through November illustrated the largest decrease in equipment during OND and we felt that this time period needed increased visibility.

As reposture the force tempo increased, it became more difficult to track the movement of ORG and TPE items because PBUSE does not delineate between Iraq and Kuwait in terms of property accountability and the location of items. Regardless if items were in Kuwait, they remained on USF-I property books and continued to show on our charts. The USF-I J4 Asset Visibility chief, in conjunction with the USF-I J4 Lean Six Sigma LNO, analyzed the property book data and extracted specific UICs from the USF-I PBUSE Task Force that had either repositioned to Kuwait, or were in the process of repositioning. In terms of reposture, these items were highlighted by the purple stacked bar as "In Kuwait" and thus removed from our on hand numbers of items remaining to reposture.

Since the decision was to proceed to "Zero", USF-I J4 did not see the need to continue briefing the progress of on hand quantities as compared to the projected slope. At this point in time, the projected line only caused confusion and concern from all interested parties. We knew where we were, and where we needed to be at end-state, so the blue line was removed.

Even after the changes above were made, USF-I J4 still felt the quantities of equipment remaining were misleading and that the true story was not portrayed effectively. By conducting further analysis of property records, we developed a method of breaking out the remaining equipment into major buckets for each of the categories of equipment. These buckets are explained, for each equipment category, in charts twelve (TPE/ORG/CMGO), thirteen (TPE), fourteen (ORG), and fifteen (CMGO).

By the end of OND, the equipment reposture charts briefed on a weekly basis included the four charts discussed above (most current data in charts twelve – fifteen below) and the two charts that USF-I originally briefed at the beginning of OND (most current data in charts sixteen and seventeen below). What started out as a couple of charts designed to paint a picture of reposture efforts for the USF-I J4 turned into a reporting standard distributed to the highest level of Army leadership. Parties receiving weekly updates on the reposture status of USF-I equipment included the USF-I J4, the USF-I CG, ARCENT, CENTCOM, and Department of the Army.

E-2-14

Chart twelve depicts the progress experienced by USF-I with regards to all three categories of equipment reposture from mid October through mid November. Additionally, the chart illustrates the changes made in the reporting method. As of 17 November 11, we had repostured approximately **2.1 million pieces of equipment since September 2010. There were approximately 403.9 thousand items remaining to reposture by 31 December. This included 11.4 thousand pieces of rolling stock and 392.5 thousand pieces of non rolling stock.

There were three main categories of the 403.9K equipment items remaining:
- 23% or 92.2K items were scheduled for redistribution within the IJOA to USM-I, the GOI (via FEPP) or to LOGCAP in support of enduring locations.

- 47% or 189.3K items were scheduled as "to accompany troops" (TAT) items such as: weapons, optics, and OCIE. As Soldiers leave, so does the equipment.

- 30% or 122.3K items will require ground movement from the IJOA. This was scheduled for movement predominantly by theater trucks or self-redeployment.

Chart 12

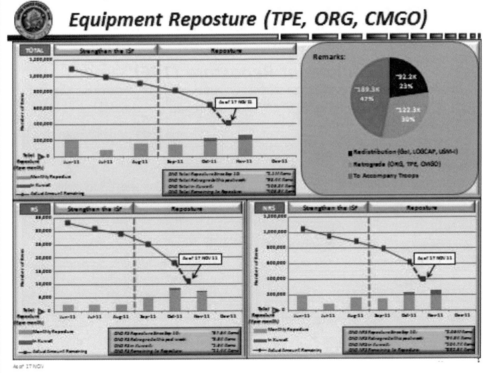

Equipment Reposture (TPE, ORG, CMGO)

**Important to note that USF-I repostured more items than beginning balance of 1.81M due to items entering the IJOA either by unit RIP/TOAs (new units brings more equipment than old unit) or equipment fieldings (XSAPI plates, MAXPRO Plus MRAPS, Tactical Wheeled Vehicles fleets).

E-2-15

Chart thirteen depicts the progress experienced by USF-I with regards to TPE reposture from mid October through mid November. Additionally, the chart illustrates the changes made in the reporting method. As of 17 November 11, we had repostured approximately 862.4 thousand pieces of equipment since September 2010. There were approximately 76.5 thousand items remaining to reposture by 31 December. This included 4.9 thousand pieces of rolling stock and 71.6 thousand pieces of non rolling stock.

There were four main categories of the 76.5K equipment items remaining:

- 56% or 43K items were GFE and other Support items that would require ground movement from the IJOA. This was scheduled for movement predominantly by theater trucks or self-redeployment.

- 29% or 22K items were XSAPI plates and were scheduled as TAT. As Soldiers leave, so does the equipment.

- 10% or 7.8K items were PLBs and were scheduled as TAT. As Soldiers leave, so does the equipment.

- 5% or 3.7K items were remaining rolling stock items that would require ground movement from the IJOA. This was scheduled for movement predominantly by theater trucks or self-redeployment.

Chart 13

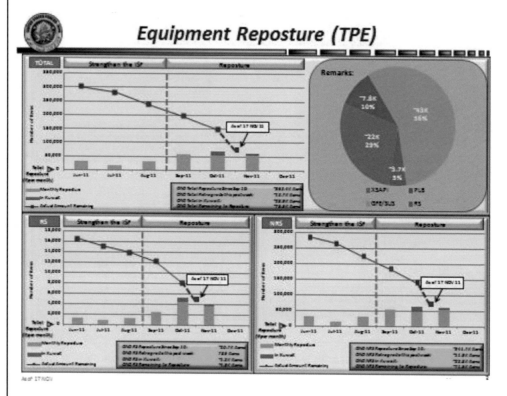

E-2-16

Chart fourteen depicts the progress experienced by USF-I with regards to ORG reposture from mid October through mid November. Additionally, the chart illustrates the changes made in the reporting method. As of 17 November 11, we had repostured approximately 644.7 thousand pieces of equipment since September 2010. There were approximately 221.4 thousand items remaining to reposture by 31 December. This included 854 pieces of rolling stock and 220.5 thousand pieces of non rolling stock.

There were four main categories of the 221.4K equipment items remaining:

- 37% or 81.1K items were weapons or optics and were scheduled as TAT. As Soldiers leave, so does the equipment.

- 28% or 61.8K items were home station and other miscellaneous items and that would require ground movement from the IJOA. This was scheduled for movement predominantly by theater trucks or self-redeployment.

- 24% or 54.3K items were OCIE equipment and were scheduled as TAT. As Soldiers leave, so does the equipment.

- 11% or 24.1K items were company level NBC type items that would require ground movement from the IJOA. This was scheduled for movement predominantly by theater trucks or self-redeployment.

Chart 14

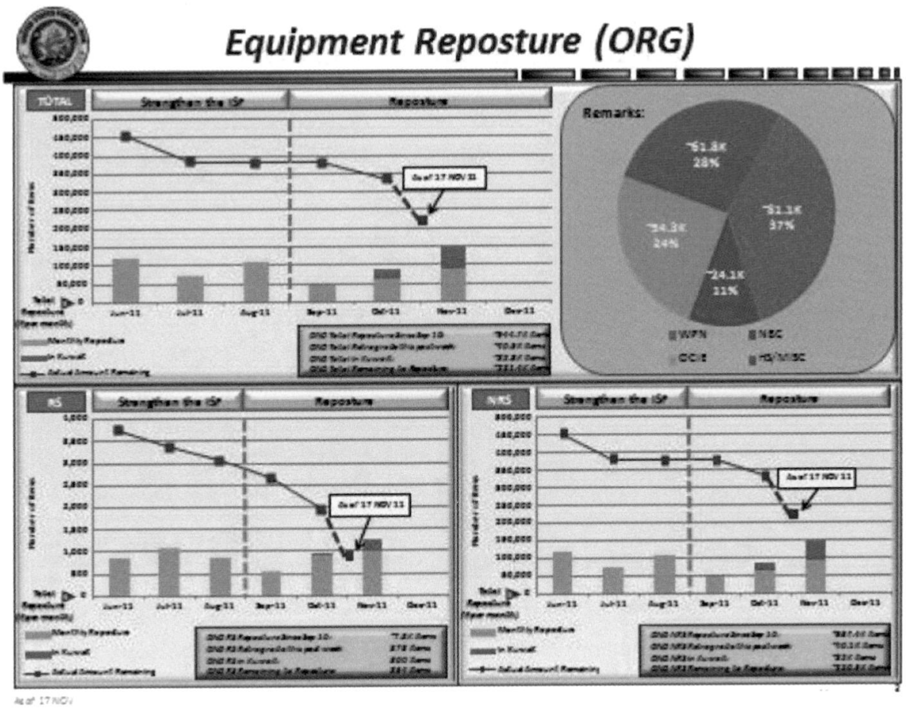

Equipment Reposture (ORG)

E-2-17

Chart fifteen depicts the progress experienced by USF-I with regards to CMGO reposture from mid October through mid November. Additionally, the chart illustrates the changes made in the reporting method. As of 17 November 11, we had repostured approximately 587.9 thousand pieces of equipment since September 2010. There were approximately 106 thousand items remaining to reposture by 31 December. This included 5.6 thousand pieces of rolling stock and 100.4 thousand pieces of non rolling stock.

There were two main categories of the 106K equipment items remaining:

- 87% or 92.2K items were scheduled for redistribution within the IJOA to either USM-I, the GOI (via FEPP) or to LOGCAP in support of enduring locations.

- 13% or 13.8K items required ground movement from the IJOA because the items were either restricted or were scheduled for redistribution to agencies outside of the IJOA. These were scheduled for movement predominantly by theater trucks or self-redeployment.

Chart 15

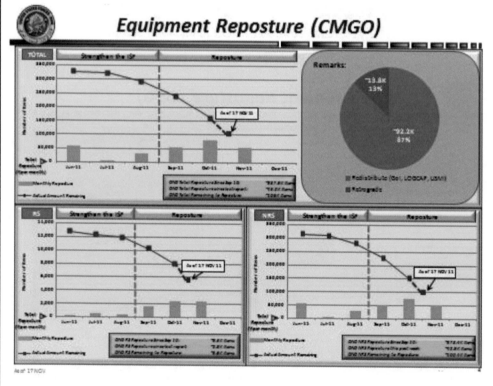

**At this time, USF-I had eight bases remaining for closure and two of those bases accounted for nearly 50% of the on hand equipment quantities remaining for reposture. The approximate 14K items requiring movement had already received TMR verification and were tracked nightly in the Equipment Drawdown Synchronization Board (EDSB).

Chart sixteen depicts the green rolling stock, white rolling stock and non-rolling stock equipment remaining in the IJOA. The collected data pertaining to all of the Theater Provided Equipment (TPE) and Organizational (ORG) equipment is pulled from the Property Book Unit Supply Enhanced (PBUSE) System. All of the data pertaining to Contractor Managed/Government Owned (CMGO) equipment was pulled from Kellogg Brown & Root's (KBR) logistical database, referred to as Strategic Tactical Enterprise Asset Management (STEAM). For all other contracted agencies, the data was consolidated and submitted twice a month by Defense Contract Management Agency (DCMA) or CENTCOM Contracting Command (C3).

As we transitioned from OIF to OND, the total pieces of equipment across the IJOA was just over 1.8 million items made up of over 40,000 pieces of rolling stock and approximately 1.77 million pieces of non-rolling stock.

The total pieces in the IJOA as of 17 November 2011 stood at ~403.9 thousand items, made up of approximately 11.4 thousand pieces of rolling stock and approximately 392.5 thousand pieces of non-rolling stock. Property in Kuwait but still on USF-I property books equaled ~106.2 thousand items, made up of approximately 1.5 thousand pieces of rolling stock and ~104.7 thousand pieces of non-rolling stock.

Chart 16

Equipment Summary (SMD)

		Bal. on 1 Sep	Bal. on 17 Nov	Progress to Date	Change from Last Week
Green RS	ORG	4,519	1,054	3,465	53
	CMGO	996	289	707	391
	TPE	13,542	4,208	9,334	39
White RS	ORG	1,027	100	927	118
	CMGO	12,936	5,345	7,591	1,943
	TPE	7,162	1,833	5,329	(639)
NRS	ORG	544,496	302,464	242,032	(2,438)
	CMGO	504,267	100,391	403,876	42,873
	TPE	219,035	94,421	124,614	9,366
	CIF/IPE	507,917	0	507,917	0
	Kuwait	N/A	106,202	106,202	N/A
Total - Kuwait		1,815,897	403,903	1,411,994	51,706

- RIP/TOAs will continue to cause adjustments

Start Point 1.81M pieces ➡ Remaining 403.9K pieces

As of 17 NOV

Chart seventeen depicts the historical trends of the on-hand balances of TPE, Organizational, and CMGO property within the IJOA.

As of 17 November 11, the organizational property trend line showed steady decreases in equipment quantities from late September through mid November attributable to units off-ramping and reposturing to Kuwait. USF-I anticipated the trend line to increase its downward velocity as units continued reposture/redeployment efforts.

The TPE trend line depicts a steady decrease in equipment quantities from mid August through mid November as units identified and retrograded non-mission essential or excess equipment. There was a significant decrease in TPE quantities in early November as units continued execution of Phase II operations.

The CMGO trend line depicts significant decreases from early August through mid November in conjunction with base closures and transfer back to the GOI. USF-I anticipated drastic decreases the rest of the month as units continued to transfer and return USF-Is largest bases.

The CIF/IPE trend line remained at zero.

Chart 17

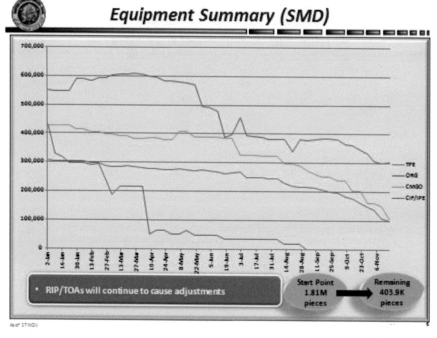

SUBJECT: Foreign Excess Personal Property (FEPP) Transfers to the Government of Iraq (Tab 3)

BACKGROUND: The Foreign Excess Personal Property (FEPP) process is the primary method of transferring excess personal property to the Government of Iraq (GOI). FEPP is declared excess by the losing military service and is redistributed in accordance with Change 2 to the CENTCOM commander's priorities for the redistribution of excess equipment from responsible drawdown in Iraq. The redistribution period is a standard 30-day process and was approved by the Office of the Secretary of Defense on 16 Oct 09.

AUTHORITIES: United States Forces-Iraq (USF-I) has worked closely with the Deputy Assistant Secretary of Defense for Logistics and Materiel Readiness (DASD L&MR) (formerly DUSD L& MR) to establish the appropriate levels of authority to transfer FEPP to the Government of Iraq. As operational requirements changed and the designated authorities were being reached, USF-I requested several increases to the current authorizations:

o   15 Jun 05 DUSD (L&MR) authorized USF-I to transfer FEPP directly to Iraqi Security Forces (ISF) MOI/MOD.
o   2 Dec 05 DUSD (L&MR) modified the authority to include all Iraqi Ministries as recipients.
o   4 Apr 08 DUSD (L&MR) established the Tiered Delegation of Authority for transferring FEPP without a base transfer.
o   6 Jun 08 DUSD (L&MR) expanded the transfer of FEPP to any approved GOI entity. Changed the terminology from "Donation" to "Transfer" of FEPP. Increased the NTV transfer to $45K.
o   19 Nov 08 DUSD (L&MR) increased property transfer from 79 FOBs to 417. Relieved USF-I of detailed accounting of barrier materiel.
o   7 Jul 09 DUSD (L&MR) Increased single FOB transfer from $5 million to $15 million. Enable the use of depreciation value rather than acquisition cost. Authorized the transfer of barrier materiel, construction material and select bridges.
o   9 Oct 09 DUSD (L&MR) increased Tiered Delegation Authority for USF-I J4 to $1 million per line item. Increased a single base transfer limit from $15 million to $30 million.
o   20 May 10 DASD (L&MR) authorized USF-I to transfer limited types of Class III (P) to the GOI
o   22 Feb 11 DASD (L&MR) authorized USF-I to exceed the $30 million base cap for the six large bases: Joint Base Balad ($200M), Victory Base Complex ($170M), Adder ($80M), Al Asad ($70M), Speicher ($65M), and Marez ($60M)
o   6 Sep 11 DASD (L&MR) authorized USF-I to transfer commercial (non-military) water purification units to the Government of Iraq
o   10 Nov 11 DASD (L&MR) authorized USF-I to increase the base caps for Al Asad and Adder to $120M each

FEPP PROCESS:

o   Unit submited FEPP list and O-6 memo nominating the property as excess to United States Government (USG) needs
o   USF-I J4 screened for USF-I requirements, U.S. Equipment Transfer to Iraq (USETTI) program and National Association of State Agencies for Surplus Property (NASASP)

o   ARCENT G4 screens for restricted items, CENTCOM requirements and declared items excess to the Army
o   Department of State (DoS) screened for internal requirements
o   Documents were prepared at USF-I J4 and packaged for signature. FEPP packet also included a legal review from USF-I Staff Judge Advocate and ARCENT G8 Cost Analysis.
o   FEPP packets were approved by the J4 and sent back to the unit to conduct the transfer

TRANSFER:

o   Transfer process was not complete until US Forces and the Government of Iraq (GOI) conducted a joint inventory and the GOI representative signed the approved FEPP inventory
o   Completed documents were maintained at USF-I J4 and sent to ARCENT G4

HISTORY: USF-I J4 (and earlier, MNC-I C4) conducted FEPP transfers to the GOI since 2007, although the data is only current from 1 Jan 09 with the inception of a Microsoft Access database tool.

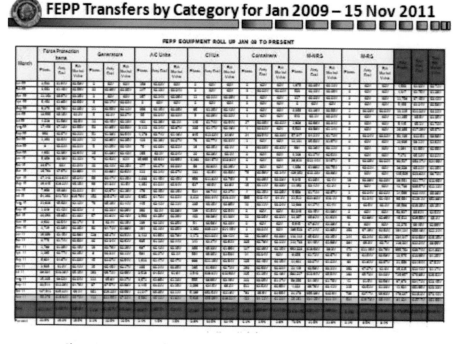

FEPP Transfers by Category for Jan 2009 – 15 Nov 2011

Chart 1. FEPP transfers by month and category, Jan 09 - present

- From January 2009 through 15 November 2011, USF-I J4 processed 954 FEPP packets and transferred over 3,976,850 items to the GoI (these items equated to an original acquisition cost of $1.3 billion and fair market value of over $439.9 million). By transferring these items to the GOI instead of shipping the items back to CONUS, the USG saved over $1.1 billion in transportation costs.

- The primary recipient of FEPP was the Ministry of Defense, although transfers were also conducted to other GOI ministries and other federal, provincial, and local Iraqi entities. Most transfers were associated with the return of fully functioning bases, but also included many tiered authority (TA) FEPP transfers not associated with base returns to the GOI. The TA FEPP process was designed as a streamlined process for USF-I to disposition property to the GOI within a specified timeline (normally within two weeks) for a specific purpose.

- During Phase 2 Reposture the Force operations, USF-I transferred 43 bases to the GOI in the final 120 days of mission of Operation New Dawn.

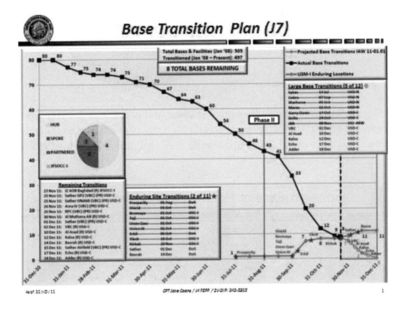

With the ability to track historical FEPP transfer data, USF-I J4 provided monthly reports to the Office of the Secretary of Defense, the Joint Staff, Department of the Army G4, CENTCOM CJ4, and ARCENT G4.

ARCENT G4, as the Title X Authority, is the theater proponent of FEPP and will maintain accountability of all FEPP records after conducting a transfer of authority with USF-I J4 on or about 15 Dec 11. They will assume FEPP processing procedures for United States Mission-Iraq at the end of Operation New Dawn. FEPP will continue to be an authorized transfer process for the USG to disposition excess property to the GOI.

The FEPP data that USF-I J4 tracked was not only reportable to Army organizations, but also in support of several other US Army Audit Agency and Government Accountability Office audits.

E-3-3

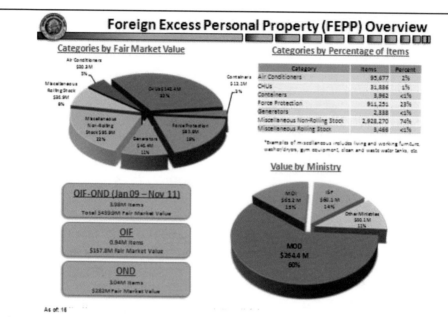

Chart 3. FEPP transfers by category (fair market value and percentage), overall fair market value transferred to each ministry, and total dollar amounts for OIF and OND, Jan 09 – present. This slide was used to brief many VIPs and USF-I senior leadership.

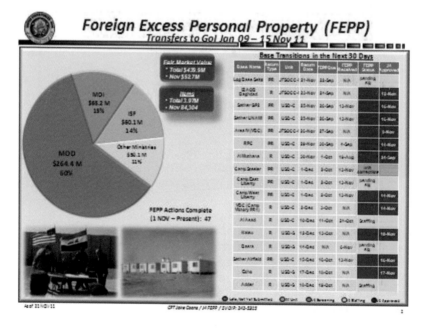

Chart 4. Weekly slide used to update USF-I J4 (and bi-monthly to the USF-I CG) on current FEPP transfers, upcoming base transitions, and the FEPP packet status for those base transitions

Open FEPP Packets

BASE/PACKET NAME	FEPP #	FEPP TYPE	BASE	J4 APPROVED	BASE RETURN/EQUIP TRANSFER DATE	REMARKS
Kalsu MUG TA #1	510	TA FEPP	USF-I J7	15-Mar-11	11-Dec-11	
Kalsu MUG TA #2	511	TA FEPP	USF-I J7	15-Mar-11	11-Dec-11	
Al Muthana AB (405)	950	Base Return	USD-C	14-Sep-11	10-Nov-11	
ISS Constitution TA #2	951	TA FEPP	USD-C	19-Oct-11	1-Nov-11	
SOF TA #1 (USD-C)	956	TA FEPP	Baghdad	26-Oct-11	9-Nov-11	
Area IV (VBC) (199.9) (Containers)	957	Partial Return	JFSOCC-I	2-Nov-11	26-Nov-11	
Kalsu (162) (Containers)	959	Base Return	USD-S	10-Nov-11	11-Dec-11	
Camp Victory PR2 (199.22) (Containers)	958	Partial Return	USD-C	14-Nov-11	1-Dec-11	
Bennett PR Taji (199.10)	961	Partial Return	JFSOCC-I	14-Nov-11	15-Nov-11	
Taji TA #15 (USD-C)	970	TA FEPP	Taji	10-Nov-11	24-Nov-11	
Z AOB Baghdad (232)	975	Base Return	CJSOTF-AP	11-Nov-11	11-Nov-11	
Taji TA #16 (USD-C)	976	TA FEPP	Taji	14-Nov-11	19-Nov-11	
Camp Victory TA #19 (USD-C)	982	TA FEPP	VBC	14-Nov-11	29-Nov-11	
Camp Steele TA #1 (USD-C)	983	TA FEPP	VBC	14-Nov-11	29-Nov-11	
Camp Victory TA #20 (USD-C)	984	TA FEPP	VBC	15-Nov-11	29-Nov-11	
Radwaniyah Palace Complex (RPC) (199.13)	987	Partial Return	JFSOCC-I	14-Nov-11	29-Nov-11	
Camp Liberty West (199.5) (Containers)	988	Partial Return	USD-C	15-Nov-11	1-Dec-11	
Echo (19) (Containers)	990	Base Return	USD-S	17-Nov-11	11-Dec-11	
Sather GP2 (199.44)	991	Partial Return	USD-C	18-Nov-11	15-Nov-11	
Sather UNAMI (199.45)	992	Partial Return	USD-C	18-Nov-11	15-Nov-11	
Sather Airfield (199.45)	993	Partial Return	USD-C	18-Nov-11	15-Dec-11	

Chart 5. Weekly slide used to update USF-I J4 on open FEPP transfers not yet completed

SUBJECT: U.S. Equipment Transfers to Iraq (USETTI) (Tab 4)

The United States Equipment Transfer to Iraq (USETTI) was one component of a comprehensive equipping strategy to enable the Iraqi Security Forces (ISF) to achieve Minimum Essential Capabilities (MEC) NLT 31DEC2011. The Program was established to transfer mostly Non-Mission Essential (NME) U.S. Forces equipment to the Government of Iraq (GoI) as an element of the responsible drawdown of U.S. Forces from Iraq.

The equipment identified for transfer under the USETTI program was governed by three Congressional Authorities as follows: (1) Section 1234 of The National Defense Authorization Act (NDAA) FY 2010 (Non Excess Defense Articles); (2) Excess Defense Articles (EDA) outlined in Section 516 of the 1961 Foreign Assistance Act (FAA), and (3) Foreign Excess Personal Property (FEPP). The FEPP items were generally Contractor Managed Government Owned (CMGO) non-standard equipments.

The USETTI program involved collaboration between the Defense Security Cooperation Agency (DSCA), Office of the Secretary of Defense, USF-I J4, HQDA G4/G8, AMC, and Iraqi Ministry of Defense and Ministry of the Interior.

The below chart 1 depicts an overview of equipment transfer process for 1234 and most 516 items. Items transferred under FEPP to USETTI were done "as is", "where is."

Chart 1

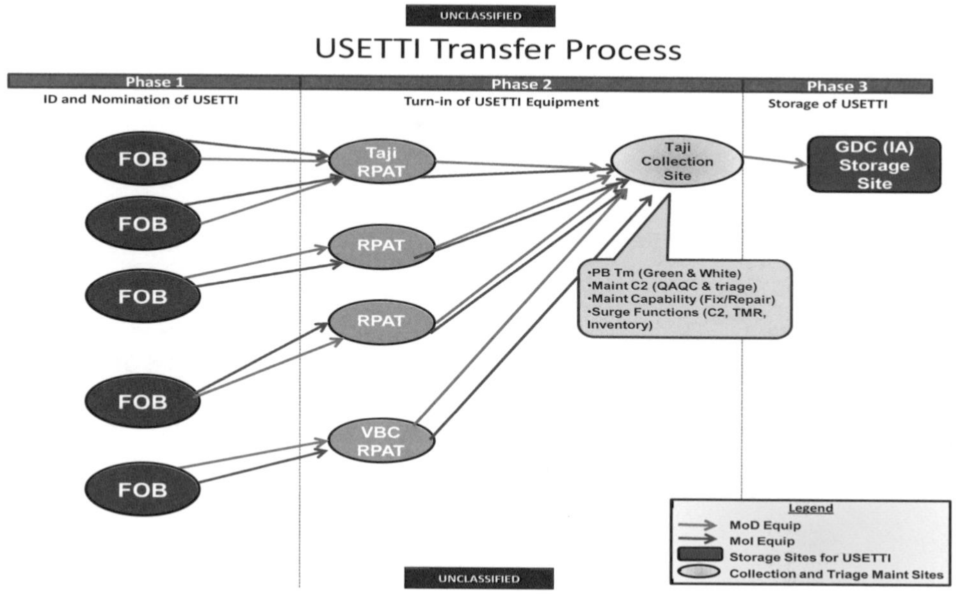

UNCLASSIFIED

USETTI Transfer Process

Phase 1	Phase 2	Phase 3
ID and Nomination of USETTI	Turn-in of USETTI Equipment	Storage of USETTI

FOB → Taji RPAT → Taji Collection Site → GDC (IA) Storage Site

FOB → RPAT

FOB → RPAT

FOB

FOB → VBC RPAT

•PB Tm (Green & White)
•Maint C2 (QAQC & triage)
•Maint Capability (Fix/Repair)
•Surge Functions (C2, TMR, Inventory)

Legend
→ MoD Equip
→ MoI Equip
▬ Storage Sites for USETTI
⬭ Collection and Triage Maint Sites

UNCLASSIFIED

E-4-1

The below chart 2 (as of Dec 2010) was used to track weekly/monthly transfer status (cumulated numbers) of Section 1234, Section 516, and FEPP. The chart shows the total threshold, number of items transferred, required remaining items, and the total transferred percentages. At this particular time, the threshold was 52,605 items and transferred total was at 14,440 items.

Chart 2

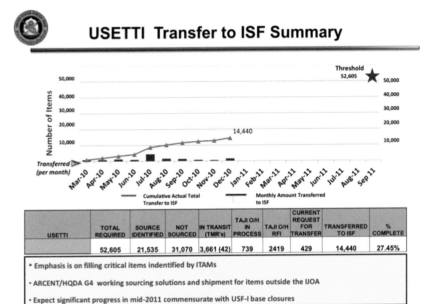

USETTI Transfer to ISF Summary

USETTI	TOTAL REQUIRED	SOURCE IDENTIFIED	NOT SOURCED	IN TRANSIT (TMR's)	TAJI O/H IN PROCESS	TAJI O/H RFI	CURRENT REQUEST FOR TRANSFER	TRANSFERRED TO ISF	% COMPLETE
	52,605	21,535	31,070	3,661 (42)	739	2419	429	14,440	27.45%

- Emphasis is on filling critical items indentified by ITAMs

- ARCENT/HQDA G4 working sourcing solutions and shipment for items outside the IJOA

- Expect significant progress in mid-2011 commensurate with USF-I base closures

The below chart 3 depicts the fill rate and the projection rate by each authority. Each authority threshold numbers were reviewed about every 90 days in order to validate the requirement by each Iraqi Training Advising Mission (ITAM) in order to discuss the possibility of reducing/deleting items based on the sourcing challenges. Once items were determined to be no longer required, a request was submitted to the Deputy Commanding General Advising and Training (DCG/A&T) for approval. The DCG/A&T is the approving authority (3 Star General Officer or equivalent) on changing the number of required threshold for the USETTI program. Once approval was obtained from DCG/A&T, it was submitted to the Defense Security Cooperation Agency (DSCA) for official notification.

Chart 3

USETTI Maximum Fill Projections by Authority

USETTI Overall Fill (Threshold = 52,605 Items)
- Current Fill 27.4% (14,440)
- Projected Final Maximum Fill: 75%

Section 1234: Project 100% Fill

Section 516:
- Project 100% Fill
- LOR in progress for 13,942 items: 12,233 items are Body Armor, Helmets, M60 Machine Guns

FEPP:
- Project Maximum Fill of 75%
- Requires continuous screening of multiple non-standard data bases and reconciliation of numerous nomenclature descriptions

Section 1234 (1,478 Items)
Current Fill: 81%
Projected Final Fill: 100%

- ■ Transferred
- ☐ Sourced
- ■ Un-sourced

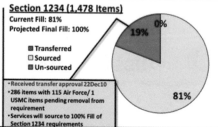

- •Received transfer approval 22Dec10
- •286 items with 115 Air Force/ 1 USMC items pending removal from requirement
- •Services will source to 100% Fill of Section 1234 requirements

Section 516 (23,168 Items)
Current Fill: 23.8% (5,504)
Projected Fill 100%

- ■ Transferred
- ☐ Sourced
- ■ Un-sourced
- ■ Sourcing Outside IJOA (HQDA G4 Working)

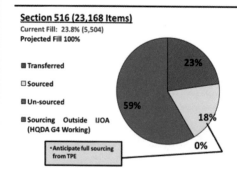

- •Anticipate full sourcing from TPE

FEPP (27,959 Items)
Current Fill 31.96% (8,936)
Projected Final Fill 75%

- ■ Transferred
- ☐ Sourced
- ■ Un-sourced

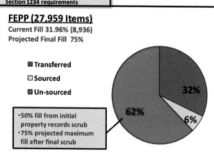

- •50% fill from initial property records scrub
- •75% projected maximum fill after final scrub

By end of October 2011, three reduction sessions had been held which removed a total of 10,091 items from the original requirement listing. These items were removed for reasons such as requirements being outdated, items no longer remaining in theater, or the Government of Iraq (GoI) no longer required the item. The total threshold was adjusted from 52,605 pieces to 42,514 pieces. The below chart 4 depicts the overall data as of Nov 2011.

Chart 4

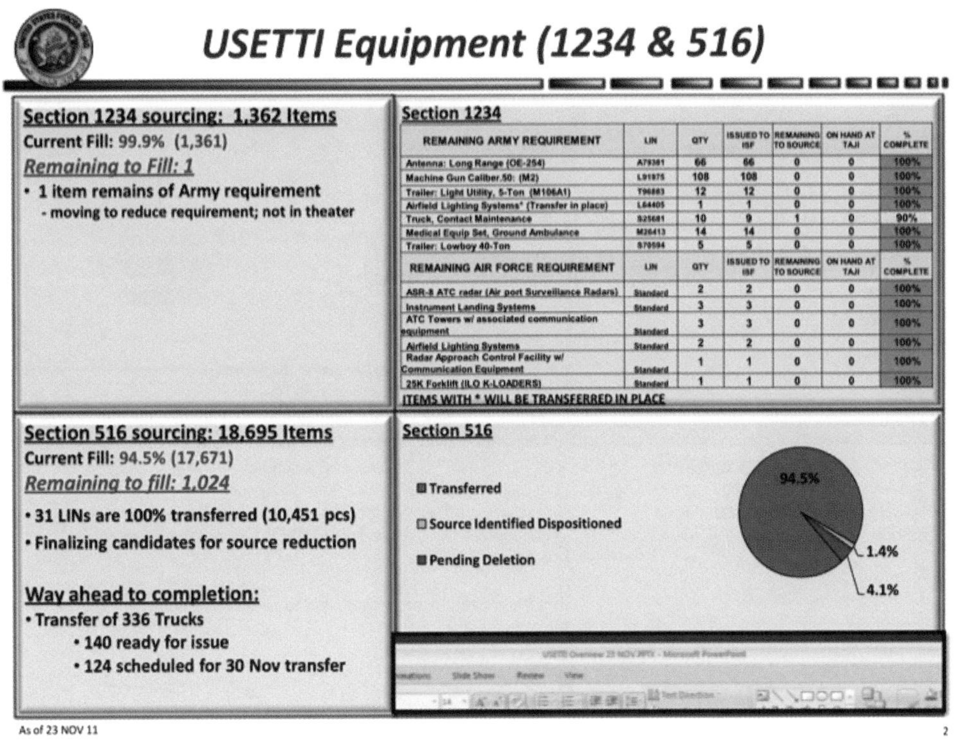

USETTI Equipment (1234 & 516)

Section 1234 sourcing: 1,362 Items

Current Fill: 99.9% (1,361)

Remaining to Fill: 1

- 1 item remains of Army requirement
  - moving to reduce requirement; not in theater

Section 1234

REMAINING ARMY REQUIREMENT	LIN	QTY	ISSUED TO IBF	REMAINING TO SOURCE	ON HAND AT TAJI	% COMPLETE
Antenna: Long Range (OE-254)	A79381	66	66	0	0	100%
Machine Gun Caliber.50: (M2)	L91875	108	108	0	0	100%
Trailer: Light Utility, 5-Ton (M106A1)	T96883	12	12	0	0	100%
Airfield Lighting Systems* (Transfer in place)	L64405	1	1	0	0	100%
Truck, Contact Maintenance	S25681	10	9	1	0	90%
Medical Equip Set, Ground Ambulance	M26413	14	14	0	0	100%
Trailer: Lowboy 40-Ton	B70594	5	5	0	0	100%
REMAINING AIR FORCE REQUIREMENT	**LIN**	**QTY**	**ISSUED TO IBF**	**REMAINING TO SOURCE**	**ON HAND AT TAJI**	**% COMPLETE**
ASR-8 ATC radar (Air port Surveillance Radars)	Standard	2	2	0	0	100%
Instrument Landing Systems	Standard	3	3	0	0	100%
ATC Towers w/ associated communication equipment	Standard	3	3	0	0	100%
Airfield Lighting Systems	Standard	2	2	0	0	100%
Radar Approach Control Facility w/ Communication Equipment	Standard	1	1	0	0	100%
25K Forklift (ILO K-LOADERS)	Standard	1	1	0	0	100%

ITEMS WITH * WILL BE TRANSFERRED IN PLACE

Section 516 sourcing: 18,695 Items

Current Fill: 94.5% (17,671)

Remaining to fill: 1,024

- 31 LINs are 100% transferred (10,451 pcs)
- Finalizing candidates for source reduction

Way ahead to completion:
- Transfer of 336 Trucks
  - 140 ready for issue
  - 124 scheduled for 30 Nov transfer

Section 516

- Transferred
- Source Identified Dispositioned
- Pending Deletion

94.5%
1.4%
4.1%

As of 23 NOV 11

2

As of Nov 2011, USF-I has transferred 17,671 items under the Section 516 (94.5%), 1,361 items under the Section 1234 (99%), and 16,130 under the FEPP authority (71.8%). The FEPP percentage will continue to rise in conjunction with closure of remaining bases. The below chart 5 depicts the data for each authority.

Chart 5

USETTI Equipment (FEPP)

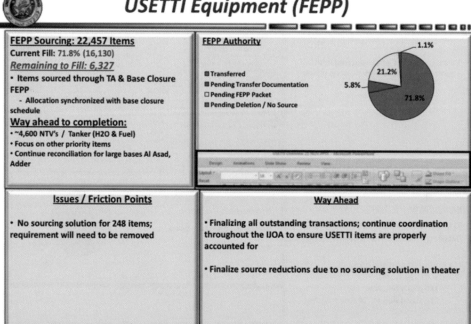

FEPP Sourcing: 22,457 Items
Current Fill: 71.8% (16,130)
Remaining to Fill: 6,327
- Items sourced through TA & Base Closure FEPP
  - Allocation synchronized with base closure schedule

Way ahead to completion:
- ~4,600 NTV's / Tanker (H2O & Fuel)
- Focus on other priority items
- Continue reconciliation for large bases Al Asad, Adder

FEPP Authority
- Transferred
- Pending Transfer Documentation
- Pending FEPP Packet
- Pending Deletion / No Source

1.1%
21.2%
5.8%
71.8%

Issues / Friction Points
- No sourcing solution for 248 items; requirement will need to be removed

Way Ahead
- Finalizing all outstanding transactions; continue coordination throughout the IJOA to ensure USETTI items are properly accounted for
- Finalize source reductions due to no sourcing solution in theater

As of 23 NOV 11   3

Subject: Total Asset Visibility and Theater Provided Equipment (TPE) Disposition Process (Tab 5)

Background: The Asset Visibility (AV) Team was required to track 100% TPE and Organizational (ORG) home station equipment and to cross-level to fill emerging requirements that was validated by the USF-I, ARCENT and HQDA. The team initially consisted of five personnel including four contractors during Phase I operation and two personnel at the beginning of Phase II operations.

The primary objective of the AV Team was to ensure the accurate and real time visibility of equipment in the IJOA, to source IJOA requirements and provide administrative input to Logistics FRAGO and to coordinate logistics actions with AMC, ARCENT and HQDA G4 on Theater policies and procedures. The use of contractors significantly increased support capabilities for the J4. This allowed Logistics J4 personnel to focus on other critical areas such as the processing of numerous Financial Liability Investigation of Property Loss's (FLIPL) and maintaining accountability of all equipment assigned to the J4 staff and to augment the J4 rear Command Post in Kuwait.

Through collaboration with the Knowledge Management Office, we were able to tailor reports and easily conduct queries for over 288,630 pieces of TPE and over 12,501,160 pieces of ORG equipment for over 1400 Derivative and ORG Unit Identification Codes (D/UICs) that span across all services and ACOMs including USAREUR, FORSCOM and USARPAC. This database was used on a regular basis by the USF-I staff and subordinate commands to quickly access data for cross leveling, reporting or accountability purposes.

Several tools used to provide asset visibility on equipment in the IJOA. The first one is the Consolidated Property Listing (CPL), Line Item Number (LIN) to provide the amount and dollar value of equipment on hand. The second tool is the Equipment Forecast Availability Tool (EFAT), which provides a tool to determine when equipment is available for redistribution within the IJOA.

Prior to Dec 09, manual excel spreadsheets and USF-I FRAGOs was used to obtain disposition for equipment and shipment out of the IJOA to fill ARCENT requirements and or for reset to the Source of Repair. The implementation of the Theater Redistributable Asset Management (TRAM) tool also known as TPE-Planner in Jan 2010 eliminated spreadsheets at the Brigade and higher level. This tool allowed greater visibility of equipment at the wholesale and retail levels and the customer the ability to view the status of their visibility as it was vetted through the process. The unit submits their non mission essential equipment listing to the next higher level and the equipment is nominated the Brigade or Separate Battalion levels, which starts the vetting process. After the vetting process the Life Cycle Management Command (LCMC) would issue disposition for turn-in to the Retrograde Property Assistance Team (RPAT) or Defense Logistics Agency (DLA) Disposition Services (DS).

The unit could then start the turn in and or lateral transfer process after receipt of the disposition instruction from TRAM/TPE Planner. After obtaining feedback to the units, it was discovered that friction points existed with the disposition process and the USF-I J4 established a deep dive to resolve these friction points.

Organization equipment was only cross leveled if authorization was received from ARCENT and or HQDA only. National Guard and Reserve equipment required HQDA approval. This disposition for this equipment was filed reset, the unit conducted reset operation at home station or Sustainment reset where the unit received shipment disposition for turn to the RPAT and follow-on shipment to the Source of Repair (SOR). Equipment with SOR disposition is dropped for the unit's property book after turn-in to the RPAT yards.

The deep dive established several key areas subjected to friction points throughout the disposition process. The submission of the excel spreadsheet from the unit it the Battalion was 5-30 days. This led to the perception that the entire disposition process could take up to a month or longer to receive disposition for equipment and commander lost faith in the vetting and disposition process. Reclaims and the constant need to cross level assets to satisfy emerging requirements at the Div, Corps and ARCENT level led to further delays. Reclaims due to incorrect disposition, to change condition codes or pull asset from turn-in to lateral transfer caused further delays at units receiving disposition. Serviceable equipment incorrectly sent to the Defense Logistics Agency-Disposition Services (DLA-DS) caused delays at the LCMC level due to reclaims to change from DLA-DS turn in to ship to Sierra Depot for future use. In August 2011 HQDA G4 issued new guidance to establish what equipment could be shipped to DLA-DS or screened in Kuwait for future use.

The lateral transfer of equipment was conducted through Phase II, up until November 2011. The biggest challenge with cross leveling TPE was the actual condition of the equipment and determining accurate availability dates of the equipment.

Property Book Accountability. 100% scrubs of Theater Provided Equipment (TPE) and organizational (ORG) property book records conducted by USF-I J4 asset visibility indicated various accountability issues such as: split property book errors (home station and deployed); serial number discrepancies; inaccurate visibility and accountability of critical equipment; poor accounting of equipment procured in theater; PM / ASA (ALT) managed TPE not on TPE hand receipts; etc. These issues did not create significant impacts during steady state operations; but if not addressed will affect phase II operations by slowing the flow of retrograde and decreasing the accuracy of property transfers and accountability.

Reconciling and correcting property records early enabled USF-I units to enter phase II operations with clean books, thus minimizing Financial Liability Investigations of Property Loss (FLIPL) and administrative corrections during peak turn-in and Reposture periods. The conduct of an Army Audit Agency (AAA) evaluation further assisted the command to obtain an outside assessment on the status of unit property books and if units had good accountability procedures in place. Also the development of a tool to track Financial Liability Investigation of Property Loss (FLIPL), Administrative Adjustment Reports (AAR) and Found on Installation (FOI) assisted with setting the conditions by issuing guidance via FRAGO to ensure compliance and issue Theater specific polices for the equipment retrograde process.

The operation to ensure accurate accountability consisted of three phases: Phase I inventory. Phase I began immediately upon receipt FRAGO 0333 and consisted of all units conducting 100% inventory of org and TPE property books and AAA simultaneously conducting property accountability audits of select units throughout the IJOA.

Phase II correcting deficiencies. Phase II consisted of units submitting adjustment documents for Found on Installation (FOI) property and any property book discrepancies identified during inventories. Unit's processed required FLIPLs through the supporting S4 and supporting property book office.

Phase III compliance and reporting. Phase III consists of units reporting compliance of USF-I directives.

Steady State & Phase II: Disposition Directive Timeline Report. Prior to Phase II, analysis was conducted to ensure that disposition request for TPE was issued within a 9 day standard from the date of nomination to the date that disposition was issued. Pending decisions from an operational standpoint, the disposition timelines increased from 3 to 9 days due to equipment held in system and not issued disposition pending decision for incoming units and mission changes.

Critically Managed Items List (CMIL). In Mar 11, it was discovered that the standing USF-I procedures to manage the release of critically managed TPE from the IJOA required further codification. There was a huge turn-over of units and key personnel involved in the process. The simultaneous efforts in support of OPORD 11.01 required a timely, well coordinated process that ensured the operational commander and USF-I staff to be involved in identifying capabilities that were required to support 11.01 Phase I and Phase II operations. After the requirements for select critically managed items across the IJOA have been satisfied the decision was made to release critical capabilities from the IJOA.

The USF-I J35, in coordination with capability proponents, maintained the Critically Managed Items List (CMIL), which identified specific TPE equipment by Line Item Number (LIN) which was deemed critical. With the rapidly approaching end of mission, replacing critical capabilities which were inadvertently released from the IJOA became increasingly difficult.

Items on CMIL generated a yellow flag in TRAM/TPE-P to notify units that the item is on the CMIL. CMIL items required a Memorandum for Record (MFR) signed by a 0-6 commander classifying the item as non-mission essential. When there was a backfilling unit within six months, the memorandum required a comment of concurrence from the incoming unit.

The unit routed the MFR through their higher headquarters to the USF-I J4. The J4 AV, received the MFR and verified that there are no unsourced requirements for this critically managed item within the IJOA. If no open requirement exists the J4 forwards the request with the recommendation to release to J35 FM.

The J35 FM confirmed no existing or emerging requirements for the capability. If no unsourced requirement exists, the J35 FM initiated a letter of release for USF-I Chief of Staff (COS) or Deputy Chief of Staff (DCOS) approval. After approval for release the J4 notifies the unit to enter the item into TRAM/TPE-P for disposition instructions.

Managing Equipment in the IJOA by Category. Due to the magnitude of equipment remaining in the IJOA, it was important to know what type of equipment remained in Theater by category. The normal means was to identify Rolling Stock (RS) by Green or White and Non Rolling Stock (NRS). The breakdown of major equipment such as Personal Locator Beacons (PLB), weapons and weapons support systems and XSAPI issued as individual equipment showed the command that although the numbers high, until the reduction of service members occurred there would not be a significant reduction of equipment maintained on hand.

The sourcing of equipment to Office of Security Cooperation-Iraq (OSC-I) and the Department of State (DoS) was a challenge, but thorough coordination with the USM-I transition cell and the USF-I unit's and contractors, over 97% of equipment on hand.

USF-I FRAGO 0984 simplified the property transfer process from USF-I to OSC-I. USF-I units located on enduring OSC-I bases were given the authority to laterally transfer equipment used to operate the base to the OSC-I primary hand receipt holder on location at the base without requesting disposition in the TRAM/TPE-P. This equipment included force protection, communications and base support equipment. HQDA EXORD support to DoS helped streamline the transfer of standard and non standard equipment to DOS under one of three authorities, loan, sale from stock or as excess defense articles.

Exception to Policy (ETP) to transfer equipment between property books.
To reduce the amount of serviceable non standard equipment sent to DLA-DS for destruction, units were given the option of transferring equipment from Theater property books to the ORG property books in order to redeploy this equipment to home station for future use without submitting a formal ETP memorandum. TPE available for this transfer consisted of items with a dollar value of 5k or less or any serviceable equipment issued disposition instructions to DLA-DS), formerly known as DRMO.

Units identified equipment on their TPE property books for transfer to the ORG property books that meets the directed retention criteria. Once identified, units will submit requests in spreadsheet format for transfer to the USF-I J4 POCs. USF-I J4 confirmed that enduring requirements do not exist and a backfill unit is not projected. USF-I J4 provided authority IAW HQDA G4 guidance to transfer the property from the TPE property book to the redeploying unit's ORG property book to the Theater Property Book office. The TPE property book officer executed the transfer using the Property Book Unit Supply Enhanced (PBUSE) system.

SUBJECT: Container Advise and Assist Teams (CAAT) (Tab 6)

Since the beginning of Operation Iraqi Freedom (OIF) and the war on terror, the United States Government has purchased and/or leased containers for use by military forces on ground. The containers mainly consisted of four different categories comprised of 20-foot dry storage; 20-foot refrigeration container; 40-foot dry storage; and 40-foot refrigeration container. The Fair Market Value (FMV) associated with these containers ranged anywhere from ~$5.2K to ~$43.3K. During the course of OIF, the Iraq Joint Operations Area (IJOA) acquired roughly 48,000 containers totaling approximately $145.8M.

As talks began on the removal of American forces from the IJOA, USF-I realized that the Container Control Officers (CCOs) assigned to each Geographical Location (GEOLOC) could use increased assistance from outside sources in regards to container management in order to account for and redistribute the vast amount of containers in a responsible manner. The personnel assigned to assist with this mission formed the Container Assessment and Assistance Team (CAAT) and arrived in the IJOA in May 2010 in preparation for the responsible withdrawal of US Forces.

Chart one below explains the mission, team composition, completed actions, and lessons learned by the CAAT assigned to the United States Forces - Iraq (USF-I) J4 during Operation New Dawn (OND).

Chart 1

UNCLASSIFIED

Container Assessment and Assistance Team (CAAT)

Mission

Establish/execute a CAAT to assist Commanders, FOB Mayors, and CCOs in accounting for containers in IJOA and facilitate removal of unknown containers with MRTs for disposition of containers/material.

Team Composition

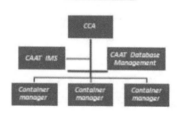

What We Did

> CAAT Members arrive 3 May 10-First Site Review 5 May 10
> Developed CAAT Standard Operating Procedures.
> Issued Multiple Frago's to Support USF-I Container Management for IJOA
> Site Visits included Sweeps of over 47,000 Containers and Re-Sweeps for over 30,000 Containers.
> Container Management Reflected an Overall Efficiency of 78% based on Containers in IBS-CMM and Containers Physically On-Site.
> Documented Results of Inventories and Provided this as a Deliverable to CCO's and Mayor Cells upon final Outbriefs.

Lessons Learned

> Container Control Officers
• Units of All Services should Assign and Train Container Control Managers as a Primary Duty Prior to Deployment
• Container Management should be part of the Joint Training & Doctrine to support CM as a Core Function.

> CAAT Personnel
• Provide Pre-deployment Orientation of the SOP for Container Sweeps.
• Train personnel on IBS-CMM prior to Deployment.
• Ensure Prior coordination is Accomplished for Computer Access at FOB's prior to Site Visits.

1

E-6-1

CAAT sweeps, and increased analysis of the information stored in the Integrated Booking System - Container Management Module (IBS-CMM) database, proved that the management of containers within the IJOA was neglected for the better part of a decade. CAAT members noticed a large discrepancy between the quantity of containers reported in IBS-CMM and the number of containers inventoried/identified during base sweeps. In an effort to correct the discrepancy, USF-I published a Fragmentary Order (FRAGO) in March 2011 outlining updated guidelines on container management for units within the IJOA.

Unit commanders at all levels are accountable and/or responsible as appropriate, for maintaining accurate inventory records of all containers under their control or in their GEOLOC regardless of ownership. The IJOA had approximately 48,000 containers listed in IBS-CMM in March that required verification with a physical hands-on inventory to ensure accuracy of the IJOA container count. The FRAGO directed that no later than (NLT) the 20th of each month, all unit commanders would account for and identify the serviceability of all containers in their possession by utilizing IBS-CMM in order to ensure the containers were fully ready for use. USF-I also published a reporting matrix, which required units to provide weekly updates, that included on hand quantities of containers (seaworthy and non-seaworthy), and projected container requirements (Organizational (ORG) and Theater Provided Equipment (TPE)) for redeployment needs.

Chart two depicts the status of containers as of 15 June 2011. The upper left quadrant illustrated the total number of containers reported in IBS-CMM and the inventory status, by United States Divisions (USDs), for the month of June. The percentage reflects the number of containers physically inventoried against the quantity listed in IBS-CMM. The percentages inventoried were low, and historically, averaged 79-82% every month. This discrepancy led USF-I to the assumption that a vast majority of containers listed in IBS-CMM were no longer physically in the IJOA and that a plan needed to be developed to resolve the issue.

The upper right quadrant identified the container type for all of the containers that were listed in IBS. The 20ft flat-racks, 40ft flat-racks, ISU-90s, 20ft flat-beds, and ROWPU were later removed from IBS-CMM and no longer tracked in the system.

The lower left quadrant portrays the container ownership as reported in IBS-CMM and highlights the delta between the IBS-CMM on hand numbers and the on hand numbers reported by USF-I units. Notice the delta of ~17.8K containers as eluded to earlier.

The lower right quadrant demonstrates the projected container disposition. Take note of the ORG/TPE requirements and the projected numbers for Foreign Excess Personal Property (FEPP) transfer. The ORG and TPE quantities were derived from the Government Owned/CMGO container quantities in the lower left quadrant. The FEPP/DRMO quantity was the remaining quantity of containers after subtracting all other quantities of containers from the IBS total. At this period of time, USF-I estimated ~27K containers would receive disposition to the Government of Iraq via FEPP or would be turned in to Defense Logistics Agency-Disposition Services (DLA-DS) as unserviceable.

Chart 2

Containers Located/Retrograde/Transferred

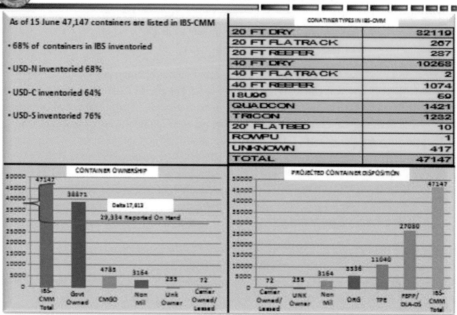

As of 15 June 47,147 containers are listed in IBS-CMM

- 68% of containers in IBS inventoried

- USD-N inventoried 68%

- USD-C inventoried 64%

- USD-S inventoried 76%

CONTAINER TYPES IN IBS-CMM	
20 FT DRY	32119
20 FT FLATRACK	207
20 FT REEFER	287
40 FT DRY	10268
40 FT FLATRACK	2
40 FT REEFER	1074
18U06	69
QUADCON	1421
TRICON	1232
20' FLATBED	10
ROWPU	1
UNKNOWN	417
TOTAL	47147

Chart three depicts updated data as of 3 November 2011. This chart also illustrates the GEOLOC closure glide path that USF-I implemented in September 2011.

The upper left quadrant highlights the hard work put forth by all container management personnel. In less than five months time, the total number of containers reported in IBS-CMM decreased by nearly 20K, and the delta with the on hand numbers reported by USF-I units decreased by nearly 7K.

The upper right quadrant shows three lines that portray the projected container disposition. The red line illustrates the IBS-CMM on hand numbers, the orange line illustrates the Unit reported on hand numbers, and the green line illustrates the number of containers required for redeployment. USF-I estimated ~11K containers would transfer to the Government of Iraq via FEPP. The remaining quantity would be put to use by United States Mission-Iraq (USM-I), Logistics Civil Augmentation Program (LOGCAP), and Army and Air Force Exchange Services (AAFES) in support of enduring presence.

The lower left chart depicts the glide path for GEOLOC closure. The blue line indicates USF-Is projection, the red line depicts the actual number of GEOLOCs open, and the black line depicts the estimated number of GEOLOCS that would remain in support of the enduring mission. The red line was elevated, but it portrayed information as of the beginning of the month and the projection reflects end of month numbers. USF-I had requested the closure of an additional 39 GEOLOCs after this report.

Chart 3

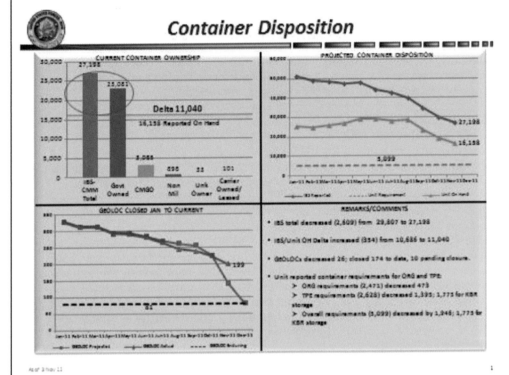

In conclusion, more than 27K containers valued at $82.2M remain in the IJOA that require proper disposition prior to the closure of OND. USF-I estimates that roughly 5.1K will be used for redeployment needs in order to retrograde the remaining ORG and TPE quantities. Approximately 8k containers will remain in the IJOA and will be used in support of the enduring US footprint, which will be managed by USM-I, AAFES, and LOGCAP. Expect approximately 3K containers to transfer to the GoI via FEPP, which will ultimately save the US Government approximately $20M in transportation costs alone. The remaining 11K containers, or the delta between IBS and on hand numbers reported by the units, will be handled in accordance with applicable regulations and policies concerning container management as a collaborative effort between Global Container Management (GCM), Container Management Element (CME), Army Intermodal and Distribution Platform Management (AIDPMO), and the Surface Deployment and Distribution Command (SDDC).

Subject: BOS-I Program Civilian and Contract Support (Tab 7)

Background: The BOS-I Program evolved as the result of a necessity to provide USDs with increased capabilities to operate and close bases as military resources declined during stability operations.

The primary objective of the BOS-I Program was to obtain a wide range of non-management, oversight and technical support services necessary to augment Mayor Cell units throughout the IJOA. The use of contractors and civilians significantly increased support capabilities for each base, and allowed the commander on the ground the ability to conduct operations and continue partnership with Iraqi units in their area of operations.

USF-I OPORD 11-01 was released 6 Jan 11 and covered USF-I operations from January 2011 through End of Mission (EOM). As a result of this OPORD, there was an increase in demand for BOS-I personnel. On 2 October, the decision was made to increase the BOS-I workforce to support both the transition to the OSC-I and DOS by 30 additional contractors to support the transition.

The BOS-I Management Support Services provided contractors to perform as operations technicians, land management specialists, information technicians, and Base Assistance and Advisory Teams (BAAT). These assets assisted in the base closure process, and provided day to day mayor cell operations support, which were either eliminated or significantly reduce as a part of the reduction of the US forces. Additionally, the Civilian Expeditionary Workforce (CEW) recruited DoD civilians to provide support to the Mayor cell on each site. These positions included the Director of Logistics (DOL), Director for Public Works (DPW), Force Protection (FP), Information Technology (IT), and Deputy Mayor for the base.

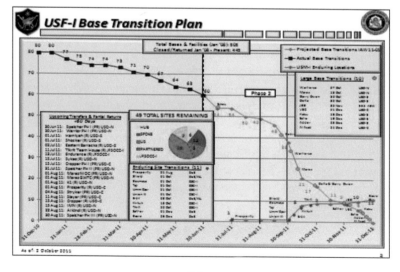

E-7-1

Slide 1 USF-I Base Transition Plan (2 October 2011)

Phase II: Reposture the Force. Phase II began on 01 September 2011 and continued until 31 December 2011. The primary mission for the BOS-I teams during this phase was to assist the USDs and Mayor Cells Reposture the Force, and identify, account, dispose, and transfer real and excess personal property. This effort comprised a large portion of the time and process required for the responsible transition of facilities and areas to the GoI.

As USF-I transitioned into Phase II, Military personnel were greatly reduced in the IJOA based on a presidential directive. Most military Base Operations Support personnel were reduced to less than 33% the previously staffed levels. The 25 BOS-I supported bases were tasked with base operations support and transitioning bases to the (Government of Iraq) GoI, Department of State (DOS), or the Office of Security Cooperation- Iraq (OSC-I). While 400 bases were already closed or transitioned, the remaining 63 bases were much larger, and required a more robust management and technical staff due to their complexity. Using a fill rate of 85%, there were 224 contractor personnel on ground, whose efforts were directed specifically toward property visibility at the mayor cell level for final equipment disposition. The majority of property on transitioning bases received disposition through the Foreign Excess Personal Property (FEPP) Process. The BOS-I teams were critical in ensuring this process included a hands on inventory, and bring to record equipment found on installation (FOI), ending with the approved transfer of real and personal property to the GoI as final base transition occurred.

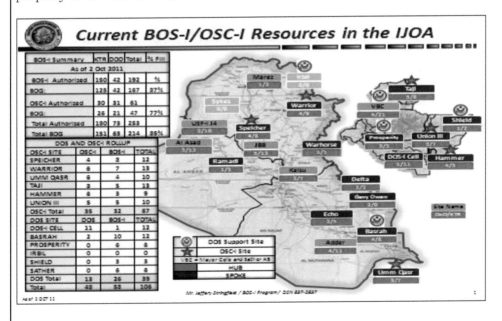

E-7-2

As Phase II USF-I transitioned into November and December, performance of the BAAT became critical, as units were required to prepare the Foreign Equipment Property Program (FEPP) documentation necessary to transfer equipment, supplies, or materiel to the GoI or IA. During Phase II (October-November time frame), the focus shifted from base operations and sustainment to preparing proper FEPP documentation. The BOS-I Program Office worked intensively with USDs to ensure mayor cells had qualified contract staff augmentation on board to assist the units with:

- Planning and synchronizing base closure property inventories, both real property and excess personnel property
- Preparing FEPP packages; inventory and supply transactions of CMGO property from the contractor to the USG property books; mayor cell operations
- I/T help desk support to mayor cells as strategic communications down sized.
- Providing asset visibility and finalizing preparation of FEPP packages prior to submission to the J4.

BOS-I Drawdown Planning. As bases were closed in the USD-N, the drawdown plan reflected contractor and DoD civilian transition requirements. During this process, the intent was to descope personnel positions at the closing sites. In addition, as the transition to GOI or DOS was completed, key personnel were redeployed, in many cases ahead of schedule to assist with the closure of problematic bases in the Baghdad or USD-S Area of Responsibility.

E-7-3

Our demobilization dates take into consideration the lead time required for the contractors to redeploy from the IJOA.

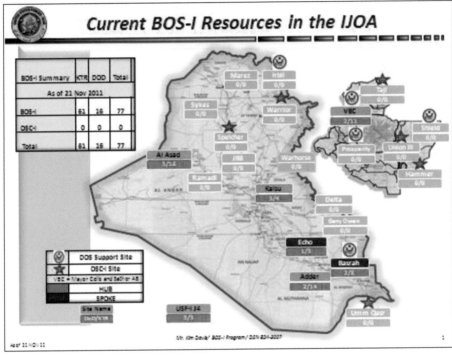

Slide 4 (Drawdown Plan)

This slide depicts the glide slope depicting off ramping of contractors from the IJOA. This demobilization plan was staffed through the USDs.

On 21 November, there were 77 personnel remaining on the BOS-I program. Our draw down projections incorporates a 10 day site release date to redeploy contractors from of the IJOA. Therefore our goal is to remove contractors 10 dates ahead of the actual base closure date.

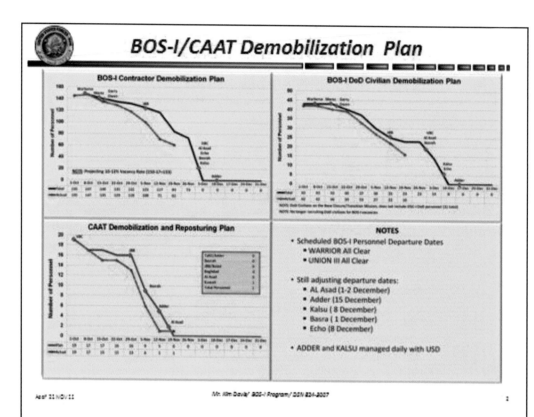

By 8 December, the Program Manager/Deputy Program Manager will have all LINC contractors out of the IJOA.

Bottom line. DoD civilians and contractors were a valuable asset to the USDs and mayor cells in executing the final base closure mission of a successful transition to DoS, OSC-I and the GoI.

Bibliography

1st Brigade Combat Team, 1st Cavalry Division, U.S. Forces–Iraq, "U.S. Forces Turn in Equipment in Support of Drawdown," press release, October 30, 2011. As of November 13, 2011: http://www.USF-Iraq.com/news/headlines/us-forces-turn-in-equipment-in-support-of-drawdown

1st Theater Sustainment Command, "1st TSC 2010 Initiatives and Highlights," fact sheet, August 2011, Not available to the general public.

1-8 Cavalry Battalion, 2d Brigade, 1st Cavalry Division, untitled after-action report, November 18, 2011.

5 Code of Federal Regulations Part 534, "Basic Pay for Employees of Temporary Organizations," *Federal Register*, Vol. 67, No. 17, January 25, 2002, p. 3581. As of April 29, 2013: http://www.opm.gov/fedregis/2002/66-003581-a.pdf

5 U.S. Code 3161, Employment and Compensation of Employees, January 3, 2012.

10 U.S. Code, Armed Forces.

22 U.S. Code, Foreign Relations and Intercourse.

22 U.S. Code 4802, Responsibility of Secretary of State, January 7, 2011.

22 U.S. Code 4805, Cooperation of Other Federal Agencies, January 7, 2011.

Abdul-Zahra, Qassim, "Al-Qaeda Making Comeback in Iraq, Officials Say," *USA Today*, October 9, 2012. As of November 8, 2012: http://www.usatoday.com/story/news/world/2012/10/09/al-qaeda-iraq/1623297/

Abdul-Zahra, Qassim, and Rebecca Santana, "Iraqis Want U.S. Trainers, Without Immunity," *Army Times*, October 4, 2011. As of November 30, 2012: http://www.armytimes.com/news/2011/10/ap-iraq-wants-trainers-without-immunity-100411/

"Abu Mahdi and Iran's Web in Iraq," United Press International, October 7, 2010. As of April 17, 2013: http://www.upi.com/Top_News/Special/2010/10/07/Abu-Mahdi-and-Irans-web-in-Iraq/UPI-52371286478366/

Ackerman, Gary L., statement, in *The Transition to a Civilian-Led U.S. Presence in Iraq: Issues and Challenges*, hearing before the Committee on Foreign Affairs, U.S. House of Representatives, 111th Cong., 2nd Sess., November 18, 2010, pp. 48–49. As of June 3, 2013: http://www.gpo.gov/fdsys/pkg/CHRG-111hhrg62399/content-detail.html

———, statement, in *Preserving Progress in Iraq, Part III: Iraq's Police Development Program*, hearing before the Subcommittee on the Middle East and South Asia, Committee on Foreign Affairs, U.S. House of Representatives, 112th Cong., 1st Sess., November 30, 2011. As of June 3, 2013: http://www.gpo.gov/fdsys/pkg/CHRG-112hhrg71400/html/CHRG-112hhrg71400.htm

Adolphson, Jason, "2nd Bn., 401st ASB Aids Drawdown," *Outpost*, May 2010, p. 4. As of August 3, 2012:
http://www.aschq.army.mil/supportingdocs/402nd_AFSB_Outpost.pdf

Ahmed, Hevidar, "Arabs Claim Kurds Aren't Implementing Article 140," *Rudaw*, December 14, 2011.

———, "Oil Companies Are Security Guarantee for Kurdistan, Experts Say," *Rudaw*, August 30, 2012.

Albert, Stewart, and Edward C. Luck, *On the Endings of Wars*, Port Washington, N.Y.: Kennikat Press, 1989.

Alkadiri, Raad, "Oil and the Question of Federalism in Iraq," *International Affairs*, Vol. 86, No. 6, 2010, pp. 1315–1328.

Allawi, Ali A., *The Occupation of Iraq: Winning the War, Losing the Peace*, New Haven, Conn.: Yale University Press, 2007.

al-Salhy, Suadad, "Iraq Lawmakers Approve 2011 Budget of $82.6 Billion," Reuters, February 20, 2011. As of April 26, 2013:
http://www.reuters.com/article/2011/02/20/us-iraq-budget-idUSTRE71J1SP20110220

———, "Iraq Blocks Syria-Bound North Korean Plane, Suspects Weapons Cargo," Reuters, September 21, 2012. As of September 21, 2012:
http://www.reuters.com/article/2012/09/21/us-syria-crisis-iraq-idUSBRE88K19720120921

Anderson, Liam, "Power-Sharing in Kirkuk: Conflict or Compromise?" paper prepared for Globalization, Urbanization, and Ethnicity Conference, Ottawa, December 3–4, 2009

Arango, Tim, "Prime Minister Puts Power-Sharing at Risk in Iraq," *New York Times*, December 21, 2011. As of April 17, 2013:
http://www.nytimes.com/2011/12/22/world/middleeast/iraqi-leader-threatens-to-abandon-power-sharing.html

———, "U.S. Planning to Slash Iraq Embassy Staff by as Much as Half," *New York Times*, February 7, 2012a. As of April 17, 2013:
http://www.nytimes.com/2012/02/08/world/middleeast/united-states-planning-to-slash-iraq-embassy-staff-by-half.html

———, "U.S. May Scrap Costly Efforts to Train Iraqi Police," *New York Times*, May 13, 2012b. As of April 17, 2013:
http://www.nytimes.com/2012/05/13/world/middleeast/us-may-scrap-costly-effort-to-train-iraqi-police.html

Arango, Tim, and Michael S. Schmidt, "Despite Difficult Talks, U.S. and Iraq Had Expected Some American Troops to Stay," *New York Times*, October 21, 2011. As of April 17, 2013:
http://www.nytimes.com/2011/10/22/world/middleeast/united-states-and-iraq-had-not-expected-troops-would-have-to-leave.html

Army National Guard, 224th Sustainment Brigade, "Guard Mobile Redistribution Team Supports Iraq Drawdown," press release, undated.

Arraf, Jane, "Iraqi Kurds Train Their Syrian Brethren," al-Jazeera, July 23, 2012. As of November 12, 2012:
http://www.aljazeera.com/indepth/features/2012/07/201272393251722498.html

Austin, III, Lloyd J., "Statement of GEN Lloyd James Austin III, Commanding General, United States Forces–Iraq," *Iraq: The Challenging Transition to a Civilian Mission*, hearing before the Committee on Foreign Relations, U.S. Senate, 112th Cong., 1st Sess., February 1, 2011a, pp. 12–13.

————, "Statement of Gen Lloyd J. Austin III, USA, Commander, U.S. Forces–Iraq," *Hearing to Receive Testimony on United States Policy Toward Iraq*, Washington, D.C.: Committee on Armed Services, U.S. Senate, February 3, 2011b. As of April 17, 2013:
http://armed-services.senate.gov/Transcripts/2011/02%20February/11-02%20-%202-3-11.pdf

Ayotte, Kelly, "Statement of Senator Kelly Ayotte," in *The Final Report of the Commission on Wartime Contracting in Iraq and Afghanistan*, hearing before the Subcommittee on Readiness and Management Support, Committee on Armed Services, U.S. Senate, 112th Cong., 1st Sess., October 19, 2011.

Aziz, Raber Younis, "Talabani Criticized for Designating Kirkuk 'Jerusalem of Kurdistan,'" *AKnews*, March 9, 2011.

Babbit, Brandon, "Webster: Drawdown and Buildup Dubbed 'Nickel II,'" *Desert Voice*, April 14, 2010.

"Bad Blood Again," *The Economist*, April 2, 2009. As of September 19, 2012:
http://www.economist.com/node/13415505

Baker, III, Fred W., "Quicker Is Better for Turkish Military Operations in Iraq, Gates Says," American Forces Press Service, February 28, 2008. As of September 26, 2012:
http://www.eucom.mil/article/20654/quicker-better-turkish-military-operations-iraq

Baron, Kevin, "Ayotte Says State Department Unequipped for Iraq Work," *National Journal*, October 20, 2011.

————, "Pentagon Extends Iraq Mission Funding," *Foreign Policy*, October 2, 2012. As of April 17, 2013:
http://e-ring.foreignpolicy.com/posts/2012/10/02/pentagon_extends_iraq_mission_funding

Bauer, Shane, "Iraq's New Death Squad," *The Nation*, June 3, 2009.

Beecroft, Robert Stephen, testimony before the Committee on Foreign Relations, U.S. Senate, September 19, 2012. As of December 6, 2012:
http://congressional.proquest.com/congressional/result/pqpresultpage.previewtitle/$2fapp-gis$2fpoltrn$2f09190003.s32/Hearing+Transcripts?accountid=25333

Bensahel, Nora, Olga Oliker, Keith Crane, Richard R. Brennan Jr., Heather S. Gregg, Thomas Sullivan, and Andrew Rathmell, *After Saddam: Prewar Planning and the Occupation of Iraq*, Santa Monica, Calif.: RAND Corporation, MG-642-A, 2008. As of April 17, 2013:
http://www.rand.org/pubs/monographs/MG642.html

Biden, Joseph, "Remarks by Vice President Biden at Event to Honor U.S. and Iraqi Servicemembers," Aw-Fal Palace, Baghdad, December 1, 2011. As of April 17, 2013:
http://www.whitehouse.gov/the-press-office/2011/12/01/remarks-vice-president-biden-event-honor-us-and-iraqi-servicemembers

Blas, Javier, "Iraq's Oil Output Overtakes Iran's," *Washington Post*, August 10, 2012. As of April 17, 2013:
http://articles.washingtonpost.com/2012-08-10/world/35493885_1_oil-output-west-qurna-oil-watchdog

Boswell, Eric, J., "Security Boost: DS Meets Challenge Posed by Iraq Drawdown," *State Magazine*, May 2011a, pp. 30–31. As of April 17, 2013:
http://digitaledition.state.gov/publication/?i=68339

———, "Statement of the Honorable Eric J. Boswell, Assistant Secretary for Diplomatic Security, U.S. Department of State," in *The Diplomat's Shield: Diplomatic Security and Its Implications for U.S. Diplomacy*, Washington, D.C.: Subcommittee on Oversight of Government Management, the Federal Workforce, and the District of Columbia, Committee on Homeland Security and Governmental Affairs, U.S. Senate, June 29, 2011b.

Bowen, Jr., Stuart, W., "Testimony of Stuart W. Bowen, Jr.," in *Transition in Iraq: Is the State Department Prepared to Take the Lead?* hearing before the Committee on Oversight and Government Reform, U.S. House of Representatives, September 23, 2010, p. 5.

———, "Statement of Stuart W. Bowen, Jr., Inspector General Office of the Special Inspector General for Iraq Reconstruction," in *U.S. Military Leaving Iraq: Is the State Department Ready?* hearing before the Subcommittee on National Security, Homeland Defense, and Foreign Operations, Committee on Oversight and Government Reform, U.S. House of Representatives, 112th Congress, 1st Sess., March 2, 2011a. As of June 4, 2013:
http://www.gpo.gov/fdsys/pkg/CHRG-112hhrg67365/html/CHRG-112hhrg67365.htm

———, "Statement of Mr. Stuart W. Bowen, Jr., Inspector General, Office of the Special Inspector General for Iraq Reconstruction," in *Preserving Progress in Iraq, Part III: Iraq's Police Development Program*, hearing before the Subcommittee on the Middle East and South Asia, Committee on Foreign Affairs, U.S. House of Representatives, 112th Cong., 1st Sess., November 30, 2011b.

———, "Statement of Stuart W. Bowen, Jr.," in *Assessment of the Transition from a Military- to a Civilian-Led Mission in Iraq*, Subcommittee on National Security, Homeland Defense, and Foreign Operations, Committee on Oversight and Government Reform, U.S. House of Representatives, 112th Cong., 2nd Sess., June 28, 2012. As of April 17, 2013:
http://oversight.house.gov/hearing/assessment-of-the-transition-from-a-military-to-a-civilian-led-mission-in-iraq-2/wp-content/uploads/2012/09/2012-06-28-SC-Natl-Sec.pdf

Bowman, Tom, "As the Iraq War Ends, Reassessing the U.S. Surge," National Public Radio, December 16, 2011. As of October 1, 2012:
http://www.npr.org/2011/12/16/143832121/as-the-iraq-war-ends-reassessing-the-u-s-surge

Branigin, William, "U.S. Shift from Iraq to Afghanistan Presents Massive Logistical Operation for Army," *Washington Post*, April 2, 2010. As of July 30, 2012:
http://www.washingtonpost.com/wp-dyn/content/article/2010/04/02/AR2010040202087.html

Brannen, Kate, "Combat Brigades in Iraq Under Different Name," *Army Times*, August 19, 2010. As of September 17, 2012:
http://www.armytimes.com/news/2010/08/dn-brigades-stay-under-different-name-081910/

Bremer, L. Paul, "Security Assessment," memorandum for Secretary Rumsfeld, Office of the Secretary Defense document number 01562-04, Rumsfeld Library, February 3, 2004.

Bremer, L. Paul, and Malcolm McConnel, *My Year in Iraq: The Struggle to Build a Future of Hope*, New York: Simon and Schuster, 2006.

Brennan, Rick, "Iran's Covert War in Iraq," *Washington Times*, March 15, 2007. As of June 3, 2013:
http://www.washingtontimes.com/news/2007/mar/15/20070315-082220-8308r

———, "USF-I Operations Thru December 31, 2011," J3, USF-I, July 5, 2010, Not available to the general public.

Broadway, Chuck, "U.S. Transfers Airspace to Iraq," U.S. Air Force website, June 6, 2011.

Brownfield, William R., Assistant Secretary of State for INL, "Subject: INL Comments on the SIGIR Draft Report 'Iraqi Police Development Program: Opportunities for Improved Program Accountability and Budget Transparency' (SIGIR 12-006, September 30, 2011)," letter to Glenn D. Furbish, Assistant Inspector General for Audits, SIGIR, October 14, 2011.

Bruno, Greg, "Finding a Place for the 'Sons of Iraq,'" backgrounder, *Council on Foreign Relations*, January 9, 2009. As of September 18, 2012:
http://www.cfr.org/iraq/finding-place-sons-iraq/p16088

Bumiller, Elisabeth, "Panetta Presses Iraq for Decision on Troops," *New York Times*, July 11, 2011. As of April 17, 2013:
http://www.nytimes.com/2011/07/12/world/middleeast/12military.html

Bush, George W., "Address to the Nation on Iraq from the U.S.S. Abraham Lincoln," The American Presidency Project website, May 1, 2003. As of April 17, 2013:
http://www.presidency.ucsb.edu/ws/index.php?pid=68675&st=iraq&st1=transition - axzz1ceriUGkp

———, "The President's News Conference with Prime Minister Tony Blair of the United Kingdom in Istanbul," The American Presidency Project website, June 28, 2004. As of April 17, 2013:
http://www.presidency.ucsb.edu/ws/index.php?pid=72676&st=iraq&st1=transition - axzz1ceriUGkp

———, "Statement on the Formation of Iraq's Government," The American Presidency Project website, May 20, 2006. As of April 17, 2013:
http://www.presidency.ucsb.edu/ws/index.php?pid=65&st=iraq&st1=transition - axzz1ceriUGkp

———, "Address to the Nation on the War on Terror in Iraq," The American Presidency Project website, January 10, 2007. As of April 17, 2013:
http://www.presidency.ucsb.edu/ws/index.php?pid=24432&st=iraq&st1=transition - axzz1ceriUGkp

Bush, George W., and Tayyip Erdogan, "President Bush and Prime Minister Tayyip Erdogan Discuss Global War on Terror," transcript, Washington, D.C.: The White House, November 5, 2007. As of September 26, 2012:
http://georgewbush-whitehouse.archives.gov/news/releases/2007/11/20071105-3.html

Carnegie Commission, *Preventing Deadly Conflict: Final Report*, New York: Carnegie Corporation of New York, 1997.

Casey, George, "Strategic Directive: Golden Mosque Bombing," memorandum to commanders and staff, February 24, 2006.

Casey, George, and Zalmay Khalilzad, "MNF-I–Embassy Joint Mission Statement: Building Success: Completing the Transition," December 6, 2005.

Casey, George, and John Negroponte, "MNF-I–Embassy Joint Mission Statement," August 18, 2004.

———, "Joint Mission Statement, A Plan for the Year Ahead: Transition to Self-Reliance," memorandum, February 7, 2005.

Cave, Damien, "A Baghdad Book Mart Tries to Turn the Page," *New York Times*, September 15, 2007.

Central Intelligence Agency, "Chiefs of State and Cabinet Members of Foreign Governments: Iraq," September 1, 2011. As of September 2011:
https://www.cia.gov/library/publications/world-leaders-1/world-leaders-i/iraq.html

Chabot, Steve, "Opening Statement of Rep. Steve Chabot," *Preserving Progress: Transitioning Authority and Implementing the Strategic Framework in Iraq, Part II*, Washington, D.C.: Subcommittee on the Middle East and South Asia, Committee on Foreign Affairs, U.S. House of Representatives, 112th Cong., 1st Sess., June 23, 2011.

Charbonneau, Louis, "Exclusive: Western Report—Iran Ships Arms, Personnel to Syria via Iraq," Reuters, September 19, 2012.

Charles, Robert B., "State Department's Air Wing and Plan Colombia," press briefing, Washington, D.C.: U.S. Department of State, October 29, 2003. As of April 17, 2013:
http://2001-2009.state.gov/p/inl/rls/prsrl/spbr/25721.htm

Chaudhry, Serena, "NATO to Continue Iraq Training Mission to End: 2013," Reuters, September 12, 2011. As of April 17, 2013:
http://www.reuters.com/article/2011/09/12/us-iraq-nato-idUSTRE78B5XU20110912

Chazan, Guy, and Javier Blas, "IEA Predicts Boom for Iraq Oil Industry," *Financial Times*, October 9, 2012.

Chiarelli, Peter W., and Patrick R. Michaelis, "The Requirements for Full-Spectrum Operations," *Military Review*, July–August 2005.

Churchill, Winston, *My Early Life: A Roving Commission*, London: Thornton Butterworth, Ltd., 1930.

Cimbala, Stephen J., ed., *Strategic War Termination*, New York: Praeger, 1986.

Clinton, Hillary Rodham, Secretary of State, letter to Senator John Kerry, January 12, 2011a.

———, "Fiscal Year 2012 State Department Budget," testimony before the Subcommittee on the Department of State, Foreign Operations and Related Programs, Committee on Appropriations, U.S. Senate, March 2, 2011b.

———, "2012 State and USAID Budget Request," testimony before the House Appropriations Committee on Foreign Operations, March 10, 2011c.

———, "America's Pacific Century," *Foreign Policy*, November 2011d. As of July 26, 2012:
http://www.foreignpolicy.com/articles/2011/10/11/americas_pacific_century?page=full

Coalition Provisional Authority, *Law of Administration for the State of Iraq for the Transitional Period*, March 8, 2004. As of July 22, 2013:
http://www.au.af.mil/au/awc/awcgate/iraq/tal.htm

Cockburn, Patrick, "Arab-Kurd Tensions Rise as US Pulls Out," *New Zealand Herald*, August 12, 2009. As of September 26, 2012:
http://www.nzherald.co.nz/world/news/article.cfm?c_id=2&objectid=10590097

Cocks, Tim, and Muhanad Mohammed, "Iraq Regains Control of Cities as U.S. Pulls Back," Reuters, June 30, 2009.

Cole, Natalie, "Seeking Sand, Birds, Ammo: Arifjan Wash Rack, Customs Ready Armored Vehicles for Redeployment," Defense Video and Imagery Distribution System, August 12, 2010. As of August 3, 2012:
http://www.dvidshub.net/news/54387/seeking-sand-birds-ammo-arifjan-wash-rack-customs-ready-armored-vehicles-redeployment#ixzz22W59wYk5

Commanding General, MNF-I, "MNF-I Campaign Action Plan for 2005—Transition to Self-Reliance," memorandum, April 22, 2005.

Commission on Wartime Contracting in Iraq and Afghanistan, "Better Planning for Defense-to-State Transition in Iraq Needed to Avoid Mistakes and Waste," CWC Special Report 3, July 12, 2010. As of June 5, 2013:
http://cybercemetery.unt.edu/archive/cwc/20110929231202/http://www.wartimecontracting.gov/index.php/reports

———, "Iraq—A Forgotten Mission?" CWC Special Report 4, March 1, 2011a. As of June 5, 2013:
http://cybercemetery.unt.edu/archive/cwc/20110929231202/http://www.wartimecontracting.gov/index.php/reports

———, *Transforming Wartime Contracting: Controlling Costs, Reducing Risks, Final Report to Congress*, August 2011b, p. 74. As of June 5, 2013:
http://wartimecontracting.gov/docs/CWC_FinalReport-lowres.pdf

Conference of the Iraqi Opposition, *Final Report on the Transition to Democracy in Iraq*, 2002.

Connable, Ben, and Martin Libicki, *How Insurgencies End*, Santa Monica, Calif.: RAND Corporation, MG-965-MCIA, 2010. As of April 17, 2013:
http://www.rand.org/pubs/monographs/MG965.html

Corbin, Michael, and Colin Kahl, "U.S. Transition in Iraq," press briefing, August 16, 2010. As of August 13, 2012:
http://www.state.gov/p/nea/rls/rm/146027.htm

Cordesman, Anthony H., *Iraqi Security Forces: A Strategy for Success*, Washington, D.C.: Center for Strategic and International Studies, 2006.

———, *Iraqi Force Development and the Challenge of Civil War: The Critical Problems the US Must Address If Iraqi Forces Are to Do the Job*, Washington, D.C.: Center for Stratgeic and International Studies, April 26, 2007.

———, *Transferring Provinces to Iraqi Control: The Reality and the Risks*, Washington, D.C.: Center for Strategic and International Studies, September 2, 2008.

Cordesman, Anthony H., and Emma R. Davies, *Iraq's Insurgency and the Road to Civil Conflict*, Washington, D.C.: Center for Strategic and International Studies, 2008.

Courts, Michael J., "State and DOD Face Challenges in Finalizing Support and Security Capabilities," written testimony in *Assessment of the Transition from a Military- to a Civilian-Led Mission in Iraq*, hearing before the Subcommittee on National Security, Homeland Defense, and Foreign Operations, Committee on Oversight and Government Reform, U.S. House of Representatives, 112th Cong., 2nd Sess., June 28, 2012. As of April 17, 2013:
http://oversight.house.gov/hearing/assessment-of-the-transition-from-a-military-to-a-civilian-led-mission-in-iraq-2/wp-content/uploads/2012/09/2012-06-28-SC-Natl-Sec.pdf

CPA—*See* Coalition Provisional Authority.

Craddock, Bantz J., "Statement of General Bantz J. Craddock, USA, Commander, United States European Command," before the Committee on Armed Services, U.S. House of Representatives March 13, 2008. As of September 26, 2012:
http://www.dod.mil/dodgc/olc/docs/testCraddock080313.pdf

Crocker, Ryan C., "Statement of Ambassador Ryan C. Crocker, United States Ambassador to the Republic of Iraq," before a Joint Hearing of the Committee on Foreign Affairs and the Committee on Armed Services, September 10, 2007.

Crawford, Neta C., "Civilian Death and Injury in Iraq 2003–2011," Costs of War Project, paper, September 2011.

CWC—*See* Commission on Wartime Contracting in Iraq and Afghanistan.

Dagher, Sam, "2 Blasts Expose Security Flaws in Heart of Iraq," *New York Times*, August 19, 2009.

———, "Standoff at U.S. Airbase in Iraq," *Wall Street Journal*, November 18, 2011. As of November 8, 2012:
http://online.wsj.com/article/SB10001424052970204517204577044441272427070.html

Dagher, Sam, and Ali Nabhan, "Iraq Dismisses Central Bank Chief Amid Investigation," *Wall Street Journal*, October 16, 2012. As of November 12, 2012:
http://online.wsj.com/article/SB10000872396390444354004578060842607619044.html

Damon, Arwa, and Mohammed Tawfeeq, "Iraq's Leader Becoming a New 'Dictator,' Deputy Warns," CNN, December 31, 2011.

Davidson, Rikeshia, "AMC Takes Lead in Task Force Aimed at Army Reset Effort," Army.mil, December 21, 2009. As of August 6, 2012:
http://www.army.mil/article/32162/amc-takes-lead-in-task-force-aimed-at-army-reset-effort/

de Rugy, Veronique, "What's the Emergency?" *Regulation*, Summer 2008.

de Rugy, Veronique, and Allison Kasic, "The Never-Ending Emergency: Trends in Supplemental Spending," Arlington, Va.: George Mason University Mercatus Center, Working Paper no. 11-30, August 2011.

Defense Institute of Security Assistance Management, *The Management of Security Cooperation (Green Book)*, January 2013. As of August 5, 2013:
http://www.disam.dsca.mil/pages/pubs/greenbook.aspx

DeYoung, Karen, "Lacking an Accord on Troops, U.S. and Iraq Seek a Plan B," *Washington Post*, October 14, 2008. As of April 18, 2013:
http://articles.washingtonpost.com/2008-10-14/
politics/36840505_1_maliki-grand-ayatollah-ali-sistani-iraqi-officials

———, "State Department Seeks Smaller Embassy Presence in Baghdad," *Washington Post*, February 8, 2012. As of April 18, 2013:
http://articles.washingtonpost.com/2012-02-08/
world/35445499_1_embassy-compound-security-concerns-contractors

DeYoung, Karen, and Ernesto Londoño, "State Dept. Faces Skyrocketing Costs as It Prepares to Expand Role in Iraq," *Washington Post*, August 11, 2010, p. Al. As of http://www.washingtonpost.com/wp-dyn/content/article/2010/08/10/AR2010081006407.html

Dobbins, James, John G. McGinn, Keith Crane, Seth G. Jones, Rollie Lal, Andrew Rathmell, Rachel Swanger, and Anga Timilsina, *America's Role in Nation Building: From German to Iraq*, Santa Monica, Calif.: RAND Corporation, MR-1753-RC, 2003. As of April 23, 2013:
http://www.rand.org/pubs/monograph_reports/MR1753.html

Dobbins, James, Seth G. Jones, Benjamin Runkle, and Siddharth Mohandas, *Occupying Iraq: A History of the Coalition Provisional Authority*, Santa Monica, Calif.: RAND Corporation, MG-847-CC, 2009. As of April 18, 2013:
http://www.rand.org/pubs/monographs/MG847.html

DoD—*See* U.S. Department of Defense.

DoD IG—*See* U.S. Department of Defense, Inspector General.

Dodge, Toby, *Iraq: From War to a New Authoritarianism*, London: International Institute for Strategic Studies, December 2012.

Donahue, Rock, USF-I J7, "Transition of Engineer Activities, United States Forces–Iraq," briefing to RAND, June 26, 2011.

DoS—*See* U.S. Department of State.

DoS DS—*See* U.S. Department of State, Bureau of Diplomatic Security.

Douthat, Ross, "In Search of a Middle Eastern Strategy," Evaluations Blog, *New York Times*, September 14, 2012. As of September 17, 2012:
http://douthat.blogs.nytimes.com/2012/09/14/in-search-of-a-middle-eastern-strategy/

Dudley, Amy, "Vice President Joe Biden: 'In America, and in Iraq, the Tide of War Is Receding,'" blog, Washington, D.C.: The White House, December 2, 2011. As of April 17, 2013:
http://www.whitehouse.gov/blog/2011/12/02/vice-president-biden-america-and-iraq-tide-war-receding

du Lac, J. Freedom, "In Iraq, the Last to Fall: David Hickman, the 4,474th U.S. Service Member Killed," *Washington Post*, December 17, 2011. As of December 19, 2011:
http://www.washingtonpost.com/world/middle_east/in-iraq-the-last-to-fall-david-hickman-the-4474th-us-service-member-killed/2011/12/15/gIQAgwl00O_story.html

Dwoskin, Elizabeth, and Gopal Ratnam, "For Sale, Cheap: The Things You Need to Invade a Nation," *Businessweek*, December 14, 2011. As of July 30, 2012:
http://www.businessweek.com/magazine/for-sale-cheap-the-things-you-need-to-invade-a-nation-12142011.html

"Economic Negativity Abounds in Iraq," Gallup poll, September 22, 2011. As of April 18, 2013:
http://www.gallup.com/poll/149702/Economic-Negativity-Abounds-Iraq.aspx

Eisenstadt, Michael, testimony before the Committee on Foreign Affairs, Subcommittee on the Middle East and South Asia, U.S. House of Representatives, June 23, 2011.

Embassy of the United States, Bosnia & Herzegovina, "Contact Us," undated. As of August 13, 2012:
http://sarajevo.usembassy.gov/contact.html

Embassy of the United States, Baghdad, and ISF-I, *Executive Core Group Briefing*, August 1, 2010, Not available to the general public.

———, "U.S. Embassy Announces New Websites for Consulates in Basrah, Erbil, and Kirkuk," press release, January 17, 2012a. As of August 13, 2012:
http://iraq.usembassy.gov/consulatewebsitespr.html

———, "U.S. Embassy: 'Police Development Program Is a Vital Part of the U.S.-Iraqi Relationship,'" press release, May 13, 2012b. As of April 18, 2013:
http://iraq.usembassy.gov/may1312poldevelop.html

Embassy of the United States, Baghdad, Knowledge Management Coordinator, "Knowledge Management Transition Plan," briefing to Executive Core Group, March 14, 2011a.

———, "USF-I/USM-I Knowledge Management Transition," briefing, June 28, 2011b.

Embassy of the United States, Yaounde, Cameroon, "Embassy Branch Office (EBO) Douala: Mission Statement," undated. As of August 13, 2012:
http://yaounde.usembassy.gov/ebo_douala.html

Enders, David, "Iraqi Tribes Reach Security Accords," *Washington Times*, July 23, 2007.

Enterline, Andrew J., J. Michael Greig, and Yoav Gortzak, "Testing Shinseki: Speed, Mass, and Insurgency in Postwar Iraq," *Defense and Security Analysis*, Vol. 25, No. 3, September 2009.

Epstein, Susan B., and Marian Leonardo Lawson, *State, Foreign Operations, and Related Programs: FY2012 Budget and Appropriations*, Congressional Research Service, R41905, January 6, 2012. As of April 18, 2013:
http://www.fas.org/sgp/crs/row/R41905.pdf

"Equipment Management During Iraq Withdrawal Presents Challenges," *AUSA News*, February 1, 2012. As of August 3, 2012:
http://www.ausa.org/publications/ausanews/archives/2012/02/Pages/EquipmentmanagementduringIraqwithdrawalpresentschallenges.aspx

Estevez, Alan F., Acting Deputy Under Secretary of Defense for Logistics and Materiel Readiness, "Authority to Transfer Foreign Excess Personal Property in Iraq," memorandum for Commanding General, Multi-National Force–Iraq, October 9, 2009. As of July 27, 2012: http://www.acq.osd.mil/log/sci/MD/Iraqi_FEPP_transfer_09Oct2009.pdf

Farrell, Stephen, "Protests in Baghdad on U.S. Pact," *New York Times*, November 21, 2008.

Farrell, Stephen, and Elisabeth Bumiller, "No Shortcuts When Military Moves a War," *New York Times*, March 31, 2010. As of July 30, 2012: http://www.nytimes.com/2010/04/01/world/01logistics.html

Felter, Joseph, and Brian Fishman, "Al Qaeda's Foreign Fighters in Iraq: A First Look at the Sinjar Records," West Point, N.Y.: Combating Terrorism Center, December 19, 2007.

Feltman, Jeffrey D., "Statement for the Record of Assistant Secretary of State for Near Eastern Affairs Jeffrey D. Feltman," in *Transition to a Civilian-Led U.S. Presence in Iraq: Issues and Challenges*, hearing before the Committee on Foreign Affairs, U.S. House of Representatives, 111th Cong., 2nd Sess., November 18, 2010a. As of June 3, 2013: http://www.gpo.gov/fdsys/pkg/CHRG-111hhrg62399/content-detail.html

———, Responses to Questions for the Record from the Honorable Russ Carnahan, in *Transition to a Civilian-Led U.S. Presence in Iraq: Issues and Challenges*, hearing before the Committee on Foreign Affairs, U.S. House of Representatives, 111th Cong., 2nd Sess., November 18, 2010b, pp. 53–63. As of June 3, 2013: http://www.gpo.gov/fdsys/pkg/CHRG-111hhrg62399/content-detail.html

Filkins, Dexter, "Back in Iraq, Jarred by the Calm," *New York Times*, September 20, 2008.

Field Manual 3-90, *Tactics,* Washington, D.C., Headquarters, Department of the Army, July 4, 2001

Francis, Bassam, "KRG Accuses Baghdad of Reducing Its Budget Share," *al-Hayat*, trans. Sami-Joe Abboud, June 14, 2012. As of September 25, 2012: http://www.al-monitor.com/pulse/business/2012/06/krg-accuses-baghdad-of-reducing.html

Franks, General Tommy, and Matthew McConnell, *American Soldier*, New York: HarperCollins, 2004.

Freedberg, Sydney J., Jr.,"In Iraq, Combat Turns into Advise and Assist," *National Journal*, December 5, 2009. As of September 17, 2012: http://www.nationaljournal.com/njmagazine/id_20091205_1454.php

Freedom House, "Iraq," *Freedom in the World 2009*, 2009. As of September 19, 2012: http://www.freedomhouse.org/report/freedom-world/2009/iraq

GAO—*See* U.S. Government Accountability Office.

Garcia-Navarro, Lourdes, "Iraq Paramilitary Group Targeted, Despite Success," NPR Morning Edition, September 4, 2008. As of September 19, 2012: http://www.npr.org/templates/story/story.php?storyId=94229791

Garramone, Jim, "Austin Gives Insight into Drawdown, Possible Aid to Iraq," American Forces Press Service, July 11, 2011.

Gatehouse, Gabriel, "Najaf's Return as a Religious Tourist Destination," BBC News, February 27, 2010.

Geisel, Harold W., "Testimony of Harold W. Geisel, Deputy Inspector General Office of Inspector General U.S. Department of State and the Broadcasting Board of Governors," in *Assessment of the Transition from a Military- to a Civilian-Led Mission in Iraq*, hearing before the Subcommittee on National Security, Homeland Defense, and Foreign Operations, Committee on Oversight and Government Reform, U.S. House of Representatives, 112th Cong., 2nd Sess., June 28, 2012. As of April 17, 2013:
http://oversight.house.gov/hearing/assessment-of-the-transition-from-a-military-to-a-civilian-led-mission-in-iraq-2/wp-content/uploads/2012/09/2012-06-28-SC-Natl-Sec.pdf

Ghazi, Yasir, and Rod Nordland, "Iraq Insurgents Kill at Least 100 After Declaring New Offensive," *New York Times*, July 23, 2012. As of April 19, 2013:
http://www.nytimes.com/2012/07/24/world/middleeast/iraqi-insurgents-kill-dozens-in-wave-of-attacks.html?_r=0

Gisick, Michael, "U.S. Base Projects Continue in Iraq Despite Plans to Leave," *Stars and Stripes*, June 1, 2010. As of October 1, 2012:
http://www.stripes.com/news/u-s-base-projects-continue-in-iraq-despite-plans-to-leave-1.105237

Glanz, James, and Alissa J. Rubin, "Iraqi Army Takes Last Basra Areas from Sadr Force," *New York Times*, April 20, 2008.

Goemans, H. E., *War & Punishment: The Causes of War Termination & The First World War*, Princeton, N.J.: Princeton University Press, 2000.

Gompert, David C., Terrence K. Kelly, and Jessica Watkins, "Security in Iraq: Emerging Threats as U.S. Forces Withdraw," Santa Monica, Calif.: RAND Corporation, RB-9481-OSD, 2010. As of April 18, 2013:
http://www.rand.org/pubs/research_briefs/RB9481.html

Gordon, Michael R., "To Stand or Fall in Baghdad: Capital Is Key to Mission," *New York Times*, October 22, 2006.

———, "Troop 'Surge' in Iraq Took Place Amid Doubt and Public Debate, *New York Times*, August 31, 2008.

———, "Iran's Master of Chaos Still Vexes U.S.," *New York Times*, October 2, 2012. As of April 18, 2013:
http://www.nytimes.com/2012/10/03/world/middleeast/qassim-suleimani-irans-master-of-iraq-chaos-still-vexes-the-us.html?hp

Gordon, Michael R., Eric Schmitt, and Stephen Farrell, "U.S. Cites Planning Gaps in Iraqi Assault on Basra," *New York Times*, April 3, 2008.

Gordon, Michael R., and Bernard E. Trainor, *The Endgame: The Inside Story of the Struggle for Iraq, from George W. Bush to Barack Obama,* New York: Pantheon, 2012.

Green, Grant S., "Statement of Grant S. Green, Commissioner, Commission on Wartime Contracting," in *U.S. Military Leaving Iraq: Is the State Department Ready?* hearing before the Subcommittee on National Security, Homeland Defense, and Foreign Operations, Committee on Oversight and Government Reform, U.S. House of Representatives, 112th Congress, 1st Sess., March 2, 2011. As of June 4, 2013:
http://www.gpo.gov/fdsys/pkg/CHRG-112hhrg67365/html/CHRG-112hhrg67365.htm

Greenberg Quinlan Rosner Research, "A Major Shift in the Political Landscape: Report on the April 2012 National Survey," June 2012. As of November 12, 2012:
http://www.ndi.org/files/NDI-Iraq%20-%20April%202012%20National%20Survey%20-%20Report.pdf

——— , "A Major Shift in the Political Landscape: Graphs for the Report on the April 2012 National Survey," May 2012. As of November 12, 2012:
http://www.ndi.org/files/NDI-Iraq%20-%20April%202012%20National%20Survey%20-%20 Presentation.pdf

Gregg, Heather S., Hy S. Rothstein, and John Arquilla, eds., *The Three Circles of War: Understanding the Dynamics of Conflict in Iraq*, Herndon, Va.: Potomac Books, 2010.

Gurney, David H., and Jeffrey D. Smotherman, "An Interview with Raymond T. Odierno," *Joint Force Quarterly*, No. 55, 4th quarter, 2009.

Gutman, Roy, Sahar Issa, and Laith Hammoudi, "Iraq's Maliki Accused of Detaining Hundreds of Political Opponents," McClatchy Newspapers, January 19, 2012.

Hacaoglu, Selcan, "Turkey to Extend Mandate for Ground Incursion Into Northern Iraq," *Bloomberg News*, September 25, 2012. As of September 26, 2012:
http://www.businessweek.com/news/2012-09-25/
turkey-to-extend-mandate-for-ground-incursion-into-northern-iraq

Hafidh, Hassan, "Iraq Says Exxon Seeks Bids on Oil-Field Stake," *Wall Street Journal*, November 8, 2012. As of November 9, 2012:
http://online.wsj.com/article/SB10001424127887324894104578107131250649640.html

Halchin, L. Elaine, "The Coalition Provisional Authority (CPA): Origin, Characteristics, and Institutional Authorities," Washington, D.C.: Congressional Research Service, RL32370, June 6, 2005.

Hanauer, Larry, Jeffrey Martini, and Omar al-Shahery, *Managing Arab-Kurd Tensions in Northern Iraq After the Withdrawal of U.S. Troops*, Santa Monica, Calif.: RAND Corporation, OP-339-USFI, 2011. As of April 18, 2013
http://www.rand.org/pubs/occasional_papers/OP339.html

Hanauer, Larry, and Laurel E. Miller, *Resolving Kirkuk: Lessons Learned from Settlements of Earlier Ethno-Territorial Conflicts*, Santa Monica, Calif.: RAND Corporation, MG-1198-USFI, 2012. As of April 18, 2013
http://www.rand.org/pubs/monographs/MG1198.html

Handel, Michael I., *War Termination—A Critical Survey*, Jerusalem: Hebrew University, 1978.

Harari, Michal, "Uncertain Future for the Sons of Iraq," backgrounder, Institute for the Study of War, August 3, 2010. As of September 18, 2012:
http://www.understandingwar.org/sites/default/files/Backgrounder_SonsofIraq_0.pdf

Haslach, Patricia M., "Iraq Transition Lessons Offsite Followup," email to Larry Hanauer, RAND Corporation, May 4, 2012

Healy, Jack, "Saudis Pick First Envoy to Baghdad in 20 Years," *New York Times*, February 21, 2012. As of April 18, 2013:
http://www.nytimes.com/2012/02/22/world/middleeast/saudi-arabia-names-ambassador-to-iraq.html

Healy, Jack, Tim Arango, and Michael Schmidt, "Premier's Acts in Iraq Raise U.S. Concerns," *New York Times,* December 13, 2011.

Higgins, Brad, "Joint Strategy, Planning & Assessments (JSPA)," memorandum to ambassador, Baghdad, November 18, 2005.

Hiltermann, Joost, "Elections and Constitution Writing in Iraq, 2005," *Mediterranean Yearbook*, 2006. As of September 19, 2012:
http://www.iemed.org/anuari/2006/aarticles/aHiltermann.pdf

Hurley, Paul C., and John J. Abbatiello, "Responsible Drawdown: Synchronizing the Joint Vision," *Joint Force Quarterly*, No. 59, 4th Quarter, 2010. As of April 18, 2013: http://www.ndu.edu/press/lib/images/jfq-59/JFQ59_127-135_Hurley-Abbatiello.pdf

Hussein, Adnan, "Six Years Later, Questions Linger About Effectiveness of Article 140 Committee," *Rudaw*, September 13, 2012.

Hylander, Devon, "Taxpayers Benefit from Army's Excess Non-Standard Equipment," Army.mil, November 24, 2010. As of July 30, 2012: http://www.army.mil/article/48581/taxpayers-benefit-from-armys-excess-non-standard-equipment/

IEA—*See* International Energy Agency.

Iklé, Fred, *Every War Must End*, New York: Columbia University Press, 1971.

Independent Commission on Security Forces in Iraq, *Report of the Independent Commission on Security Forces in Iraq*, September 7, 2007.

International Crisis Group, *Unmaking Iraq: A Constitutional Process Gone Awry*, Middle East Briefing No. 19, September 26, 2005. As of September 19, 2012: http://www.crisisgroup.org/~/media/Files/Middle%20East%20North%20Africa/Iraq%20Syria%20Lebanon/Iraq/B019%20Unmaking%20Iraq%20A%20Constitutional%20Process%20Gone%20Awry.pdf

———, *Iraq and the Kurds: Confronting Withdrawal Fears*, Middle East Report No. 103, March 28, 2011. As of September 25, 2012: http://www.crisisgroup.org/~/media/Files/Middle%20East%20North%20Africa/Iraq%20Syria%20Lebanon/Iraq/103%20Iraq%20and%20the%20Kurds%20--%20Confronting%20Withdrawal%20Fears

———, *Iraq's Secular Opposition: The Rise and Decline of Al-Iraqiya*, Middle East Report No. 127, July 31, 2012. As of November 12, 2012: http://www.crisisgroup.org/en/regions/middle-east-north-africa/iraq-iran-gulf/iraq/127-iraqs-secular-opposition-the-rise-and-decline-of-al-iraqiya.aspx

International Energy Agency, *Iraq Energy Outlook*, October 9, 2012. As of November 9, 2012: http://www.iea.org/publications/freepublications/publication/WEO_2012_Iraq_Energy_OutlookFINAL.pdf

"Iraq Attacks Kill 110 in Deadliest Day in 2 Years," Associated Press, July 23, 2012. As of November 8, 2012: http://usatoday30.usatoday.com/news/world/iraq/story/2012-07-23/Iraq-Baghdad-suicide/56424314/1

Iraq Civil Aviation Authority, "Iraq Aeronautical Information Publication (AIP)," November 17, 2011.

"Iraq Deaths from Violence," Iraq Body Count website, January 1, 2013. As of February 7, 2012: http://www.iraqbodycount.org/analysis/numbers/2012/

"Iraq Election Chief Gets Prison Sentence for Graft," Agence France-Presse, August 28, 2012.

"Iraq Ex-Electoral Chief Says Cleared of Graft," Agence France-Presse, October 16, 2012. As of November 12, 2012: http://english.ahram.org.eg/News/55736.aspx

"Iraq: Procurement," *Jane's Sentinel Security Assessment—The Gulf States*, October 15, 2012.

"Iraq Releases Suspected Hezbollah Operative Daqduk," Reuters, November 16, 2012. As of ____: http://www.reuters.com/article/2012/11/16/us-iraq-daqduq-release-idUSBRE8AF0SS20121116

"Iraq Stops Paying Salary of Convicted Fugitive VP," Associated Press, October 10, 2012. As of November 12, 2012:
http://news.yahoo.com/iraq-stops-paying-salary-convicted-fugitive-vp-135224240.html

Iraq Study Group, *The Iraq Study Group Report*, December 2006.

"Iraqi Army Not Ready to Take Over Until 2020, Says Country's Top General," *The Guardian*, August 12, 2010.

Iraqi Constitution, 2005. As of July 22, 2013:
http://www.wipo.int/wipolex/en/text.jsp?file_id=230000

"Iraqi Pilots Begin Flight Training in U.S.," *Defense World*, September 3, 2012. As of April 18, 2013:
https://defenseworld.net/news/7438/Iraqi_Pilots_Begin_Flight_Training_In_U_S

"Iraqi VP Tariq al-Hashemi Sentenced to Death," BBC News, September 9, 2012. As of September 21, 2012:
http://www.bbc.co.uk/news/world-middle-east-19537301

"Iraq's Anti-US Cleric Considers US Embassy Employees Occupiers," *China Daily*, October 23, 2011. As of September 28, 2012:
http://www.chinadaily.com.cn/xinhua/2011-10-23/content_4146358.html

"Iraq's Sadr Says to Halt Attacks on U.S. Troops," Thomson/Reuters, September 11, 2011. As of September 28, 2012:
http://www.newsmax.com/Newsfront/Iraq-Sadr-US-troops/2011/09/11/id/410450

Jakes, Lara, "Wave of Bombings Kills 26 Across Iraq," Associated Press, October 1, 2012.

Jeffrey, James F., "Statement by Ambassador James F. Jeffrey: Senate Foreign Relations Committee," July 20, 2010. As of April 17, 2013:
http://www.foreign.senate.gov/imo/media/doc/Jeffrey,%20James%20Franklin1.pdf

———, "Statement of Hon. James F. Jeffrey, Ambassador to Iraq, U.S. Department of State, Washington, DC," *Iraq: The Challenging Transition to a Civilian Mission*, hearing before the Committee on Foreign Relations, U.S. Senate, 112th Cong., 1st Sess., February 1, 2011a, pp. 6–8.

———, "Statement of Hon. James F. Jeffrey, U.S. Ambassador to Iraq," *Hearing to Receive Testimony on United States Policy Toward Iraq*, Washington, D.C.: Committee on Armed Services, U.S. Senate, February 3, 2011b. As of April 17, 2013:
http://armed-services.senate.gov/Transcripts/2011/02%20February/11-02%20-%202-3-11.pdf

Jeffrey, James F., and Lloyd J. Austin, III, "Prepared Joint Statement of Ambassador James F. Jeffrey and Gen Lloyd J. Austin," *Iraq: The Challenging Transition to a Civilian Mission*, hearing before the Committee on Foreign Relations, U.S. Senate, 112th Cong., 1st Sess., February 1, 2011, pp. 8–12.

JLOC—*See* USF-I J-4 Joint Logistics Operation Center.

Joint Publication 3-24, *Counterinsurgency Operations*, Washington, D.C.: Joint Staff, October 5, 2009.

Jones, James L., chairman, *The Report of the Independent Commission on the Security Forces of Iraq*, September 6, 2007.

Joyner, James, "Rumsfeld Bans Word 'Insurgents,'" Outside the Beltway blog, November 30, 2005.

"June Deadliest Month for U.S. Troops in 2 Years," *USA Today*, June 30, 2011. As of September 28, 2012:
http://www.usatoday.com/news/world/iraq/2011-06-30-us-troops-iraq_n.htm

Kahl, Colin, "Statement of Colin Kahl, Ph.D, Deputy Assistant Secretary of Defense for the Middle East, U.S. Department of Defense," and testimony, in *The Transition to a Civilian-Led U.S. Presence in Iraq: Issues and Challenges*, hearing before the Committee on Foreign Affairs, U.S. House of Representatives, 111th Cong., 2nd Sess., November 18, 2010. As of June 3, 2013: http://www.gpo.gov/fdsys/pkg/CHRG-111hhrg62399/content-detail.html

Katulis, Brian, "Standoff in Khanaquin: Trouble Brews Between Arabs and Kurds in a Volatile Corner of Iraq," Center for American Progress, August 29, 2008. As of September 26, 2012: http://www.americanprogress.org/issues/military/news/2008/08/29/4794/standoff-in-khanaqin/

Katzman, Kenneth, *Iraq: Reconciliation and Benchmarks*, Washington, D.C.: Congressional Research Service, 2005.

———, *Iraq: Post-Saddam Governance and Security*, Washington, D.C.: Congressional Research Service, July 8, 2009.

———, *Iraq: Politics, Elections, and Benchmarks*, Washington, D.C.: Congressional Research Service, March 1, 2011a.

———, *Iraq: Politics, Governance, and Human Rights*, Washington, D.C.: Congressional Research Service, April 1, 2011b.

———, *Iraq: Politics, Governance, and Human Rights*, Washington, D.C.: Congressional Research Service, RS21968, August 21, 2012.

Kelemen, Michele, "Huge Embassy Keeps U.S. Presence in Iraq," National Public Radio, December 18, 2012. As of July 10, 2012: http://www.npr.org/2011/12/18/143863722/with-huge-embassy-u-s-still-a-presence-in-iraq

Kendall, Frank, "Frank Kendall, Principal Deputy Under Secretary of Defense for Acquisition, Technology and Logistics, U.S. Department of Defense," in *U.S. Military Leaving Iraq: Is the State Department Ready?* hearing before the Subcommittee on National Security, Homeland Defense, and Foreign Operations, Committee on Oversight and Government Reform, U.S. House of Representatives, 112th Congress, 1st Sess., March 2, 2011a. As of June 4, 2013: http://www.gpo.gov/fdsys/pkg/CHRG-112hhrg67365/html/CHRG-112hhrg67365.htm

———, "Statement of Hon. Frank Kendall, Principal Deputy Under Secretary of Defense dor Acquisition, Technology, and Logistics," in *The Final Report of the Commission on Wartime Contracting in Iraq and Afghanistan*, hearing before the Subcommittee on Readiness and Management Support, Committee on Armed Services, U.S. Senate, 112th Cong., 1st Sess., October 19, 2011b.

Kendall, Frank, and Brooks L. Bash, "Joint Prepared Statement by Frank Kendall and Lt. Gen. Brooks L. Bash, USAF," in *The Final Report of the Commission on Wartime Contracting in Iraq and Afghanistan*, hearing before the Subcommittee on Readiness and Management Support, Committee on Armed Services, U.S. Senate, 112th Cong., 1st Sess., October 19, 2011.

Kennedy, Patrick F., Under Secretary of State for Management, letter to the Department of Defense, April 7, 2010a.

———, Under Secretary of State for Management, letter to Ashton Carter, Under Secretary of Defense (Acquisition, Technology, and Logistics), December 20, 2010b.

———, "Statement of Ambassador Patrick Kennedy, Under Secretary of State for Management, U.S. Department of State," in *U.S. Military Leaving Iraq: Is the State Department Ready?* hearing before the Subcommittee on National Security, Homeland Defense, and Foreign Operations, Committee on Oversight and Government Reform, U.S. House of Representatives, 112th Congress, 1st Sess., March 2, 2011a. As of June 4, 2013: http://www.gpo.gov/fdsys/pkg/CHRG-112hhrg67365/html/CHRG-112hhrg67365.htm

―――, "Statement by Patrick F. Kennedy, Department of State, Under Secretary for Management Department of State," statement for the record, for *Department of Statev Contracting: CWC's Second Interim Report, the QDDR, and Iraq Transition*, hearing before the Commission on Wartime Contracting, June 6, 2011b. As of June 5, 2013:
http://www.wartimecontracting.gov/docs/hearing2011-06-06_testimony-Kennedy.pdf

―――, "Statement of the Honorable Patrick F. Kennedy," in *Assessment of the Transition from a Military- to a Civilian-Led Mission in Iraq*, hearing before the Subcommittee on National Security, Homeland Defense, and Foreign Operations, Committee on Oversight and Government Reform, U.S. House of Representatives, 112th Cong., 2nd Sess., June 28, 2012. As of April 17, 2013:
http://oversight.house.gov/hearing/assessment-of-the-transition-from-a-military-to-a-civilian-led-mission-in-iraq-2/wp-content/uploads/2012/09/2012-06-28-SC-Natl-Sec.pdf

Kerry, John, *Testimony of John Kerry, Legislative Proposals Relating to the War in Southeast Asia, Hearings Before the Committee on Foreign Relations*, U.S. Senate, 92 Cong., 1st Sess., Washington, D.C.: Government Printing Office, April 22, 1971.

―――, "Opening Statement of Senator John Kerry," *Iraq: The Challenging Transition to a Civilian Mission*, hearing before the Committee on Foreign Relations, U.S. Senate, 112th Cong., 1st Sess., February 1, 2011.

Keskin, Funda, "Turkey's Trans-Border Operations in Northern Iraq: Before and After the Invasion of Iraq," *Research Journal of International Studies*, No. 8, November 2008.

Khalilzad, Zalmay, and George W. Casey, "Joint Campaign Plan: Operation Iraqi Freedom, Transition to Iraqi Self-Reliance," April 28, 2006a.

―――, "2006 Joint Campaign Action Plan 'Unity, Security, Prosperity,'" June 9, 2006b.

Kline, David, "Mobile RPAT Mission at COB Basra," *Outpost*, Vol. 1, No. 4, September 2010.

Knickmeyer, Ellen, and Jonathan Finer, "Insurgent Leader Al-Zarqawi Killed in Iraq," *Washington Post*, June 8, 2006. As of April 18, 2013:
http://www.washingtonpost.com/wp-dyn/content/article/2006/06/08/AR2006060800114.html

Knights, Michael, "The Evolution of Iran's Special Groups in Iraq," Combating Terrorism Center, November 1, 2010. As of April 18, 2013:
http://www.ctc.usma.edu/posts/the-evolution-of-iran%E2%80%99s-special-groups-in-iraq

―――, "The JRTN Movement and Iraq's Next Insurgency," Combating Terrorism Center, July 1, 2011. As of April 18, 2013:
http://www.ctc.usma.edu/posts/the-jrtn-movement-and-iraq%E2%80%99s-next-insurgency

Kramer, Andrew E., "Leaving Camp Victory in Iraq, the Very Name a Question Mark," *New York Times*, November 10, 2011. As of November 12, 2011:
http://www.nytimes.com/2011/11/11/world/middleeast/United-States-Prepares-for-Moving-Day-at-Camp-Victory-in-Iraq.html?ref=iraq

―――, "Fate of a U.S. Base Heightens Tensions in Iraq's Contested North," *New York Times*, November 18, 2011. As of July 12, 2012:
http://www.nytimes.com/2011/11/19/world/middleeast/fate-of-us-base-raises-tensions-around-iraqi-kurdistan.html

―――, "U.S. Leaving Iraqi Comrades-in-Arms in Limbo," *New York Times*, December 13, 2011. As of September 18, 2012:
http://www.nytimes.com/2011/12/14/world/middleeast/united-states-leaving-sunni-awakening-comrades-in-iraq-in-limbo.html?pagewanted=all

Lake, Jason, "Iraqi Air Force Leaders Thank U.S. Advisors, Celebrate 80th Anniversary," 321st Air Expeditionary Wing Public Affairs, April 22, 2011. As of August 9, 2012:
http://www.afcent.af.mil/news/story.asp?id=123252922

Landler, Mark, "Biden in Iraq to Prepare for Postwar Relations," *New York Times*, November 30, 2011.

Lardner, Richard, "State Department Wants a Mini-Army in Iraq," *Army Times*, June 14, 2010. As of July 26, 2012:
http://www.armytimes.com/news/2010/06/ap_state_department_iraq_security_061410/

"Last US Troops Withdraw from Iraq," BBC News, December 18, 2011. As of December 19, 2011:
http://www.bbc.co.uk/news/world-middle-east-16234723

Latif, Ali, "Iraq's Central Bank Governor Is Removed Under Cloud," *Azzaman* (Iraq), trans. Sahar Ghoussoub and Naria Tanoukhi, October 15, 2012. As of November 12, 2012:
http://www.al-monitor.com/pulse/business/2012/10/warrant-issued-for-iraqs-central-bank-governor.html

Lavey, A. M., "South Dakota Medevac Team Ready for Action," press release, U.S. Forces–Iraq, October 23, 2011.

Layton, Lyndsey, "The Story Behind the Iraq Study Group," *Washington Post*, November 21, 2006.

Levinson, Charles, and Ali A. Nabhan, "U.S., Iraq Negotiate on Troop Immunity," *USA Today*, October 7, 2008.

Lew, Jacob, "The Next Phase in America's Relationship with Iraq," Statesmen's Forum video, Center for Strategic and International Studies, August 5, 2010. As of April 19, 2013:
http://csis.org/event/statesmens-forum-james-b-steinberg-deputy-secretary-state-and-jacob-j-lew-deputy-secretary-sta

Licklider, Roy, ed., *Stopping the Killing: How Civil Wars End*, New York: New York University Press, 1993.

Londoño, Ernesto, "Operation Iraqi Freedom Ends as Last Combat Soldiers Leave Baghdad," *Washington Post*, August 19, 2010. As of September 17, 2012:
http://www.washingtonpost.com/wp-dyn/content/article/2010/08/18/AR2010081805644.html

Loney, Jim, "Iraqi Air Defense: A Work in Progress," Reuters, September 27, 2011. As of August 9, 2012:
http://www.reuters.com/article/2011/09/27/us-iraq-security-airforce-idUSTRE78Q2RH20110927

Lute, Jane Holl, *From the Streets of Washington to the Roofs of Saigon: Domestic Politics and the Termination of the Vietnam War*, dissertation, Stanford, Calif.: Stanford University, 1989.

Lynn, William, "Statement of William J. Lynn, III, Deputy Secretary of Defense," before the Senate Budget Committee, March 10, 2011.

Madhani, Aamer, "U.S. Urges Kurdish Leader to Repair Ties to Iraqi Government," *USA Today*, April 5, 2012. As of April 19, 2013:
http://www.usatoday.com/news/washington/story/2012-04-04/kurdistan-barzani-visit/54015702/1

Mahmoud, Nawzad, "Property Claims Law Fails Thousands of Kurdish Families," *Rudaw*, August 27, 2012.

Mansour, Renad, "Iraqi Kurdistan & the Syrian-Kurd Pursuit of Autonomy," Al-Jazeera Center for Studies, September 24, 2012. As of November 12, 2012:
http://studies.aljazeera.net/en/reports/2012/09/201291910402907471.htm

Mardini, Ramzi, "Iraq's First Post-Withdrawal Crisis," Washington, D.C.: Institute for the Study of War, December 19, 2012.

Markey, Patrick, "Analysis: Syrian Border Standoff a New Front in Iraq-Kurdish Rift," Reuters, August 8, 2012. As of August 29, 2013:
http://in.reuters.com/article/2012/08/08/us-syria-crisis-iraq-kurdistan-idINBRE8770IX20120808

———, "Iraq Says Signs Contract for 18 F-16 Fighter Jets," Reuters, October 18, 2012. As of November 12, 2012:
http://www.reuters.com/article/2012/10/18/us-iraq-military-jets-idUSBRE89H14B20121018

Mason, Chuck R., *U.S.-Iraq Withdrawal/Status of Forces Agreement: Issues for Congressional Oversight*, Washington, D.C.: Congressional Research Service, July 13, 2009.

———, *Status of Forces Agreement (SOFA): What Is It, and How Has It Been Utilized?* Washington, D.C.: Congressional Research Service, January 5, 2011.

Mason, Raymond, "Statement of Lieutenant General Raymond V. Mason, Deputy Chief of Staff, Logistics, G4 , U.S. Army," in *Army and Marine Corps Materiel Reset*, hearing before the Subcommittee on Readiness, Committee on Armed Services, U.S. House of Representatives, 112th Cong., 2nd Sess., March 28, 2012.

McChrystal, Stanley, *My Share of the Task: A Memoir*, New York: Penguin Group, 2013.

McClelland, Robert, letter to Arch Bevis, Chair, Parliamentary Joint Committee on Intelligence and Security, document 08/11412, October 21, 2008. As of September 18, 2012:
http://aph.gov.au/Parliamentary_Business/Committees/House_of_Representatives_Committees?url=pjcis/asg_jua_aqi/subs.htm

McCrummen, Stephanie, "Protesters Say Maliki Is Using Special Security Forces to Shut Down Demonstrations in Iraq," *Washington Post*, March 3, 2011.

McDonald, Lori K., "Drawdown Brings Non-Standard Equipment Mission to Sierra," Army.mil, January 18, 2011. As of July 30, 2012:
http://www.army.mil/article/50497/

McGurk, Bret H., "Statement of Brett H. McGurk, Visiting Scholar, Columbia University School of Law," in *Security Issues Relating to Iraq*, hearing before the Committee on Armed Services, U.S. Senate, 112 Cong., 1st Sess., November 15, 2011.

Migdalovitz, Carol, *Turkey: Selected Foreign Policy Issues and U.S. Views*, Congressional Research Service Report RL34642, November 28, 2010. As of September 26, 2012:
http://www.fas.org/sgp/crs/mideast/RL34642.pdf

Miles, Donna, "Brigade Tests New Concept in Iraq," American Forces Press Service, November 4, 2009. As of September 17, 2012:
http://www.defense.gov/news/newsarticle.aspx?id=56522

———, "Panetta Concludes Iraq Mission Noting Service, Sacrifice," American Forces Press Service, December 15, 2011. As of 16 December, 201:
http://www.defense.gov/news/newsarticle.aspx?id=66483

"Military Spending: 11,175,067," *Time*, August 14, 2012.

MNF-I—*See* Multi-National Force–Iraq.

MNSTC-I—*See* Multi-National Security Transition Command–Iraq.

Muhammad, Nehro, "Peshmerga Budget Next on Agenda in Baghdad-Erbil Talks," *Rudaw* (Erbil), September 26, 2012.

Al-Mullah, Haidar, Iraqiya spokesman media conference (in Arabic), Al-Iraqiya News, June 2011.

Multi-National Force–Iraq, "Command Briefing," undated, Not available to the general public.

———, *Campaign Plan: Operation Iraqi Freedom; Partnership: From Occupation to Constitutional Elections*, Baghdad, 2004a.

———, *Building Legitimacy and Confronting Insurgency in Iraq: End States and Bottom Line*, Baghdad, Iraq, July 15, 2004b.

———, *Five Month Campaign Progress Review*, Baghdad, December 12, 2004c.

———, *The September Assessment*, September 23, 2005a.

———, *Campaign Progress Review, June 2005–December 2005*, Baghdad, December 20, 2005b.

———, "ISF Terms of Reference," briefing slide, October 26, 2010.

Multi-National Security Transition Command–Iraq, "Shaping the Long-Term Security Partnership with Iraq," briefing to General Odierno, Commander, MNF-I, February 24, 2009a.

———, "Terms of Reference, Iraqi Security Force (ISF) Capabilities," information paper, October 15, 2009b.

National Association of State Agencies for Surplus Property, "NASASP Mission Statement," website, 2009. As of April 29, 2013:
http://nasasp.org/site/pages/organization_missionstatement.htm

National Counterterrorism Center, "Al-Qa'ida in Iraq (AQI)," *Counterterrorism 2013 Calendar*, 2012. As of April 17, 2013:
http://www.nctc.gov/site/groups/aqi.html

National Security Archive, "New State Department Releases on the 'Future of Iraq' Project," Electronic Briefing Book No. 198, September 1, 2006. As of July 12, 2013:
http://www.gwu.edu/~nsarchiv/NSAEBB/NSAEBB198/

National Security Council, *National Strategy for Victory in Iraq*, November 2005. As of April 19, 2013:
http://www.washingtonpost.com/wp-srv/nation/documents/Iraqnationalstrategy11-30-05.pdf

National Security Decision Directive 38, "Staffing at Diplomatic Missions and Their Overseas Constituent Posts," Office of Management Policy, Rightsizing, and Innovation, June 2, 1982.

National Security Presidential Directive 24, "Post-War Iraq Reconstruction," January 20, 2003.

National Security Presidential Directive 36, "United States Government Operations in Iraq," May 11, 2004.

"NATO to End Training Mission in Iraq," ABC News (Australia), December 13, 2011. As of December 13, 2011:
http://www.abc.net.au/news/2011-12-13/nato-ends-training-mission-in-iraq/3727614

"NATO Will Not Extend Iraq Training Mission Beyond 2011," *The National*, December 12, 2011. As of July 26, 2012:
http://www.thenational.ae/news/world/middle-east/
nato-will-not-extend-iraq-training-mission-beyond-2011

Neil, Rob, "Targeting Terror's Roots—Part 1," *Pacific Wings*, Vol. 75, No. 8, August 2007.

Nickmeyer, Ellen, "Tikrit Palace Complex Allegedly Picked Clean," *Seattle Times*, January 13, 2006. As of November 12, 2011:
http://community.seattletimes.nwsource.com/archive/?date=20060113&slug=tikrit13

Nides, Thomas, "Rightsizing U.S. Mission Iraq," briefing, Washington, D.C.: U.S. Department of State, February 8, 2012. As of April 19, 2013:
http://m.state.gov/md183598.htm

NSDD—*See* National Security Decision Directive.

NSPD—*See* National Security Presidential Directive.

Nuland, Victoria, State Department Daily Press Briefing, Washington, February 7, 2012. As of April 19, 2013:
http://www.state.gov/r/pa/prs/dpb/2012/02/183489.htm

O'Bagy, Elizabeth, and Stephen Wicken, "Fact Sheet: Ali Mussa Daqduq," Institute for the Study of War, May 14, 2012. As of April 19, 2013:
http://www.understandingwar.org/reference/fact-sheet-ali-mussa-daqduq

Obama, Barack, "Responsibly Ending the War in Iraq," remarks, Camp Lejeune, N.C., February 27, 2009. As of April 19, 2013:
http://www.whitehouse.gov/the_press_office/
Remarks-of-President-Barack-Obama-Responsibly-Ending-the-War-in-Iraq

————,"Remarks by the President on Ending the War in Iraq," Washington, D.C.: The White House, Office of the Press Secretary, October 21, 2011. As of April 17, 2013:
http://www.whitehouse.gov/the-press-office/2011/10/21/remarks-president-ending-war-iraq

Obama, Barack, and Michelle Obama, "Remarks by the President and First Lady on the End of the War in Iraq," Fort Bragg, N.C., December 14, 2012. As of December 7, 2012:
http://www.whitehouse.gov/the-press-office/2011/12/14/remarks-president-and-first-lady-end-war-iraq

Obama, Barack, and Nuri al-Maliki, "Remarks by President Obama and Prime Minister al-Maliki of Iraq in a Joint Press Conference," transcript, Washington, D.C.: The White House, December 12, 2011. As of December 13, 2011:
http://www.whitehouse.gov/photos-and-video/video/2011/12/12/
president-obama-s-press-conference-prime-minister-maliki

"Obama and Maliki Back Iraq Post-War Future," BBC News, December 12, 2011. As of April 17, 2013:
http://www.bbc.co.uk/news/world-middle-east-16134259

Odierno, Raymond, "Statement of Gen. Ray Odierno, USA, Commanding General, Multi-National Force–Iraq," in *Status of Ongoing U.S. Efforts in Iraq*, hearing before the Committee on Armed Services, U.S. House of Representatives, 111 Cong., 1st Sess., September 30, 2009. As of September 17, 2012:
http://www.gpo.gov/fdsys/pkg/CHRG-111hhrg55277/pdf/CHRG-111hhrg55277.pdf

————, "Operation New Dawn: Building a Long-Term Strategic Partnership Through Stability Operations," *Army Magazine*, October 2010. As of September 27, 2012:
http://www.ausa.org/publications/armymagazine/archive/2010/10/Documents/Odierno_1010.pdf

Office of Management and Budget, *Section 2207 Report on Iraq Relief and Reconstruction, 3rd Quarter, Fiscal Year 2004*, Appendix 1, "Sectoral Descriptions," July 2, 2004. As of December 3, 2012:
http://www.whitehouse.gov/sites/default/files/omb/assets/omb/legislative/appdx_1_sectors_final.pdf

Office of Security Cooperation in Iraq, "OSC-I Transition Plan," briefing to RAND, June 27, 2011.

OIG—See U.S. Department of State, Office of the Inspector General.

Olive Group, "Weekly Security Update: Regional Activity (Iraq) 22 Jan 2012–4 Nov 2012," *Iraq Business News*, November 7, 2012.

Ollivant, Douglas A., "Producing Victory: Rethinking Conventional Forces in COIN Operations," *Military Review*, October 2006.

Operations Order (OPORD) 09-01, "Operations Order for the Responsible Drawdown of U.S. Forces from Iraq," MNF-I, May 23, 2009.

OSC-I—*See* Office of Security Cooperation in Iraq.

Oubre, Michael S., "Equipment Once Used in Iraq Is Now Helping State and Local Governments," *Army AL&T Magazine*, January 6, 2012. As of July 30, 2012: http://armyalt.va.newsmemory.com/rss.php?date=20120106&edition=Army+AL%26T+Magazine&subsection=LOGISTICS&page=49alt2012jan-mar_logistics.pdf.0&id=art_0.xml&device=

Owles, Eric, "Then and Now: A New Chapter for Baghdad Book Market," *New York Times*, December 18, 2008.

Palmieri, Paco, "Update to Core Group: MOI/Police Advise and Train Mission—Transition to INL," Baghdad, June 27, 2011.

Panetta, Leon E., "Secretary of Defense Leon E. Panetta Submitted Testimony on Iraq," Committee on Armed Services, U.S. Senate, November 15, 2011. As of June 5, 2013: http://www.armed-services.senate.gov/statemnt/2011/11%20November/Panetta%2011-15-11.pdf

Panter, Jr., Frank, "Statement of Lieutenant General Frank A. Panter, Deputy Commandant, Installations and Logistics," in *Army and Marine Corps Materiel Reset*, hearing before the Subcommittee on Readiness, Committee on Armed Services, U.S. House of Representatives, 112th Cong., 2nd Sess., March 28, 2012.

Parrish, Karen, "Command's Relocation Aids 'Strategic Velocity,'" American Forces Press Service, October 28, 2010. As of August 6, 2013: http://www.defense.gov/news/newsarticle.aspx?id=61466

———, "Forces in Iraq Pursue 'Methodical, Careful' Exit Plan," press release, U.S. Central Command, November 3, 2011. As of November 12, 2011: http://www.centcom.mil/news/forces-in-iraq-pursue-methodical-flexible-exit-plan

Perito, Robert, *The Interior Ministry's Role in Security Sector Reform*, Washington, D.C.: United States Institute of Peace, May 2009.

Perry, Walter L., Stuart E. Johnson, Keith Crane, David C. Gompert, and John Gordon IV, Robert E. Hunter, Dalia Dassa Kaye, Terrence K. Kelly, Eric Peltz, and Howard J. Shatz, *Withdrawing from Iraq: Alternative Schedules, Associated Risks, and Mitigating Strategies*, Santa Monica, Calif: RAND Corporation, MG-882-OSD, 2009. As of April 18, 2013: http://www.rand.org/pubs/monographs/MG882.html

Peter, Tom A., "US Withdraws from Iraq Cities," *Global Post*, June 29, 2009.

Peterson, Stephen W., "Central but Inadequate: The Application of Theory in Operation Iraqi Freedom," research paper, Washington, D.C.: National War College, 2004.

Petraeus, David, "DoD Briefing on NATO Training Mission in Iraq," February 10, 2005. As of August 30, 2013: http://www.defense.gov/Transcripts/Transcript.aspx?TranscriptID=1707

———, "Report to Congress on the Situation in Iraq," September 10–11, 2007.

Pillar, Paul, *Negotiating Peace: War Termination As a Bargaining Process*, Princeton, N.J.: Princeton University Press, 1983.

Pincus, Walter, "State Department Could Buy Local Food in Iraq," *Washington Post*, November 15, 2011. As of August 30, 2013:
http://articles.washingtonpost.com/2011-11-15/
world/35282444_1_convoys-private-security-local-food

———, "Troops Have Withdrawn from Iraq, but U.S. Money Hasn't," *Washington Post*, June 27, 2012a. As of April 19, 2013:
http://articles.washingtonpost.com/2012-06-27/
world/35459625_1_baghdad-embassy-water-system-compound

———, "Iraq Transition Raises Thorny and Expensive Questions," *Washington Post*, July 2, 2012b. As of April 19, 2013:
http://articles.washingtonpost.com/2012-07-02/
world/35489248_1_iraq-security-forces-iraq-transition-iraq-government

Pirnie, Bruce R., and Edward O'Connell, *Counterinsurgency in Iraq (2003–2006): RAND Counterinsurgency Study—Volume 2*, Santa Monica, Calif.: RAND Corporation, MG-595/3-OSD, 2008. As of July 25, 2013:
http://www.rand.org/pubs/monographs/MG595z3.html

Pollack, Kenneth M., "Pollack: Iraq May End Up Worse Off Than Under Saddam," interview with Bernard Gwerzman, Council on Foreign Relations, October 13, 2006. As or April 19, 2013:
http://www.cfr.org/middle-east/pollack-iraq-may-end-up-worse-off-than-under-saddam/p11727

———, "Iraq Back on the Brink: Maliki's Sectarian Crisis of His Own Making," *Daily Beast*, December 23, 2011. As of 27 September 2012:
http://www.thedailybeast.com/articles/2011/12/23/iraq-back-on-the-brink-maliki-s-sectarian-crisis-of-his-own-making.html

Pollack, Kenneth M., and Michael E. O'Hanlon, "Iraq's Ban on Democracy," *New York Times*, January 17, 2010.

Pregent, Richard, "Rule of Law Capacity Building in Iraq," *International Law Studies*, Vol. 86, 2010. As of December 3, 2012:
http://www.usnwc.edu/getattachment/4af0e517-d0e3-4600-a3ba-a7bde66b31e5/
Rule-of-Law-Capacity-Building-in-Iraq

"President's Chief of Staff: Kurds Cannot Decide to Replace Maliki Alone," *Rudaw*, May 15, 2012.

Public Law 87-195, U.S. Foreign Assistance Act of 1961, September 4, 1961.

Public Law 108-106, Emergency Supplemental Appropriations Act for Defense and for the Reconstruction of Iraq and Afghanistan, 2004, November 6, 2003.

Public Law 110-417, The Duncan Hunter National Defense Authorization Act for Fiscal Year 2009, October 14, 2008.

Public Law 111-84, National Defense Authorization Act for Fiscal Year 2010, October 28, 2009.

Public Law 112-81, National Defense Authorization Act for Fiscal Year 2012, December 31, 2011.

Quinlivan, James T., "Coup-Proofing: Its Practice and Consequences in the Middle East," *International Security*, Vol. 24, No. 2, Fall 1999, pp. 131–165.

Rabasa, Angel, John Gordon IV, Peter Chalk, Christopher S. Chivvis, Audra K. Grant, K. Scott McMahon, Laurel E. Miller, Marco Overhaus, and Stephanie Pezard, *From Insurgency to Stability*, Vol. 1: *Key Capabilities and Practices*, Santa Monica, Calif.: RAND Corporation, MG-1111/1-OSD, 2011. As of April 29, 2013:
tp://www.rand.org/pubs/monographs/MG1111z1.html

Rathmell, Andrew, *Fixing Iraq's Internal Security Forces: Why Is Reform of the Ministry of Interior So Hard?* Washington, D.C.: Center for Strategic and International Studies, November 2007.

Rathmell, Andrew, Olga Oliker, Terrence K. Kelly, David Brannan, and Keith Crane, *Developing Iraq's Security Sector: The Coalition Provisional Authority's Experience*, Santa Monica, Calif.: RAND Corporation, MG-365-OSD, 2005. As of October 25, 2012:
http://www.rand.org/pubs/monographs/MG365.html

Raufoglu, Alakbar, "Turkey Looks to KRG for Help Against PKK," *SES Türkiye*, 4 July 4 2012. As of September 26, 2012:
http://turkey.setimes.com/en_GB/articles/ses/articles/features/departments/world/2012/07/04/feature-01

Rawya, Rageh, "Iraq Tensions Rise over Kirkuk," Al-Jazeera, April 17, 2011.

Rayburn, Joel D., "Rise of the Maliki Regime," *Journal of International Security Affairs*, Spring/Summer 2012. As of November 12, 2012:
http://www.securityaffairs.org/issues/2012/22/rayburn.php

Rice, Condoleezza, "Dr. Condoleezza Rice Discusses Iraq Reconstruction," press briefing, April 4, 2003. As of April 23, 2013:
http://georgewbush-whitehouse.archives.gov/news/releases/2003/04/20030404-12.html

Rice, Condoleezza, and Robert Gates, "What We Need Next in Iraq," *Washington Post*, February 13, 2008.

Richardson, Thomas, "The Role of U.S. Forces–Iraq with Force Reposturing and the Status of Government-Owned Equipment in Iraq," roundtable with defense bloggers, August 3, 2011. As of August 6, 2012:
http://www.dodlive.mil/files/2011/08/0803rich.pdf

Ricks, Tom, *Fiasco: The American Military Adventure in Iraq*, New York: Penguin Books, 2006.

———, *The Gamble: General Petraeus and the American Military Adventure in Iraq*, New York: Penguin Books, 2009a.

———, "Iraq, the Unraveling (XXIV): U.S. Embassy vs. U.S. Military, Again," *Foreign Policy*, September 28, 2009b. As of April 25, 2013:
http://ricks.foreignpolicy.com/posts/2009/09/28/iraq_the_unraveling_xxiv_us_embassy_vs_us_military_again

Ridolfo, Kathleen, "Iraq: Kurdish Official Says Kirkuk Normalization to Proceed," RFE/RL, June 21, 2007. As of September 20, 2012:
http://www.rferl.org/content/article/1077257.html

Risen, James, "End of Immunity Worries U.S. Contractors in Iraq," *New York Times*, November 30, 2008.

Robinson, Linda, *Tell Me How This Ends: General David Petraeus and the Search for a Way Out of Iraq*, New York: PublicAffairs, 2008.

Rogers, Harold, "Opening Statement of Rep. Harold Rogers," in Fiscal Year 2012 Budget Request for State Department and Foreign Assistance Programs, hearing before the Subcommittee on State and Foreign Operations, Committee on Appropriations, U.S. House of Representatives, March 10, 2011.

Rogin, Josh, "Romney Campaign Links Embassy Attacks to Obama's Failed Middle East Policies," Foreign Policy (The Cable), September 12, 2012. As of September 17, 2012
http://thecable.foreignpolicy.com/posts/2012/09/12/romney_campaign_links_embassy_attacks_to_obama_s_failed_middle_east_policies

———, "Congress Fails to Reauthorize the Pentagon's Mission in Iraq," *Foreign Policy*, October 1, 2012. As of November 30, 2012:
http://thecable.foreignpolicy.com/posts/2012/10/01/
congress_fails_to_reauthorize_the_pentagon_s_mission_in_iraq

Rose, Gideon, *How Wars End: Why We Always Fight the Last Battle*, New York: Simon & Schuster, 2012.

"The Routes: Iraq to Afghanistan," *New York Times*, March 31, 2010. As of July 30, 2012:
http://www.nytimes.com/imagepages/2010/03/31/world/01logistics-map.html?ref=world.

Rudd, Gordon W., *Reconstructing Iraq: Regime Change, Jay Garner, and the ORHA Story*, Lawrence, Kan.: University of Kansas Press, 2011.

Rumsfeld, Donald, "DoD News Briefing with Secretary Rumsfeld and North Korean Minister of National Defense Yoon Kwang-Ung at the Pentagon," October 20, 2006.

Ryan, Mike, and Jason Coats, "The U.S.-Iraqi Security Agreement and the Changing Nature of U.S. Military Operations in Iraq," *Military Review*, September–October 2009. As of September 18, 2012:
http://usacac.army.mil/CAC2/MilitaryReview/Archives/English/MilitaryReview_20091031_art009.pdf

SA—*See* United States of America and the Republic of Iraq, 2008a.

Sanger, David E., "With Korea as Model, Bush Team Ponders Long Support Role in Iraq," *New York Times*, June 3, 2007. As of October 1, 2012:
http://www.nytimes.com/2007/06/03/washington/03assess.html

Santana, Rebecca, "'Bottomless Pit': Watchdog Slams US Training of Iraqi Police," Associated Press, October 24, 2011. As of August 30, 2013:
http://www.msnbc.msn.com/id/45011157/ns/world_news-mideast_n_africa/t/
bottomless-pit-watchdog-slams-us-training-iraqi-police/

Schinasi, Katherine, testimony, in *The Final Report of the Commission on Wartime Contracting in Iraq and Afghanistan*, hearing before the Subcommittee on Readiness and Management Support, Committee on Armed Services, U.S. Senate, 112th Cong., 1st Sess., October 19, 2011, p. 23.

Schmidt, Michael, "Attacks in Iraq Heighten Political Tensions," *New York Times*, August 16, 2011. As of August 30, 2013:
http://www.nytimes.com/2011/08/17/world/middleeast/17iraq.html

Schmidt, Michael S., and Eric Schmitt, "Flexing Muscle, Baghdad Detains U.S. Contractors," *New York Times*, January 15, 2012:
http://www.nytimes.com/2012/01/16/world/middleeast/asserting-its-sovereignty-iraq-detains-american-contractors.html?scp=4&sq=michael%20s.%20schmidt%20and%20eric%20
schmitt&st=cse

Schmitt, Eric, and Michael S. Schmidt, "U.S. Drones Patrolling Its Skies Provoke Outrage in Iraq," *New York Times*, January 29, 2012:
http://www.nytimes.com/2012/01/30/world/middleeast/iraq-is-angered-by-us-drones-patrolling-its-skies.html?pagewanted=all

Schreck, Adam, "Iraq's Ousted Bank Chief Professes His Innocence," Associated Press, November 9, 2012.

Secretary-General of the United Nations, "An Agenda for Peace: Preventative Diplomacy, Peacemaking, and Peacekeeping," New York, 1992.

Secretary of State, *Congressional Budget Justification*, Vol. 1: *Department of State Operations, Fiscal Year 2012*, Washington, D.C.: U.S. Department of State, February 14, 2011a.

———, *Congressional Budget Justification*, Vol. 2: *Foreign Operations, Fiscal Year 2012*, Washington, D.C.: U.S. Department of State, April 2011b.

SFA—*See* United States of America and the Republic of Iraq, 2008b.

Shadid, Anthony, "Ambassador Leaves Iraq with Much Still Unsettled," *New York Times*, August 12, 2010.

Shafran, Stacie, "Air Force Assists in Historic Final Transfer of Iraq's Airspace," U.S. Air Forces Central, Baghdad Media Outreach Team, 2011. As of November 25, 2011: http://www.dvidshub.net/news/79492

Shanker, Thom, Michael S. Schmidt, and Robert F. Worth, "In Baghdad, Panetta Leads Uneasy Moment of Closure," *New York Times*, December 15, 2011. As of July 30, 2012: http://www.nytimes.com/2011/12/16/world/middleeast/panetta-in-baghdad-for-iraq-military-handover-ceremony.html?pagewanted=all

Sharp, Jeremy M., and Christopher M. Blanchard, *Post-War Iraq: Foreign Contributions to Training, Peacekeeping, and Reconstruction*, Washington, D.C.: Congressional Research Service, June 6, 2005.

Sherlock, Richard, "DoD Operational Update Briefing with Maj. Gen. Sherlock from the Pentagon Briefing Room, Arlington Va.," September 6, 2007. As of September 12, 2013: http://www.defense.gov/transcripts/transcript.aspx?transcriptid=4035

SIGIR—*See* Special Inspector General for Iraq Reconstruction.

Slovensky, Andrew, "New Radar Paints Iraq's Air Defense Picture," 362nd Mobile Public Affairs Detachment, November 3, 2011. As of August 9, 2012: http://www.dvidshub.net/news/79487/new-radar-paints-iraqs-air-defense-picture

———, "Heavy Metal Takes a Ride to Kuwait," press release, U.S. Forces–Iraq, November 10, 2011.

Smith, Austin B., "Experts: State Dept. Unprepared to Oversee Contracting in Iraq," Medill News Service for *Military Times*, November 9, 2011.

Special Inspector General for Iraq Reconstruction, "Status of the Provincial Reconstruction Team Program in Iraq," Washington, D.C., SIGIR-06-034, October 29, 2006.

———, "Status of the Provincial Reconstruction Team Program Expansion in Iraq," Washington, D.C., SIGIR-07-014, July 25, 2007a.

———, *Quarterly and Semiannual Report to the United States Congress*, July 30, 2007b. As of June 7, 2013: http://www.sigir.mil/publications/quarterlyreports/

———, "Commander's Emergency Response Program in Iraq Funds Response Program in Iraq Funds Many Large-Scale Projects," Washington, D.C., SIGIR-08-006, January 25, 2008.

———, *Hard Lessons: The Iraq Reconstruction Experience*, Washington, D.C.: U.S. Government Printing Office, 2009a.

———, "Iraq Commander's Emergency Response Program Generally Managed Well, but Project Documentation and Oversight Can Be Improved," Washington, D.C., 10-003, October 27, 2009b.

———, *Quarterly Report to the United States Congress*, April 30, 2009c. As of June 7, 2013: http://www.sigir.mil/publications/quarterlyreports/

———, "Iraqi Security Forces: Police Training Program Developed Sizable Force, but Capabilities Are Unknown," SIGIR 11-003, October 25, 2010a. As of September 17, 2012: http://www.sigir.mil/files/audits/11-003.pdf

———, "Iraqi Security Forces: Special Operations Force Program Is Achieving Goals, but Iraqi Support Remains Critical to Success," SIGIR 11-004, October 25, 2010b.

———, *Quarterly Report to the United States Congress*, October 30, 2010c. As of August 30, 2013:
http://www.sigir.mil/publications/quarterlyreports/October2010.html

———, "Sons of Iraq Program: Results Are Uncertain and Financial Controls Were Weak," SIGIR 11-010, January 28, 2011a. As of September 18, 2012:
http://www.sigir.mil/files/audits/11-010.pdf

———, *Quarterly Report and Semiannual Report to the United States Congress*, January 30, 2011b. As of June 7, 2013:
http://www.sigir.mil/publications/quarterlyreports/

———, *Quarterly Report to the United States Congress,* April 30, 2011c, As of June 7, 2013:
http://www.sigir.mil/publications/quarterlyreports/

———, *Quarterly Report and Semiannual Report to the United States Congress*, July 30, 2011d. As of June 7, 2013:
http://www.sigir.mil/publications/quarterlyreports/

———, *Iraqi Police Development Program: Opportunities for Improved Program Accountability and Budget Transparency*, SIGIR 12-006, October 24, 2011e, p. 1. As of August 10, 2012:
http://www.sigir.mil/files/audits/12-006.pdf#view=fit

———, *Quarterly Report to the United States Congress*, October 30, 2011f. As of June 7, 2013:
http://www.sigir.mil/publications/quarterlyreports/

———, *Quarterly Report and Semiannual Report to the United States Congress*, January 30, 2012a. As of June 7, 2013:
http://www.sigir.mil/publications/quarterlyreports/

———, "Interim Report on Spend Plans for Fiscal Years 2011–2012 Iraq Security Forces Funds (ISFF)," SIGIR 12-015, April 26, 2012b. As of August 30, 2013:
http://www.sigir.mil/files/audits/12-015-F.pdf

———, *Quarterly Report to the United States Congress*, April 30, 2012c. As of June 7, 2013:
http://www.sigir.mil/publications/quarterlyreports/

———, *Iraq Police Development Program: Lack of Iraqi Support and Security Problems Raise Questions About the Continued Viability of the Program*, SIGIR 12-020, July 30, 2012d, p. 4. As of June 7, 2013:
http://www.sigir.mil/files/audits/12-020.pdf

———, *Quarterly Report and Semiannual Report to the U.S. Congress*, July 30, 2012e. As of June 7, 2013:
http://www.sigir.mil/publications/quarterlyreports/

———, *Sustaining the Progress Achieved by U.S. Rule of Law Programs in Iraq Remains Questionable*, SIGIR 13-001, October 25, 2012f. As of December 3, 2012:
http://www.sigir.mil/files/audits/13-001.pdf#view=fit

———, *Quarterly Report to the United States Congress*, October 30, 2012g. As of June 7, 2013:
http://www.sigir.mil/publications/quarterlyreports/

Spratt, Pierre A., "402nd AFSB Provides Retrograde Support," Army.mil, June 17, 2011. As of August 3, 2012:
http://www.army.mil/article/59923/402nd_AFSB_provides_retrograde_support/

"Standing Up the IqAF: King Air 350s," *Defense Industry Daily*, August 25, 2009. As of August 13, 2012:
http://www.defenseindustrydaily.com/Standing-Up-the-IqAF-King-Air-350s-05101/

Stanford University, "Al-Qaeda in Iraq (AQI)," Mapping Militant Organizations website, August 8, 2012a. As of April 17, 2013:
http://www.stanford.edu/group/mappingmilitants/cgi-bin/groups/view/1

———, "Promised Day Brigades," Mapping Militant Organizations website, August 27, 2012b. As of April 25, 2013:
http://www.stanford.edu/group/mappingmilitants/cgi-bin/groups/view/249

Stanley, Elizabeth, *Paths to Peace: Domestic Coalition Shifts, War Termination and the Korean War*, Stanford, Calif.: Stanford University Press, 2009.

"State Dept.: Al-Qaeda in Iraq Fighting in Syria," CBS/Associated Press, July 31, 2012. As of November 12, 2012:
http://www.cbsnews.com/8301-202_162-57483500/state-dept-al-qaeda-in-iraq-fighting-in-syria/

Stratman, Henry W., "Orchestrating Instruments of Power for Nationbuilding," *Joint Force Quarterly*, No. 41, 2nd quarter, 2006.

Strouse, Thomas, "Kata'ib Hezbollah and the Intricate Web of Iranian Military Involvement in Iraq," *Terrorism Monitor*, Vol. 8, No. 9, March 4, 2010. As of June 11, 2013:
http://www.jamestown.org/single/?no_cache=1&tx_ttnews[tt_news]=36109

Stuart, Jeffrey, "Operations Transitions Working Group," briefing to RAND, Baghdad, June 26, 2011.

Sullivan, Marisa Cochrane, "Obama's Iraq Abdication," *Wall Street Journal*, July 28, 2011. As of November 30, 2012:
http://online.wsj.com/article/SB10001424052702303661904576456220275931918.html

Swedberg, Claire, "U.S. Army Deploys 'Soldier-Friendly' System to Track Thousands of Vehicles in Kuwait," *RFID Journal*, November 8, 2011:
http://www.rfidjournal.com/article/view/8946/

Tan, Michelle, "Logisticians Shuffle Troops' Equipment," *Army Times*, March 28, 2010. As of July 30, 2012:
http://www.armytimes.com/news/2010/03/army_drawdown_032810w/

TAL—*See* Coalition Provisional Authority, 2004.

Tavernise, Sabrina, "U.S. Agrees to Lift Immunity for Contractors in Iraq," *New York Times*, July 2, 2008.

Thompson, Mark, "With U.S. Pullout, Iraq Takes Ownership of Its War," *Time*, June 30, 2009.

Transparency International, "Corruption Perception Index 2012," 2012. As of March 10, 2013:
http://www.transparency.org/cpi2012/results#myAnchor1

United Nations Security Council, "The Situation Between Iraq and Kuwait," Resolution 1546, 2004.

United States of America and the Republic of Iraq, "Agreement Between the United States of America and the Republic of Iraq on the Withdrawal of United States Forces from Iraq and the Organization of Their Activities During Their Temporary Presence in Iraq," November 17, 2008a. As of April 17, 2013:
http://graphics8.nytimes.com/packages/pdf/world/20081119_SOFA_FINAL_AGREED_TEXT.pdf

United States of America and the Republic of Iraq, "Strategic Framework Agreement for a Relationship of Friendship and Cooperation Between the United States of America and the Republic of Iraq," November 17, 2008b. As of April 17, 2013: http://www.acq.osd.mil/log/ps/p_vault/se_sfa.pdf

United States of America and the Republic of Iraq Higher Coordinating Committee, "Joint Statement by the United States of America and the Republic of Iraq Higher Coordinating Committee," Washington, D.C.: The White House, Office of the Press Secretary, November 30, 2011. As of April 17, 2013: http://www.whitehouse.gov/the-press-office/2011/11/30/ joint-statement-united-states-america-and-republic-iraq-higher-coordinat

U.S. Army Regulation 350-1, *Army Training and Leader Development*, 2011

U.S. Army and U.S. Marine Corps, *Counterinsurgency Field Manual*, FM 3-24/MCWP 3-33.5, Chicago: University of Chicago Press, 2007.

USCENTCOM—*See* U.S. Central Command.

U.S. Central Command, planning slides, August 2002. As of January 17, 2012: http://www.gwu.edu/~nsarchiv/NSAEBB/NSAEBB214/index.htm

——, *Iraq Country Plan*, June 10, 2011a, Not available to the general public.

——, *Iraq Country Plan, Appendix 3 to Annex O (Security Cooperation), Iraqi Army Security Cooperation Roadmap*, Headquarters, October 12, 2011b, Not available to the general public.

——, *Iraq Country Plan, Appendix 5 to Annex O (Security Cooperation), Iraqi Air Force (IqAF) Security Cooperation Roadmap*, October 12, 2011c, Not available to the general public.

USCENTCOM Regulation 200-2, Contingency Environmental Guidance, 2012.

U.S. Congress, "Joint Resolution Making Continuing Appropriations for Fiscal Year 2012, and for Other Purposes," 111th Cong., 1st Sess., H.J. Res. 79, Sec. 116, September 14, 2011.

U.S. Department of Defense, "2009–10 GOI Budget Shortfall Impact on Security," undated, Not available to the general public.

——, *Measuring Stability and Security in Iraq*, July 2005.

——, *Measuring Stability and Security in Iraq*, November 2006.

——, *Measuring Stability and Security in Iraq*, September 2008.

——, *Measuring Stability and Security in Iraq*, March 2009a.

——, *Measuring Stability and Security in Iraq*, December 2009b.

——, *Measuring Stability and Security in Iraq*, June 2010.

——, *Sustaining U.S. Global Leadership: Priorities for 21st Century Defense*, January 2012a. As of July 26, 2012: http://www.defense.gov/news/Defense_Strategic_Guidance.pdf

——, "List of Casualties in Operation Iraqi Freedom, Operation New Dawn, and Operation Enduring Freedom," as of July 30, 2012b.

——, "Contracts," press release no. 659-12, August 7, 2012c. As of November 12, 2012: http://www.defense.gov/contracts/contract.aspx?contractid=4849

U.S. Department of Defense, Inspector General, *Assessment of U.S. Government Efforts to Develop the Logistics Sustainment Capability of the Iraq Security Forces*, Report No. SPO-20 11-001, November 17, 2010, pp. 13, 17. As of April 26, 2013:
http://www.dodig.mil/SPO/Reports/ISF10Nov10.pdf

———, *Assessment of Planning for Transitioning the Security Assistance Mission in Iraq from Department of Defense to Department of State Authority*, Report No. SPO-2011-008, August 25, 2011. As of August 7, 2012:
http://www.dodig.mil/SPO/Reports/ISATReport082511.pdf

———, *Assessment of the DOD Establishment of the Office of Security Cooperation–Iraq*, Report No. DODIG-2012-063, March 16, 2012. As of August 7, 2012:
http://www.dodig.mil/spo/Reports/DODIG-2012-063.pdf

U.S. Department of Labor, Bureau of Labor Statistics, *Occupational Outlook Handbook*, 2010–2011.

U.S. Department of Labor, Office of Workers' Compensation Programs, "Defense Base Act Case Summary by Nation," web page, September 1, 2001–March 31, 2013.

U.S. Department of State, "11 02 01 State FY 2010–FY12 Iraq Funding_FINAL.docx," undated a.

———, Biography of Joseph Pennington, former chief of EBO Mostar (2000–2001), undated b. As of August 13, 2012:
http://prague.usembassy.gov/joe-pennington.html

———, "Iraq Jobs: Benefits," web page, undated c.

———, "Solicitation SAQMMA11-R-0010: Department of State Health Care Service Support in Iraq," February 9, 2001.

———, "The Future of Iraq Project: Overview," Washington, D.C., May 12, 2003. As of July 12, 2013:
http://www.gwu.edu/~nsarchiv/NSAEBB/NSAEBB198/FOI%20Overview.pdf

———, *Section 2207 Report on Iraq Relief and Reconstruction*, January 5, 2005a. As of June 7, 2013:
http://2001-2009.state.gov/s/d/rm/rls/2207/jan2005/index.htm

———, *Section 2207 Report on Iraq Relief and Reconstruction*, April 6, 2005b. As of June 7, 2013:
http://2001-2009.state.gov/s/d/rm/rls/2207/apr2005/index.htm

———, "Designation of Kata'ib Hizballah as a Foreign Terrorist Organization," press release, July 2, 2009a. As of March 11, 2013:
http://www.state.gov/r/pa/prs/ps/2009/july/125582.htm

———, "U.S. Official Visits Kazakstan, Opens U.S. Consulate General in Almaty," December 11, 2009b. As of August 13, 2012:
http://almaty.usconsulate.gov/pr-12-11-09.html

———, "Treasury Targets Iran's Islamic Revolutionary Guard Corps," press release, February 10, 2010a. As of March 11, 2013:
http://www.state.gov/r/pa/prs/ps/2010/02/136595.htm

———, Response to Question for the Record #2 Submitted by the Subcommittee on State, Foreign Relations, and Related Programs, Committee on Appropriations, U.S. House of Representatives, drafted August 20, 2010b.

———, "FY2010 Supplemental Appropriations Spending Plan for State Department Operations in Iraq," accompanying Assistant Secretary of State for legislative Affairs Richard Verma's letters to multiple Chairs of Authorizing and Appropriations Committees, October 20, 2010c.

———, "State FY 2010 and 2011 Iraq Funding," February 1, 2011a.

———, *Executive Budget Summary: Function 150 and Other International Programs, Fiscal Year 2012*, February 14, 2011b. As of August 6, 2013:
http://www.state.gov/documents/organization/156214.pdf

———, "2012 Iraq Service Recognition Package (ISRP)," cable to all diplomatic and consular posts, SecState #052041, May 26, 2011c.

———, "Country Reports in Human Rights Practices for 2011," website, April 24, 2012a.

———, "Foreign Terrorist Organizations," in *Country Reports on Terrorism 2011*, July 31, 2012b. As of September 28, 2012:
http://www.state.gov/j/ct/rls/crt/2011/195553.htm#kh

———, "Rest and Recuperation (R&R) Travel," *Foreign Affairs Manual*, Vol. 3, *Personnel*, 3 FAM 3720, November 8, 2012c.

U.S. Department of State, Bureau of Near Eastern Affairs, Iraq, "Policy and Mission Overview—Iraq," briefing, April 2011.

U.S. Department of State, Bureau of Diplomatic Security, "Vigilant in an Uncertain World: Diplomatic Security 2010 Year in Review," March 2011. As of April 29, 2013:
http://www.state.gov/documents/organization/158786.pdf

U.S. Department of State, Office of Acquisitions, "Baghdad Operation and Maint Support," Solicitation Number SAQMMA-12-R-0012," January 17, 2012. As of August 6, 2013:
https://www.fbo.gov/index?s=opportunity&mode=form&tab=core&id=c10b2476f574316583aec1131 1a67a38&cview=O

U.S. Department of State, Office of Inspector General, "Inspection of Rule-of-Law Programs, Embassy Baghdad," OIG Report No. ISP-IQO-06-01, October 2005. As of December 3, 2012:
http://oig.state.gov/documents/organization/103473.pdf

———, "Report of Inspection: Embassy Sarajevo, Bosnia and Herzegovina," Report No. ISP-I-09-55A, September 2009. As of August 13, 2012:
http://oig.state.gov/documents/organization/132896.pdf

———, "The Bureau of International Narcotics and Law Enforcement Affairs Air Wing Program in Afghanistan and Pakistan: Performance Audit," Report Number MERO-A-10-03, March 2010a.

———, "Compliance Follow-Up Review of Embassy Baghdad, Iraq," Report ISP-C-11-0SA, October 2010b.

———, "Iraq Transition," briefing, October 6, 2010c.

———, "DOS Planning for the Transition to a Civilian-Led Mission in Iraq," Report No. MERO-I-11-08, May 2011. As of 10 July 2012:
http://oig.state.gov/documents/organization/165037.pdf

U.S. Department of State, Office of Logistics Management, "Baghdad Operation and Maintenance Support," presolicitation notice, Solicitation Number SAQMMA-12-R-0012, November 3, 2011.

U.S. Department of State and U.S. Agency for International Development, *Leading Through Civilian Power: The First Quadrennial Diplomacy and Development Review*, Washington, D.C., 2010a. As of July 26, 2013:
http://www.state.gov/documents/organization/153108.pdf.

———, "Supplemental Budget Justification, Fiscal Year 2010," 2010b. As of June 10, 2013:
http://www.state.gov/documents/organization/136356.pdf

U.S. Department of State and the Broadcasting Board of Governors, Office of Inspector General, *Inspection of Rule-of-Law Programs*, report of inspection, Report Number ISP-IQO-06-01, October 2005.

U.S. Deparment of the Army, *Commander's Guide to Money as a Weapons System*, Center for Army Lessons Learned, April 2009. As of November 22, 2011:
http://usacac.army.mil/cac2/call/docs/09-27/09-27.pdf

———, "Biography of USF-I Spokesman and Director for Strategic Effects Major General Stephen R. Lanza," April 2010. As of September 24, 2012:
http://www.dodlive.mil/files/2010/04/MG-Stephen-Lanza-Bio.pdf

———, "Audit of the U.S. Equipment Transfer to Iraq Program—Phase II," 2011a.

———, "Limited Source Justification for Exception to Fair Opportunity Ordering Process Under Multiple Award Indefinite Delivery Contracts," Control Number 12-003, October 27, 2011b. As of June 11, 2013:
http://lignesdedefense.blogs.ouest-france.fr/files/olive%20group.pdf

———, *The Iraq War: 2003–2011*, U.S. Government Printing Office, May 10, 2012.

U.S. Department of the Army, Judge Advocate General's Legal Center and School, Center for Law and Military Operations, *Rule of Law Handbook: A Practitioner's Guide for Judge Advocates*, 2010. As of December 3, 2012:
http://www.loc.gov/rr/frd/Military_Law/pdf/rule-of-law_2010.pdf

USF-I—*See* U.S. Forces–Iraq.

U.S. Forces–Iraq, *Joint Campaign Plan*, Annex C, "Rule of Law," November 23, 2009.

———, "Joint Campaign Plan Overall Goal Assessment," in *Iraq Transition Senior Leader Conference*, briefing, Washington, July 22–23, 2010a, Not available to the general public.

———, "Strategic Transition Overview," in *Iraq Transition Senior Leader Conference*, briefing, Washington, D.C., July 22–23, 2010b, slide 36.

———, "CS/TS LOO Update to GEN Lloyd J. Austin, III," September 29, 2010c.

———, "Combined Security Mechanism," press guidance, September 13, 2010d.

———, *Quarterly Command Report, 4th Quarter, FY10 (1 July–30 September 2010)*, October 31, 2010e, Not available to the general public.

———, *OIF/OND Chronology*, 2011a.

———, OPORD 11-01, January 6, 2011b, Not available to the general public.

———, "ExSum 11-01 Operations ROC Drill," memorandum, Baghdad, February 16, 2011c.

———, OPORD 11-01, Change 1 to Appendix 4 to Annex V (Establishment of the Office of Security Cooperation–Iraq), May 5, 2011d.

———, "Arab-Kurd Transition Plan," update, April 4, 2011e.

———, *Responsible Redeployment of United States Armed Forces from Iraq*, draft report for Congress, October 2011f, Not available to the general public.

———, "JCP Activity Transition," briefing to RAND Corporation, June 25, 2011g, Not available to the general public.

———, "Iraq Training and Assistance Mission–Army, Transition Plan," briefing to RAND, Baghdad, June 27, 2011h, Not available to the general public.

———, Iraq Training and Advisory Mission–Navy, "Transition Plan," briefing to RAND, June 28, 2011i.

———, "Advising and Training and NATO Training Mission—Iraq," briefing, July 12, 2011j.

———, "Commanders' Update," briefing, October 17, 2011k, Not available to the general public.

———, "FMS Synchronization," briefing, October 6, 2011m.

U.S. Forces–Iraq, J4 Joint Logistics Operation Center (JLOC), "J4 Transition to DoS Plan Presentation to RAND Corporation," briefing, Baghdad, June 25, 2011a.

———, "BOS-I Program Civilian and Contract Support," December 13, 2011b.

———, "Foreign Excess Personal Property (FEPP) Transfers to the Government of Iraq," December 13, 2011c.

———, "Joint Logistics Operation Center (JLOC) Overview," December 13, 2011d.

———, "J-4 Summary of Equipment Drawdown," December 13, 2011e.

———, memo to RAND, December 13, 2011f.

———, "Operation New Dawn Commodity Drawdown," December 13, 2011g.

———, "USETTI Background Paper," December 13, 2011h.

———, "USF-I J4 DOS Transition Cell History Report," December 13, 2011i.

———, "Road Show," briefing, Kuwait, December 18, 2011j.

U.S. Forces–Iraq, J5 Strategy, "Joint Assessment Board," briefing to RAND, Baghdad, June 30, 2011.

U.S. Forces–Iraq, J7, "Base Transition to Enduring U.S. Entities," September 29, 2010.

———, *Base Closure Smartbook*, February 2011.

U.S. Forces–Iraq, J35, "J35 Transition Plan Presentation to RAND Corporation," Baghdad, July 1, 2011, slide 4 Not available to the general public.

U.S. Forces–Iraq, Partnership Strategy Group–Iraq (PSG-I), "Office of Security Cooperation–Iraq (OSC-I)," July 9, 2010.

U.S. Forces–Iraq, Public Affairs Office, "Besmaya Combat Training Center Transferred to Iraq Army," August 1, 2010a.

———, "Iraqi Armor School Moves to Besmaya," December 12, 2010b.

U.S. Forces–Iraq and U.S. Embassy Baghdad, *2010 Joint Campaign Plan*, November 23, 2009.

———, "Arab-Kurd Relations Transition Plan," April 5, 2011, Not available to the general public.

U.S. Forces–Iraq and the U.S. Department of State, "Iraq Transition Senior Leader Conference," briefing, National Defense University, July 23, 2010.

U.S. Government Accountability Office, *Supplemental Appropriations: Opportunities Exist to Increase Transparency and Provide Additional Controls*, GAO-08-314, January 2008a.

———, *Military Operations: Actions Needed to Better Guide Project Selection for Commander's Emergency Response Program and Improve Oversight in Iraq*, Washington, D.C., GAO-08-736R, June 23, 2008b.

———, *Operation Iraqi Freedom: Actions Needed to Enhance DOD Planning for Reposturing of U.S. Forces from Iraq*, Washington, D.C., September 2008c.

———, *Provincial Reconstruction Teams in Afghanistan and Iraq*, Washington, D.C., GAO-09-86R, October 1, 2008d. As of June 11, 2013:
http://www.gao.gov/assets/100/95824.pdf

———, *Iraq Drawdown: Opportunities Exist to Improve Equipment Visibility, Contractor Demobilization, and Clarity of Post-2011 DOD Role*, Washington, D.C., GAO-11-774, September 2011. As of August 20, 2013:
http://www.gao.gov/new.items/d11774.pdf

U.S. Mission–Iraq and Multi-National Force–Iraq, *Campaign Progress Review, June 2006–December 2006*, Baghdad, 2006a.

———, *Campaign Progress Review, June 2004–June 2006*, June 14, 2006b.

U.S. Senate, Committee on Appropriations, "Making Emergency Supplemental Appropriations for Disaster Relief and Summer Jobs for the Fiscal Year Ending September 30, 2010, and for Other Purposes," Conf. Rpt. 111-188, 111th Cong., 2nd Sess., May 14, 2010. As of August 30, 2013:
http://gpo.gov/fdsys/pkg/CRPT-111srpt188/pdf/CRPT-111srpt188.pdf

———, "Department of State, Foreign Operations, and Related Programs Appropriations Bill, 2013: Report to Accompany S. 3241," Report 112-172, 112th Cong., 2nd Sess., May 24, 2012.

U.S. Senate, Committee on Armed Services, "Report to Accompany S. 3001, the National Defense Authorization Act for Fiscal Year 2009," May 12, 2008. As of August 30, 2013:
http://www.gpo.gov/fdsys/pkg/CRPT-110srpt335/pdf/CRPT-110srpt335.pdf

U.S. Senate, Committee on Foreign Relations, "Iraq: The Transition from a Military Mission to a Civilian-Led Effort," S. Prt. 112-3, January 31, 2011.

———, *Iraq Report: Political Fragmentation and Corruption Stymie Economic Growth and Political Progress; A Minority Staff Trip Report*, S. Prt. 112-34, 112th Cong., 2nd Sess., April 30, 2012. As of August 14, 2012:
http://www.gpo.gov/fdsys/pkg/CPRT-112SPRT74162/pdf/CPRT-112SPRT74162.pdf

Verga, Peter F., Chief of Staff to the Under Secretary of Defense for Policy, "Statement of the Department of Defense," in *Assessment of the Transition from a Military- to a Civilian-Led Mission in Iraq*, Subcommittee on National Security, Homeland Defense, and Foreign Operations, Committee on Oversight and Government Reform, U.S. House of Representatives, 112th Cong., 2nd Sess., June 28, 2012. As of April 17, 2013:
http://oversight.house.gov/hearing/assessment-of-the-transition-from-a-military-to-a-civilian-led-mission-in-iraq-2/wp-content/uploads/2012/09/2012-06-28-SC-Natl-Sec.pdf

Visser, Reidar, "SOFA Issues," in *historiae.org*, October 20, 2008.

———, "Another Batch of Security Ministry Nominees: Turning the Clock Back to 2006?" *Iraq and Gulf Analysis*, May 6, 2011. As of August 30, 2013:
http://gulfanalysis.wordpress.com/2011/05/06/
another-batch-of-security-ministry-nominees-turning-the-clock-back-to-2006/

———, "Sadr Demands Resistance Against the US Embassy in Baghdad," *Iraq and Gulf Analysis*, October 22, 2011. As of August 30, 2013:
http://gulfanalysis.wordpress.com/2011/10/22/
sadr-demands-resistance-against-the-us-embassy-in-baghdad/

Vogt, Richard, "Warhorse Transfers COS Garry Owen to IA, Closes Historic Chapter," press report, October 27, 2011.

Walt, Vivienne, "Can Petraeus Salvage Iraq?" *Time*, June 19, 2004.

Walter, Emily, "Base Transfer Marks Progress," *Expeditionary Times*, July 28, 2010. As of November 12, 2011:
http://www.hood.army.mil/4sus/news/ExpTimes/et100728.pdf

"We Are Ready to Die to Protect Kurds of Iraq's Disputed Areas: Kurdistan Interior Ministry Official," *AK News*, July 21, 2012. As of September 20, 2012:
http://www.ekurd.net/mismas/articles/misc2012/7/kurdsiniraq171.htm

Wilbanks, Mark, and Efraim Karsh, "How the 'Sons of Iraq' Stabilized Iraq," *Middle East Quarterly*, Fall 2010. As of September 18, 2012:
http://www.meforum.org/meq/pdfs/2788.pdf

Williams, Timothy, and Duraid Adnan, "Sunnis in Iraq Allied with U.S. Rejoin Rebels," *New York Times*, October 16, 2010. As of September 18, 2012:
http://www.nytimes.com/2010/10/17/world/middleeast/17awakening.html?ref=awakeningmovement

Woodward, Bob, *Plan of Attack*, New York: Simon & Schuster, 2004.

———, *The War Within: A Secret White House History 2006–2008*, New York: Simon & Schuster, 2008.

Wright, Robin, "Kurdish Eyes on Iraq's Future," *Los Angeles Times*, November 24, 2002.

Zachary, Stacia, "Iraqi Air Force Builds ISR Foundation with Help from Airmen," Armed Forces News Service, June 16, 2009. As of August 30, 2013:
http://www.afcent.af.mil/news/story.asp?id=123154108

———, "Iraqi Air Force Brings ISR Capabilities Online," Armed Forces News Service, August 24, 2009. As of August 30, 2013:
http://www.afcent.af.mil/news/story.asp?id=123164549

Zane, Anthony, "Equipment Leaving Iraq Saves Tax Dollars While Soldiers Return Home, Materials Are Redistributed," press release, U.S. Forces–Iraq, November 5, 2011a.

———, "Last Stop for Fuel in Iraq: Bulk Fuel Farm Consolidates Fuel as Drawdown Continues in Iraq," press release, U.S. Forces–Iraq, November 7, 2011b.

Zartman, I. William, *Ripe for Resolution: Conflict and Intervention in Africa*, New York: Oxford University Press, 1989.

Ziezulewica, Geoff, "USAF General: Iraqi Air Defenses to Have Two-Year 'Gap,'" *Stars and Stripes*, November 7, 2011. As of August 13, 2012:
http://www.stripes.com/news/usaf-general-iraqi-air-defenses-to-have-two-year-gap-1.160030

Zogby, James, "Iraq: The War, Its Consequences & the Future," Zogby Research Services, LLC, November 18–20, 2011.

Index